Alexandre C. Dimian and
Costin Sorin Bildea
Chemical Process Design

Related Titles

S. Engell (Ed.)

Logistic Optimization of Chemical Production Processes

2008

ISBN 978-3-527-30830-9

L. Puigjaner, G. Heyen (Eds.)

Computer Aided Process and Product Engineering

2006

ISBN 978-3-527-30804-0

K. Sundmacher, A. Kienle, A. Seidel-Morgenstern (Eds.)

Integrated Chemical Processes

Synthesis, Operation, Analysis, and Control

2005

ISBN 978-3-527-30831-6

Alexandre C. Dimian and Costin Sorin Bildea

Chemical Process Design

Computer-Aided Case Studies

WILEY-VCH Verlag GmbH & Co. KGaA

The Authors

Prof. Alexandre C. Dimian
University of Amsterdam
FNWI/HIMS
Nieuwe Achtergracht 166
1018 WW Amsterdam
The Netherlands

Prof. Costin Sorin Bildea
University "Politehnica" Bucharest
Department of Chemical Engineering
Str. Polizu 1
011061 Bucharest
Romania

All books published by Wiley-VCH are carefully produced. Nevertheless, authors, editors, and publisher do not warrant the information contained in these books, including this book, to be free of errors. Readers are advised to keep in mind that statements, data, illustrations, procedural details or other items may inadvertently be inaccurate.

Library of Congress Card No.: applied for

British Library Cataloguing-in-Publication Data
A catalogue record for this book is available from the British Library

Bibliographic information published by the Deutsche Nationalbibliothek
Die Deutsche Nationalbibliothek lists this publication in the Deutsche Nationalbibliografie; detailed bibliographic data are available in the Internet at http://dnb.d-nb.de

© 2008 WILEY-VCH Verlag GmbH & Co. KGaA, Weinheim

All rights reserved (including those of translation into other languages). No part of this book may be reproduced in any form – by photoprinting, microfilm, or any other means – nor transmitted or translated into a machine language without written permission from the publishers. Registered names, trademarks, etc. used in this book, even when not specifically marked as such, are not to be considered unprotected by law.

Printed in the Federal Republic of Germany
Printed on acid-free paper

Cover design wmx design, Heidelberg
Typesetting SNP Best-set Typesetter Ltd., Hong Kong
Printing Strauss GmbH, Mörlenbach
Bookbinding Litges & Dopf GmbH, Heppenheim

ISBN: 978-3-527-31403-4

Contents

Preface XV

1 Integrated Process Design 1
1.1 Motivation and Objectives 1
1.1.1 Innovation Through a Systematic Approach 1
1.1.2 Learning by Case Studies 2
1.1.3 Design Project 3
1.2 Sustainable Process Design 5
1.2.1 Sustainable Development 5
1.2.2 Concepts of Environmental Protection 5
1.2.2.1 Production-Integrated Environmental Protection 6
1.2.2.2 End-of-pipe Antipollution Measures 7
1.2.3 Efficiency of Raw Materials 7
1.2.4 Metrics for Sustainability 9
1.3 Integrated Process Design 13
1.3.1 Economic Incentives 13
1.3.2 Process Synthesis and Process Integration 14
1.3.3 Systematic Methods 15
1.3.3.1 Hierarchical Approach 16
1.3.3.2 Pinch-Point Analysis 16
1.3.3.3 Residue Curve Maps 16
1.3.3.4 Superstructure Optimization 17
1.3.3.5 Controllability Analysis 17
1.3.4 Life Cycle of a Design Project 17
1.4 Summary 19
References 20

2 Process Synthesis by Hierarchical Approach 21
2.1 Hierarchical Approach of Process Design 22
2.2 Basis of Design 27
2.2.1 Economic Data 27
2.2.2 Plant and Site Data 27
2.2.3 Safety and Health Considerations 28

Chemical Process Design: Computer-Aided Case Studies. Alexandre C. Dimian and Costin Sorin Bildea
Copyright © 2008 WILEY-VCH Verlag GmbH & Co. KGaA, Weinheim
ISBN: 978-3-527-31403-4

2.2.4	Patents	28
2.3	Chemistry and Thermodynamics	28
2.3.1	Chemical-Reaction Network	28
2.3.2	Chemical Equilibrium	31
2.3.3	Reaction Engineering Data	31
2.3.4	Thermodynamic Analysis	32
2.4	Input/Output Analysis	32
2.4.1	Input/Output Structure	33
2.4.1.1	Number of Outlet Streams	34
2.4.1.2	Design Variables	35
2.4.2	Overall Material Balance	35
2.4.3	Economic Potential	36
2.5	Reactor/Separation/Recycle Structure	41
2.5.1	Material-Balance Envelope	41
2.5.1.1	Excess of Reactant	43
2.5.2	Nonlinear Behavior of Recycle Systems	43
2.5.2.1	Inventory of Reactants and Make-up Strategies	43
2.5.2.2	Snowball Effects	44
2.5.2.3	Multiple Steady States	45
2.5.2.4	Minimum Reactor Volume	45
2.5.2.5	Control of Selectivity	45
2.5.3	Reactor Selection	45
2.5.3.1	Reactors for Homogeneous Systems	46
2.5.3.2	Reactors for Heterogeneous Systems	46
2.5.4	Reactor-Design Issues	47
2.5.4.1	Heat Effects	47
2.5.4.2	Equilibrium Limitations	48
2.5.4.3	Heat-Integrated Reactors	48
2.5.4.4	Economic Aspects	49
2.6	Separation System Design	49
2.6.1	First Separation Step	50
2.6.1.1	Gas/Liquid Systems	50
2.6.1.2	Gas/Liquid/Solid Systems	51
2.6.2	Superstructure of the Separation System	51
2.7	Optimization of Material Balance	54
2.8	Process Integration	55
2.8.1	Pinch-Point Analysis	55
2.8.1.1	The Overall Approach	56
2.8.2	Optimal Use of Resources	58
2.9	Integration of Design and Control	58
2.10	Summary	58
	References	60

3	**Synthesis of Separation System** 61	
3.1	Methodology 61	
3.2	Vapor Recovery and Gas-Separation System 64	
3.2.1	Separation Methods 64	
3.2.2	Split Sequencing 64	
3.3	Liquid-Separation System 71	
3.3.1	Separation Methods 72	
3.3.2	Split Sequencing 73	
3.4	Separation of Zeotropic Mixtures by Distillation 75	
3.4.1	Alternative Separation Sequences 75	
3.4.2	Heuristics for Sequencing 76	
3.4.3	Complex Columns 77	
3.4.4	Sequence Optimization 78	
3.5	Enhanced Distillation 79	
3.5.1	Extractive Distillation 79	
3.5.2	Chemically Enhanced Distillation 79	
3.5.3	Pressure-Swing Distillation 79	
3.6	Hybrid Separations 79	
3.7	Azeotropic Distillation 84	
3.7.1	Residue Curve Maps 84	
3.7.2	Separation by Homogeneous Azeotropic Distillation 88	
3.7.2.1	One Distillation Field 88	
3.7.2.2	Separation in Two Distillation Fields 89	
3.7.3	Separation by Heterogeneous Azeotropic Distillation 95	
3.7.4	Design Methods 98	
3.8	Reactive Separations 99	
3.8.1	Conceptual Design of Reactive Distillation Columns 100	
3.9	Summary 101	
	References 101	
4	**Reactor/Separation/Recycle Systems** 103	
4.1	Introduction 103	
4.2	Plantwide Control Structures 106	
4.3	Processes Involving One Reactant 108	
4.3.1	Conventional Control Structure 108	
4.3.2	Feasibility Condition for the Conventional Control Structure 111	
4.3.3	Control Structures Fixing Reactor-Inlet Stream 112	
4.3.4	Plug-Flow Reactor 114	
4.4	Processes Involving Two Reactants 115	
4.4.1	Two Recycles 115	
4.4.2	One Recycle 117	
4.5	The Effect of the Heat of Reaction 118	
4.5.1	One-Reactant, First-Order Reaction in PFR/Separation/Recycle Systems 118	

4.6	Example–Toluene Hydrodealkylation Process	122
4.7	Conclusions	126
	References	127

5 Phenol Hydrogenation to Cyclohexanone 129

5.1	Basis of Design	129
5.1.1	Project Definition	129
5.1.2	Chemical Routes	130
5.1.3	Physical Properties	131
5.2	Chemical Reaction Analysis	132
5.2.1	Chemical Reaction Network	132
5.2.2	Chemical Equilibrium	133
5.2.2.1	Hydrogenation of Phenol	133
5.2.2.2	Dehydrogenation of Cyclohexanol	135
5.2.3	Kinetics	137
5.2.3.1	Phenol Hydrogenation to Cyclohexanone	137
5.2.3.2	Cyclohexanol Dehydrogenation	139
5.3	Thermodynamic Analysis	140
5.4	Input/Output Structure	141
5.5	Reactor/Separation/Recycle Structure	144
5.5.1	Phenol Hydrogenation	144
5.5.1.1	Reactor-Design Issues	145
5.5.2	Dehydrogenation of Cyclohexanol	151
5.5.2.1	Reactor Design	151
5.6	Separation System	152
5.7	Material-Balance Flowsheet	153
5.7.1	Simulation	153
5.7.2	Sizing and Optimization	155
5.8	Energy Integration	156
5.9	One-Reactor Process	158
5.10	Process Dynamics and Control	161
5.10.1	Control Objectives	161
5.10.2	Plantwide Control	162
5.11	Environmental Impact	166
5.12	Conclusions	170
	References	172

6 Alkylation of Benzene by Propylene to Cumene 173

6.1	Basis of Design	173
6.1.1	Project Definition	173
6.1.2	Manufacturing Routes	173
6.1.3	Physical Properties	175
6.2	Reaction-Engineering Analysis	176
6.2.1	Chemical-Reaction Network	176
6.2.2	Catalysts for the Alkylation of Aromatics	178

6.2.3	Thermal Effects	180
6.2.4	Chemical Equilibrium	181
6.2.5	Kinetics	181
6.3	Reactor/Separator/Recycle Structure	183
6.4	Mass Balance and Simulation	185
6.5	Energy Integration	187
6.6	Complete Process Flowsheet	192
6.7	Reactive Distillation Process	195
6.8	Conclusions	199
	References	200

7	**Vinyl Chloride Monomer Process**	**201**
7.1	Basis of Design	201
7.1.1	Problem Statement	201
7.1.2	Health and Safety	202
7.1.3	Economic Indices	202
7.2	Reactions and Thermodynamics	202
7.2.1	Process Steps	202
7.2.2	Physical Properties	205
7.3	Chemical-Reaction Analysis	205
7.3.1	Direct Chlorination	206
7.3.2	Oxychlorination	208
7.3.3	Thermal Cracking	210
7.4	Reactor Simulation	212
7.4.1	Ethylene Chlorination	212
7.4.2	Pyrolysis of EDC	212
7.5	Separation System	213
7.5.1	First Separation Step	213
7.5.2	Liquid-Separation System	215
7.6	Material-Balance Simulation	216
7.7	Energy Integration	219
7.8	Dynamic Simulation and Plantwide Control	222
7.9	Plantwide Control of Impurities	224
7.10	Conclusions	229
	References	229

8	**Fatty-Ester Synthesis by Catalytic Distillation**	**231**
8.1	Introduction	231
8.2	Methodology	232
8.3	Esterification of Lauric Acid with 2-Ethylhexanol	235
8.3.1	Problem Definition and Data Generation	235
8.3.2	Preliminary Chemical and Phase Equilibrium	236
8.3.3	Equilibrium-based Design	238
8.3.4	Thermodynamic Experiments	239
8.3.5	Revised Conceptual Design	240

8.3.6	Chemical Kinetics Analysis	*241*
8.3.6.1	Kinetic Experiments	*241*
8.3.6.2	Selectivity Issues	*242*
8.3.6.3	Catalyst Effectiveness	*243*
8.3.7	Kinetic Design	*244*
8.3.7.1	Selection of Internals	*245*
8.3.7.2	Preliminary Hydraulic Design	*246*
8.3.7.3	Simulation	*248*
8.3.8	Optimization	*250*
8.3.9	Detailed Design	*251*
8.4	Esterification of Lauric Acid with Methanol	*251*
8.5	Esterification of Lauric Acid with Propanols	*254*
8.5.1	Entrainer Selection	*255*
8.5.2	Entrainer Ratio	*257*
8.6	Conclusions	*258*
	References	*259*

9	**Isobutane Alkylation** *261*	
9.1	Introduction	*261*
9.2	Basis of Design	*263*
9.2.1	Industrial Processes for Isobutane Alkylation	*263*
9.2.2	Specifications and Safety	*263*
9.2.3	Chemistry	*264*
9.2.4	Physical Properties	*265*
9.2.5	Reaction Kinetics	*265*
9.3	Input–Output Structure	*267*
9.4	Reactor/Separation/Recycle	*268*
9.4.1	Mass-Balance Equations	*268*
9.4.2	Selection of a Robust Operating Point	*272*
9.4.3	Normal-Space Approach	*274*
9.4.3.1	Critical Manifolds	*274*
9.4.3.2	Distance to the Critical Manifold	*275*
9.4.3.3	Optimization	*277*
9.4.4	Thermal Design of the Chemical Reactor	*278*
9.5	Separation Section	*280*
9.6	Plantwide Control and Dynamic Simulation	*281*
9.7	Discussion	*284*
9.8	Conclusions	*285*
	References	*285*

10	**Vinyl Acetate Monomer Process** *287*	
10.1	Basis of Design	*287*
10.1.1	Manufacturing Routes	*287*
10.1.2	Problem Statement	*288*
10.1.3	Health and Safety	*289*

10.2	Reactions and Thermodynamics	289
10.2.1	Reaction Kinetics	289
10.2.2	Physical Properties	293
10.2.3	VLE of Key Mixtures	294
10.3	Input–Output Analysis	294
10.3.1	Preliminary Material Balance	294
10.4	Reactor/Separation/Recycles	296
10.5	Separation System	298
10.5.1	First Separation Step	299
10.5.2	Gas-Separation System	300
10.5.3	Liquid-Separation System	300
10.6	Material-Balance Simulation	302
10.7	Energy Integration	304
10.8	Plantwide Control	305
10.9	Conclusions	310
	References	311

11	**Acrylonitrile by Propene Ammoxidation**	313
11.1	Problem Description	313
11.2	Reactions and Thermodynamics	314
11.2.1	Chemistry Issues	314
11.2.2	Physical Properties	317
11.2.3	VLE of Key Mixtures	318
11.3	Chemical-Reactor Analysis	319
11.4	The First Separation Step	321
11.5	Liquid-Separation System	324
11.5.1	Development of the Separation Sequence	324
11.5.2	Simulation	324
11.6	Heat Integration	328
11.7	Water Minimization	332
11.8	Emissions and Waste	334
11.8.1	Air Emissions	334
11.8.2	Water Emissions	334
11.8.3	Catalyst Waste	335
11.9	Final Flowsheet	335
11.10	Further Developments	337
11.11	Conclusions	337
	References	338

12	**Biochemcial Process for NO_x Removal**	339
12.1	Introduction	339
12.2	Basis of Design	341
12.3	Process Selection	341
12.4	The Mathematical Model	343
12.4.1	Diffusion-Reaction in the Film Region	343

12.4.1.1 Model Parameters *346*
12.4.2 Simplified Film Model *348*
12.4.3 Convection-Mass-Transfer Reaction in the Bulk *351*
12.4.3.1 Bulk Gas *351*
12.4.3.2 Bulk Liquid *352*
12.4.4 The Bioreactor *354*
12.5 Sizing of the Absorber and Bioreactor *355*
12.6 Flowsheet and Process Control *357*
12.7 Conclusions *358*
References *360*

13 PVC Manufacturing by Suspension Polymerization *363*
13.1 Introduction *363*
13.1.1 Scope *363*
13.1.2 Economic Issues *363*
13.1.3 Technology *365*
13.2 Large-Scale Reactor Technology *365*
13.2.1 Efficient Heat Transfer *367*
13.2.2 The Mixing Systems *369*
13.2.3 Fast Initiation Systems *370*
13.3 Kinetics of Polymerization *371*
13.3.1 Simplified Analysis *374*
13.4 Molecular-Weight Distribution *376*
13.4.1 Simplified Analysis *377*
13.5 Kinetic Constants *378*
13.6 Reactor Design *378*
13.6.1 Mass Balance *379*
13.6.2 Molecular-Weight Distribution *382*
13.6.3 Heat Balance *383*
13.6.4 Heat-Transfer Coefficients *384*
13.6.5 Physical Properties *385*
13.6.6 Geometry of the Reactor *385*
13.6.7 The Control System *385*
13.7 Design of the Reactor *388*
13.7.1 Additional Cooling Capacity by Means of an External Heat Exchanger *389*
13.7.2 Additional Cooling Capacity by Means of Higher Heat-Transfer Coefficient *390*
13.7.3 Design of the Jacket *390*
13.7.4 Dynamic Simulation Results *390*
13.7.5 Additional Cooling Capacity by Means of Water Addition *392*
13.7.6 Improving the Controllability of the Reactor by Recipe Change *393*
13.8 Conclusions *396*
References *396*

14	**Biodiesel Manufacturing** *399*	
14.1	Introduction to Biofuels *399*	
14.1.1	Types of Alternative Fuels *399*	
14.1.2	Economic Aspects *401*	
14.2	Fundamentals of Biodiesel Manufacturing *402*	
14.2.1	Chemistry *402*	
14.2.2	Raw Materials *404*	
14.2.3	Biodiesel Specifications *405*	
14.2.4	Physical Properties *406*	
14.3	Manufacturing Processes *409*	
14.3.1	Batch Processes *409*	
14.3.2	Catalytic Continuous Processes *411*	
14.3.3	Supercritical Processes *413*	
14.3.4	Hydrolysis and Esterification *414*	
14.3.5	Enzymatic Processes *415*	
14.3.6	Hydropyrolysis of Triglycerides *415*	
14.3.7	Valorization of Glycerol *416*	
14.4	Kinetics and Catalysis *416*	
14.4.1	Homogeneous Catalysis *416*	
14.4.2	Heterogeneous Catalysis *419*	
14.5	Reaction-Engineering Issues *420*	
14.6	Phase-Separation Issues *422*	
14.7	Application *423*	
14.8	Conclusions *426*	
	References *427*	
15	**Bioethanol Manufacturing** *429*	
15.1	Introduction *429*	
15.2	Bioethanol as Fuel *429*	
15.3	Economic Aspects *431*	
15.4	Ecological Aspects *433*	
15.5	Raw Materials *435*	
15.6	Biorefinery Concept *437*	
15.6.1	Technology Platforms *437*	
15.6.2	Building Blocks *439*	
15.7	Fermentation *440*	
15.7.1	Fermentation by Yeasts *440*	
15.7.2	Fermentation by Bacteria *441*	
15.7.3	Simultaneous Saccharification and Fermentation *441*	
15.7.4	Kinetics of Saccharification Processes *442*	
15.7.5	Fermentation Reactors *444*	
15.8	Manufacturing Technologies *445*	
15.8.1	Bioethanol from Sugar Cane and Sugar Beets *445*	
15.8.2	Bioethanol from Starch *446*	
15.8.3	Bioethanol from Lignocellulosic Biomass *447*	

15.9 Process Design: Ethanol from Lignocellulosic Biomass *449*
15.9.1 Problem Definition *449*
15.9.2 Definition of the Chemical Components *450*
15.9.3 Biomass Pretreatment *450*
15.9.4 Fermentation *452*
15.9.5 Ethanol Purification and Water Recovery *456*
15.10 Conclusions *458*
References *459*

Appendix A Residue Curve Maps for Reactive Mixtures *461*
Appendix B Heat-Exchanger Design *474*
Appendix C Materials of Construction *483*
Appendix D Saturated Steam Properties *487*
Appendix E Vapor Pressure of Some Hydrocarbons *489*
Appendix F Vapor Pressure of Some Organic Components *490*
Appendix G Conversion Factors to SI Units *491*

Index *493*

Preface

> "I hear and I forget. I see and I remember. I do and I understand."
> *Confucius*

Chemical process design today faces the challenge of sustainable technologies for manufacturing fuels, chemicals and various products by extended use of renewable raw materials. This implies a profound change in the education of designers in the sense that their creativity can be boosted by adopting a systems approach supported by powerful systematic methods and computer simulation tools. Instead of developing a single presumably good flowsheet, modern process design generates and evaluates several alternatives corresponding to various design decisions and constraints. Then, the most suitable alternative is refined and optimized with respect to high efficiency of materials and energy, ecologic performance and operability.

This book deals with the conceptual design of chemical processes illustrated by case studies worked out by computer simulation. Typically, more than 80% of the total investment costs of chemical plants are determined at the conceptual design stage, although this activity involves only 2–3% of the engineering costs and a reduced number of engineers. In addition, a preliminary design allows critical aspects in research and development and/or in searching subcontractors to be highlighted, well ahead of starting the actual plant design project.

The book is aimed at a wide audience interested in the design of innovative chemical processes, especially chemical engineering undergraduate students completing a process and/or plant design project. Postgraduate and PhD students will find advanced and thought-provoking process-design methods. The information presented in the book is also useful for the continuous education of professional designers and R&D engineers.

This book uses ample case studies to teach a generic design methodology and systematic design methods, as explained in the first four chapters. Each project starts by analysing the fundamental knowledge about chemistry, thermodynamics and reaction kinetics. Environmental problems are highlighted by analysing the detailed chemistry. On this basis the process synthesis is performed. The result is the generation of several alternatives from which the most suitable is selected for refinement, energy integration, optimization and plantwide control. Computer

Chemical Process Design: Computer-Aided Case Studies. Alexandre C. Dimian and Costin Sorin Bildea
Copyright © 2008 WILEY-VCH Verlag GmbH & Co. KGaA, Weinheim
ISBN: 978-3-527-31403-4

simulation is intensively used for data analysis, supporting design decisions, investigating the feasibility, sizing the equipment, and finally for studying process dynamics and control issues. The results are compared with flowsheets and performance indices of industrial licensed processes. Complete information is given such that the case studies can be reproduced with any simulator having adequate capabilities.

The distinctive feature of this book is the emphasis on integrating process dynamics and plant wide control, starting with the early stages of conceptual design. Considering the reaction/separation/recycle structure as the architectural framework and employing kinetic modelling of chemical reactors render this approach suited for developing flexible and adaptive processes. Although the progress in software technology makes possible the use of dynamic simulation directly in the conceptual design phase, the capabilities of dynamic simulators are largely underestimated, because little experience has been disseminated. From this perspective the book can be seen as a practical guide for the efficient use of dynamic simulation in process design and control.

The book extends over fifteen chapters. The first four chapters deal with the fundamentals of a modern process design, while their application is developed in the next eleven case studies.

Chapter 1 *Introduction* presents the concepts and metrics of sustainable development, as well as the framework of an integrated process design by means of two interlinked activities, process synthesis and process integration.

The conceptual design framework is developed in Chapter 2 *Process Synthesis by Hierarchical Approach*. An efficient methodology is proposed aiming to minimize the interactions between the synthesis and integration steps. The core activity concentrates on the reactor/separation/recycle structure as defining the process architecture, by which the reactor design and the structure of separations are examined simultaneously by considering the effect of recycles on flexibility and stability. By placing the reactor in the core of the process, the separators receive clearly defined tasks of plantwide perspective, which should be fulfilled later by the design of the respective subsystems. The heat and material balances built upon this structure supply the key elements for sizing the units and assessing capital and operation costs, and on this basis establish the process profitability.

Chapter 3 deals with the *Synthesis of the Separation System*. A task-oriented approach is proposed for generating close-to-optimum separation sequences for which both feasibility and performance of splits are guaranteed. Emphasis is placed on the synthesis of distillation systems by residue curve map methods.

Chapter 4 deals in more detail with the analysis of the *Reactor/Separation/Recycle Systems*. Undesired nonlinear phenomena can be detected at early conceptual stages through steady-state sensitivity and dynamic stability analysis. This approach, developed by the authors, allows better integration between process design and plantwide control. Two different approaches to plantwide control are discussed, namely controlling the material balance of the plant by using the self-regulation property or by applying feedback control.

The first case study of Chapter 5 *Cyclohexanone by Phenol Hydrogenation* developed in a tutorial manner, allows the reader to navigate through the key steps of the methodology, from thermodynamic analysis to reactor design, flowsheet synthesis and simulation. The key issue is designing a plant that complies with flexibility and selectivity targets. The initial design of the plant contains two reaction sections, but selective catalyst and adequate recycle policy allow an efficient and versatile single reactor process to be developed. In addition, the case study deals with waste reduction by design, with both economical and ecological benefits.

Chapter 6 on *Alkylation of Benzene by Propene to Cumene* illustrates the design of a modern process for a petrochemical commodity. The process employs a zeolite catalyst and an adiabatic reactor operated at higher pressure. Large benzene recycle limits the formation of byproducts, but implies considerable energy consumption. Significant energy saving can be achieved by heat integration by using double-effect distillation and recovering the reaction heat as medium-pressure steam. The performance indices of the designed process are in agreement with the best technologies. A modern alternative is catalytic reactive distillation. While appealing at first sight, this method raises a number of problems. Reactive distillation can bring benefits only if a superior catalyst is available, exhibiting much higher activity and better selectivity than the liquid-phase processes.

Chapter 7 *Vinyl Chloride Monomer Process* emphasizes the complexity of designing a large chemical plant with multireactors and an intricate structure of recycles. The raw materials efficiency is close to reaction stoichiometry such that only the VCM product leaves the plant. Because a large spectrum of chloro-hydrocarbon impurities is formed, the purification of the intermediate ethylene di-chloride becomes a complex design and plantwide control problem. The solution implies not only the removal of impurities accumulating in recycle by more efficient separators, but also their minimization at source by improving the reaction conditions. In particular, the yield of pyrolysis can be enhanced by making use of initiators, some being produced and recycled in the process itself. In addition, the chemical conversion of impurities accumulating in recycle prevents the occurrence of snowball effects that otherwise affect the operation of reactors and separators. Steady-state and dynamic simulation models can greatly help to solve properly this integrated design and control problem.

Chapter 8 deals with the manufacturing of *Fatty Esters by Reactive Distillation* using superacid solid catalyst. The key constraint is selective water removal to shift the chemical equilibrium and to ensure a water-free organic phase. Because the catalyst manifests similar activities for several alcohols, the study investigates the possibility of designing a multiproduct reactive distillation column by slightly adjusting the operation conditions. The residue curve map analysis brings useful insights. The esterification with propanols raises the problem of breaking the alcohol/water azeotrope. The solution passes by the use of an entrainer. The equipment is simple and efficient. The availability of an active and selective catalyst remains the key element in technology.

Chapter 9 *Isobutane/Butene Alkylation* illustrates in detail the integration of design and plantwide control. Special attention is paid to the reaction/separation/

recycle structure, showing how plantwide control considerations are introduced during the early stages of conceptual design. Thus, a simplified plant mass balance based on a kinetic model for the reactor and black-box separation models is used to generate plantwide control alternatives. Nonlinear analysis reveals unfavourable steady state behavior, such as high sensitivity and state multiplicity. An important part is devoted to robustness study in order to ensure feasible operation when operation variables change or the design parameters are uncertain.

The case study on *Vinyl Acetate Process*, developed in Chapter 10, demonstrates the benefit of solving a process design and plantwide control problem based on the analysis of the reactor/separation/recycles structure. In particular, it is demonstrated that the dynamic behavior of the chemical reactor and the recycle policy depend on the mechanism of the catalytic process, as well as on the safety constraints. Because low per pass conversion of both ethylene and acetic acid is needed, the temperature profile in the chemical reactor becomes the most important means for manipulating the reaction rate and hence ensuring the plant flexibility. The inventory of reactants is adapted accordingly by fresh reactant make-up directly in recycles.

Chapter 11 *Acrylonitrile by Ammoxidation of Propene* illustrates the synthesis of a flowsheet in which a difficult separation problem dominates. In addition, large energy consumption of both low- and high-temperature utilities is required. Various separation methods are involved from simple flash and gas absorption to extractive distillation for splitting azeotropic mixtures. The problem is tackled by an accurate thermodynamic analysis. Important energy saving can be detected.

Chapter 12 handles the design of a *Biochemical Process for NO_x Removal* from flue gases. The process involves absorption and reaction steps. The analysis of the process kinetics shows that both large G/L interfacial area and small liquid fraction favor the absorption selectivity. Consequently, a spray tower is employed as the main process unit for which a detailed model is built. Model analysis reveals reasonable assumptions, which are the starting point of an analytical model. Then, the values of the critical parameters of the coupled absorber–bioreactor system are found. Sensitivity studies allow providing sufficient overdesign that ensures the purity of the outlet gas stream when faced with uncertain design parameters or with variability of the input stream.

Chapter 13 *PVC Manufacturing by Suspension Polymerization* illustrates the area of batch processes and product engineering. The central problem is the optimization of a polymerization recipe ensuring the highest productivity (shortest batch time) of a large-scale reactor with desired product-quality specifications defined by molecular weight distribution. A comprehensive dynamic model is built by combining detailed reaction kinetics, heat transfer and process-control system. The model can be used for the optimization of the polymerization recipe and the operation procedure in view of producing different polymer grades.

The last two chapters are devoted to problems of actual interest, manufacturing biofuels from renewable raw materials. Chapter 14 deals with *Biodiesel Manufacturing*. This renewable fuel is a mixture of fatty acid esters that can be obtained from vegetable or animal fats by reaction with light alcohols. A major aspect in technology is getting a composition of the mixture leaving the reactor system that matches the fuel specifications. This is difficult to achieve in view of the large variety of raw materials. On the basis of kinetic data, the design of a standard biodiesel process based on homogeneous catalysis is performed. The study demonstrates that employing heterogeneous catalysis can lead to a much simpler and more efficient design. The availability of superactive and robust catalysts is still an open problem.

Bioethanol Manufacturing is handled in Chapter 15. The case study examines different aspects of today's technologies, such as raw materials basis, fermentation processes and bioreactors. The application deals with the design of a bioethanol plant of the second generation based on lignocellulosic biomass. Emphasis is placed on getting realistic and consistent material and energy balances over the whole plant by means of computer simulation in order to point out the impact of the key technical elements on the investment and operation costs. To achieve this goal the complicated biochemistry is expressed in term of stoichiometric reactions and user-defined components. The systemic analysis emphasizes the key role of the biomass conversion stage based on simultaneous saccharification and fermentation.

The book is completed with *Annexes* on the analysis of reactive mixtures by residue curve maps, design of heat exchangers, selection of construction materials, steam tables, vapor pressure of typical chemical components and conversion table for the common physical units.

The authors acknowledge the contribution to this book of many colleagues and students from the University of Amsterdam and Delft University of Technology, The Netherlands. Special thanks go to the Dutch Postgraduate School for Process Technology (OSPT) for supporting our postgraduate course in Advanced Process Integration and Plantwide Control, where the integration of design and control is the main feature. The authors express their appreciation to the software companies AspenTech and MathWorks for making available for education purposes an outstanding simulation technology.

And last but not the least we express our gratitude and love to our families, for continuous support and understanding.

January 2008

Alexandre C. Dimian
Costin Sorin Bildea

1
Integrated Process Design

1.1
Motivation and Objectives

1.1.1
Innovation Through a Systematic Approach

Innovation is the key issue in chemical process industries in today's globalization environment, as the best means to achieve high efficiency and competitiveness with sustainable development. The job of a designer is becoming increasingly challenging. He/she has to take into account a large number of constraints of technical, economical and social nature, often contradictory. For example, the discovery of a new catalyst could make profitable cheaper raw materials, but needs much higher operating temperatures and pressures. To avoid the formation of byproducts lower conversion should be maintained, implying more energy and equipment costs. Although attractive, the process seems more expensive. However, higher temperature can give better opportunities for energy saving by process integration. In addition, more compact and efficient equipment can be designed by applying the principles of process synthesis and intensification. In the end, the integrated conceptual design may reveal a simpler flowsheet with lower energy consumption and equipment costs.

The above example is typical. Modern process design consists of the optimal combination of technical, economic, ecological and social aspects in highly *integrated processes*. The conceptual approach implies the availability of effective cost-optimization design methods aided by powerful computer-simulation tools.

Creativity is a major issue in process design. This is not a matter only of engineering experience, but above all of adopting the approach of *process systems*. This consists of a systemic viewpoint in problem analysis supported by systematic methods in process design.

A systematic and systems approach has at least two merits:

1. Provides guidance in assessing firstly the *feasibility* of the process design as a whole, as well as its flexibility in operation, before more detailed design of components.

Chemical Process Design: Computer-Aided Case Studies. Alexandre C. Dimian and Costin Sorin Bildea
Copyright © 2008 WILEY-VCH Verlag GmbH & Co. KGaA, Weinheim
ISBN: 978-3-527-31403-4

2. Generates not only one supposed optimal solution, but several good *alternatives* corresponding to different *design decisions*. A remarkable feature of the systemic design is that quasioptimal targets may be set well ahead detailed sizing of equipment. In this way, the efficiency of the whole engineering work may improve dramatically by avoiding costly structural modifications in later stages.

The motivation of this book consists of using a wide range of case studies to teach generic creative issues, but incorporated in the framework of a technology of industrial significance. Computer simulation is used intensively to investigate the feasibility and support design decisions, as well as for sizing and optimization. Particular emphasis is placed on thermodynamic modeling as a fundamental tool for analysis of reactions and separations. Most of the case studies make use of chemical reactor design by kinetic modeling.

A distinctive feature of this book is the *integration of design and control* as the current challenge in process design. This is required by higher flexibility and responsiveness of large-scale continuous processes, as well as by the optimal operation of batchwise and cyclic processes for high-value products.

The case studies cover key applications in chemical process industries, from petrochemistry to polymers and biofuels. The selection of processes was confronted with the problem of availability of sufficient design and technology data. The development of the flowsheet and its integration is based on employing a systems viewpoint and systematic process synthesis techniques, amply explained over three chapters. In consequence, the solution contains elements of originality, but in each case this is compared to schemes and economic indices reported in the literature.

1.1.2
Learning by Case Studies

Practising is the best way to learn. "I see, I hear and I forget", says an old adage, which is particularly true for passive slide-show lectures. On the contrary, "I see, I do and I understand" enables effective education and gives enjoyment.

There are two types of active learning: problem-based and project-based. The former addresses specific questions, exercises and problems, which aim to illustrate and consolidate the theory by varying data, assumptions and methods. On the contrary, the project-based learning, in which we include case studies, addresses complex and open-ended problems. These are more appropriate for solving real-life problems, for which there is no unique solution, but at least a good one, sometime "optimal", depending on constraints and decisions. In more challenging cases a degree of uncertainty should be assumed and justified.

The principal merits of learning by case studies are that they:

1. bridge the gap between theory and practice, by challenging the students,
2. make possible better integration of knowledge from different disciplines,

3. encourage personal involvement and develop problem-solving attitude,
4. develop communication, teamwork skills and respect of schedule,
5. enable one to learn to write professional reports and making quality presentations,
6. provide fun while trying to solve difficult matters.

There are also some disadvantages that should be kept in mind, such as:

1. frustration if the workload is uneven,
2. difficulties for some students to maintain the pace,
3. complications in the case of failure of project management or leadership,
4. possibility of unfair evaluation.

The above drawbacks, merely questions of project organization, can be reduced to a minimum by taking into account the following measures:

1. provide clear definition of content, deliverables, scheduling and evaluation,
2. provide adequate support, regular evaluation of the team and of each member. If possible, separate support end evaluation, as customer/contractor relation,
3. evaluate the project by public presentation, but with individual marks,
4. propose challenging subjects issued from industry or from own research,
5. attract specialists from industry for support and evaluation.

1.1.3
Design Project

Teaching modern chemical process design can be organized at two levels:

- Teach a systems approach and systematic methods in the framework of a *process design and integration* introductory course. A period of 4–6 weeks fulltime (160 to 240 h) should be sufficient. Here, a first *process-integration project* is proposed, which can be performed individually or in small groups.

- Consolidate the engineering skills in the framework of a larger *plant design project*. A typical duration is 10–12 weeks full time with groups of 3–5 students.

Although dissimilar in extension and purpose, these projects largely share the content, as illustrated by Fig. 1.1. The main points of the approach are as follows:

1. Provide clear *definition* of the design problem. Collect sufficient engineering data. Get a comprehensive picture of chemistry and reaction conditions, thermal effects and chemical equilibrium, as well as about safety, toxicity and environmental problems. Examine the availability of physical properties for components and mixtures of significance. Identify azeotropes and key binaries. Define the key constraints.

2. The basic flowsheet structure is given by the reactor and separation systems. Alternatives can be developed by applying *process-synthesis methods*. Use computer simulation to get physical insights into different conceptual issues and to evaluate the performance of different alternatives.

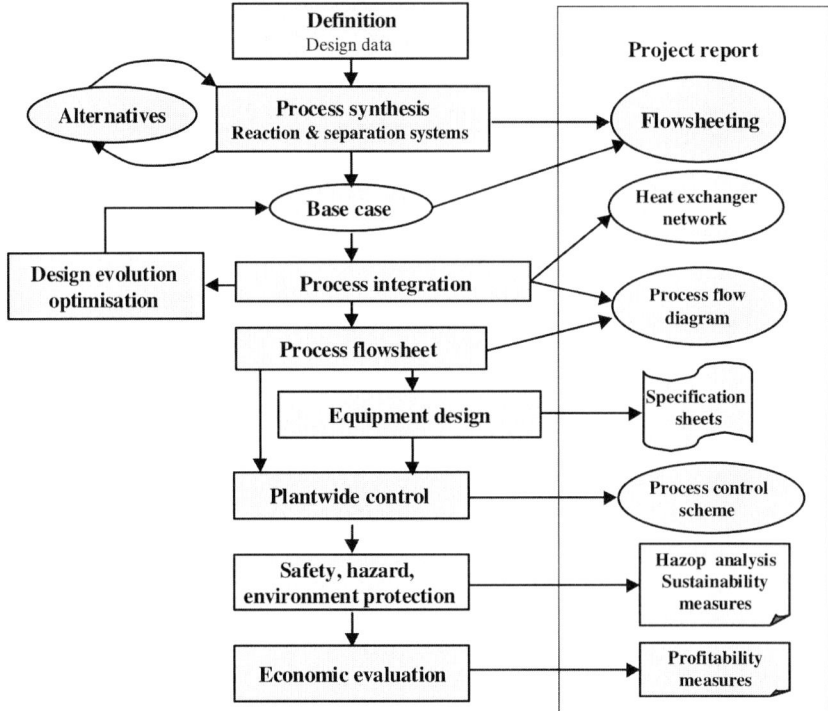

Figure 1.1 Outline of a design project.

3. Select a good base case. Determine a consistent material balance. Improve the design by using *process-integration* techniques. Determine targets for utilities, water and mass-separation agents. Set performance targets for the main equipment. Optimize the final flowsheet.

4. Perform *equipment design*. Collect the key equipment characteristics as *specification sheets*.

5. Examine *plantwide control* aspects, including safety, environment protection, flexibility with respect to production rate, and quality control.

6. Examine measures for *environment protection*. Minimize waste and emissions. Characterize process sustainability.

7. Perform the *economic evaluation*. This should be focused on profitability rather than on an accurate evaluation of costs.

8. Elaborate the *design report*. Defend it by public presentation.

In the process-integration project the goal is to encourage the students to produce original processes rather than imitate proven technologies. The emphasis is on learning a systemic methodology for flowsheet development, as well as suitable systematic methods for the design of subsystems. The emphasis is on generating flowsheet alternatives. The student should understand why several competing technologies can coexist for the same process, and be able to identify the key design decisions in each case. Thus, stimulating the creativity is the key issue at this level.

A more rigorous approach will be taught during the plant-design project. Here, the objective is to develop professional engineering skills, by completing a design project at a level of quality close to an engineering bureau. The subject may be selected from existing and proven technologies, but the rationale of the flowsheet development has to be retraced by a rigorous revision of the conceptual levels and of design decisions at each step. This time the efficiency in using materials and energy, equipment performance and the robustness of the engineering solution are central features. The quality of report and of the public presentation plays a key role in final mark. More information about this approach may be found elsewhere [1].

1.2
Sustainable Process Design

1.2.1
Sustainable Development

Sustainable development designates a production model in which fulfilling the needs of the present society preserves the rights of future generations to meet their own needs. Sustainable development is the result of an equilibrium state between economic success, social acceptance and environmental protection. Ecological sustainability demands safeguarding the natural life and aiming at zero pollution of the environment. Economic sustainability aspires to maximize the use of renewable raw materials and of green energies, and saving in this way valuable fossil resources. Social sustainability has to account of a decent life and respect of human rights in the context of the global free-market economy.

An efficient use of scarce resources by nonpolluting technologies is possible only by a large innovation effort in research, development and design. Sustainability aims at high material yield by the minimization of byproducts and waste. The same is valid for energy, for which considerable saving may be achieved by the heat integration of units and plants.

A systemic approach of the whole supply chain allows the designer to identify the critical stages where inefficient use of raw materials and energy takes place, as well as the sources of toxic materials and pollution. Developing sustainable processes implies the availability of consistent and general accepted sustainability measures. A comprehensive analysis should examine the evolution of sustainability over the whole life cycle, namely that raised by the dismantling the plant.

1.2.2
Concepts of Environmental Protection

In general, a manufacturing process can be described by the following relation [2]:

1 Integrated Process Design

$$(A + B + I) + (M + C + H) \xrightarrow{E} P + S + R + W + F$$

The inputs – main reactants A, coreactants B and impurities I – shape the generic category of *raw materials*. In addition, *auxiliary materials* are needed for technological reasons, as reaction medium M, catalyst C, and helping chemicals H. The process requires, naturally, an amount of energy E. The outputs are: main products P, secondary products S, residues R and waste W. The term *residue* signifies all byproducts and impurities produced by reaction, including those generated from the impurities entered with the raw materials. Impurities have no selling value and are harmful to the environment. On the contrary, the secondary products may be sold. The term *waste* means materials that cannot be recycled in the process. Waste can originate from undesired reactions involving the raw materials, as well as from the degradation of the reaction medium, of the catalyst, or of other helping chemicals. The term *F* accounts for gas emissions, as CO_2, SO_2 or NO_x, produced in the process or by the generation of steam and electricity.

There are two approaches for achieving minimum waste in industry, as illustrated by Fig. 1.2 [2], briefly explained below.

1.2.2.1 Production-Integrated Environmental Protection

By this approach, the solution of the ecological problems results fundamentally from the conceptual process design. Two directions can be envisaged:

- Intrinsically protection, by eliminating at source the risk of pollution.
- Full recycling of byproducts and waste in the manufacturing process itself.

In an ecologically integrated process only saleable products should be found in outputs. Inevitably a limited amount of waste will be produced, but the overall yield of raw materials should be close to the stoichiometric requirements. By applying heat-integration techniques the energy consumption can be optimized. The economic analysis has to consider penalties incurred by greenhouse gases (GHG), as well as for the disposal of waste and toxic materials.

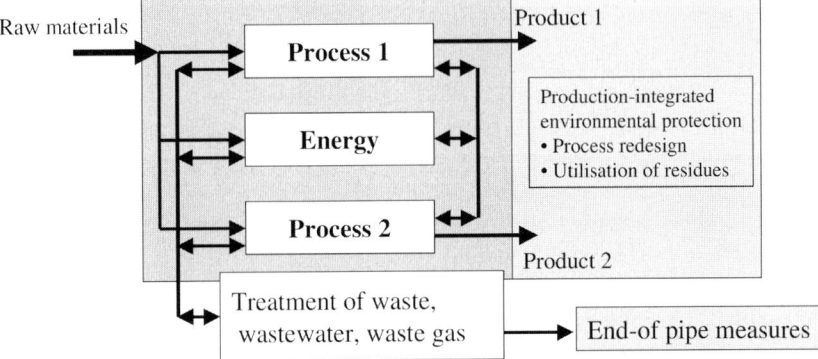

Figure 1.2 Approaches in environmental protection [2].

1.2.2.2 End-of-pipe Antipollution Measures

When a production-integrated approach cannot be applied and the amount of waste is relatively small, then end-of-pipe solutions may be employed. Examples are:

- Transformation of residues in environmental benign compounds, as by incineration or solidification.
- Cleaning of sour gases and toxic components by chemical adsorption.
- Treatment of volatile organic components (VOC) from purges.
- Wastewater treatment.

Obviously, the end-of-pipe measures can fix the problem temporarily, but not remove the cause. Sometimes the problem is shifted or masked into another one. For this reason, an end-of-pipe solution should be examined from a plantwide viewpoint and beyond. For example, sour-gas scrubbing by chemical absorption may cut air pollution locally, but involves the pollution created by the manufacture of chemicals elsewhere. In this case, physical processes or using green (recyclable) solvents are more suitable. The best way is the reduction of acid components by changing the chemistry, such as for example using a more selective catalyst.

End-of-pipe measures are implemented in the short term and need modest investment. In contrast, production-integrated environmental protection necessitates longer-term policy committed towards sustainable development.

Summing up, the following measures can be recommended for improving the environmental performances of a process:

- If possible, modify the chemical route.
- Improve the selectivity of the reaction step leading to the desired product by using a more selective catalyst. Make use primarily of heterogeneous solid catalysts, but consider pollution incurred by regeneration. If homogeneous catalysis is more efficient then developing a recycle method is necessary.
- Optimize the conversion that gives the best product distribution. Low conversion gives typically better selectivity, but implies higher recycle costs. Recycle costs can be greatly reduced by employing energy-integration and process-intensification techniques.
- Change the reaction medium that generates pollution problem. For example, replace water by organic solvents that can be recovered and recycled.
- Purify the feeds to chemical reactors to prevent the formation of secondary impurities, which are more difficult to remove.
- Replace toxic or harmful solvents and chemicals with environmentally benign materials.

1.2.3 Efficiency of Raw Materials

Measures can be used to characterize a chemical process in term of environmental efficiency of raw materials, as described below [2]. Consider the reaction:

$$v_A A + v_B B + \ldots \to v_P P + v_R R + v_S S$$

A is the reference reactant, B the coreactant, P the product, R the byproduct (valuable) and S the waste product.

Stoichiometric yield RY is defined as the ratio of the actual product to the theoretical amount that may be obtained from the reference reactant:

$$RY = \frac{v_A}{v_P} \frac{M_A}{M_P} \frac{m_P}{m_A} \tag{1.1}$$

This measure is useful, but gives only a partial image of productivity, since it ignores the contribution of other reactants and auxiliary materials, as well as the formation of secondary valuable products.

The next measures are more adequate for analyzing the efficiency of a process by material-flow analysis (MFA). Two types of materials can be distinguished:

1. Main reaction materials, which are involved in the main reaction leading to the target product. All or a part of these can be found in secondary products and byproducts in the case of more complex reaction schemes, or in residues if some are in excess and nonrecycled.

2. Secondary materials, as those needed for performing the reactions and other physical operations, as catalysts, solvents, washing water, although not participating in the stoichiometric reaction network.

The following definitions are taken from Christ [2] based on studies conducted in Germany by Steinbach (www.btc-steinbach.de).

Theoretical balance yield BA_t is given by the ratio between the moles of the target product and the total moles of the primary raw materials (PRM), including all reactants involved in the stoichiometry of the synthesis route.

$$BA_t = \frac{\text{moles target product}}{\text{moles of primary raw materials}} = \frac{n_P M_P}{\sum (n_A M_A + n_B M_B + \ldots)_{PRM}} \tag{1.2}$$

This measure considers always an ideal process, but in contrast with the stoichiometric yield, takes into account the quantitative contribution of other molecules. For this reason it is equivalent to an "atomic utilization". This parameter is constant over a synthesis route and as a result a measure of material utilization. Thus, it is the maximum productivity to be expected. A lower BA_t value means more waste in intermediate synthesis steps and a signal to improve the chemistry, by fewer intermediate steps or better selectivity.

Real balance yield BA is the ratio of the target product to the total amount of materials, including secondary raw materials (SRM) as solvents and catalysts, and given by:

$$BA = \frac{\text{amount target product}}{\text{amount primary and secondary materials}} = \frac{m_P}{\sum m_{PRM} + \sum m_{SRM}} \tag{1.3}$$

BA is a measure of productivity, which should be maximized by design.

The ratio of the above indices, called *specific balance yield*, is a measure of the raw material efficiency:

$$sp_{BA} = \frac{BA}{BA_t} \qquad (1.4)$$

The same index can be calculated by the following relation:

$$sp_{BA} = F \times RY \times EA_p \qquad (1.5)$$

The factor EA_p characterizes the efficiency of primary raw materials:

$$EA_p = \frac{\text{amount of primary raw materials}}{\text{amount of primary and secondary raw materials}}$$
$$= \frac{m_{PRM}}{\sum m_{PRM} + \sum m_{SRM}} \qquad (1.6)$$

The factor F expresses the excess of primary raw materials, and is defined as:

$$F = \frac{\text{stoichiometric raw materials}}{\text{excess of primary raw materials}} \leq 1 \qquad (1.7)$$

From Eqs. (1.4) and (1.5) one gets:

$$BA = BA_t \times (F \times RY) \times EA_p \qquad (1.8)$$

The crossexamination of the above measures can suggest means for improving the technology, in the first place the real balance yield BA. For example, the use of an excess of reactant can give higher stoichiometric yield RY, but lower real balance yield BA, if the reactant is not recycled. Hence, increasing the efficiency of primary raw materials EA_p to the theoretical limit of one is an objective of the process design. This can be achieved by replacing steps involving unrecoverable reactants and chemicals with operations where their recycle is possible. Thus, recovery and recycle of all materials inside the process is the key to sustainability from the viewpoint of material efficiency.

1.2.4
Metrics for Sustainability

The measure for assessing the sustainability of a process design should consider the complete manufacturing supply chain over the predictable plant life cycle. The metrics should be simple, understandable by a larger public, useful for decision-making agents, consistent and reproducible. The metrics described below [3] have

Example 1.1: Production of Phenone by Acetylation Reaction [2]

Phenone is produced by the acetylation of benzyl chloride with o-xylene via a Friedel–Crafts reaction. Table 1.1 presents the elements of the material balance. Calculate the efficiency of raw materials.

The stoichiometric equation is:

$$C_6H_5\text{-}COCl + (C_6H_4)\text{-}(CH_3)_2 + AlCl_3 + 3H_2O =$$
$$\phantom{C_6H_5\text{-}}140.6 106.2 133.4 48$$

$$(C_6H_5)\text{-}CO\text{-}(C_6H_4)\text{-}(CH_3)_2 + Al(OH)_3 + 4HCl$$
$$210.2 78 146$$

From the relations (1.1) to (1.8) the following values result for the:

$$RY = \frac{n_P}{n_A} = \frac{4.76}{4.98} = 0.956$$

$$BA_t = \left(\frac{m_P}{\sum m_{reactants}}\right)_{theoretical} = \frac{210.2}{(140.6 + 106.2 + 133.4 + 48)} = 0.484$$

$$BA = \left(\frac{m_P}{\sum m_{reactants}}\right)_{real} = \frac{1000}{3300} = 0.303 \quad sp_{BA} = \frac{0.303}{0.484} = 0.626$$

$$EA_p = \frac{PRM}{\sum m_{reactants\,real}} = \frac{(700 + 550 + 700 + 258)}{3300} = 0.669$$

$$F = \frac{n_A \cdot \sum M_{w,reactants}}{PRM} = \frac{4.98(140.6 + 106.2 + 133.4 + 48)}{(700 + 550 + 700 + 258)} = 0.979$$

The calculation shows that the stoichiometric yield RY is acceptable, but the theoretical balance yield BA_t poor, because catalyst complex lost after reaction. A significant improvement would be the use of solid catalyst. Other alternative is regeneration of $AlCl_3$ complex by recycling. The two solutions would lead to the same theoretical yield, but with different costs. Therefore, a deeper investigation should take into account a cost flow analysis too. More details can be found in Christ [2].

these properties. They refer to the same unit of output, the value-added monetary unit,

Value-added dollar ($VA) = Revenues – Costs of raw materials and utilities

that are consistent in the sense that the lower the value the more effective the process, and indicate the same direction. A short description is given below:

Table 1.1 Material balance for the Example 1.1.

	Input				Output		
	Mw	Mass	Moles		Mw	Mass	Moles
PRM				**Target product**			
R-COCl	140.6	700	4.98	Phenone	210.2	1000	4.76
o-Xylene	106.2	550	5.18				
$AlCl_3$	133.4	700	5.25	**Wastewater**			
H_2O	18	258	14.33	$Al(OH)_3$	78	410	5.26
				HCl	36.5	600	16.44
				Other		123	
SRM				**Waste**			
Toluene		900		Toluene		900	
H_2SO_4		192		Other		267	
Total		3300				3300	

PRM: primary raw materials; SRM: secondary raw materials.

1. *Material intensity* is given by the mass of waste per unit of output. Waste is calculated by subtracting the mass of products and saleable subproducts from the raw materials. Water and air are not included unless incorporated in the product.

2. *Energy intensity* is the energy consumed per unit of output. It includes natural gas, fuel, steam and electricity, all converted in net-fuel or the same unit for energy. For consistent calculations 80% average efficiency is considered for steam generation and 31% for electricity generation, corresponding to 3.138 MJ/kg steam and 11.6 MJ/kWh electricity. This metric captures in a synthetic manner the energy saving not only by heat integration, reflected by low steam and fuel consumption, but also by more advanced techniques, as cogeneration of heat and power. Negative values would mean export of energy to other processes. This situation is likely for processes involving high exothermic reactions, where the heat developed by reaction should be added as negative term in the energy balance.

3. *Water consumption* gives the amount of fresh water (excluding rainwater) per unit of output, including losses by evaporation (7% from the recycled water) and by waste treatment.

4. *Toxic emissions* consider the mass of toxic materials released per unit of output. The list of toxic chemicals can be retrieved from the website of the Environmental Protection Agency (USA).

5. *Pollutant emissions* represent the mass of pollutants per unit of output. The denominator is calculated as equivalent pollutant rather than effective mass. This topic is more difficult to quantify, but the idea is to use a unified measure.

6. *Greenhouse gas emissions* are expressed in equivalent carbon dioxide emitted per unit of output. Besides the CO_2 from direct combustion, this metric should include other sources, such as the generation of steam and electricity.

The advantage of using these measures in design is that the comparison of alternatives on a unique basis allows the designer to identify the best chemistry and flowsheet leading to the lowest resources and environmental impact. Usually the objective function is profit maximization. Including the above measures, at least as constraints, could contribute to conciliating the economic efficiency with the environmental care, a concept designated today by the label *ecoefficiency*.

A distinctive feature of these metrics is that they can be stacked along the whole product supply chain. In this way, ecological bottlenecks can be identified readily. For example, a chemical product that might appear as benign for the environment, could involve, in reality, highly toxic materials in some intermediate steps of manufacturing.

As an illustration, Table 1.2 shows values for some representative chemical processes. The output units refer to the added-value dollar $VA explained before. It can be seen that phosphoric acid has very unfavorable indices on the whole line, being very intensive as material, energy and water consumption. Acrylonitrile produced by ammonoxidation has also poor environmental performance with respect to toxics and pollutants. Note also the large amount of CO_2 produced by the methanol process. The best process in the list is the acetic acid made by the carbonylation of methanol.

Table 1.2 Sustainability metrics for some processes [3].

Process	Material kg/$	Energy MJ/$	Water m³/$	Toxics g/$	Pollutants g/$	CO_2 kg/$
Methanol (natural gas reforming)	0.2721	165.12	0.161	5.90	0	8.80
Acetic acid (MeOH carbonylation)	0.1769	16.76	0.029	0.313	0	1.10
Terephtalic acid (p-xylene oxidation)	0.4264	47.34	0.085	35.38	2.721	3.05
Acrylonitrile (Propene ammonoxidation)	2.1228	62.74	0.121	63.50	99.789	6.22
Phosphoric acid (Wet process)	144.3	267.4	0.788	1909.62	0	17.10

Sustainability metrics can be used as decision-support instruments. Among the most important tools in *life-cycle analysis* of processes we mention:

- Practical minimum-energy requirements (PME) set reference values for the intensive-energy steps and suggests energy-reduction strategies.

- Life-cycle inventory (LCI) deals with the material inventories of each phase of a product life, namely by tracking the variation between input and output flows.

- Life-cycle assessment (LCA) consists of determining the impact on the environment of each phase of a life cycle, as material and energy intensity, emissions and toxic releases, greenhouse gases, *etc.*
- Total cost assessment (TCA) provides a comparison of costs of sustainability, and by consequence, a consistent evaluation of alternative processes.

1.3
Integrated Process Design

The principles of the systematic and systemic design of chemical-like processes have been set by the works of Jim Douglas and coworkers, largely disseminated by his book from 1988 [4]. In the field of energy saving fundamental contributions have been made by Linnhoff and coworkers [5]. Several books addressing the design by systematic methods, but from different perspectives and professional backgrounds, have been published more recently, such as by Biegler et al. [6], Seider et al. [7], Dimian [1] and Smith [8].

The assembly of the systematic methods applied to the design of chemical processes are captured today in the paradigm of *integrated process design*. The application on modern design methods becomes possible because of *process-simulation* software systems, which encode not only sophisticated computational algorithms but also a huge amount of data. Combining design and simulation allows the designer to understand the behavior of complex system and explore design alternatives, and on this basis to propose effective innovative solutions.

1.3.1
Economic Incentives

Conceptual design designates that part of the design project dealing with the basic elements defining a process: flowsheet, material and energy balances, equipment specification sheets, utility consumption, safety and environmental issues, and finally economic profitability. Therefore, in conceptual design the emphasis is on the behavior of the process as a system rather than only sizing the equipment.

It is important to note that conceptual design is responsible for the major part of the investment costs in a process plant, even if its fraction in the project's fees is rather small. An erroneous decision at the conceptual level will propagate throughout the whole chain up to the detailed sizing and procurement of equipment. Moreover, much higher costs are necessary later in the operation to correct misconceptions in the basic design. Figure 1.3 shows typical cost-reduction opportunities in a design project (Pingen [9]). It can be seen that the conceptual phase takes only a very modest part, about 2% of the total project cost, although it contributes significantly in cost-reduction opportunities, with more than 30%. In the detailed design phase the cost of engineering rises sharply to 12%, but saving opportunities goes down to only 15%. In contrast, the cost of procurement and

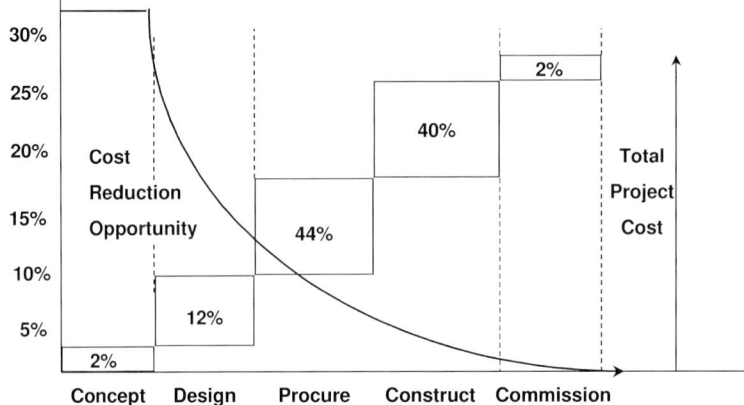

Figure 1.3 Economic incentives in a project.

construction are more than 80%, but the savings are below 10%. At the commissioning stage the total project cost is frozen.

1.3.2
Process Synthesis and Process Integration

In this book we consider the paradigm of *integrated process design* as the result of two complementary activities, process synthesis and process integration [1]. Figure 1.4 depicts the concept by means of a representation similarly with the onion diagram proposed originally by Linnhoff et al. [5]. *Process synthesis* focuses on the structural aspects that define the material-balance envelope and the flowsheet architecture. The result is the solution of the layers regarding the reaction (R) and the separation (S) systems, including the recycles of reactants and mass-separation agents. *Process integration* deals mainly with the optimal use of heat (H) and utilities (U), but includes two supplementary layers for environmental protection (E), as well as for controllability, safety and operability (C).

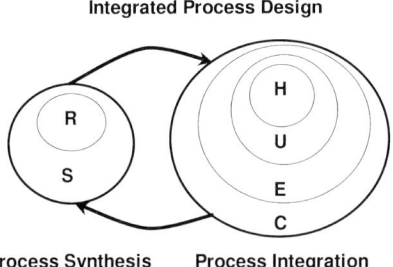

Figure 1.4 Integrated process design approach.

The key features of an integrated process design are:

1. The main objective of design is the *flowsheet architecture*. We mean by this type of units, performance and connections by material and energy streams. Systemic techniques are capable of calculating optimal targets for subsystems and components without the need of the detailed sizing of equipment.

2. The approach consists of developing alternatives rather than a unique flowsheet. The selected solution is the best cost-effective means only for the assumed constraints of technological, ecological, economical and social nature.

3. Computer simulation is the key tool for analysis, synthesis and evaluation of designs. The efficiency in using the software depends on the capacity of the designer to integrate generic capabilities with particular engineering knowledge.

4. The methodology addresses new design, debottlenecking and retrofit projects, and it can be applied to any type of process industries.

We stress again the importance of developing alternatives in which design targets are set well ahead of the detailed sizing of equipment. The last feature indicates a qualitative change that is removed from the concept of unit operations in favor of a more generic approach based on generic *tasks*. Using tasks instead of standard unit operations facilitates the invention of nonconventional equipment that can combine several functionalities, such as reaction and separations. This approach is designated today by *process intensification*. Moreover, the task-oriented design is more suited for applying modern process-synthesis techniques based on the optimization of superstructures.

1.3.3
Systematic Methods

The long road from an idea to a real process can be managed at best by means of a systemic approach. A design methodology consists of a combination of analysis and synthesis steps. *Analysis* is devoted to the knowledge of the elements of a system, such as for example the investigation of physical properties of species and mixtures, the study of elements characterizing the performance of reactors and unit operations, or the evaluation of profitability. *Synthesis* deals with activities aiming to determine the architecture of the system, as the selection of suitable components, their organization in the frame of a structure, as well as with the study of connections and interactions.

A design problem is always *underdefined*, either by the lack of data or insufficient time and resources. Moreover, a design problem is always *open-ended* since the solution depends largely on the *design decisions* taken by the designer at different stages of project development, for example to fulfil technical or economical constraints, or to avoid a license problem.

The systematic *generation of alternatives* is the most important feature of the modern conceptual design. The best solution is identified as the optimal one in the context of constraints by using consistent evaluation and ranking of alternatives. In the last two decades, a number of powerful systematic techniques have emerged to support the integrated process design activities. These can be classified roughly as:

- heuristics-based methods,
- thermodynamic analysis methods,
- optimization methods.

Note that so-called heuristics does not mean necessarily empirical-based rules. Most heuristics are the results of fundamental studies or extensive computer simulation, but may be formulated rather as simple decisional rules than by means of mathematical algorithms.

Today, the field of integrated process design is an active area of scientific research with immediate impact on the engineering practice. Methods accepted by the process-engineering community are described briefly below.

1.3.3.1 Hierarchical Approach

The hierarchical approach is a generic methodology for laying out the conceptual flowsheet of a process. The methodology consists of decomposing a complex problem into simpler subproblems. The approach is organized in "levels" of design decisions and flowsheet refinement. Each level makes use of heuristics to generate alternatives. Consistent evaluation eliminates unfeasible alternatives, keeping only a limited number of schemes for further development. Finally, the methodology allows the designer to develop a good "base case", which can be further refined and optimized by applying process-integration techniques. Chapters 2 to 4 present a revisited approach with respect to a previous presentation [1].

1.3.3.2 Pinch-Point Analysis

Pinch-point analysis deals primarily with the optimal management of energy, as well as with the design of the corresponding heat-exchanger network. The approach is based on the identification of the *pinch point* as the region where the heat exchange between the process streams is the most critical. The pinch concept has been extended to other systemic issues, as process water saving and hydrogen management in refineries. More details about this subject can be found in the monograph by Linnhoff et al. [5], as well as in the recent book by Smith [8].

1.3.3.3 Residue Curve Maps

The feasibility of separations of nonideal mixtures, as well as the screening of mass-separation agents for breaking azeotropes can be rationalized by means of thermodynamic methods based on residue curve maps. The treatment was extended processes with simultaneous chemical reaction. Two comprehensive books have been published recently by Stichlmair and Frey [10], as well as by Doherty and Malone [11].

1.3.3.4 Superstructure Optimization

A process-synthesis problem can be formulated as a combination of tasks whose goal is the optimization of an economic objective function subject to constraints. Two types of mathematical techniques are the most used: mixed-integer linear programming (MILP), and mixed-integer nonlinear programming (MINLP).

Process synthesis by superstructure optimization consists of the identification of the best flowsheet from a superstructure that considers many possible alternatives, including the optimal one. A substantial advantage is that integration and design features may be considered simultaneously. At today's level of software technology the superstructure optimization is still an emerging technique. However, notable success has been achieved in numerous applications. The reference in this field is the book of Biegler et al. [6].

1.3.3.5 Controllability Analysis

Plantwide control can be viewed as the strategy of fulfilling the production objectives of a plant, such as keeping optimal the material and energy balance, while preserving safety and waste minimization. Plantwide control means also that the global control strategy of the plant has to be compatible with the local control of units, for which industry proven solutions exist. Controllability analysis consists of evaluating the capacity of a process to be controlled. The power of manipulated variables should be sufficient (this is a design problem) to effectively keep the controlled variables on setpoints for predictable disturbances, or to move the plant onto new setpoints when changing the operation procedure. Controllability analysis and plantwide control can be handled today by a systematic approach. For a deeper study see the books of Luyben and Tyreus [12], Skogestad and Postlewaite [13], Dimian [1], as well as the recent monograph edited by Seferlis and Georgiadis [14].

1.3.4
Life Cycle of a Design Project

Life-cycle models can be used to manage the elaboration of complex projects [1]. A simple but efficient model can be built up on the basis of a waterfall approach. This indicates that the project sequencing should be organized so as to avoid excessive feedback between phases, and in particular to upset the architectural design. More sophisticated approaches, such as V-cycle or spiral models, could be used to handle projects requiring more flexibility and uncertainty, as in the case of software technology.

As a general approach by systems engineering, the phases of a project must be clearly defined such as the output of one stage falls cleanly into the input of the next stage. Complete definition of goals and requirements comes first. Systemic (architectural) design always precedes the detailed design of components. The modeling of units should be at the level of detail capable of capturing the behavior of the system, not more. After solving appropriately the conceptual phase, the

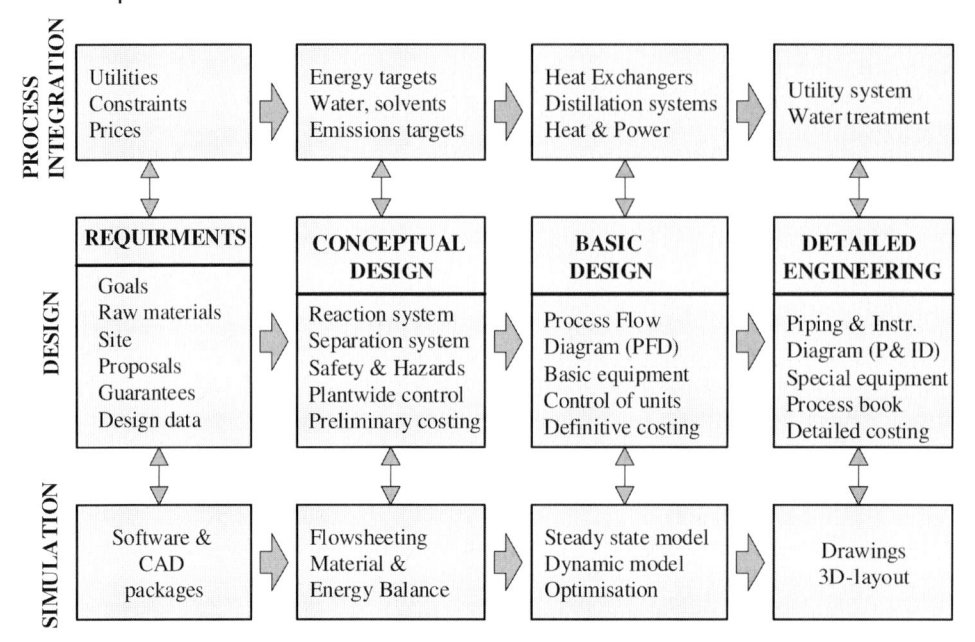

PROCESS INTEGRATION	Utilities Constraints Prices	Energy targets Water, solvents Emissions targets	Heat Exchangers Distillation systems Heat & Power	Utility system Water treatment
DESIGN	**REQUIREMENTS** Goals Raw materials Site Proposals Guarantees Design data	**CONCEPTUAL DESIGN** Reaction system Separation system Safety & Hazards Plantwide control Preliminary costing	**BASIC DESIGN** Process Flow Diagram (PFD) Basic equipment Control of units Definitive costing	**DETAILED ENGINEERING** Piping & Instr. Diagram (P& ID) Special equipment Process book Detailed costing
SIMULATION	Software & CAD packages	Flowsheeting Material & Energy Balance	Steady state model Dynamic model Optimisation	Drawings 3D-layout

Figure 1.5 Life cycle of an integrated design project [1].

project may proceed with the implementation and test of units, and finally with the test of the system, in most cases by computer simulation.

The development of an idealized process design project can be decomposed into four major phases: *requirements, conceptual design, basic design*, and *detailed engineering*, as shown in Fig. 1.5. Typical integration and simulation activities are listed. For example, the flowsheet developed during the conceptual design consists mainly of the reaction and separation subsystems. Other issues solved at this level are safety and hazards, environmental targets, plantwide control objectives and preliminary economic evaluation. By process-integration techniques targets for utilities, water and solvents are assessed. Several alternatives are developed, but only one base case is selected for further refinement. In this phase, process simulation is a key activity for getting consistent material and energy balances.

The development of the selected alternative is continued in the *basic design* phase, by the integration of subsystems, which leads to final process flow diagram (PFD). Specific integration activities regard the design of the heat-exchanger network, the energy saving in distillations, or combined heat and power generation.

Completing the flowsheet allows the generation of a steady-state simulation model. A dynamic simulation model may be developed for supporting process control implementation and for the assessment of operation strategies.

In *detailed engineering*, the components of the project are assembled before commissioning.

In practice, the workflow of a project may be different from the idealized frame presented above. For example, parallel engineering may be used to improve the overall efficiency. However, recognizing the priority of conceptual tasks and minimizing the structural revisions remain key factors.

1.4 Summary

Innovation is the key issue in today's chemical process industries. The main directions are *sustainability* and *process intensification*. Sustainability means in the first place the efficient use of raw materials and energy close to the theoretical yields. By process intensification the size of process plants is considerably reduced. The integration of several tasks in the same unit, as in reactive separations, can considerably simplify the flowsheet and decrease both capital and operation costs.

Production-integrated environmental protection implies that ecological issues are included in the conceptual design at very early stages. This approach should prevail over the end-of-pipe measures, which shift but do not solve the problem.

Increasing recycling of materials and energy results in highly integrated processes. Saving resources and preserving flexibility in operation could raise conflicts. These can be prevented by integrating flowsheet design and plantwide control.

By a systems approach, a process is designed as a complex system of interconnected components so as to satisfy agreed-upon measures of performance, such as high economic efficiency of raw materials and energy, down to zero waste and emissions, together with flexibility and controllability faced with variable production rate.

Integrated process design is the paradigm for designing efficient and sustainable processes. Key features are:

- Integrated flowsheet architecture for a cost-effective process is the main objective. Appropriate systemic techniques are capable of determining close-to-optimum targets for components without the need for detailed design and sizing.

- The conceptual design consists of developing several alternatives rather than a single flowsheet. The reason for alternatives is that every development step is controlled by design decisions. The selected solution among the alternatives should fulfil at best the optimization criteria within the environment of constraints.

- Process simulation is the main conceptual tool, both for analysis and synthesis purposes. Today, the traditional art-of-engineering is replaced by accurate computer simulation. Modern steady-state and dynamic simulation techniques make possible the investigation of complex processes close to the real situation.

- Getting accurate data, namely for thermodynamic and kinetic modeling, remains a challenge. For this reason, confronting the predictions by simulation with industrial reality is necessary, each time when this is possible.
- The systematic methods and the analysis tools of integrated process design can be applied to any type of chemical process industries, from refining to biotechnologies, as well as to new or revamped projects.
- For assessing the sustainability of process design consistent measures should be applied, such to minimize the material waste, energy, water consumption, toxic and pollutant emissions per unit of added value.
- The management of design project can be ensured by adopting the life-cycle modeling approach, in which the key elements are recognizing the priority of conceptual tasks and minimizing the structural revisions.

References

1 Dimian A.C., Integrated Design and Simulation of Chemical Processes, Computer-Aided Chemical Engineering 13, Elsevier, Amsterdam, The Netherlands, 2003
2 Christ C. (ed.), Production-Integrated Environmental Protection and Waste Management in the Chemical Industry, Wiley-VCH, Weinheim, Germany, 1999
3 Schwartz J., Beloff B., Beaver, E., Chem. Eng. Progress, July, pp. 58–64, 2002
4 Douglas J.M., Conceptual Design of Chemical Processes, McGraw-Hill, New York, USA, 1988
5 Linnhoff B., Townsend D.W., Boland D., Hewitt G.F., Thomas B., Guy A.R., Marsland R.H., User Guide on Process Integration, The Institution of Chemical Engineers, UK, 1994
6 Biegler L., Grossmann I., Westerberg A., Systematic Methods of Chemical Process Design, Prentice Hall, Upper Saddle River, NJ, USA, 1998
7 Seider W.D., Seader J.D., Lewin D.R., Product and Process Design Principles, Wiley, New York, USA, 2003
8 Smith R., Chemical Process Design and Integration, Wiley, New York, USA, 2004
9 Pingen J., A vision of future needs and capabilities in process modeling, simulation and control, ESCAPE-11 Proceedings, Elsevier, Amsterdam, The Netherlands, 2001
10 Stichlmair J.G., Frey J.R., Distillation, Principles and Practice, Wiley-VCH, Weinheim, Germany, 1999
11 Doherty M.F., Malone, M., Conceptual Design of Distillation Systems, McGraw-Hill, New York, USA, 2001
12 Luyben W.L., Tyreus B., Plantwide Process Control, McGraw-Hill, New York, USA, 1999
13 Skogestaad S., Postlewaite I., Multivariable Feedback Control, Wiley, NewYork, USA, 1998
14 Seferlis P., Georgiadis M.C. (eds.), The Integration of Process Design and Control, CACE 17, Elsevier, Amsterdam, The Netherlands, 2004

2
Process Synthesis by Hierarchical Approach

The flowsheet synthesis of continuous chemical-like process can be performed following a systematic strategy known as the *hierarchical approach*. The procedure, initially proposed by Jim Douglas and coworkers in the decade 1980–90 [1, 2], describes the conceptual design process as a logical sequence of analysis and synthesis steps grouped in levels. Each level involves a flowsheet development mechanism based on *design* decisions. The result is not a unique solution but a collection of alternative flowsheets from which an evaluation procedure eliminates the less attractive ones.

In the original formulation, the evaluation and ranking of alternatives relies on the computation of an economic potential (EP) as representing the difference between revenues and manufacturing fees. Following the strategy known as the "depth-first approach", at each level only the best alternative is kept for further development. A single base-case flowsheet is generated, which serves further for refinement, energy integration and optimization.

Another option is merging the alternatives into a large superstructure that can be submitted to global optimization by appropriate MINLP techniques. However, this approach is not workable at present by using commercial packages.

We should mention that the modern process-synthesis methods can ensure intrinsically the optimization of subsystems, such as reactions, separations and heat exchange. Therefore, by searching in the first stage the structural optimization of the flowsheet and by performing later the optimization of units should lead quite soon to the true overall optimum in a large number of situations.

Computer simulation based on rigorous modeling may be applied today at any level of the methodology for replacing shortcut methods or design heuristics. On the other hand, even the most sophisticated software cannot cover the richness of physical situations. Therefore, adapting the modeling environment of a simulator to a particular technology and embedding the engineering expertise is the strategic approach to solve a challenging design problem.

We may define the objectives of conceptual process design as follows:

- finding the optimal architecture of the flowsheet with respect to:
  - efficiency of raw materials and energy,
  - minimal impact on the environment,
  - flexibility in throughput and quality of raw materials.

- setting feasible and optimal tasks for units,
- evaluating the effect of interactions through recycles of mass and energy,
- selecting and sizing the main equipment units.

The detailed design of components remains for the most part a downstream activity.

A process-design problem is always "open-ended". To put it in a nutshell, the central problem in design is generating relevant alternatives in an environment of specific constraints and decisions of economic, technical and social nature.

However, only the static evaluation of process performance around an operation point is not sufficient. The flexibility in operation and the dynamic behavior should be explored by dynamic simulation, including the implementation of the main features of the control system. Therefore, process dynamics and plantwide control issues are largely addressed in this book.

2.1
Hierarchical Approach of Process Design

The classical hierarchical approach of conceptual design proposed by Douglas is organized in eight levels [2], as presented in Figure 2.1:

0. Input information.
1. Number of plants.
2. Input/output structure and connection of plants.
3. Recycle structure of simple plants.
4. Separation systems of simple plants:
    4a. General architecture: identify specific separation subsystems.
    4b. Vapor and gases recovery and separation system.
    4c. Solid recovery: getting valuable solids from solutions.
    4d. Liquid-separation system: separate products from liquid mixtures.
    4e. Solid-separation system: separate solid products.
    4f. Combine the separation systems and study interactions.
5. Energy integration:
    5a. Pinch-point analysis for optimal heat and power saving.
    5b. Water minimization: design an efficient system for water recycling.
    5c. Solvent minimization: design an efficient system for solvent recycling.
6. Design alternatives.
7. Hazop analysis:
    7a. Identify the sources of hazards and risks.
    7b. Perform hazard and operability study.
8. Control-system synthesis:
    8a. Plantwide control.
    8b. Control structure of units.

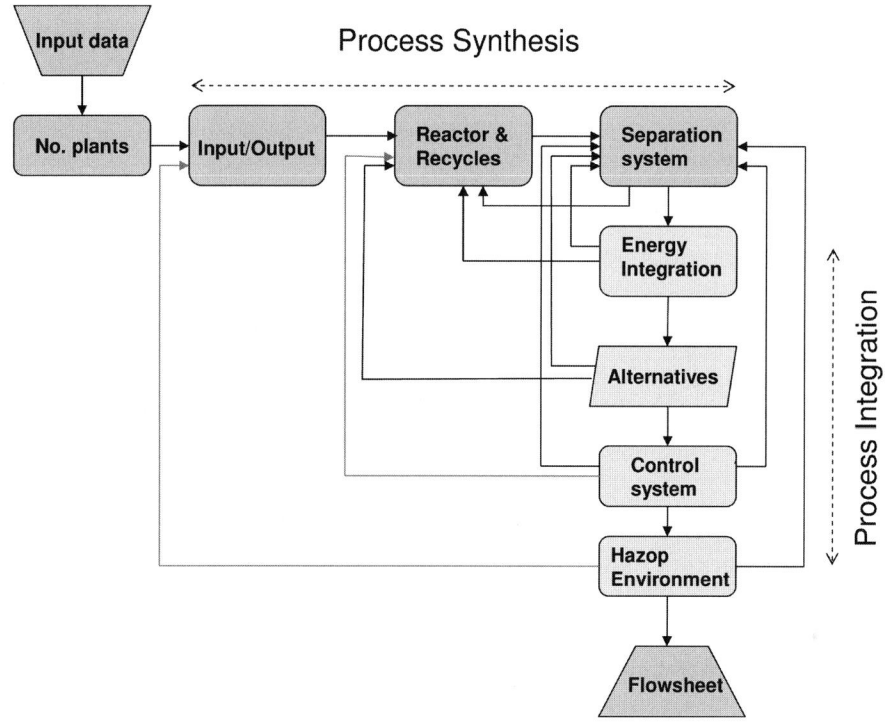

Figure 2.1 Classical hierarchical approach of conceptual process design [2].

The first four levels belong to the activity known as *process synthesis* dealing with the flowsheet architecture: the nature of units and their assigned performance, as well as the connection of units by flows of materials and energy. Based on this, a preliminary material and energy balance is computed. The remaining *process-integration* stages handle the optimal use of energy and of additional resources, as well as hazard and environmental problems. The design and sizing of units are consolidated. The implementation of the process-control system follows as the final activity.

The decomposition of the design activity into two phases, synthesis and integration, is now well established. It may be observed, however, that considerable backflow of information could occur between different conceptual levels, particularly between energy integration on the one hand and reaction and separation systems on the other hand. Alternatives implying structural flowsheet modifications are clearly not desirable.

For this reason, this book proposes an *improved hierarchical approach*, as depicted in Figure 2.2. The goal is to make the design methodology more efficient by reducing the interactions between different levels of synthesis and integration. In this improved approach, the emphasis is on developing the *reactor/separation/recycle* structure as defining the essential of the flowsheet architecture. This level also

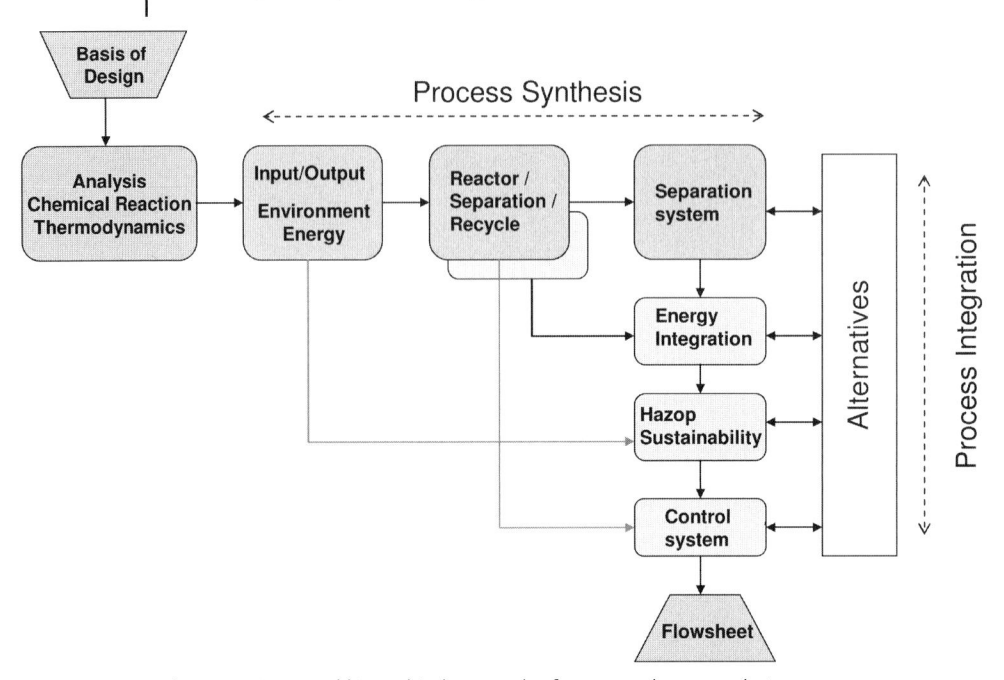

Figure 2.2 Improved hierarchical approach of conceptual process design.

considers some key features of the energy integration, namely around the chemical reactor. In addition, the content of levels is redefined putting more emphasis on the analysis tools supported by computer simulation. The whole approach is described below.

Level 0: Basis of Design. This step consists of gathering fundamental technological and economic data needed for performing a conceptual design, including health, safety and environmental risks.

Level 1: Chemistry and Thermodynamics. This level deals with the analysis of the fundamental knowledge needed for performing the conceptual process design. A detailed description of chemistry is essential for designing the chemical reactor, as well as for handling safety and environmental issues. Here, the constraints set by chemical equilibrium or by chemical kinetics are identified. The non-ideal behavior of key mixtures is analyzed in view of separation, namely by distillation.

Level 2: Input/Output Analysis. This stage sets the framework of the overall material balance delimited by raw materials at input, and products, byproducts and waste at output. The key design decision regards the performance of the reaction system. On this basis, the initial feasibility is evaluated by means of an economic potential or by any other measure for the added value. The analysis should include the negative costs incurred by handling environmental problems, namely by the treatment of effluents. If the reaction is highly endothermic, an estimation of energy costs based on existing processes may be included for a more realistic analysis. Note that the economic potential at the I/O level should be high enough in order to accept further reductions when the operating and capital costs are accounted for.

Level 3: Reactor/Separation/Recycle. This level deals with the key elements defining the process architecture, namely the chemical reactor interacting with the separators through recycles. The emphasis is placed on the reactor design in recycles on a kinetic basis. Secondary reactions and formation of impurities are considered, at least quantitatively. The design is performed not only around one presumed optimal operating point, but within an "operation window" defined by the flexibility in the production rate and by the variability of raw materials. A more rigorous analysis makes use of bifurcation analysis. In this way, the designer may examine plantwide control issues, such as reactor stability and makeup of reactants, well ahead to control system implementation. Chapter 4 is devoted to the theoretical explanation of this new approach.

Another feature is the early analysis of heat integration issues regarding the chemical reactor, before applying pinch-point analysis to the whole flow-sheet. High exothermic reactions are of particular interest with respect to (1) stability of the chemical reaction system faced to feedback of materials and energy, and (2) optimum use of energy for covering own needs and exporting the surplus. On the other hand, the endothermic reactions are constrained by the availability of utilities, as well as by costly devices for generating heat and power.

If the Level 3 is solved properly, the flowsheet development should follow a nearly sequential track consisting of synthesis of subsystems and solving local integration problems. In particular, the energy saving may bring some modifications regarding the separation system, but without affecting neither the chemical reactor design nor the structure of recycles. Thus, the reactor/separator/recycle level appears as being the most important in the hierarchy of conceptual design.

Level 4: Separation System. As opposed to the Douglas' procedure, the synthesis of the separation system makes use of a task-oriented methodology. The concept, originally proposed by Barnicki and Fair [15–17], will be developed in Chapter 3. After solving the *first split*, the synthesis problem is divided into subproblems for treating homogeneous fluids, which in turn generates *separation subsystems* for gases, liquids and solids. The objective is finding a near-optimum *separation sequence* in each subsystem. The approach consists of identifying the separation task by means of logical *selectors*, having as an effect a significant reduction in the searching space. The ranking of separation techniques is based on the identification of a *characteristic property* among the components of a mixture. The generation of the separation sequence relies for the most part on *heuristics*, although it may include *optimization methods*.

At the end of the Level 4, the result is a close-to-optimum process flowsheet together with a consistent material balance. The next levels will have as a goal the solution of the problems related with the optimal use of energetic resources and material utilities, as well as with waste minimization and plantwide process control.

Level 5: Energy Integration. This level involves a large spectrum of design activities dealing with the minimization of energy and material utilities. These can be classified as follows:

5a. Pinch-point analysis for optimal heat and power consumption.
5b. Design of energy-integrated separations.
5c. Design of refrigeration systems.
5d. Water minimization: design an efficient system for water recycling.
5e. Solvent minimization: design an efficient system for solvent recycling.
5f. Site integration of energetic and material utilities.

Note that the evolution of design between the levels 4 and 5 can generate a number of *alternatives*, but these should not affect the basic flowsheet structure defined at the reactor/separations/recycle level. In addition, employing *complex units* and *process-intensification tech*niques can produce more compact flowsheet and cheaper hardware.

Level 6: Hazop and Environment. Since the factors leading to hazard and environmental problems are handled at earlier stages, this level should imply only a quantitative evaluation of effects with limited consequences on conceptual design.

Level 7: Process-Control System. The key issues of process dynamics and control, namely fresh feed policy and stability in operation of the reaction/separation/recycle system, are solved at Level 3. Consequently, the implementation of a process-control system may be realized without affecting the basic flowsheet structure, but taking into account fundamental process control principles, as proposed in the methodology developed by Luyben and Tyreus [20].

The next section will explain in more detail the content of different levels. The approach will be applied throughout the whole book, although not in a repetitive manner. Actually, each case study will emphasize only generic conceptual elements. The case study regarding the manufacturing of cyclohexanone by phenol hydrogenation is a kind of leading example. Since the book focuses on technical aspects, there is no place for cost evaluation and profitability analysis.

The above methodology is applicable to any type of chemical process industry. The best results are obtained when the user goes through all the steps, avoiding the temptation of reproducing existing flowsheets. The approach is valuable not only for new processes, but also for the improvement of existing processes (revamping and retrofitting), where starting from "scratch" is often the best way to stimulate innovative ideas.

2.2
Basis of Design

2.2.1
Economic Data

(a) *Product price versus purity specifications*
Express the desired specifications of the final product, namely by identifying critical impurities that affect specifications and price. Search prices for valuable byproducts.

(b) *Raw materials prices*
The prices for raw materials are a function of purity. For commodities and intermediate chemicals, it is more advantageous to integrate the process in an existing industrial complex. Consider the prices for storage and transport.

(c) *Utilities*
Process utilities are fuel, steam, cooling water, chilled water, brines, electricity and refrigeration. Prices on transactional basis are lower. Specify limitations in supply.

(d) *Costs of waste disposal*
Estimate the cost of waste disposal, as well as the treatment of volatile organic components (VOCs), polychlorinated biphenyls (PCBs), and any other materials forbidden by the environmental regulations.

2.2.2
Plant and Site Data

(a) *Location*
Proximity to the source of raw materials is preferable. The integration on an existing site is an advantage, as well as the availability of shipping facilities.

(b) *Storage facilities*
The costs for the storage of raw materials, products and intermediates can be significant, namely for large-scale commodities. The storage of toxic and hazard chemicals should be avoided.

(c) *Climate*
Climate conditions are determinant for the selection of utilities. Get data about minimum and maximum temperatures, humidity, wind and meteorological variability.

(d) *Utility system*
Select the type of utilities, namely the steam pressures and the temperature of recycled water, as well as the use of inert gas and refrigeration fluids.

(e) *Environmental legislation*
List the specific requirements that the process must fulfil with respect to the environmental legislation, such as gaseous emissions, soil and water pollution.

2.2.3
Safety and Health Considerations

Safety and hazard problems can justify important design decisions. The following elements deserve attention:

(a) *Explosions risks*
Identify potential explosive mixtures in the chemical reactors and storage facilities, particularly the mixtures with air. Specify concentration and temperature range.

(b) *Fire risks*
Find information about flash point, autoignition temperature and flammability limits.

(c) *Toxicity*
Specify the toxic or nontoxic character of the main chemicals involved in the process. Information on toxicity and hazard effects can be found on the websites of agencies for public environment and health, such as for example the US Environmental Protection Agency (EPA) and the European Environmental Agency. A good introduction to environmental engineering is the book of Allen and Shonnard [4]. In the field of process safety, the book of Crowl and Louvar [5] is still popular. The book of Kletz [6] covers the topics of hazard and operability, as well as hazard analysis.

2.2.4
Patents

Patents provide valuable technology information for designers. Firstly, information about process feasibility may be collected with respect to chemistry, catalyst, safety and operation conditions. Qualitative data regarding the reaction engineering, such as conversion and selectivity, as well as the productivity and residence time are useful for the selection of the chemical reactor. Even more important are data regarding the reaction-mixture composition for the assessment of separations, namely with respect to byproducts and impurities.

Some patents address process-design issues, as separation techniques and energy-recovery methods. Their reliability can be checked against the predictions offered by computer simulation.

2.3
Chemistry and Thermodynamics

2.3.1
Chemical-Reaction Network

Finding the *independent chemical reactions* provides consistency and proper specification for both material balance and chemical kinetics. By definition, a set of reactions is independent if any given reaction cannot be express as a combination of the remaining ones. There are two identification methods:

1. The reduction of an extensive set of reactions supposed to take place.
2. Knowing the species involved in the chemistry.

To explain the first approach let us consider S chemical involved in r chemical reactions, from which only R are independent. The stoichiometric relations give a set of linear algebraic equations:

$$\sum_{j=1}^{S} v_{i,j} A_j = 0, \quad i = 1, \ldots, r \quad (2.1)$$

The number of independent reactions R can be found simply as the rank of the matrix of stoichiometric coefficients $\lfloor v_{i,j} \rfloor$ with dimension $S \times r$ such that $R \leq r$. Different methods can be applied, such as reduction to triangular matrix by Gaussian elimination for small-size matrices, or computer methods for larger problems.

Following the second approach, only the knowledge of the chemical species involved in the chemistry is sufficient. The problem can be formulated as follows: given a list of *chemical species S* find a proper set of *independent chemical equations R* that ensures the conservation of *atomic species N* in terms of the molecular formulas of the system species. A proper set allows that any other chemical equation can be obtained from this by algebraic operations. A chemical species is characterized by formula, isomeric form and phase, and it may be neutral molecules, ionics and radicals. The atomic species can be atoms and charge (+ or −).

Let us consider a mixture $\{A_j\}$ of S chemical species involving N elementary components $\{E_k\}$. By setting the atoms in rows and the species in columns, a formula matrix B can be built:

$$B = \begin{pmatrix} & A_1\ldots & A_j\ldots & A_S \\ E_1 & b_{11}\ldots & b_{1j}\ldots & b_{1S} \\ E_k & b_{j1}\ldots & b_{kj}\ldots & b_{jS} \\ E_N & b_{N1}\ldots & b_{Nk}\ldots & b_{NS} \end{pmatrix} \quad (2.2)$$

The number of independent chemical equations R can be found simply by the relation:

$$R = S - \text{rank}(B) \quad (2.3)$$

The final visualization of the reduced B matrix allows finding the basic set of independent chemical equations. Note that $C = \text{rank}(B)$ gives the number of "component species" that may form all the other noncomponent species by a minimum of independent chemical reactions. The procedure can be applied by hand calculations for simple cases, or by using computer algebra tools for a larger number of species. More details can be found in the book of Missen et al. [7], or at www.chemical-stoichiometry.net.

At this point, we should mention the difference between independent chemical equations and independent chemical reactions. The former are of mathematical significance, being helpful to carry out consistent material balance. The latter are useful for describing the chemical steps implied in a chemical-reaction network. They may be identical with the independent stoichiometric equations, or derived by linear combination. This approach is useful in formulating consistent kinetic models.

From the perspective of process synthesis, the analysis of process chemistry should include the aspects examined below.

(a) *Main reactions*
- Identify the number of independent chemical reactions associated with each reaction step in the manufacture of the desired product. Include intermediates that can be separated and recycled. At each step indicates the phase(s), as well as the range of feasible temperatures and pressures.

- List the thermal effect for each chemical reaction. Identify highly exothermic reactions, as well as temperature-sensitive reactions with a large activation energy.

- List technological constraints, such as the ratio of reactants at reactor inlet, pressure and temperature, maximum allowable concentration, flammability and explosion limits.

(b) *Secondary reactions*
- List secondary reactions leading to byproducts and impurities, in the range of temperatures and pressures mentioned above.

- Find data about selectivity and its variation with conversion. This information is essential in conceptual design. The distribution of chemical species in different conditions is helpful for evaluating the selectivity pattern, at least qualitatively.

- Pay special attention to the formation of impurities in chemical reactors, but also in some physical operations, because of long residence time or high temperature.

- Consider reactions involving impurities entered with the feed of raw materials.

(c) *Catalyst*
- List alternative catalysts and note the following properties:
  - activity per unit of reaction volume,
  - selectivity towards the desired products at the required conversion, and the achievable per-pass yield,
  - possibility for regeneration,
  - thermal and chemical stability,
  - geometry and its effect on activity,
  - mechanical strength and resistance with respect to attrition.
- Examine the formation of byproducts and impurities specific to each catalyst.

- Check the effect of temperature and of potential impurities on catalyst activity.
- Revise the environmental problems raised by regeneration and disposal, as well as the need for solvents or special chemicals.
- Select the most suitable catalyst by taking into account activity, operation time, purchasing costs, and fees for regeneration and disposal.

2.3.2
Chemical Equilibrium

Chemical-equilibrium analysis allows finding the maximum achievable per-pass conversion and the composition of the reaction mixture at equilibrium. Accordingly, it may suggest measures for improving both conversion and selectivity.

Gibbs free-energy minimization offers an elegant computational manner without the need to specify the stoichiometry. In addition, phase equilibrium is accounted for. Since occasionally the method fails, considering explicit reactions is safer.

Note that the chemical-equilibrium computation is very sensitive with respect to errors involved in the thermodynamic data because of the exponential form $K = \exp(-\Delta G^0/RT)$. The effect depends both on the errors in Gibbs free energy and on the temperature level. For example, RT is about 4 kJ/mol at 500 K and 6.5 kJ/mol at 800 K, while component ΔG_f^0 values are in the order of 100 kJ/mol. Since the calculation of ΔG^0 involves addition/subtraction of large numbers, the accuracy in estimating ΔG_f^0 should be better than 5 kJ/mol or 1 kcal/mol. At the limit of the practical range [−4,4] for $\Delta G^0/RT$ an error of 15% gives a relative error of about 180% in estimating K. Therefore, checking the accuracy of thermochemical data should be done systematically when calculating the chemical equilibrium.

2.3.3
Reaction Engineering Data

(a) *Reactor design*
The following aspects are of significance:
- alternative reactor types, geometrical characteristics, mixing pattern, operating conditions, residence time and productivity,
- technological constraints, such as minimum feed temperature, maximum temperature and pressure,
- safety issues,
- mechanical problems related to high pressures or temperatures, sophisticated mixing or distribution devices, use of special materials.

(b) *Kinetics of the main reactions*
Kinetic data are necessary for sizing the chemical reactor and for assessing the key features of process dynamics. However, the absence of kinetic data does not prevent the development of a process flowsheet, although the reactor will be described as a black-box steady-state unit, on stoichiometric or yield basis.

(c) Formation of byproducts and impurities

The knowledge of selectivity is crucial for developing a realistic process. Preferably, the formation of byproducts should be expressed by kinetic equations, or by reference to the main species. Because in most cases this information is hardly available, the user should consider realistic estimations for impurities that might cause troubles in operation and/or affect the product quality. A good approach is the examination of patents.

2.3.4
Thermodynamic Analysis

(a) Physical properties of key components

The minimum information covers chemical formula, molecular weight, normal boiling point, freezing point, liquid density, water solubility and critical properties. Additional properties are enthalpies of phase transitions, heat capacity of ideal gas, heat capacity of liquid, viscosity and thermal conductivity of liquid. Computer simulation can estimate missing values. The use of graphs and tables of properties offers a wider view and is strongly recommended.

(b) Phase equilibrium

VLE and VLLE diagrams of representative binaries should be plotted for representative binaries in the range of operating pressures and temperatures. The formation of azeotropes should be checked against experimental data. Partial solubility of liquids, as well as the solubility of gases is difficult to predict. In the first case suitable liquid-activity models should be used, namely Uniquac and NRTL. For describing gas solubility, three methods can be employed: (1) Henry coefficients for diluted solutions, (2) Equations of state, and (3) Asymmetric definition of component fugacity with liquid-activity modeling. The reliability of binary interaction parameters is a key issue. Regression of parameters from experimental data should be performed systematically when accuracy is needed. The use of group contribution methods, such as UNIFAC, should be restricted to screening. Evaluating several thermodynamic options is recommended. Models and methods are presented in detail elsewhere [3].

(c) Residue curve map

Plotting residue curves maps (RCM) allows the designer to anticipate problems by the separation of nonideal mixtures, namely when dealing with homogeneous and heterogeneous azeotropes. By reactor selection, it may foresee problems incurred by the recycle of some reactants.

2.4
Input/Output Analysis

The main steps involved in the chemical-reaction network determine the number of reactors and consequently, the number of simple plants. The following approach is useful for the synthesis of a complex plant:

1. Consider individual plants around each chemical reactor.
2. If chemical species are produced in different reactors, consider the possibility of handling these components in a common separation system.
3. In a first attempt, do not consider large recycle loops due to the recovery of some auxiliary materials, as water, organic solvents and hydrogen.

2.4.1
Input/Output Structure

The input/output structure defines the material balance boundary of the flowsheet. Often it is referred as the *inside battery limit* envelope. A golden rule requires that the total mass flow of all components entering the process must be equal with the total mass flow of all components leaving it. It should be kept in mind that the recycles affect only the internal process streams, but not the input/output material balance.

Figure 2.3 displays the input/output structure, which includes inlet and outlet streams, as well as large external recycles that requires separate plants and storage facilities. Design decisions formulated as heuristics, are given in Table 2.1, mostly suggested by Douglas [1].

Some comments are helpful. The first decision with strong implications regards the *feed purification*. If the impurity will affect neither the reaction nor the separations, this should enter the process. Otherwise, its removal is imperative.

The second key decision regards the *recycling of reactants and auxiliary materials*. Here, we may cite recycling gases such as hydrogen or nitrogen, or water and mass-separation agents, as well as the regeneration of catalyst. Recycling implies substantial costs in treatment, storage and transportation, which should be included from the beginning in the economic analysis.

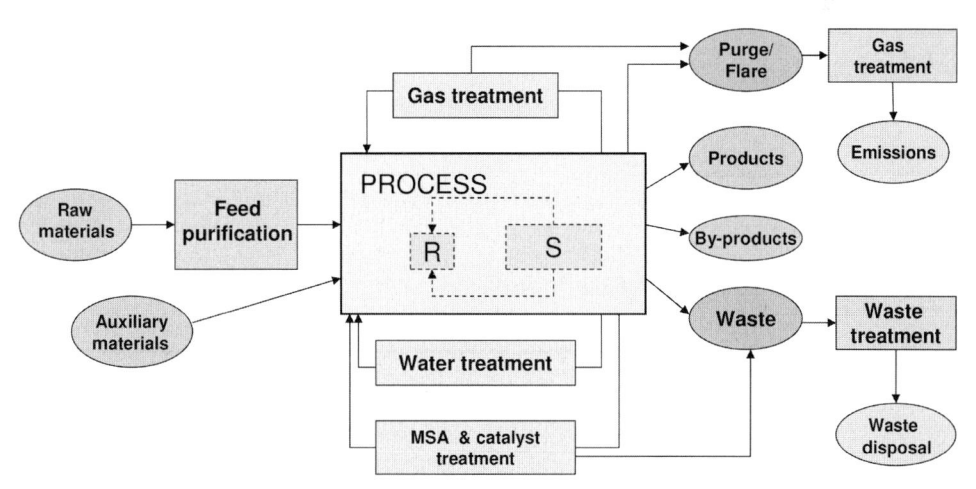

Figure 2.3 Input/output structure of a flowsheet.

Table 2.1 Heuristics at input/output level.

Heuristics 2A: Feed purification
1. Examine the product specification. Consider secondary reactions involving impurities in feed that lead to undesired impurities in product. Purify the feed or switch to appropriate raw materials.
2. If an impurity is not an inert but in significant amount, remove it.
3. If an impurity is present in a gas feed, as first choice let the impurity enter the process.
4. If an impurity in a liquid feed stream is also a byproduct or product component usually it is better to process the feed through the separation system.
5. If an impurity is present as azeotrope with the reactant, often is better to process it.
6. If a feed impurity is an inert that can be separated easier from the product than from the feed, it is better to let it pass through the process.
7. Impurities affecting the catalyst must be removed. Evaluate the cost of an extra purification system for feeds, as well as the cost of recycling harmful impurities, including equipment fouling and maintenance.

Heuristics 2B: Recycling of reactants and auxiliary materials
1. In a first attempt consider full recycling of reactants and no losses in products.
2. Do not recycle very cheap reactants, such as air, but examine implication on gaseous emissions. Consider using oxygen for minimizing emissions.
3. Recycling of water is imperative. Because water treatment implies substantial expense, it must be considered as a negative term in the economic potential. If the cost is excessive then alternative reactors should be considered.
4. In a first attempt consider the removal of byproducts in reversible reactions.

Heuristics 2C: Purge and bleed streams
1. Consider gaseous purges or liquid bleeds if some components tend to accumulate in recycles. Minimize purge and bleeding streams.
2. Examine the post-treatment of purges and other emissions, for example by physical operations (ex: adsorption) or by chemical conversion (ex: combustion).
3. Consider the transformation of light impurities by chemical conversion in more heavy components that can be easy eliminated in waste streams.
4. Consider new separation techniques, such as membranes, to recover valuable components from purges and bleeds.

A third important decision regards the *post-treatment of emissions and waste*. Dumping toxic emissions and waste is forbidden. Often, CO_2 is undesirable and involves negative costs.

In summary, the design decisions at the input/output level aim to obtain the most efficient material balance and helps in defining ecological targets.

2.4.1.1 Number of Outlet Streams

The correct assignment of outlet streams ensures the consistency of the material balance. Remember that all the species – products, byproducts and impurities – should leave the process in outlet streams! Some guidelines are given below:

1. Examine carefully the composition of the outlet reaction mixture.
2. Order the components by their normal boiling point.

Table 2.2 Destination code and component classification.

	Component classification	Destination code
1	Reactant (liquid)	Liquid recycle (exit)
2	Reactant (solid)	Recycle or waste
3	Reactant (gas)	Gas recycle & purge, vent
4	Byproduct (gas)	Fuel or flare
5	Byproduct (reversible reaction)	Recycle or exit
6	Reaction intermediate	Recycle (exit)
7	Product	Product storage
8	Valuable byproduct	Byproduct storage
8	Fuel byproduct	Fuel supply
9	Waste byproducts	
	• aqueous waste	Biological treatment
	• incineration waste	Incinerator
	• solid waste	Landfill
10	Feed impurity	Same as byproduct
11	Homogeneous catalyst	Recycle
12	Homogeneous catalyst activator	Recycle
13	Reactor or Product solvent	Recycle (exit)

3. Assign destination code to each component, as shown in Figure 2.3 and Table 2.2.
4. Group neighboring components with the same destination.
5. The number of all groups minus the recycle streams gives the number of the outlet streams. Azeotropes or solid components may change the rule.

2.4.1.2 Design Variables

The design variables originate from the design decisions. At the input/output level, the design variables define the degrees of freedom of the overall material balance. That is why it is impossible to develop a unique material balance for a process, even with the same chemistry. Table 2.3 lists some possible choices.

2.4.2
Overall Material Balance

The overall material balance gives the relation between the input streams, essentially raw materials, and the output streams, such as products, byproducts, purge and waste. In the first attempt, a simplified approach is useful for framing the main characteristics. Usually the assumptions are:

- Consider only products and byproducts, neglect subproducts and impurities.
- Consider 100% recovery of all recycled components.

Table 2.3 Design variables at the input/output level.

Reaction system	• level of conversion and selectivity
	• molar ratio of reactants
	• reaction temperature and/or pressure
Reactants	• need for feed purification
	• reactants not recovered
	• use of gas recycle and purge
Byproducts	• separation or recycling
	• waste treatment

The *preliminary material balance* sets the limits for the variables of significance for process performance, namely the minimum material consumption and the maximum yield in products, and from this viewpoint is a kind of ideal material balance.

The procedure below requires only a spreadsheet. The steps are:

1. Identify the inlet and outlet streams, as described before.
2. Express the production rate in convenient units.
3. Determine the input/output partial flow rates for every main component.
4. Express the formation of byproducts in terms of design variables (conversion, molar ratio, *etc.*).
5. Determine the flow rates for reactants in excess and not recycled, and include them in outlet streams.
6. Determine the inlet flow rates for impurities entered with some reactant streams.
7. Calculate outlet flow rates for impurities in purge or bleed streams.

2.4.3
Economic Potential

At each level of the design the feasibility of alternatives may be evaluated by means of an *economic potential* (EP). This index has the significance of an *added value* representing the difference between earnings and expenses on yearly basis. Since at the *input/output* level the process configuration is not known, only material costs, including auxiliary resources and ecological costs, are considered:

$$EP_{I/O} = \{\text{Product value}\} + \{\text{Byproduct value}\} - \{\text{Raw materials costs}\}$$
$$- \{\text{Auxiliary materials costs}\} - \{\text{Ecological costs}\} \quad (2.4)$$

$EP_{I/O}$ is useful for assessing the feasibility of alternative chemistries and sources of raw materials before undertaking a more detailed flowsheet development. Clearly, the economic potential should be largely positive to accept further reductions when the capital and operation costs are taken into account.

2.4 Input/Output Analysis

An inconvenience of the above approach is its high sensitivity to the uncertainties in costs. In general, one may presume that the selling prices of products should be better known, being regulated by the market. On the contrary, the purchasing prices of raw materials are much more uncertain. These should consider commercial fees and transportation costs. Using spot prices is not reliable for the long-term assessment of a process plant. For example, for some commodities the difference between the prices of raw materials and products might be occasionally negative. When the raw materials are intermediate commodities, as in petrochemistry, the best solution is integrating several plants on the same site.

Therefore, the economic potential at the I/O level could be seen merely as a tool for selecting the chemical route, and at the same time for setting targets when purchasing raw materials. As a rule of thumb, the ratio of selling prices of products to the purchasing prices of raw materials should be a minimum of two when the payback time of the total capital investment is greater than five years. Preferably, this ratio should be about three for a payback time of three years [3]. More accurate calculations may be carried out easily by means of profitability analysis tools.

After deciding the chemical route and selecting the raw materials, the assessment of alternatives during the evolution of the process design may be done solely on the basis of cumulative capital and operating costs, including additional ecological costs, here called *processing costs*:

$$\text{Processing costs} = \{\text{Equipment costs/Payback time}\} + \{\text{Cost of utilities}\} \\ + \{\text{Additional environmental costs}\} \quad (2.5)$$

The economic potential at a downstream process design level n becomes:

$$EP_n = EP_{I/O} - \{\text{Processing Costs}\}_n \quad (2.6)$$

Example 2.1: Acrylonitrile by Ammoxidation of Propylene: Input/Output Analysis

The manufacture of acrylonitrile is based today on the ammoxidation of propylene:

$$CH_2\text{=}CH\text{–}CH_3 + NH_3 + 3/2 O_2 \rightarrow CH_2\text{=}CH\text{–}CN + 3H_2O$$

The reaction is highly exothermic ($\Delta H = -123$ kcal/mol) taking place in gaseous phase over a suitable catalyst at temperatures between 300–500 °C and pressures of 1.5 to 3 bar. Modern catalysts achieve a yield in acrylonitrile of 80–82%. A recent patent (US 6,595,896, 2003) gives activity data for a molybdenum/bismuth-modified catalyst that may be used in revamping plants employing fluid-bed reactor technology (Table 2.4). A pressure of 2 bar seems optimum. Based on these data evaluate a preliminary input/output material balance. Estimate the key features ensuring the economic feasibility including the ecological impact of emissions and waste.

Table 2.4 Outlet reaction mixture composition in molar percentage for a feed of propylene/ammonia/air 1:1.2:9.8, reaction temperature 440 °C and mass load WWH 0.085 h^{-1} (ton propylene per ton catalyst per hour).

P (MPa)	Conv. propene	AN	ACT	HCN	Acrolein + Acrylic acid	CO_2 + CO
0.18	97.8	79.6	2.1	2.3	4.1	9.6
0.2	98.3	80.1	2.1	2.7	2.7	10.7
0.25	97.5	78.2	3.2	2.5	2.1	11.2

AN: acrylonitrile; ACT: acetonitrile

Consider the following prices in $/kg: propylene (0.7), ammonia (0.2), acrylonitrile (1.8), hydrogen cyanide (1.5), acetonitrile (1.3), ammonia sulfate (0.15). Environmental fees are estimated as: emissions (0.001), wastewater per kg organic loading (0.5), acid water per kg acid (0.05), incineration organics/water (0.6), incineration organic solids (1.2) and landfill (0.20). In addition, consider a CO_2 penalty of 0.10 $/kg.

Solution The data regarding the species distribution are converted in stoichiometric equations. Table 2.5 gives the independent chemical reactions. Note the presence of a reaction for describing the formation of heavies, lumped as dinitrile succinate. The material balance around the chemical reactor can be easily calculated by using a spreadsheet. Table 2.6 presents the results for 1 kmol propylene, 1.2 kmol NH_3 and 9.5 kmol air. Propylene conversion is 0.983 and the selectivity in acrylonitrile 79.6%.

Table 2.5 Chemical reaction network and selectivity data.

	Reactions	Extent
1	$CH_2 = CH-CH_3 + NH_3 + 3/2 O_2 \rightarrow CH_2 = CH-CN$ (AN) $+ 3H_2O$	0.801
2	$2CH_2 = CH-CH_3 + 3NH_3 + 3/2 O_2 \rightarrow 3CH_3-CN$ (ACT) $+ 3H_2O$	0.021
3	$CH_2 = CH-CH_3 + 3NH_3 + 3O_2 \rightarrow 3HCN + 6H_2O$	0.027
4	$CH_2 = CH-CH_3 + 3/2 O_2 \rightarrow 3CO_2 + 3H_2O$	0.107
5	$CH_2 = CH-CH_3 + 1/2 O_2 \rightarrow CH_2 = CH-CHO$ (ACR) $+ H_2O$	0.027
6	$CH_2 = CH\text{-}CN + HCN \rightarrow NC-CH_2\text{-}CH_2\text{-}CN$ (SCN)	0.005

Table 2.6 Material balance around the chemical reactor for 1 kmol/h propylene.

	$-C3$	NH_3	O_2	N_2	AN	HCN	ACN	ACR	SCN	CO_2	H_2O
I	1	1.2	1.9	7.6	–	–	–	–	–	–	–
O	0.017	0.286	0.077	7.6	0.796	0.076	0.031	0.027	5E-3	0.321	3.07

The above results can be extrapolated to the industrial scale. Next, we consider a production rate of 340 kmol/h AN, which corresponds to about 100 kton/yr. Figure 2.4 shows the input/output flow diagram. The inputs consist of raw materials, such as propylene, ammonia and air. The outputs are product (acrylonitrile), byproducts (HCN and acetonitrile), gaseous emissions, lights, heavies, wastewater and eventually solids.

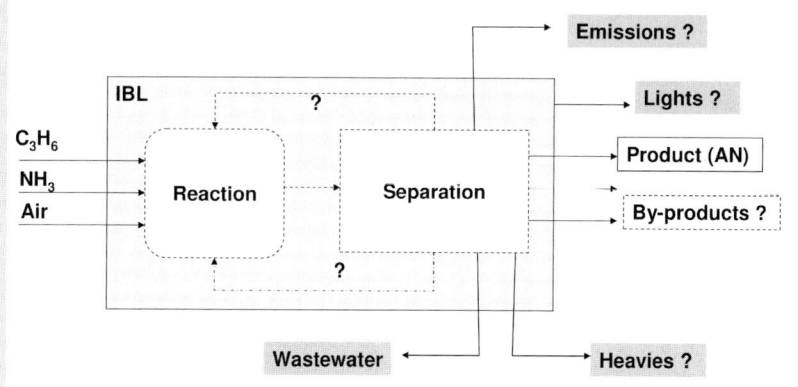

Figure 2.4 Input/output structure of the acrylonitrile manufacturing.

The process flowsheet inside the battery limits (IBL) is at this stage unknown. However, the recycle of reactant may be examined. The patent reveals that the catalyst ensures very fast reaction rate. Conversion above 98% may be achieved in a fluid-bed reactor for residence time of seconds. Thus, recycling propylene is not economical. The same conclusion results for ammonia. The small ammonia excess used is to be neutralized with sulfuric acid (30% solution) giving ammonium sulfate. Oxygen supplied as air is consumed in the main reaction, as well as in the other undesired combustion reactions.

The preliminary input/output material balance built with the above elements is illustrated by the Table 2.7. The gaseous species are found in emissions. The lumped fraction "lights" consists of acroleine, while "heavies" described by di-nitrile succinate.

Some interesting conclusions may be drawn:

1. The mass of gaseous emissions is five times the acrylonitrile production rate, because a large nitrogen amount is carried out with air. CO_2 is of significance too.

2. Wastewater forms by reaction at the same rate as the main product. A supplementary amount comes from ammonia neutralization, as well as the recycled water for the extractive distillation of acetonitrile. Hence, the wastewater sent to treatment is at least three times the acrylonitrile amount.

3. About 8% valuable byproducts (HCN and acetonitrile) are formed.

Table 2.7 Input/output material balance for acrylonitrile manufacturing.

	M_w	Input		Output			Destination
		kmol/h	kg/h	kmol/h	kg/h		
$C_3^=$	42	340	14 307.4	5.78	243.2		
O_2	32	646	20 671.2	26.35	843.2		
N_2	28	2584	72 386.8	2584	72 386.8		
CO_2	44	0	0.0	109.14	4803.2	78 276.5	Emissions
NH_3	17	408	6948.5	97.41	1658.9		
H_2SO_4	98	50	4900.0	50	4900.0		
H_2O solv.	18		32 666.7		32 666.7	44 125.6	Waste
AN	53	0	0.0	270.34	14 345.2	14 345.2	Product
HCN	27	0	0.0	25.54	690.2		
ACN	41	0	0.0	10.71	439.7	1129.9	Byproducts
ACR	56	0	0.0	9.18	514.7	514.7	Lights
SCN	80	0	0.0	2	160.2	160.2	Heavies
H_2O	18	30	540.5	1041.84	18 769.0	18 769.0	Wastewater
Total			157 321.1		157 321.1	157 321.1	

4. The amounts of lights and heavies are substantial. These should be minimized.

A preliminary economic analysis can be carried out, as follows:

$$\text{Raw materials} = (14\,307.4 \times 0.7 + 6948.5 \times 0.2 + 1658.9 \times 132/34 \times 0.35)$$
$$= 11\,404.9\ \$$$
$$\text{Sales} = (14\,345.2 \times 1.8 + 690.2 \times 1.5 + 439.7 \times 1.3$$
$$+ 1658.9 \times 132/34 \times 0.35) = 29\,682.5\ \$$$
$$\text{Sales/raw materials (without environmental costs)} = 2.60$$

A first part of water contamination originates from dissolved organics. Let us assume a yield for AN separation of 98.5% with 1.5% losses in water. Sour water concerns the neutralization of ammonia with sulfuric acid. The lights and heavies are incinerated. Summing up, the following environmental costs should be included:

$$\text{Removal of organics from wastewater} = 14\,345.2 \times 0.015 \times 0.5 = 116\ \$$$
$$\text{Treatment of sour water} = 16\,333.3 \times 0.03 = 490\ \$$$
$$\text{Incineration of organics} = (514.7+160.2) \times 0.8 = 539.9\ \$$$
$$\text{Treatment of gaseous emissions} = 78\,276.5 \times 0.02 = 156.5\ \$$$

In total, the environmental fees are estimated at 1302.5 $ per hour, or 4.38% from sales, which seems realistic [4]. If the CO_2 penalty is considered then 480.3 $ should be added, raising the environmental cost to 1782.8 $ and the ratio to sales at 6%.

The economic potential is 29 682.5 − (11 404.9 + 1782.8) = 16 494.8 $, or about 1.5 times the price of the raw materials. Thus, the added value is 1.5 times the fees incurred by purchasing the raw materials.

Obviously, the above calculations should be seen as order-of-magnitude. In this case, the I/O analysis highlights the possibilities for significantly improving the economics and the ecological performance of an acrylonitrile process. The following aspects should deserve attention:

1. The immediate measure would be replacing propylene by propane, which is at least half the price. The cost of raw materials could drop by 44% raising the ratio sales/raw materials from 2.6 to 4.6, evidently much more profitable.

   Catalysts for ammoxidation of alkanes were investigated intensively in the last two decades. Recent formulations can achieve selectivity around 50–60% but at lower conversions around 50% with more CO_2 and acetonitrile formation [21, 22].

2. Using pure oxygen instead of air and recycling the inert should diminish considerably the amount of emissions. A good inert could be CO_2, a reaction product, which has a higher thermal conductivity than nitrogen. Indeed, some recent patents indicate the possibilities of using CO_2 at dilution rate below 40%.

3. Employing a reusable solvent for ammonia recovery could suppress the treatment of a large amount of wastewater. For an ROI of two years the investment in treating the sour water would be of 490 × 8400 × 2 = 8.23 M$.

4. Conversion of byproducts in more valuable chemicals is profitable. For example acetonitrile can be converted to acrylonitrile by oxidative methylation with CH_4, while HCN in acetone cyanhydrine, methacrylic acid, methionine, *etc.*

2.5
Reactor/Separation/Recycle Structure

2.5.1
Material-Balance Envelope

From the material-balance viewpoint a chemical process consists of two subsystems, *reactions* and *separations*, linked by *recycles*. Figure 2.5 illustrate the basic flowsheet for the reaction A(g) + B(l) = P(l) + R(g). The feeds of the chemical reactor are fresh and recycled reactants via a conditioning device. The separation section consisting of two subsystems, for gases and liquids, that deliver products, subproducts and waste, as well as recycles of unconverted reactants. The figure emphasizes two key issues regarding the plantwide material balance. First, there is an optimal *ratio of reactants* at the inlet of the chemical reactor. This value should comply with the experimental investigation in laboratory or pilot plant regarding the reaction engineering, and as a consequence can be very different with respect to stoichiometry. In the second place the *makeup of fresh reactants* must respect

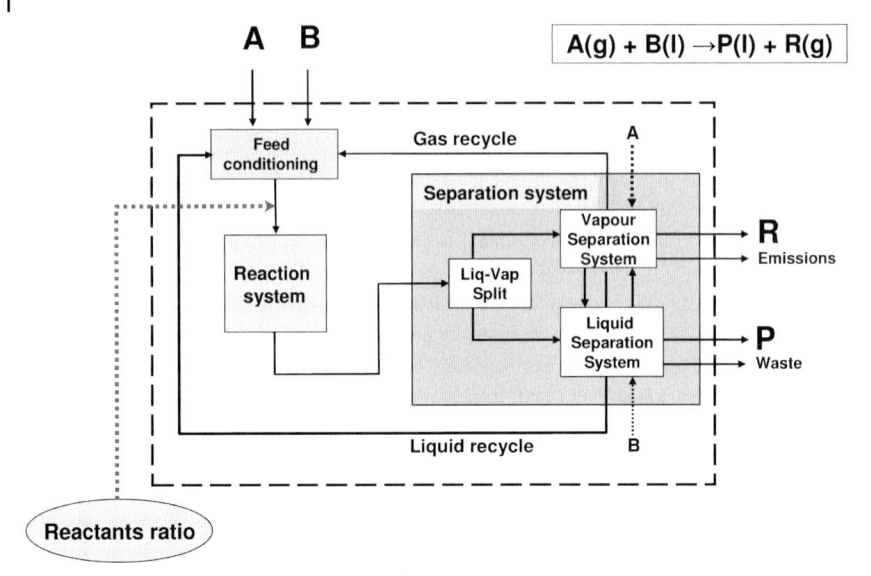

Figure 2.5 Generic reactor/separation/recycle structure.

the overall material balance, meaning that all entered materials have to leave the plant in products, subproducts and waste.

The above constraints can be fulfilled only by building up an appropriate structure of recycles. At the industrial level this implies the integration of design and control of the units implied in the plantwide material balance. Hence, we may speak about the reactor/separation/recycle (RSR) as the major architectural structure defining a chemical process.

It is clear that the reaction and separation systems are interrelated and, in principle, their design should be examined simultaneously. In practice, this approach turns to be very difficult. It is possible to apply, however, a simplified approach. Indeed, from the systemic viewpoint the functions and the connections of units have priority on their detailed design. Therefore, a functional analysis based on the RSR structure only is valuable. This analysis should primarily show that the flowsheet architecture is feasible and appropriate for stable operation. On this basis design targets for the subsystems may be assigned.

In the RSR approach the chemical reactor is the key unit, designed and simulated in terms of productivity, stability and flexibility. From the systemic viewpoint the key issue is the quality and dynamics of flows entering the reactor and less how they have been produced. Obviously, these flows include fresh reactants and recycle streams. The dynamics of flows must respect the overall material balance at steady state, as well as the process constraints. For this reason, the chemical-reactor analysis should be based on a kinetic model.

A sensible assumption is that the separation units are well controlled. They may be viewed as "black-boxes" supplying constant purity recycle flows. However, the flows rates of recycles may exhibit large variations, some with periodical character, some out-of-control with increasing or decreasing tendency, or some chaotic. In these cases, one speaks of the occurrence of *nonlinear* phenomena, which are not desired from the operation viewpoint. Performing a comprehensive nonlinear analysis of a recycle system is not trivial, but a simplified approach may be found, as presented in Chapter 4.

2.5.1.1 Excess of Reactant

Employing an excess of reactant can bring important advantages, as it:

- shifts the maximum achievable conversion for equilibrium-controlled reactions,
- shifts the product distribution by means of kinetic effects,
- helps complete conversion of a reactant for which recycling is not convenient,
- increases the rate of heat and mass transfer inside the reactor,
- prevents the formation of undesired byproducts,
- protects the catalyst,
- helps in solving safety problems.

The excess of reactant at the reactor inlet should be realized by means of recycles, and not by feeding fresh reactant above the stoichiometric ratio. If an excess of reactant is imposed by technological reasons, this part must be removed in purge or bleed streams and becomes an optimization variable.

Table 2.8 presents heuristics for two reactants involved in series/parallel reactions. First, the hypothesis of complete consumption of the reactant that could raise separation problems should be investigated. However, the recycling of both reactants should be envisaged when high selectivity is the aim. The optimization should examine both reactor design and recycle policy.

Table 2.8 Heuristics for dealing with complex reactions A+B→ products.

1. If the selectivity is not affected consider the total conversion of one reactant and recycle the other one.
2. If selectivity is affected consider recycling both reactants. Control the selectivity by optimizing the reaction temperature or by means of recycle policy.

2.5.2
Nonlinear Behavior of Recycle Systems

2.5.2.1 Inventory of Reactants and Make-up Strategies

Estimating the inventory of reactants and anticipating their dynamic effects is fundamental in the design and control of chemical plants. The occurrence of nonlinear phenomena is often interrelated with the method of controlling the *makeup of fresh reactants* [8]. There are two methods for controlling the component inventory in a plant. By *self-regulation* the fresh reactant is set on flow control at a value given by the desired production rate. No attempt is made to measure or evaluate the inventory of the unconverted amount. On the contrary, in the mode named *regulation-by-feedback* the inventory of each reactant is evaluated by direct or indirect measurements, and adjusted by manipulating the fresh feed. Note that adding fresh reactant can take place anywhere in the recycle path, usually on level control for liquids and pressure control for gases. Self-regulation and regulation-by-feedback modes may be combined to produce alternative control structures in the case of multiple reactants, as will be shown in Chapter 4.

The self-regulation offers the advantage of setting directly the production rate by means of the fresh reactants. In the regulation-by-feedback the production rate is set indirectly. A good practice is setting the flow of reactant(s) entering the reactor. The two strategies reflect different viewpoints in plantwide control, based on unit-by-unit and by a systemic approach of the whole plant, respectively.

2.5.2.2 Snowball Effects

Figure 2.6 illustrates qualitatively major nonlinear phenomena that could occur in recycle systems, namely *high sensitivity* and *state multiplicity*. High sensitivity of plants with recycles is manifested typically by the so-called *snowball effect*, which consists of exhibiting a large nonlinear response in flows due to small variations in some parameters of units. Figure 2.6 (left-hand) shows that small changes in the fresh feed of reactants give much larger variations in the flows sent to separations. Snowball may cause severe troubles in operation, for example causing flooding in distillation columns, typically designed close to maximum vapor load. Luyben [9] demonstrated that snowball effects are responsible for difficulties in controlling plants with recycle. He developed a useful rule called the *fixed recycle flow rate*. Accordingly, the flow rate of a reactant in recycle should be fixed in order to get better plantwide control of the material balance.

Snowball is in itself not a dynamic but a steady-state phenomenon characterizing systems employing the self-regulation of inventory. As shown by Bildea and Dimian [10] the occurrence of snowball is rather a problem of unit design than of process control. For example, snowball is caused by too small a reactor volume giving insufficient reactant conversion when higher production is the aim. Accordingly, lower conversion needs to be compensated by larger recycle flow. Reactor design by steady-state optimization may lead to high sensitivity. When the reactor

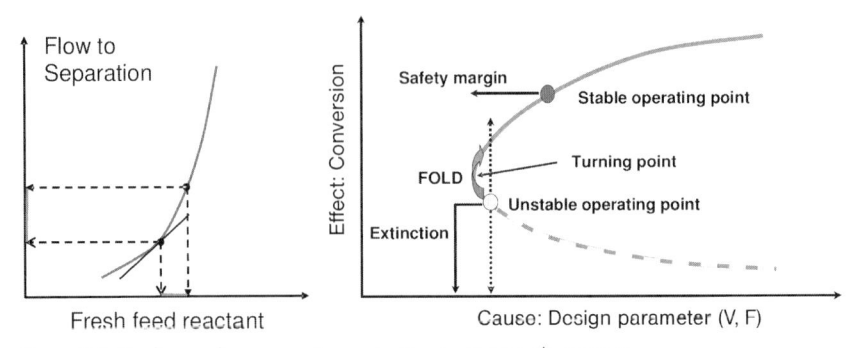

Figure 2.6 Nonlinear phenomena in reactor/separator/recycle systems.

2.5 Reactor/Separation/Recycle Structure

is large enough the classical control structures may be employed. But a better approach is using a regulation-by-feedback strategy, such as fixing the reactant flow at the reactor inlet. The topic will be discussed in more detail in Chapter 4.

2.5.2.3 Multiple Steady States

It is well known that multiple steady states can appear in the case of high exothermic chemical reactions taking place in a standalone CSTR. On the contrary, standalone PFR has a unique state, the nonlinearity manifesting as the occurrence of hotspot and parametric sensitivity. The behavior of a chemical reactor placed in a recycle system can be very different from its standalone counterpart. Figure 2.6 (right-hand) presents a plot of conversion versus reactor volume showing that *multiple steady states* may occur in a recycle system due solely to the feedback effect of material recycles. The C-shape curve has two branches: stable states at high conversions and unstable states at low conversions. The transition takes place at the *turning point*. To the left of the turning point there is no feasible state. Obviously, the designer prefers the operation in a stable state, if possible at high conversion.

2.5.2.4 Minimum Reactor Volume

Figure 2.6 (right-hand) emphasizes that a *minimum reactor volume* is necessary for operation in a recycle system. On the other hand, the steady-state optimization of reactor volume is likely to lead to minimum volume, and as a result is characterized by the high sensitivity mentioned before. Therefore, in contrast to the standalone view, ensuring stable operation in a recycle system asks that the reactor volume must be larger than a minimum value. The demonstration was performed initially for simple reactions by Bildea et al. [11], and later extended to more complex reactions by Kiss et al. [12]. Similar diagrams have been obtained for CSTR and PFR in the case of simple and complex kinetics. The differences are quantitatively important in design, but the trend remains. More details are given in Chapter 4.

2.5.2.5 Control of Selectivity

Selectivity is a key topic for the design of reactors in recycles. From the standalone viewpoint the means to influence selectivity are reactor type, conversion level and mixing method. In the standalone view low conversion and PFRs are recommended for achieving good selectivity. By contrast, following the results of RSR analysis, the recycle policy and the plantwide control of reactant feeds can play the determinant role; the reactor type or the conversion level is less important [13].

2.5.3
Reactor Selection

The selection of a chemical reactor should ensure safe operation, high productivity and yield, low capital and operating costs, environmental acceptability, and acceptable flexibility with respect to production rate and raw materials composition.

Table 2.9 Heuristics: Selection of reactor for homogeneous systems.

1. For single reactions minimize the reaction volume.
   (a) For positive reaction orders CSTR always requires a larger volume than PFR. The difference increases with higher conversions and higher reaction orders.
   (b) Using a series of CSTRs drastically reduces the total reaction volume. For more than ten units the performance of a PFR is achieved.
   (c) CSTR followed by PFR may be an interesting alternative.
   (d) At low conversions the difference between CSTR a PFR is not relevant. The selection can be motivated on mechanical technology, controllability and safety.

2. For series reactions as A→P→R when the goal is the maximization of the intermediate, do not mix reactant and intermediates. PFR gives the highest yield.

3. For parallel reactions as A→P, A→R the objective is a desired product distribution.
   (a) Low concentrations favor the reaction of lowest order, while high concentrations favor the reaction of highest order.
   (b) For similar reaction orders the product distribution is not affected by concentration, the only solution being to use a suitable catalyst.

4. Complex reactions can be analyzed by means of simple series and parallel reactions. For first-order series-parallel reactions the behavior as series reactions dominates. PFR is more advantageous for maximizing an intermediate component.

5. High temperatures favor reactions with larger activation energy. Reactions with small activation energy are slightly affected, so low temperature is preferred.

2.5.3.1 Reactors for Homogeneous Systems

The selection of a chemical reactor can be formulated as guidelines in term of relative performances of the two basic types, CSTR and PFR, as functions of stoichiometry and kinetics, as shown in Table 2.9. In most cases PFR offers better productivity and yield, particularly at high conversion. However, the CSTR is superior from the viewpoint of heat- and mass-transfer operations. The performance of CSTR may be improved by building a cascade of perfect mixing zones. The combination of zones of perfect mixing and plug flow is advantageous for carrying out complex reactions with autocatalytic pattern, such as in polymerization and biological processes. The guidelines hold also for heterogeneous reactions described by pseudohomogeneous models.

2.5.3.2 Reactors for Heterogeneous Systems

The selection of reactors dealing with a heterogeneous reaction should take into account three aspects: catalyst selection, reactant injection and dispersion, choice of hydrodynamic flow regime, as illustrated by Figure 2.7 [14].

Catalyst selection involves two features: productivity and selectivity. The process rate is a subtle combination of four limiting steps: adsorption/desorption of reactants/product, surface reaction between species, diffusion through pores and diffusion through external film. Pore structure, surface area, nature and distribution of active sites play a crucial role in forming the process rate at the level of catalyst

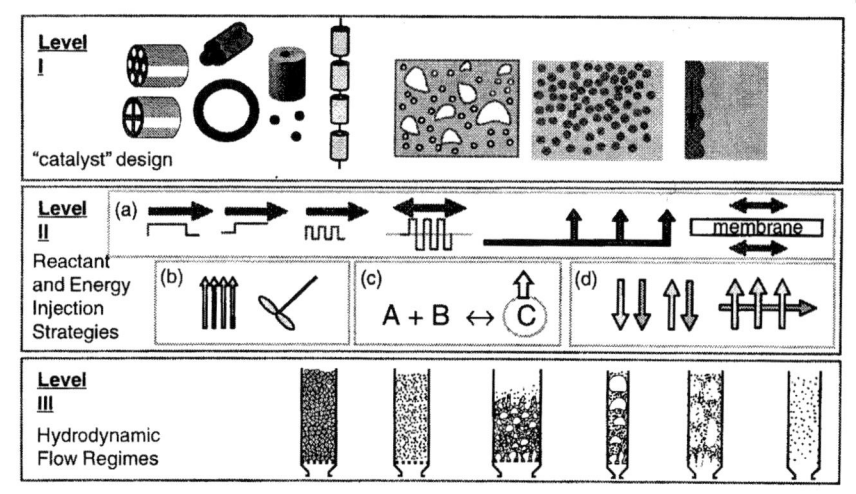

Figure 2.7 Three-level strategy for multiphase reactor selection [14].

grain. If the chemical surface reaction is very fast, some physical steps may become rate controlling. In this case, the catalyst is more efficient if coated on a surface.

Contact of reactants involves design decisions regarding the following aspects:

(a) Reactant injection strategy, such as one-shot continuous pulsed injection, reversed flow, staged injection and use of membrane.
(b) Choice of the optimum mixedness of fluid phases.
(c) Use of *in-situ* product separation or of energy removal.
(d) Phase contact as co-, counter-, and crosscurrent flows.

The reversible reactions deserve particular attention. The *in-situ* removal of a product by reactive distillation, reactive extraction or by using selective membrane diffusion should be investigated.

Hydraulic design aims at the realization of an intensive heat and mass transfer. For two-phase gas-liquid or gas-solid systems, the choice is between different regimes, such as dispersed bubbly flow, slug flow, churn-turbulent flow, dense-phase transport, dilute-phase transport, *etc*.

2.5.4
Reactor-Design Issues

2.5.4.1 Heat Effects

The adiabatic temperature change ΔT_a helps to evaluate the importance of heat effects in a design of a chemical reactor, even if the reactor itself is not an adiabatic one. Table 2.10 presents some useful heuristics. The use of an inert to remove or add heat is the most employed in low-cost adiabatic reactors. Considering heat-transfer devices is more expensive, but better for energy integration.

Table 2.10 Heuristics for the thermal design of chemical reactors.

1. Exothermic reactions. If $\Delta T_{ad} > 0$ is too high, then:
   - Increase the flow rate and/or lower the per-pass conversion.
   - Use a heat carrier to remove heat (example excess inert).
   - Consider a device with heat transfer to a cooling agent.

2. Endothermic reactions. If $\Delta T_{ad} < 0$ is too high, then:
   - Preheat the reactants at sufficiently high temperature.
   - Use direct heat carrier to supply heat (for example steam).
   - Consider a device with heat transfer from a hot agent.
   - Consider the possibility of internal heat generation by exothermic reactions.

2.5.4.2 Equilibrium Limitations

For a highly exothermic reaction the optimization of the temperature profile is the key factor in maximizing the reactor productivity. For endothermic reactions maximizing the reaction temperature and employing a heat carrier is often the best solution. Table 2.11 summarizes the guidelines.

Table 2.11 Heuristics for chemical-equilibrium-controlled reactions.

1. Reversible exothermic reactions
Optimize the temperature profile. Higher temperature is advantageous for the reaction rate, but not for equilibrium conversion. Lower temperature has the opposite effect. The reaction should start at higher temperature and finish at the lower one.

2. Reversible endothermic reactions
The temperature should the highest permitted by the technological constraints.
Use of heat carriers may be considered, such as steam, hot gas or solid inert. If the number of moles increase by reaction, the dilution with inert shifts the equilibrium conversion to higher values. However, more energy is needed for inert recycling. Therefore, the molar ratio inert/reactant is an optimization variable.

2.5.4.3 Heat-Integrated Reactors

Highly exothermic reactions are excellent candidates for energy integration. Adiabatic reactors are often preferred as they are robust and cheap. Figure 2.8 presents a typical flowsheet structure designated as *heat-integrated reactor*. Important energy saving may be achieved by means of a feed–effluent heat exchanger, often abbreviated as FEHE. Additional units are usually included, such as the following:

- Heater for start-up. Because positive feedback due to heat integration may lead to state multiplicity, the heater duty can be manipulated in a temperature control loop to ensure stable operation.

- Steam generator. The energy introduced by a heater has to be removed, for example by raising steam. Placing the steam generator after the reactor but before FEHE allows heat recovery at higher temperature, preferable from the viewpoint of exergy.

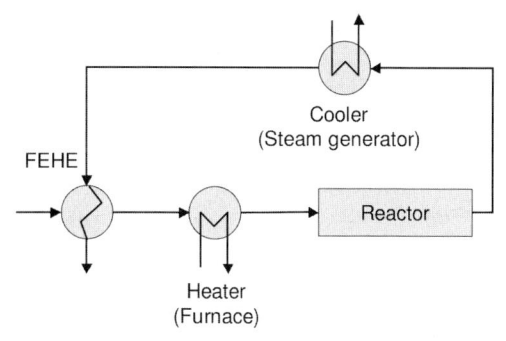

Figure 2.8 Generic flowsheet structure of a heat-integrated reactor.

The above flowsheet is generic for industrial applications. Since the heat source is a furnace and the excess of energy is rejected to a heat sink (steam generator), the reactor can be viewed as a heat pump. The properties of this structure have been studied by Bildea and Dimian from the perspective of state multiplicity, stability and controllability [12]. There is a close relation between design and controllability. A surprising conclusion is that high-performance design of each individual unit might be counterproductive for the controllability of the whole. For example, minimizing the furnace load by maximizing the FEHE efficiency is not appropriate when the reactor temperature is controlled by the furnace duty, the main disturbance being the feed flow. On the contrary, the best dynamic performance is given by a design with moderate furnace duty, small steam generator and efficient FEHE.

2.5.4.4 Economic Aspects

From the economic viewpoint three aspects are of interest in reactor selection:

(a) Consider alternative reactor types, such as for example, a gas-phase catalytic reactor against a slurry gas/liquid reactor.
(b) Evaluate alternative reactor design. For example, one may have interest in comparing a heat carrier versus a heat-transfer system.
(c) Define the optimality range for the design variables, usually temperature, pressure, conversion and reactants ratio.

Because at the RSR level the separators are not yet known, the cost of recycles may account only for transport and conditioning of streams. Transporting gases involves high capital and operation costs for compressors. Similarly, thermal feed conditioning may involve expensive equipment, such as evaporators and furnaces, as well as the cost of heat carriers.

2.6
Separation System Design

The structure of separations is determined by the composition and the thermodynamic properties of the mixture leaving the reactor. The *first separation step* is essential, because it helps to divide a large separation problem of a multiphase mixture into smaller subproblems for separating monophase mixtures that may be handled further by specific systematic methods.

2.6.1
First Separation Step

2.6.1.1 Gas/Liquid Systems

For the first separation step the following techniques can be employed, such as simple flash or sequence of flashes, physical or chemical absorption, or reboiled stripping. Simple flash is suitable when the ratio of K values of light and heavy components is larger than 10 (Figure 2.9a). Pressure and temperature are optimization variables. Better separation can be obtained in more stages with intermediate heating/cooling (Figure 2.9b). Good recovery can be realized by means of a gas-absorption device. The absorbent can be a process stream or a recycled solvent (Figure 2.9c). Vapor recompression could be used to improve the separation after a first flash (Figure 2.9d).

Reboiled stripping is efficient for mixtures containing a significant amount of light and intermediate components. An example is the separation of C2 and C3 fractions from a hydrocarbon mixture issued from fluid catalytic cracking. The initial precooled mixture is sent to the top of a distillation column provided only with reboiler. The top product contains gases and light components stripped out

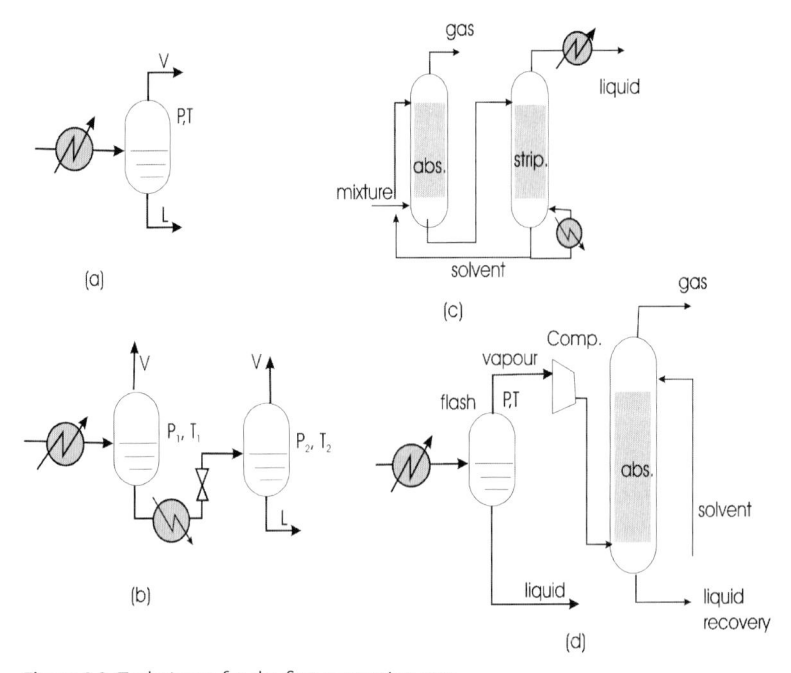

Figure 2.9 Techniques for the first separation step.

by the vapor produced in the reboiler, while the bottom collects the remaining heavy components. Note that the temperature of the inlet mixture stream should be low enough to prevent entraining heavier components at the top. Recycled heavy solvent may be used to increase the separation efficiency, the initial mixture being fed in at a lower position.

2.6.1.2 Gas/Liquid/Solid Systems

Solid particles may be present in the reactor effluent or generated by deeper cooling. Another method is precipitation by means of a suitable mass-separation agent. This technique is more expensive than simple cooling, because it introduces a new recycle loop. Adding water might lead to waste-treatment problem. After precipitation, the suspension is sent to a filter or centrifuge. In general, two liquid recovery systems are generated, for reactant and solvent, respectively.

2.6.2
Superstructure of the Separation System

The separation system can be described as a "superstructure" of subsystems [2], as illustrated by Figure 2.10, corresponding to the dominant physical state during processing, respectively as gas and vapor, liquid and solid. Inspection of Figure 2.10 emphasizes the role of the first separation step in generating separation subsystems. The subsystems are interconnected by recycles. Because recycling is

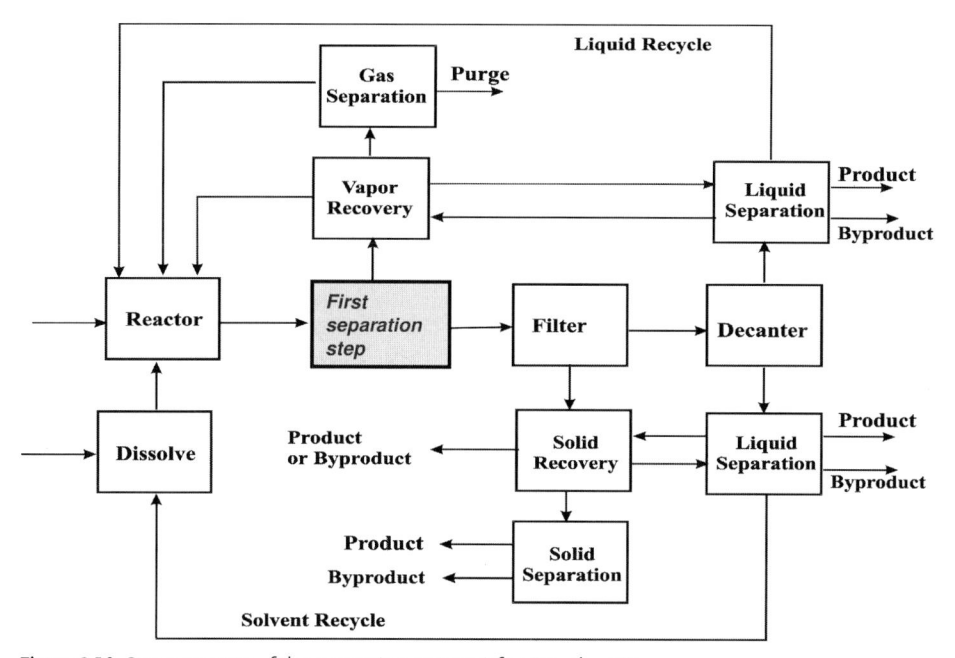

Figure 2.10 Superstructure of the separation system (after Douglas [2]).

not economical, the challenge for the designer is to minimize the number and the flow rates of recycles. Special attention should be paid to the accumulation of byproducts and impurities in recycles.

As a strategy, the synthesis procedure should start with vapor recovery and gas separations, from which some components are sent to liquid separations. For the same reason, the solid-separation system should be placed in the second place. Note that the subsystems of gas and solid separations are largely uncoupled. As a result, the liquid-separation system should is handled the last.

At this point, we pause the problem of synthesis of subsystems of separations. This will be resumed in Chapter 3, where a systematic methodology based on a task-oriented approach will be described in more detail.

Example 2.2: Structure of Separations for the Hydrodealkylation of Toluene

The hydrodealkylation of side-chain aromatics to nonsubstituted parents is a major process in petrochemistry. A typical example is the conversion of toluene to benzene:

$$C_6H_5-CH_3 + H_2 \rightarrow C_6H_6 + CH_4 \tag{1}$$

The reaction may be carried out at 30–50 bar, either thermally at 550–700 °C, or over suitable catalyst at somewhat lower temperatures. Higher boiling sub-products form, such as biphenyl and fluorene. In thermal processes the conversion is usually 60–80%, with a selectivity of about 95%. In catalytic processes the selectivity is significantly higher, which can compensate the cost of the catalyst. Sketch the structure of separations by considering as feeds pure toluene and hydrogen with 5% methane.

Solution: Firstly, we solve the problem of the input/output material balance. Besides the main reaction a secondary reaction leading to the subproduct diphenil takes place:

$$2C_6H_6 \rightarrow C_{12}H_{10} + H_2$$

The first design decision is that incomplete conversion of toluene take place, for example 70%. In addition, the selectivity to benzene is of 95%. The second decision is using a purge to avoid the accumulation of methane. Consequently, the hydrogen should be introduced in excess with respect to stoichiometry. The third decision is fixing the ratio hydrogen/toluene at the reactor inlet at a suitable value, for example five, for minimizing the formation of carbon by hydrocarbon decomposition. This gives a supplementary constraint on the flow rate and composition of the gas recycle.

The flowsheet is presented in Figure 2.11. This can be submitted to simulation by using a stoichiometric reactor and a black-box separator. Note that a

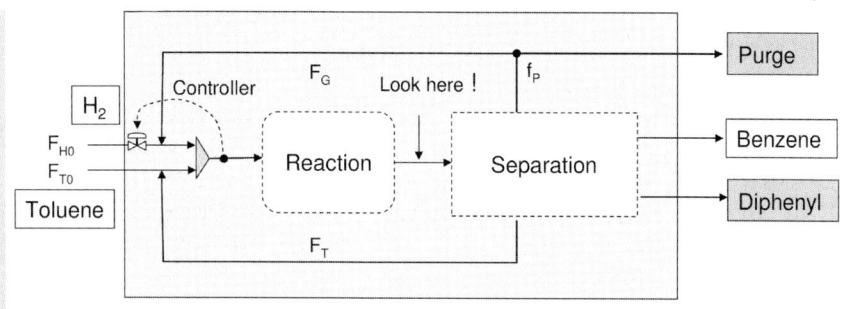

Figure 2.11 Input/output structure by toluene hydrodealkylation process.

controller (design specification in Aspen) is employed to fulfil the third design decision. The simulation of this flowsheet is a simple exercise. If the inputs of both hydrogen and toluene are fixed, as well as the purge ratio f_P, the convergence fails, unless correct values are matched by trial and error. In addition, the third condition will not be fulfilled. The solution is to use a controller to adjust an input, for example the hydrogen fresh feed. Instead of specifying the purge fraction better is fixing the recycle flow F_G. This makes more sense from a plantwide control viewpoint too. The flowsheet controller simulates the behavior of a process control loop.

Table 2.12 shows the table stream calculated with Aspen Plus for toluene fresh feed of 100 kmol/h, purge fraction of 0.06 and ratio of hydrogen/toluene in the inlet reactor mixture of 5. In these conditions, the gas recycle rate is about ten times the molar flow rate of the inputs.

For the assessment of the first separation step, the behavior of the outlet reactor mixture is of interest. A flash at 40 bar and 303 K with Peng–Robinson EOS gives the K values listed in the last column. It may be observed that sharp separation of gases from condensable species is possible just by a single flash. Hydrogen and toluene, much more volatile by a factor of 1000, separate very easily in the gas phase, while benzene, toluene and diphenyl pass completely in the liquid phase. This simple solution is applicable in a large number of cases, as illustrated in this book by the case studies referring to the manufacturing of cyclohexanone and vinylacetate.

Table 2.12 Stream table by toluene hydrodealkylation process.

Mole flow kmol/h	Input	Output	Reactor in	Reactor out	K values
H_2	134.5	37.0	714.3	616.8	7.23E + 01
CH_4	7.1	107.1	1684.8	1784.8	1.07E + 01
C_6H_6	0.0	95.0	0.0	95.0	8.94E – 03
C_7H_8	100.0	0.0	142.9	42.9	3.03E – 03
$C_{12}H_{10}$	0.0	2.5	0.0	2.5	9.11E – 06
Total flow	241.6	241.6	2541.9	2541.9	

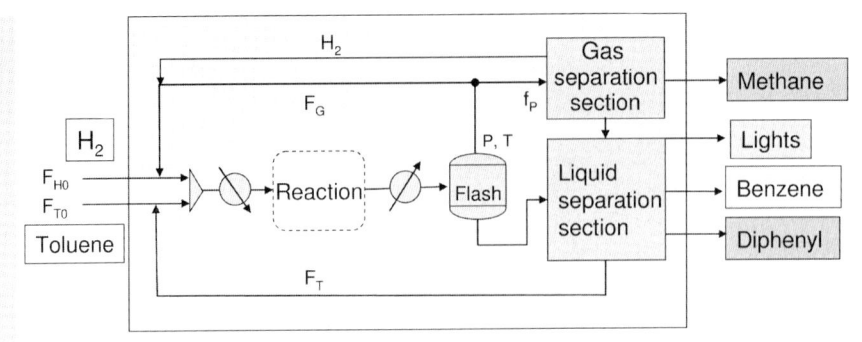

Figure 2.12 Structure of separations for the toluene hydrodealkylation (HDA) process.

Since the flash produces two monophase streams, the separation system will contain both gas and liquid-separation sections (Figure 2.12). On the gas side, the separation of methane as a useful byproduct and the recycling of hydrogen is today economically viable by using membranes. The operation would greatly reduce the cost of the gas compression, as well as the size of the chemical reactor. An already obsolete alternative would be sending the purge stream directly to combustion. On the liquid side, as indicated by the K values, the separation of the three components benzene, toluene and di-phenyl can be realized easily, either in a sequence of two distillation columns, or in a single column with side stream. Since small amounts of methane and hydrogen remain dissolved in the liquid after flash, these are removed as lights.

2.7
Optimization of Material Balance

The economic potential may be used to optimize the flowsheet with respect to the material balance. After Level 4 the mass-balance envelope is closed by completing the synthesis of reaction and separation systems. The economic potential EP_4 becomes:

$$EP_4 = EP_2 - \{\text{Reactor costs/Payback time}\} - \{\text{Cost of separations}\} \\ - \{\text{Utilities costs}\} - \{\text{Additional environmental costs}\} \qquad (2.7)$$

Put in this way, the approach consists of a complex multivariable optimization task subject to many uncertainties. A simpler method is by assuming that the chemical reactor and the cost of recycles determine the overall optimum, the key variable being the conversion of the reference reactant. Lower conversion gives in general better selectivity, but higher costs of recycles. Higher conversion gives more subproducts and impurities, sharply increasing the cost of separations. Performing a more accurate optimization consists of finding simple cost relations as functions of throughput and performance of separation units.

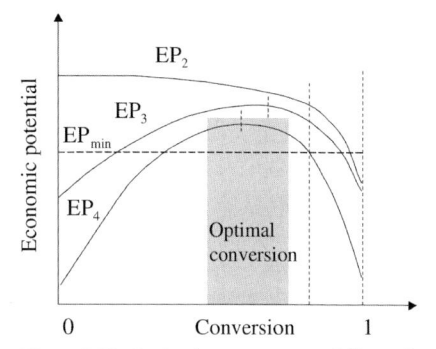

Figure 2.13 Optimal conversion at different levels of the process synthesis by the hiearchical approach.

Figure 2.13 illustrates the variation of the economic potential during flowsheet synthesis at different stages as a function of the dominant variable, reactor conversion. EP_{min} is necessary to ensure the economic viability of the process. At the input/output level EP_2 sets the upper limit of the reactor conversion. On the other hand, the lower bound is set at the reactor/separation/recycle level by EP_3, which accounts for the cost of reactor and recycles, and eventually of the separations. In this way, the range of optimal conversion can be determined. This problem may be handled conveniently by means of standard optimization capabilities of simulation packages, as demonstrated by the case study of a HDA plant [3].

2.8
Process Integration

2.8.1
Pinch-Point Analysis

After developing the basic flowsheet and mass balance the next step is the integration of energy. Systematic methods are now well established in the framework of pinch-point analysis (PPA), developed for the most part by the contributions of Linnhoff and coworkers [18]. It is useful to remember that *pinch* designates the location among process streams where the heat transfer is the largest constraint. The pinch can be identified in an enthalpy–temperature plot as the nearest distance between the hot and cold composite curves. Accordingly, the energy-management problem is split into two parts: above and below the pinch. In principle only heat exchange between streams belonging to the same region are energetically efficient. Moreover, heat should be supplied only above and removed only below the pinch. When the pinch principle is violated energy penalties are incurred. The designer should be aware of it and try to find measures that limit the transfer of energy across the pinch.

The essential merit of pinch-point analysis is that makes possible the identification of key targets for energy saving with minimum information about the performance of heat-exchange equipment. The key results of are:

1. Computation of minimum energy requirements (MER).
2. Generation of an optimal heat exchangers network.
3. Identification of opportunities for combined heat and power production.
4. Optimal design of the refrigeration system.

2.8.1.1 The Overall Approach

Figure 2.14 illustrates the overall approach by pinch-point analysis. The first step is extraction of *stream data* from the process synthesis. This step involves the simulation of the material-balance envelope by using appropriate models for the accurate computation of enthalpy. On this basis *composite curves* are obtained by plotting the temperature T against the cumulative enthalpy H of streams selected for analysis, hot and cold, respectively. Two aspects should be taken into account:

- Proper selection of streams with potential for energy integration.
- Adequate linearization of T–H relation by *segmentation*.

The next step is the selection of utilities. Additional information regards the partial heat-transfer coefficients of streams and utilities, as well as the price of utilities and the cost laws of heat exchangers.

After completing the input of data one can proceed with the assignment of tasks for heat recovery by a *supertargeting* optimization procedure. In the first place the *minimum difference temperature* ΔT_{min} is determined as a trade-off between energy and capital costs. If the economic data are not reliable, selecting a practical ΔT_{min} is safer. Next, initial design targets are determined as:

1. *minimum energy requirements* (MER) for hot and cold utilities,
2. overall heat exchange area, and
3. number of units of the *heat-exchanger network* (HEN).

The approach continues by *design evolution*. This time, the design of units is examined in more detail versus optimal energy management. Thus, the "appropriate placement" of unit operations against pinch is checked. This may suggest design modifications by applying the "plus/minus principle". The options for utility are revisited. Capital costs are the trade-off again against energy costs. The procedure may imply several iterations between targeting and design evolution. Significant modifications could require revisiting the flowsheet simulation.

The iterative procedure is ended when no further improvement can be achieved. Note that during different steps of the above procedure the individual heat exchangers are never sized in detail, although information about the heat-transfer coefficients of streams is required. Only after completing the overall design targets can the detailed sizing of units take place. Optimization methods can be used to refine the design. Then, the final solution is checked by rigorous simulation.

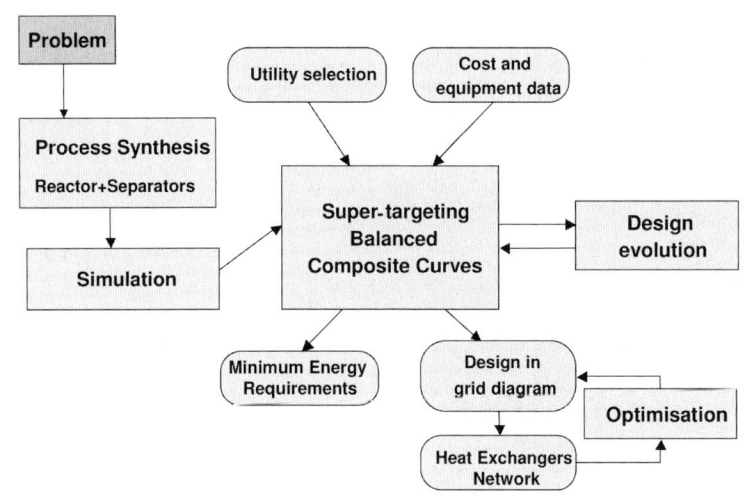

Figure 2.14 Overal approach by pinch-point analysis.

Appendix B presents useful information in preliminary conceptual design of heat exchange systems. This concerns the selection of heat transfer fluids and of heat transfer coefficients, the design of shell-and-tube and air cooled heat exchangers, as well as of heat exchangers for special applications (plate, compact and spiral). Appendix C handles the selection of material of constructions, determinant for equipment costing. Since steam is the most used utility in process engineering, Appendix D gives a comprehensive table of thermodynamic properties of water.

An important feature of the methodology is determining the appropriate placement of unit operations with respect to pinch. The analysis can find which changes in the design of units are necessary and perform a quantitative evaluation of these changes. The strongest impact has the design of the chemical reactor, namely the pressure and temperature. It is useful to know that higher reaction temperatures give better opportunities for heat integration. Another important source of energy saving is the integration of distillation columns by thermal coupling or by integrated devices, such as the divided wall column (DWC).

However, very tight energy integration might be detrimental for controllability and operability, by removing some degrees of freedom. Thus, the analysis of heat integration should investigate the consequences on process control.

Summing up, pinch-point analysis consists of a systematic screening of the maximum energy saving that can be obtained in a plant by internal process/process exchange, as well as by the optimal use of the available utilities. The method is capable of assessing optimal design targets for the heat-exchanger network well ahead of detailed sizing of the equipment. Furthermore, the method may suggest design improvements capable of significantly enhancing the energetic performance of the whole process.

2.8.2
Optimal Use of Resources

Beside energy, the optimal use of other resources is of highest interest for process integration. Similar methods based on the pinch principle have been developed for optimal recycling and management of water, hydrogen and solvents. However, the analogy is restricted by the difficulty of finding representative properties and simple graphical methods. More appropriate seems the direct optimization methods based on the allocation of resources. An advanced treatment of this topic can be found in the newest book of Smith [19].

2.9
Integration of Design and Control

Conceptual process design must guarantee good controllability characteristics. Therefore, design and control should be integrated as early as possible. Much experience has been accumulated over the years in the control of individual unit operations, or controlling groups of units by traditional SISO or more advanced MIMO methods. However, the search of appropriate strategies of controlling the plant as a whole, so-called plantwide control, are relatively of recent date [20]. The need for such approach originates from three reasons:

1. The increase of material and energy recycles in modern plants because of tight integration, increasing the interactions between units too.
2. The suppression or the limitation of intermediate storage tanks in order to improve the overall dynamics and/or increase the safety. The result is that the control of chemical reactors and separators is submitted to more frequent and larger disturbances.
3. Flexible plants, with stable behavior and good responsiveness both at lower and higher throughput.

The advent of powerful and user-friendly dynamic simulation software makes it possible to handle the plantwide control strategy directly with nonlinear plant model. This advantage will be greatly exploited in this book. Chapter 4 will treat in more detail the problem of nonlinear analysis of plants with recycle with consequences on the plantwide control strategies.

2.10
Summary

The hierarchical approach is a simple but powerful methodology for the development of process flowsheets. It consists of a top-down analysis organized as clearly defined sequence of tasks aggregated in levels. Each level handles a fundamental conceptual problem: input/output structure, reactor design, structure of separations and recycles, design of separation subsystems, energy and resources integration, protection of environment, safety and hazard problems, as well as plantwide control issues. At each level, systematic methods can be applied for the analysis and synthesis of subsystems.

In the first place, the designer has to identify dominant design variables and take appropriate design decisions, in turn determined by economic, technological, safety and environmental constraints. In this way, the procedure generates not only one presumed good flowsheet, but also a number of alternatives. These are submitted to an evaluation procedure, based on economic and technologic criteria. Following a depth-first approach only one alternative is retained at each level. Another possibility would be the optimization of a superstructure of alternatives. When there is no obvious distinction, the few best ones can be kept for further analysis.

Formally, the hierarchical design procedure can be divided in two steps: process synthesis and process integration. The first deals with the synthesis of the process flowsheet and basic material balance. The second step handles the problem of efficient use of energy and helping material resources (water, solvents, hydrogen). Clearly, the two steps are interrelated.

This chapter proposed an improved methodology that aims to minimize the interactions between synthesis and integration steps by putting more effort on the structural elements of the flowsheet architecture. The core is the level reactor/separation/recycle in which the reactor design and the structure of separations are examined simultaneously by considering the effect of recycles. A major issue is the feed policy of reactants and its implications on plantwide control of reactant inventory and separators. By placing the reactor in the core of the process, the separators receive clearly defined design tasks from plantwide perspective. This level also tackles energy-integration problems around the chemical reactor.

For the synthesis of separation systems the strategy is decomposing the overall problem into subsystems handling essentially gases, liquids and solids, for which systematic methods exist. Environmental protection can also be handled by taking into account the tasks defined early in the above procedure.

The conceptual flowsheet with heat and material balance built upon supplies the key elements for sizing the units and assessing capital and operation costs, and on this basis establish the process profitability.

Thus, the major merit of the hierarchical approach is its generic conceptual value, offering a consistent framework for developing alternatives rather than a single presumed optimal design. The solution is never unique, depending on assumed design decisions and constraints. Conceptual shortcomings that may be identified are valuable insights for further improvement of the process design, as well as new ideas in research and development.

References

1 Douglas, J. M., Conceptual Design of Chemical Processes, McGraw-Hill, New York, USA, 1988
2 Douglas, J. M., Synthesis of separation system flowsheets, *AIChEJ*, 41, 252, 1995
3 Dimian, A. C., Integrated Design and Simulation of Chemical Processes, CAPE series No. 13, Elsevier, Amsterdam, The Netherlands, 2003
4 Allen, D. T., Shonnard, D. R., Green Engineering – Environmentally Conscious Design of Chemical Processes, Prentice Hall, Upper Saddle River, NJ, USA, PTR, 2002
5 Crowl, D., Louvar, J. F. Chemical Process Safety: Fundamentals with Applications, Prentice Hall, Upper Saddle River, NJ, USA, 1989
6 Kletz, T., Hazop and Hazan: Identifying and Assessing Process Industry Hazards, Hemisphere Publ., 1999
7 Missen, R. W., Mims, C. A., Saville B. A., Chemical Reaction Engineering and Kinetics, Wiley, New York, USA, 1999
8 Dimian, A. C., Bildea, C. S., Component inventory control, in Integrating Design and Control (eds P. Seferlis and M. Georgiadis), CACE series No.17, Elsevier, Amsterdam, The Netherlands, pp. 401–430, 2004
9 Luyben, W. L. Snowball effects in reactor/separator processes with recycle, *Ind. Eng. Chem. Res.*, 33, 299–305, 1994
10 Bildea, C. S., Dimian A. C., Fixing flow rates in recycle systems: Luyben's rule revisited, *Ind. Eng. Chem. Res*, 42, 4578–4588, 2003
11 Bildea C. S., Dimian, A. C. Iedema, P. D. Nonlinear Behavior of Reactor-Separator-Recycle Systems, *Comp. Chem. Eng.*, Vol. 24, Nos. 2–7, 209–217, 2000
12 Bildea, C. S., Dimian, A. C. Stability and multiplicity approach to the design of heat-integrated PFR, *AIChE J.*, Vol. 44, 703–2712, 1998
13 Kiss, A. A., Bildea, C. S. Dimian, A. C. *Comp. Chem. Eng.*, 31, 601–611, 2007
14 Krishna, R., Sie, S. T. Strategies for multiphase reactor selection, *Chem. Eng. Sci.*, Vol. 49, 24A, 4029–406, 1994
15 Barnicki, S.D., Fair, J.R. Separation System Synthesis: A Knowledge Based Approach: 1. Liquid Mixture Separations, *Ind. Eng. Chem. Res.*, 29, 431–439, 1990
16 Barnicki, S.D., Fair, J.R. Separation System Synthesis: A Knowledge Based Approach: 2. Gas/Vapor mixtures, *Ind. Eng. Chem. Res.*, 31, 1679–1694, 1992
17 Siirola, J.J., Industrial applications of chemical process synthesis, in Advances in Chemical Engineering, Vol. 23, Process Synthesis, Academic Press, 1996
18 Linnhoff, B., Townsed, D. W. Bolland, D. Hewitt, G. F. Thomas, B. Guy, A.R. Marsland, R.H. User Guide on Process Integration, The Institution of Chemical Engineers, UK, 1994
19 Smith, R., Chemical Process Design and Integration, Wiley, New York, USA, 2004
20 Luyben, W. L., Tyreus, B. Plantwide Process Control, McGraw-Hill, New York, USA, 1999
21 Grasselli, R. K., Selectivity issues in (amm)oxidation catalysis, *Catal. Today*, 99, 23–31, 2005
22 Centi, G., Grasselli, R. K., Trifiro, F., Propane ammoxidation to acrylonitrile, an overview, *Catal. Today*, 661–666, 1992

3
Synthesis of Separation System

3.1
Methodology

The overall picture of the synthesis problem as a superstructure of separations has been presented in Chapter 2. For generating separation sequences inside subsystems we adopt the formalism of task-oriented approach proposed by Barnicki and Fair [1, 2], but with modifications issued from own experience [5]. Interesting ideas for developing a systematic treatment of a separation problem can be found also in the paper of Barnicki and Siirola [3].

The goal in process synthesis of separations is the development of close-to-optimum flowsheets, in which both the feasibility and the performance of splits are guaranteed. The detailed design of units is left as a downstream activity that can be treated by specialized algorithms or ensured by an equipment supplier. Since the global optimal solution is very difficult or impossible, a workable strategy consists of splitting the overall synthesis problem in smaller and relatively independent problems for which systematic methods are available. The approach consists of a building a hierarchy of separations, as displayed in Figure 3.1. This can be worked out manually, or implemented as a knowledge-based computerized tool.

The first step is the split of the initial mixture in essentially monophase submixtures, as gas, liquid and solid. This operation, called *the first separation step*, can employ simple flash or a sequence of flashes, adsorption/desorption and reboiled stripping, or the combination of the above techniques. Next, the process synthesis activity can be further handled by specialized *managers*, namely gas split manager (GSM), liquid split manager (LSM) and solid split manager (SSM).

Handling the separation problem in each subsystem is further aided by *selectors*. A selector designates a generic *separation task* for which separation techniques are available. The use of selectors has as result a significant reduction in the searching space of all possible separations takes place. By applying the selectors to successive mixtures allows a *separation sequence* to be generated.

The task assigned to a selector can be executed by several separation techniques. Specialized *designers* bring the expertise necessary to design the unit. From the practical viewpoint these can be identified with the standard unit operations, for which computation methods are available, but also with novel techniques that can

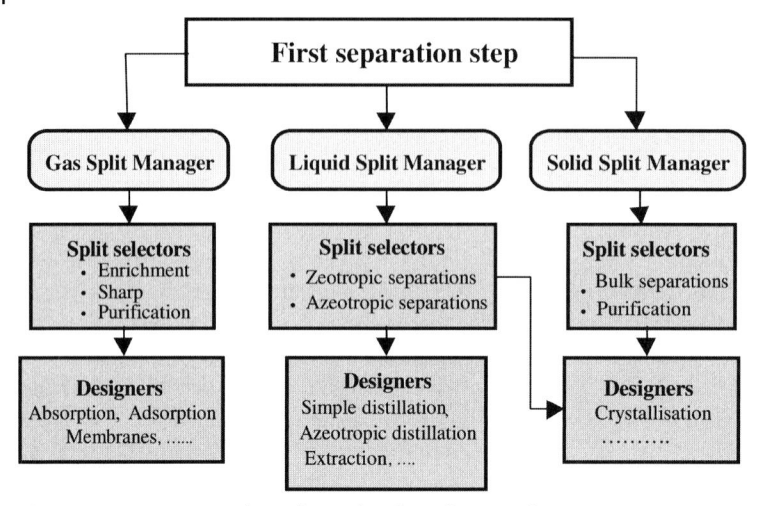

Figure 3.1 Separation synthesis hierarchy of a task-oriented system.

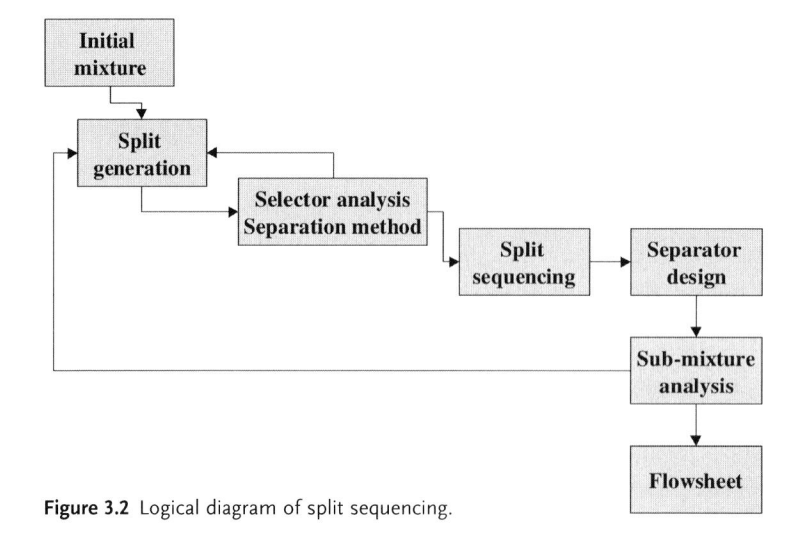

Figure 3.2 Logical diagram of split sequencing.

be assumed as feasible. Details about design can be found in specialized books, as for distillation [8, 14], membranes [15] and separations in general [10].

The generation of a *separation sequence* follows the logical diagram described in Fig. 3.2. The steps are:

1. Split generation

Possible splits are proposed taking into account the mixture composition and the specifications of products to achieve by applying appropriate *heuristics*. Then, the selected splits are placed into the corresponding selector.

Table 3.1 General heuristics for split generation.

1. Remove corrosive, hazardous and toxic materials in the first place.
2. Remove troublesome impurities.
3. Favor separations that match directly the products.
4. Remove the most plentiful component.
5. Prefer separations dividing the feed as equally as possible.

2. Selector analysis

In each selector, a logical diagram will guide the identification of a suitable separation method for the split proposed above. This is done by ranking the mixture components versus a *characteristic property*. For example, the relative volatility is a characteristic property for separation by simple distillation (see Appendices E and F). This approach, however, is not applicable when azeotropes are involved and other characteristic properties should be investigated, such as, for example, the chemical structure. A split becomes potential if complete by at least one method.

3. Split sequencing

The potential splits identified above are compared in order to decide the best one to perform next. The evaluation may be assisted by *heuristics* or by a more sophisticated approach. When distillation is applicable, computer simulation can be used for comparing alternatives.

4. Separator design

Each potential split selected at the step 3 can be the subject of a design procedure, which can be of shortcut type or based on rigorous simulation. For some particular or innovative separators the sizing algorithm might not be available, but this fact should not hinder the procedure. It suffices that the feasibility of the task can be guaranteed by an equipment supplier, or this is put on the list of equipment for which suppliers are searched.

5. Submixture analysis

The composition of intermediate streams generated in each split is compared to the desired product specifications. The steps 1 to 4 are repeated until the generation of all products is accomplished.

Table 3.1 shows general heuristics for split generation applicable to all managers. The removal of toxic, hazardous and corrosive materials has the highest priority. Next, troublesome impurities should be dealt with. Matching directly the products by the shortest sequence of splits is optimal in most cases. When no choice is obvious, dividing the feed as equal as possible is often the best strategy. More specific heuristics will be formulated later in this chapter.

We may remark that the above procedure does not necessarily lead to a unique solution, but rather to a number of alternatives. The optimal solution may be identified by a deep-first search in which the best alternative is kept at each step, or by the optimization of a superstructure.

3.2
Vapor Recovery and Gas-Separation System

Gas-separation manager includes both vapor recovery and gas-separation systems. Vapor recovery handles the recovery of valuable condensable components from a gas stream or the removal of undesired components since they are corrosive, toxic, polymerizable, have a bad odor, *etc*. Gas separation deals with the recovery of recycled gaseous reactants, as well as with the delivery of purified products and byproducts. Douglas [6] recommends the following heuristics for placing the vapor-recovery system:

- on the purge, if significant amount of product is being lost,
- on the gas-recycle stream, if impurities could affect the reactor operation,
- on the vapor stream after flash, if both items 1 and 2 are valid,
- does not use vapor recovery if neither item 1 nor 2 are important.

3.2.1
Separation Methods

The selection of a separation method is based on the identification of a suitable characteristic property, whose variation should be important for the component(s) to be separated. Table 3.2 presents characteristic properties for gas separations. A first group of methods relies on physical properties, such as boiling point, relative volatility, solubility, *etc.*, which generates separation techniques such as condensation, distillation, physical absorption, *etc*. The second category exploits the reactivity of some functional groups, as in chemical absorption, catalytic oxidation, catalytic hydrogenation and chemical treatment.

3.2.2
Split Sequencing

Gas-separation manager makes use of three selectors: enrichment, sharp separation and purification. In the original treatment of Barnicki and Fair logical diagrams are used for the selection of suitable separation methods. A simpler procedure can be imagined as a multiple choice matrix [5], as presented in Table 3.3.

1. *Enrichment* consists of a significant increase in the concentration of one or several species in the desired stream, although by this operation neither high recovery nor purity is achieved. Condensation, physical absorption, membrane permeation, cryogenic distillation, and adsorption are convenient separation techniques.

Table 3.2 Methods used in vapor recovery and gas separations.

Method	Characteristic property	Condition	Observation
Condensation	Boiling points Relative volatility	Difference in boiling points >40 °C or $\alpha_{ij} > 7$	Optimize pressure and temperature
Cryogenic distillation	Boiling points Relative volatility	$\alpha_{ij} > 2$	Large-scale processes Remove first freezable components
Physical absorption	Solubility	$K_i > 4$	Optimize P and T Recycle the solvent
Chemical absorption	Reactive function as acid or base groups	Reversible process	Optimize the solvent ratio
Molecular sieving	Size/shape	Significant differences	Remove first fouling components
Equilibrium adsorption	Adsorption coefficient	Favorable adsorption	Remove first fouling components
Membrane permeation	Perselectivity	Perselectivity greater than 15	Remove first fouling components
Catalytic oxidation	Chemical family	Impurities below 10% of the flammability point	Danger of dioxine, not for halogenated organics
Catalytic hydrogenation	Chemical family	Components containing double bound	Develop selective catalyst
Chemical treatment	Chemical family	Selective reaction	Dry treatment preferred Recovery of chemical agent

Table 3.3 Selectors and methods for gas separations.

Separation method	Enrichment	Sharp separation	Purification
Condensation	Yes	No	No
Cryogenic distillation	Yes	Yes	No
Physical absorption	Yes	Yes	No
Chemical scrubbing	No	Yes	Yes
Molecular-sieve adsorption	Yes	Yes	Yes
Equilibrium-limited adsorption	Yes	Yes	Yes
Membranes	Yes	Yes	No
Catalytic oxidation	No	No	Yes
Catalytic hydrogenation	No	No	Yes
Chemical treatment	No	No	Yes

2. *Sharp separation* consists of obtaining splitting of the mixture into products with a high recovery of target components. The sharpness is defined as the ratio of key component concentrations in products. This should be better than 10. Potential techniques are: physical absorption, cryogenic distillation, molecular sieving, as well as equilibrium adsorption when the molar fraction of the adsorbate is less than 0.1. Chemical absorption may also be applicable when the component concentration is low.

3. *Purification* deals with the removal of impurities with the goal of achieving very high concentration of the dominant component. The initial concentration of impurity in the mixture should be lower than 2000 ppm, while the final concentration of impurity in the product should be less than 100 ppm. Suitable separation methods are equilibrium adsorption, molecular-sieve adsorption, chemical absorption and catalytic conversion.

The generation and sequencing of splits can be managed by means of heuristics. It starts by trying sequentially the generic rules presented in Table 3.1. Firstly, corrosive, hazardous and other troublesome species must be removed. For example, water or CO_2 that can freeze will foul the equipment in cryogenic distillation. Next, should be placed the split ensuring the direct separation of a product. The removal of the most plentiful component has the same priority, diminishing significantly the cost of downstream separations. A 50/50 split is recommended when there is no clear choice.

Table 3.4 presents more specific heuristics for gas separations. The condensation of subcritical components at suitable pressure by cooling with water is often practical. Before applying low-temperature separations or membranes the removal of water by glycol absorption or by adsorption is compulsory. In the case of impurities accumulating in recycles a powerful method is their chemical conversion by selective catalysis. Before treatment the impurities should be concentered by an enrichment operation. Catalytic conversion is also recommended for handling volatile organic components (VOC).

Table 3.4 Special sequencing heuristics for gas separations.

1. Favor condensation for removing high boilers from noncondensable when cooling water can be used as thermal agent.
2. Favor glycol absorption for large-scale desiccation operations requiring dew-point depression of 27 °C or less.
3. Favor adsorption for small-scale desiccation operations. This is the cheapest alternative for processing small amounts of gas.
4. Favor adsorption for processes that require essentially complete removal of water vapor. This is capable of achieving dew-point depression of more than 44 °C.
5. Favor catalytic conversion when impurities may be converted into desired product, or when they accumulate in recycles, or when they produce other impurities by side reactions.

Example 3.1: Treatment of a Landfill Gas

This example, adapted from [2], illustrates the application of a task-oriented methodology in developing a separation sequence. The object is the treatment of a landfill gas (LFG) and the recovery of useful products. The transformation of municipal solid waste (MSW) in landfill gas by anaerobic decomposition is an ecological technique used for energy recovery from waste, while reducing significantly the impact of greenhouse gases on the environment. Table 3.5 shows the raw gas composition, as well as the specifications for products, which are biogas and high-purity CO_2. The output gas capacity is $10^5\,N\,m^3$ per day, which is typical for a medium-size plant.

The transformation of municipal, agricultural and industrial waste in landfill gas is a viable solution for saving fossil fuels and preventing climate warming. Typically, the raw gas contains, in equal proportions, methane and CO_2 with small amounts of organic components, as well as H_2S and siloxanes. The US Environmental Protection Agency (www.epa.gov) estimates that generalizing the transformation of municipal waste into biogas could deliver power for more than 1 million homes, or reduce emissions equivalent to 12 million

Table 3.5 Gas composition and product specifications.

Component	LFG	CO_2	Biogas
Methane	57.5	–	96–99 mol%
Carbon dioxide	37.0	99.9 mol%	0.1–1
Nitrogen	3.7	–	1–3
Oxygen	0.95	–	0.01–0.1
Hydrogen sulfide	0.05	0–0.3 ppm	10–50 ppm
Aromatics (benzene)	0.30	0–5 ppm	100–200 ppm
Halo-hydrocarbons (chloro-ethane)	0.50	0–1 ppm	5–15 ppm

vehicles. The biogas can replace the natural gas whenever needed in industry as fuel, or for manufacturing various chemicals.

The cleaning technology is not unique, depending on the gas composition, local conditions and end-use specifications. More details about industrial projects and methods may be found in a guide published by the Scottish Environment Agency [18]. The sequence below is only an illustration.

1. First split

The components are ordered by their relative amount, as in Table 3.6. Methane and CO_2 largely dominate, but other species keep the attention, even in very small amounts. Particularly harmful is H_2S, highly toxic for humans. Its level, as well as for benzene and halo-hydrocarbons, should be of a few ppm in the end products.

For split generation we make use of heuristics, as given in Table 3.1. The removal of troublesome impurities is suggested in the first place, here H_2S, benzene and chloro-ethane. Then the split is placed in an appropriate selector, in this case of type "purification". Table 3.3 indicates that six separation methods could be applied to perform this task: chemical absorption, molecular-sieve adsorption, physical adsorption, catalytic oxidation, catalytic hydrogenation and chemical treatment.

The next action is to estimate the separation targets and allocate the components in products. For the assessment of a separation a good practice is by setting up a *recovery matrix* as shown in Table 3.6. This expresses the split of a component between feed and products. Note that this information is available in Aspen Plus when simulating separators.

Let us consider H_2S. The total amount is $10^5 \times 0.05 \times 34/22.4 = 75.89$ kg/day. Since H_2S should pass only in the biogas, the recovery in the first split is determined by its specifications: 50 ppm as fuel for power station, but only 10 ppm in a municipal network. The amount of biomethane is $10^5 \times 0.575 \times 16/22.4 = 41\ 071.43$ kg/day. Thus, the H_2S allowed in biogas for power stations is $41\ 071.43 \times 50E{-}6 = 2.05$ kg/day. Accordingly, the minimum recovery faction of H_2S is $1-2.05/71.42 = 0.970$. Similarly, the maximum recovery fraction for domestic use is 0.994. In the same way recovery targets can be determined for

Table 3.6 Recovery matrix for the first separation split.

Component	Gas mol%	Product 1	Product 2
Methane	57.5	0.01	0.99
Carbon dioxide	37.0	0.05	0.95
Nitrogen	3.70	none	none
Oxygen	0.99	none	none
Halo-hydrocarbons (chloroethane)	0.50	0.999	0.01
Aromatics (benzene)	0.30	0.999	0.01
Hydrogen sulfide	0.05	0.973–0.994	0.027–0.006

Table 3.7 Ranked list of properties for purification selector.

Chemical absorption		Molecular sieving		Equilibrium adsorption	
Component	Chemical family	Component	Kinetic diameter (Å)	Component	Loading (mol/g ads)
CO_2	Acid gas	Oxygen	<3	Oxygen	0
H_2S	Acid gas	Nitrogen	<3	Nitrogen	0
Nitrogen	Inert gas	H_2S	<4	Methane	0.0005
Oxygen	Inorg. gas	CO_2	<4	CO_2	0.0035
Chloroethane	Chloride	Methane	<4	Benzene	0.0046
Benzene	Aromatic	Chloroethane	<5	H_2S	0.0069
Methane	n-Alkane	Benzene	<8	Chloroethane	0.0070

other species. Note that recoveries for benzene and halo-hydrocarbons come from purity requirements for food-quality CO_2.

Next, characteristic properties of components are listed to select appropriate separation method (Table 3.7). Because the trace components belong to different chemical families, we eliminate gas-phase catalytic oxidation or hydrogenation. More specific chemical-based techniques remain. A first one is reversible chemical absorption. As solvents we may enumerate liquid redox systems (chelated iron), caustic washing solutions, amines or special formulations, as Selexol™. Since H_2S and CO_2 both have an acid character, we may expect that a certain amount of CO_2 will pass in the off-gas stream. Dry chemical treatment could also be used, as reaction of H_2S with iron-sponge or impregnated wood chips.

To evaluate the applicability of molecular sieves, we compare the kinetic diameter of components. Low-size sieves would filter all, except chloro-ethane and benzene. Then, the remaining H_2S may be handled by wet scrubbing.

Finally, we examine the feasibility of equilibrium adsorption. The loading capacity of adsorbent (active charcoal) indicates that all the trace components can be selectively adsorbed, and therefore the equilibrium adsorption can be applied for their removal.

2. Second split

The remaining gas mixture now has the composition: nitrogen 3.7 mol%, oxygen 1.0 mol%, methane 47.9 mol%, CO_2 47.4 mol%. The third heuristic in Table 3.1 applies: try to match the products. The appropriate selector is "sharp split". Table 3.8 present lists of characteristic properties. Potential methods are: absorption, cryogenic distillation, molecular sieving, membranes and equilibrium adsorption.

Absorption is indeed a standard technique for CO_2 removal. Physical absorption in water under pressure is applicable for recovery below 90%. Chemical solvents are preferred for advanced recovery, such as amine solutions, hot carbonate, cold methanol, Selexol, *etc*.

Table 3.8 Ranked list of properties for sharp split-separation selector.

Absorption		Cryogenic distillation		Molecular sieve adsorption		Equilibrium adsorption	
Comp.	Family	Comp.	Relative volatility	Comp.	Kinetic diam. (Å)	Comp.	Loading (mol/g)
Oxygen	Inorg gas	Nitrogen	1.13	Oxygen	<3	Nitrogen	≈0
Nitrogen	Inorg gas	Oxygen	2.73	Nitrogen	<3	Oxygen	≈0
CO_2	Acid gas	Methane	1	CO_2	<4	Methane	0.005
Methane	n-Alkane	CO_2	Freeze	Methane	<4	CO_2	0.0035

The second technique for CO_2/methane separation is membrane permeation, now standard in industry. Cryogenic distillation may be applied for methane-enrichment purposes with some technical precautions because of CO_2 freezing. Equilibrium adsorption on activated carbon would give poor separation. Molecular sieving would permit only the removal of oxygen and nitrogen.

3. Third split

After CO_2 removal, the remaining mixture has the composition 91 mol% CH_4, 7.1 mol% nitrogen and 1.9 mol% oxygen. The appropriate selector is again "purification". Examination of Table 3.8 indicates that molecular sieving may be applied to remove both nitrogen and oxygen. Another possibil-

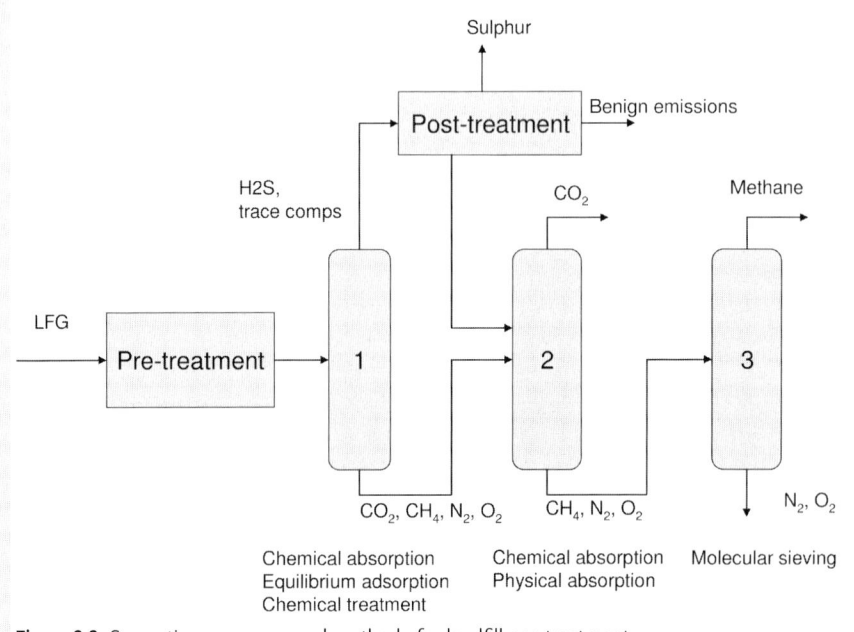

Figure 3.3 Separation sequence and methods for landfill gas treatment.

ity would be using cryogenic distillation, but this is economical for large-scale plants.

With these elements we can sketch a conceptual scheme. Figure 3.3 presents the flowsheet and alternative separation methods. Firstly, the LFG should be pretreated against contaminants, such as water and particulates, which could harm the downstream separations. Then the gas enters the main treatment. The first split purifies the gas of trace troublesome impurities as H_2S, aromatics and halo-hydrocarbons. Appropriate techniques are chemical absorption, equilibrium adsorption, as well as dry chemical treatment. The second split consists of a sharp separation of CO_2 from methane-rich gas. Suitable methods are based on chemical absorption in various solvents including cryogenic absorption in methanol. In the third split, if needed, the amount of nitrogen and oxygen in biogas may be reduced by means of molecular sieving. Note that the pressure and temperature level at each step should be adapted to fulfil optimal operation conditions.

3.3
Liquid-Separation System

Removing light and heavy-end impurities is compulsory before other separations. Figure 3.4 presents some common methods for light-ends removal:

1. Series of flashes, by reducing the pressure or/and by increasing the temperature.

2. Distillation with vapor product. When a *partial condenser* is used, the flash drum plays the role of a vapor/liquid separator. In the setup known as a *stabilizer* there is only vapor distillate, while the liquid is returned as reflux. The column has a *pasteurization* section when a gaseous stream leaves at the top, while the

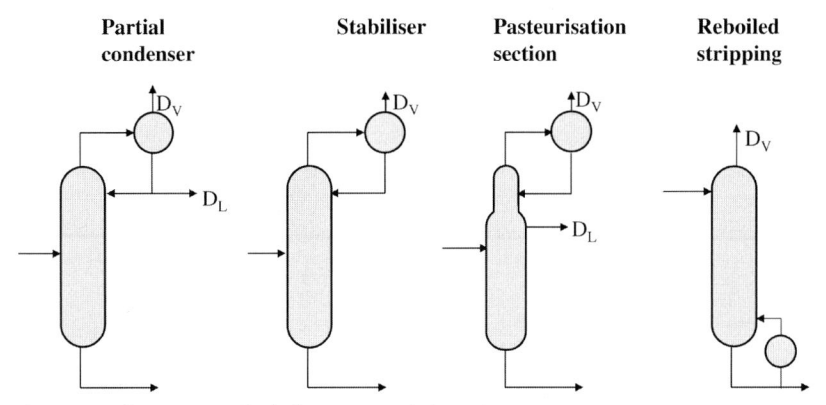

Figure 3.4 Alternative methods for removing light-ends.

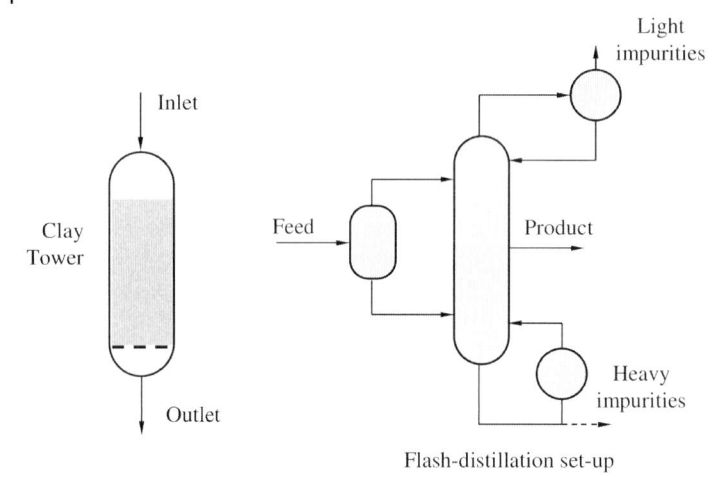

Figure 3.5 Alternative methods for removing heavy-ends.

purified liquid product is obtained as a side stream a few stages below. The condenser temperature is determined by the available temperature of the cooling agent. The column pressure should be optimized against losses of the useful component.

3. Reboiled stripping. In this case, the vapor needed for stripping is produced internally. This method may replace conveniently steam distillation. Note that the initial mixture should be fed at sufficiently low temperature on the top stage.

Some techniques for the removal of heavy-ends are shown in Figure 3.5. A first one is *washing with solvent*, which can be filtered, distilled and recycled. Making use of an adsorption device, such as for example *clay tower*, is widespread in industry, with the inconvenience that the retention capacity diminishes in time and the adsorbent has to be regenerated. A simple but effective solution is the use of a *flash-distillation* device. The flash serves as a preseparator of light from heavies, while the purification of the bulk component takes place in a second column provided with a side stream. The heavies are removed as bleed from the reboiler, while the lights distillate at the top as a vapor distillate. This device is suited for the simultaneous removal of lights and heavies, when their amount is very small compared with a large medium-boiling fraction. This situation can be encountered typically in hydrocarbon oxidation processes. The above device prefigures more sophisticated arrangements for the distillation of mixtures with lights/intermediate/heavies, as the divided-wall column.

3.3.1
Separation Methods

Characteristic physical properties and corresponding separation methods are presented in Table 3.9. Industrial experience demonstrated that when the distillation is feasible, it should be adopted immediately, particularly for large-scale processes.

Table 3.9 Separation methods for liquid mixtures.

Separation method	Characteristic property	Observation
Simple distillation	Relative volatility α	Use heuristics for sequencing. Not feasible if $\alpha < 1.1$
Simple and azeotropic distillations	Vapor-pressure variation	Check thermal stability of components
Stripping, L–L extraction	Boiling point	Use stripping and L–L extraction for thermal sensitive components
Melt crystallization	Freezing point	Differences larger than 20 °C
Adsorption chromatography	Polarity	Pay attention to adsorbent regeneration
Membrane permeation	Shape and size	Emerging technology
Azeotropic distillation, Extractive distillation, L–L extraction	Chemical family	MSA selection is the main issue. Recycling of MSA increases the costs
L–L extraction, stripping, adsorption, crystallization	Temperature sensitivity	Recycle of MSA increases the costs

In a second attempt other alternative separation methods could be rated against distillation.

3.3.2
Split Sequencing

The liquid split manager deals with two selectors: *zeotropic* and *azeotropic* mixtures. A second decomposition may take place as a function of mixture composition and sensitivity to temperature. The first criterion generates *dilute* and *bulk* separations, while the second one leads to *temperature-sensitive* and *temperature-insensitive* separations.

A separation is *diluted* when the distillate or the bottom product is less than 5 wt% with respect to the feed. The distillation (simple, extractive or azeotropic) might not be the most economical, but other methods, such as liquid–liquid extraction, stripping, crystallization, adsorption, or membrane permeation, should be tried. The decision depends on the mixture composition and the nature of the components.

In *bulk* separations, the desired product is more than 5% in the initial mixture. Distillation is the most economical separation method and should be tried in the first place. It is important to note that the split generation follows different patterns for zeotropic and azeotropic mixtures.

For zeotropic mixtures the problem is not the feasibility of separation, because a suitable column can always be designed, but the *optimal sequencing* of splits. The optimality criterion should be the total cost of separations, in term of investment and operation.

In contrast, when dealing with azeotropic mixtures the feasibility of separation is not guaranteed. Entrainer selection and feasibility is the central problem. The

Table 3.10 Liquid-separation manager: selectors and separation methods.

Separation method	Dilute separations	Bulk separations	Azeotropic mixtures	Temperature sensitive
Simple distillation	Yes	Yes	Yes	Yes[a]
Complex distillation	No	Yes	No	No
Stripping	Yes	Yes	No	Yes
Extractive distillation	No	Yes	Yes	No
Azeotropic distillation	No	No	Yes	No
L–L extraction	Yes	Yes	Yes	Yes
Adsorption	Yes	Yes	Yes	Yes
Molecular sieving	Yes	Yes	Yes	Yes
Membrane permeation	Yes	No	Yes	Yes
Melt crystallization	No	Yes	No	Yes

a) Vacuum distillation.

number of splits is less important. For example, in the case of a binary mixture typically a minimum of two and a maximum of three columns are employed for separation, with solvent recycle (see later in this chapter).

In the case of *temperature-sensitive* separations the column temperature profile is constrained. Appropriate methods are stripping, liquid–liquid extraction, adsorption and crystallization, as well as vacuum distillation.

The classification of separation methods suitable for the four mentioned selectors is summarized in Table 3.10. Note that azeotropic distillation, membrane permeation and melt crystallization are the most expensive, but unavoidable in handling more complex mixtures.

For sequencing the separation of liquid mixtures, there are general heuristics, as shown in Table 3.11. More specific rules for the separations of zeotropic mixtures by distillation will be discussed later in this chapter.

Table 3.11 General heuristics for separation sequencing of liquid mixtures.

1. Remove first corrosive, hazardous, fouling, reactive and any troublesome components. Consider also in the first place the removal of light-ends.
2. Deliver high-purity products as top distillate. The same is valid for reactants sent to reactors sensitive to impurities.
3. When separation by distillation is feasible, prefer it in a first attempt.
4. Isolate zeotropic and azeotropic mixtures.
5. Perform difficult zeotropic separations later, but before azeotropic separations. Examine other options, such as extractive distillation, L–L extraction, crystallization, adsorption, or molecular sieves.
6. Examine the separation of azeotropic mixtures last.
7. Remove the components in order of decreasing percentage of the feed. This operation will reduce the cost of the next separation.
8. Favor 50/50 splits.

Some comments about the above heuristics are useful:

1. Remove troublesome species as early as possible, with the priority of lights.

2. Obtaining a high-purity component as the top product in the final distillation is the "golden rule" in industry. For recycle streams sent to reactors the purity should be not necessarily high, but constant. Obviously, harmful impurities for the catalyst should be removed.

3. Distillation remains the most reliable separation technique, and is the most efficient in a large number of cases. Therefore, it should be tried first and other separation methods should be rated against it.

4. The partition of a mixture in zeotropic and azeotropic submixtures tremendously simplifies the separation problem. Some components may be distributed between the two subsystems, in which case these are coupled by recycles and bridge separations.

5. Difficult zeotropic separations should not be appear in the middle of a separation sequence, because they can be a bottleneck, but in the last place before azeotropic separations. Actually, a low relative volatility separation demands a large number of stages and high reflux ratio. Energy saving is a central issue. Employing other separation methods should be investigated, such as crystallization, adsorption and molecular sieving.

6. The separation of azeotropic mixtures is complicated. Homogeneous azeotropic distillation is appealing but finding a suitable entrainer is difficult. In a number of cases, such as breaking water/alcohol azeotropes, the heterogeneous azeotropic distillation gives much better results. Extractive distillation is a good alternative when high-boiler solvent may be found. Often, combining distillation with other separation techniques is recommended.

7. Removing systematically the most plentiful component in intermediate mixtures will drastically reduce the cost of separation, both as investment and energy consumption. This heuristics typically leads to a direct sequence.

8. When the components are evenly distributed, a 50/50 split is recommended since it leads to a drastic reduction of the number of downstream separations.

3.4
Separation of Zeotropic Mixtures by Distillation

3.4.1
Alternative Separation Sequences

The typical problem is the separation of a ternary mixture ABC, with components ranked by decreasing volatility. Figure 3.6 shows two basic alternatives employing simple columns (one feed two products). In the *direct sequence*, the components

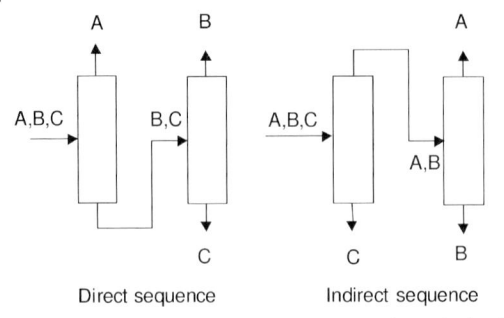

Figure 3.6 Direct and indirect sequences of simple distillation columns.

Table 3.12 Sequences for separating four-component mixture by simple columns.

No.	Type	First split	Second split	Third split
1	Direct	A/BCD	B/CD	C/D
2	Equal split	AB/CD	A/B	C/D
3	Indirect	ABC/D	AB/C	A/B
4	Direct/indirect	A/BCD	BC/D	B/C
5	Indirect/direct	ABC/D	A/BC	B/C

are taken as the overhead product in the order of their relative volatilities, first A then B. In the *indirect sequence*, the components are separated in the inverse order of volatility, firstly the heaviest C as bottoms from the first column, followed by the A/B separation in the second column. Direct sequence is mostly applied in industry, but indirect sequence may found also application in some complex schemes.

In the case of a four-component mixture ABCD there are five possible sequences, each of three columns, as shown in Table 3.12. In the *direct sequence* all the components, except the heaviest, are taken as top products. In the *indirect sequence* all the components are obtained as bottoms, except the lightest. In *equal split* both A and C are obtained as overhead, while B and D as bottoms. There also two mixed sequences, such as "direct/indirect" and "indirect/direct", the second split making the difference.

3.4.2
Heuristics for Sequencing

The procedure known as the *list-processing method* allows the designer to identify feasible sequences for separating ideal or slightly nonideal zeotropic mixtures by means of simple distillation columns. Table 3.13 presents a list of more specific heuristics for sequencing, supplementary to those in Table 3.7. The first one shows

Table 3.13 Heuristics for separation sequencing of zeotropic mixture.

1. Perform difficult separations the last, but before separations of azeotropes.
2. Remove firstly the lightest component one by one as overhead products.
3. Remove components in order of decreasing percentage of the feed.
   This operation will reduce the cost of the next separation.
4. Favor near 50/50 splits.

that the most difficult separation should be pushed to the end, but before handling azeotropes. Secondly, the sequential separation of the lightest component leading to the direct sequence is often the best. The third rule recommends the separation in the first place of the most plentiful component, which will reduce accordingly the cost of downstream separations. When none of the above applies, a 50/50 split ensures optimality in most cases.

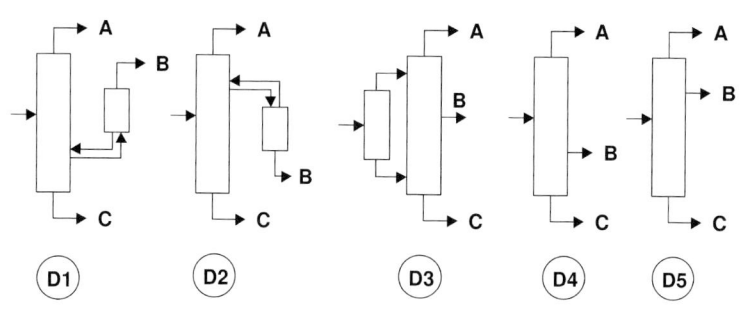

Figure 3.7 Types of complex columns.

3.4.3
Complex Columns

Complex columns are distillation devices that can handle a mixture of minimum three components and deliver more than two products. A complex column consists of a main tower surrounded by additional columns, as prefractionator, side strippers and side rectifiers. As illustration, Figure 3.7 presents five alternatives for separating a ternary mixture ABC:

1. Side-stream rectifier: A and C from main column, B as top product of side rectifier.
2. Side-stream stripper: A and C as before, but B as bottoms of side stripper.
3. Prefractionator: separate AB and BC mixtures in first column by sloppy separation, then take pure components A, B, and C in a side-stream second column.
4. Side-stream low position: take B as side stream below the feed.
5. Side-stream high position: take B as side stream above the feed.

The key advantage of a complex column over a sequence of simple columns is more compact equipment with substantial reduction of investment, instrumentation, manpower and maintenance costs. In some cases substantial energy saving is obtained.

3.4.4
Sequence Optimization

The heuristics mentioned above lead to a feasible but not always optimal sequence. In principle, the sequences can be evaluated and ranked against operation and capital costs. Near-optimal sequences are those minimizing the total vapor rate of key components [8, 12]. However, when the relative volatility of keys differs considerably this method might be in large error. A true optimal sequencing can be found by mathematical programming. Earlier attempts proposed the optimization of a superstructure including all the possible sequences and using detailed modeling of columns. However, finding the solution by mixed-integer nonlinear programming (MINLP) is hindered by various nonlinear and nonconvex elements. More efficient is a generic tree-like representation of the separations based on assigned tasks. The sequencing can be formulated more conveniently as a mixed-integer linear programming (MILP) optimization problem for which robust algorithms are available. The tasks may consist of simple or hybrid columns modeled by appropriate shortcut or semirigorous methods. For deeper analysis of process synthesis based on an optimization approach we recommend the book of Biegler et al. [4].

By using computer simulation accurate near-optimal solutions can be found with relatively reduced effort quickly by the sequencing of zeotropic mixtures. The following strategy can be applied:

1. Generate a (large) list of candidates by means of general and specific heuristics.
2. Produce a shortlist of candidates by ranking the alternatives following the total vapor rate. A minimum reflux calculation design based on Fenske–Underwood–Gilliland method should be sufficiently accurate.
3. Evaluate the best candidates by paying attention to heat integration and technological constraints.
4. Select the optimal sequence. Use rigorous simulation for the final assessment.

The above method cannot be applied for highly nonideal mixtures involving azeotropes and distillation boundaries. In this case reducing the separation to handling of ternary mixtures is recommended, for which two or three columns are normally sufficient, either by direct or indirect sequence. Since the entrainer plays an important role in economics, the sequences with the entrainer recycle as bottoms are preferred.

3.5
Enhanced Distillation

Enhanced distillation designates special techniques for separating nonideal azeotropic mixtures, or zeotropic mixtures with very low relative volatility, based on the use of a dedicated mass-separation agent that has to be recycled. Sometimes, a component already present in the mixture can play this role.

3.5.1
Extractive Distillation

The extractive distillation profits from the capacity of an entrainer (solvent) to modify selectively the relative volatility of species. Normally, the entrainer is the highest boiler, while the component to be separated becomes heavier, being carried out in bottoms. For this reason, this operation may be regarded as an extractive absorption. Extractive distillation can be used for separating both zeotropic and azeotropic mixtures. The entrainer is fed near the top for a zeotropic mixture or a minimum-boiling azeotrope, or mixed with the feed for a maximum-boiling azeotrope. The separation sequence normally has two columns, for extraction and solvent recovery [5].

3.5.2
Chemically Enhanced Distillation

Chemically enhanced distillation is similar to extractive distillation, but this time the solvent is an ionic salt or a complex organic molecule. As an example, acetone separates easier from a solution with methanol in the presence of a concentrated solution of calcium chloride. More modern solvents are based on ionic liquids.

3.5.3
Pressure-Swing Distillation

When the composition of an azeotrope is sensitive to moderate changes in pressure, then pressure-swing distillation may be used to separate almost pure components. The first column separates pure A in top or in bottoms, if this is a minimum or maximum boiler, respectively. Then the second column makes possible the separation of pure B while recycling the A/B azeotrope.

3.6
Hybrid Separations

By a hybrid separation, distillation is combined with other separation methods, such as L–L extraction, adsorption, crystallization and membranes. It is mainly employed when distillation is unfeasible, or is very costly because of an unfavorable vapor-pressure difference, or because azeotropes form between the key components.

The combination with L–L extraction is appropriate for breaking azeotropes, such as for example cyclohexane/benzene by acetone, in turn extracted selectively with water, distilled and recycled.

The combination with molecular-sieve adsorption or membrane permeation allows breaking an azeotrope without the need of a contaminating solvent. This is advantageous when separating ultrapure components, such as medical grade ethanol from its azeotrope with water. More details can be found elsewhere [5, 10, 14].

Example 3.2: Separation of Aromatics

The aromatics are of key importance for the petrochemical industry. Major sources of aromatics are *pyrolysis of gasoline* and *catalytic reforming*. The first is richer in benzene and toluene, while the second is richer in xylenes. These mixtures of C6–C8 aromatics, commonly designated by BTX, are raw materials for producing valuable chemicals, such as styrene, cumene, phenol, phtalic anhydride, terephtalic acid, *etc.* Therefore, a separation process of aromatics has to be adaptable to the market requirements. This requirement is fulfilled in industry by combining physical separations and chemical-conversion processes, such as dealkylation, transalkylation and isomerization. Since large amounts of energy are involved, saving energy is a central design objective. Details about recent technologies can be found in the monograph of Meyers [19].

Figure 3.8 presents an efficient separation scheme widely used in industry. This exercise will try to trace back the logic of such complicated scheme taking into consideration the generic methodology developed before.

A representative mixture has the composition C6-C8 nonaromatics (20), benzene (12), toluene (18), ethylbenzene (7), p-xylene (10), m-xylene (20), o-xylene (13), all in wt%. The goal is the separation of benzene and p-xylene of high purity. The other aromatics should be recovered at convenient purity in view of selling or for further conversion into the main products.

The strategy of flowsheet synthesis will consist of generating splits on the basis of characteristic properties and heuristics, placing them in selectors, choosing suitable separation methods, evaluating them and selecting the most appropriate, as described by the logical diagram in Figure 3.2.

(a) *First split:* mixture partition
In the absence of troublesome components the appropriate heuristic is "favor separations that match directly the products". Accordingly, we try the isolation of aromatics from nonaromatics. The selector is *bulk separation*. To assess the

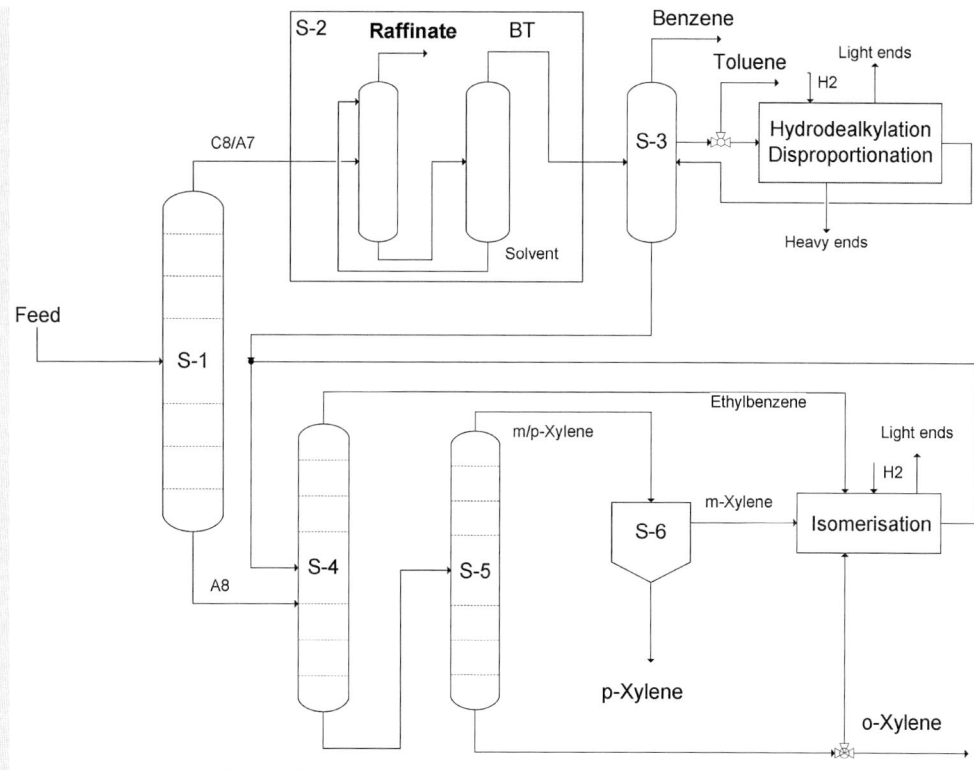

Figure 3.8 Separation scheme of aromatics from catalytic reforming feedstock.

Table 3.14 Normal boiling points (°C) of key hydrocarbon components.

C7	C8	C9[a)]	A6	A7	A8
79.2–98.4	106.3–125.7	124.1–150.8	80.1	110.6	136.2–144.4

a) Given as illustration, but not allowed in the initial mixture.

feasibility we examine the list of boiling points of representative components, as shown in Table 3.14. C6 to C9 denotes alkanes, A6 to A8 aromatics, the boiling interval reflecting the isomers. It may be observed that lighter aromatics, benzene and toluene, overlaps with C7 and some C8, while C9 overlaps with A8. In addition, numerous azeotropes form between components, such as for example cyclohexane and heptane with benzene and toluene. Therefore, sharp split of the mixture in aromatics and nonaromatics is not possible. However, cutting before ethylbenzene ensures good separation of A8, leaving a BT fraction with nonaromatics. Because the presence of C9 would complicate the xylenes separation the naphta fraction submitted to catalytic reforming should be chosen in the range C6–C8. Indeed, high severity (higher conversion into aromatics) can be achieved in modern processes.

As a second alternative, we may assume that the first split ensures the recovery of the whole aromatics fraction. By inspecting Table 3.9 three separation methods may be envisaged: 1) azeotropic distillation, 2) extractive distillation and 3) liquid–liquid extraction. They are commented briefly:

1. Azeotropic distillation is appropriate only at high content in aromatics, over 90%. Nonpolar entrainers, such as methanol and acetone, can increase the volatility of nonaromatics by forming low-boiling azeotropes. This method depends strongly on the charge composition, and therefore is not used in industry.

2. Extractive distillation employs high-boiler solvents to selectively extract the aromatics, letting the nonaromatics in at the top. As solvents we may cite N-metyl-pyrolidone (NMP), N-formylmorpholine (NFM), sulfolane and glycols. As illustration, the volatility of benzene vesus n-heptane, which is normally 0.5, may be inversed to about 2. Since the strongest effect on relative volatility is for benzene and toluene the isolation of a BT fraction seems rational. A key advantage of extractive distillation is a simple flowsheet with only two columns: extractive absorption and solvent recovery with stripping of extracted component(s). Since the solvent is recycled as bottoms the energy consumption is not sensitive to the solvent ratio. As recently reported, the flowsheet may be reduced to a single divided-wall column (DWC), with supplementary investment advantage and energy saving.

3. Liquid–liquid extraction is applicable over a large mixture composition. A supplementary advantage is low energy costs. All the solvents cited for extractive distillation may be used, plus dimethyl sulfoxide (DMSO) with the advantage of easy recovery. An extract-reflux is used to increase the yield of the extraction operation. In addition water is used for solvent recovery from raffinate and extract streams. As result, a typical liquid–liquid extraction for aromatics has four columns: extractor, extractive stripper, extract recovery column and water-washing column.

Summing up, for the first split there are two alternatives:

- Cutting a rich benzene/toluene fraction at the top with the remaining A_8 aromatics in bottoms. Simple distillation is suitable.
- Complete recovery of aromatics. Liquid–liquid extraction is the first choice, but extractive distillation with high-performance solvents is also applicable.

We select the first alternative as representative for modern plants.

(b) *Second split:* BT isolation

This operation concerns the separation of mixture with the weight composition nonaromatics (40), benzene (24) and toluene (36). The selector is again bulk separation. The appropriate method is extractive distillation, as discussed previously. Modern processes are based on solvents such as sulfolane (Shell-UOP), NFM (Uhde-Edeleanu) or mixtures (GTC). The nonaromatics pass completely in raffinate, while the BT fraction passes in the extract.

(c) *Third split:* BT separation
The BT mixture can be easily separated by simple distillation in a two-column sequence, or in a single column with side stream. The last alternative is preferred because it is advantageous energetically. The side-stream toluene is usually sent to *hydrodealkylation* or *transalkylation* units to increase the yield in benzene and xylenes. The bottom product goes to higher aromatics treatment.

(d) *Fourth and fifth splits:* isolation of m/p-xylenes
The mixture has a mass composition of ethylbenzene (14), p-xylene (20), m-xylene (40) and o-xylene (26). Examination of the boiling points in Table 3.15 reveals that the separation by distillation is very difficult, the largest temperature difference ethylbenzene/o-xylene being of only 8.2 °C at a relative volatility of 1.25. Using vacuum does not help since the vapor-pressure variation with the temperature is similar for all isomers.

The chemical structure of A8 isomers suggests that the appropriate property for separation may be the shape of the molecule, reflected in the melting point, kinetic diameter and dipole moment. The large variation of the melting points can be explained by the differences in the spatial structure that minimizes the energy of the solid state. P-xylene consists of flat thin molecules stacked together, which leads to a higher freezing point. On the contrary, the shape of m/o-xylene and ethylbenzene is more spherical, reflected in looser packing and lower melting points. The same trend is observed in the variation of kinetic diameter and Debye moment. Hence, crystallization is a suitable method for p-xylene separation. In addition, other methods based on shape and size might be workable, such as selective adsorption and molecular sieving.

When crystallization applies care should be paid to the fixed-composition points (eutectics) as with the azeotropes. In this case both ethylbenzene and o-xylene have to be removed before since they give eutectics with p-xylene. Although demanding, the distillation can be used. A direct sequence scheme is appropriate with ethylbenzene separated in the top of the first split and then o-xylene in bottoms of the second split.

Table 3.15 Property variation of A8 aromatic hydrocarbons.

Components	B.p., °C	Change b.p. °C/Pa × 10^4	Melting point, °C	Kinetic diameter, nm	Debye Dipole
Ethyl benzene	136.2	3.68	−95.0	0.600	0.4
p-Xylene	138.4	3.69	13.3	0.585	0.1
m-Xylene	139.1	3.68	−47.9	0.680	0.3
o-Xylene	144.4	3.73	−25.2	0.680	0.5

(e) *Sixth split:* m/p-xylene separation
The remaining mixture has now the weight composition p-xylene (27) and m-xylene (54), thus a ratio 1:2 optimum for separation by crystallization. The binary p-xylene/m-xylene gives a minimum eutectic at −52 °C and 13% p-xylene, indicating that simple separation of pure components is not possible. After water removal by drying the raw p-xylene is submitted to purification in several stages by melting and recrystallization. The final purity can be of 99.5%. Further p-xylene is separated as crystal sludge and then as solid cake by centrifugation. The liquid filtrate contains m-xylene and nonrecovered p-xylene. This stream, to which ethylbenzene and o-xylene streams can be added, is sent to chemical conversion by *isomerization*. The near-equilibrium mixture of xylenes thus obtained is sent back to separation. In this way the overall yield can be adapted to market requirements.

Today the melt crystallization can be advantageously replaced by a more challenging separation method known as *simulated moving bed* (SMB) technology. The method exploits the differences in affinity of zeolitic adsorbents for p-xylene with respect to other A8 components. Despite the name, the adsorbent phase is stationary and only fluid phase is distributed in a cyclic manner by a multivalve system. Operation parameters are temperatures of 125 to 200 °C and pressures up to 15 bar. Lighter (toluene) or heavier solvents (p-diethylbenzene) may be used as a desorbent. The Parex process working on this principle today has many applications.

The final inspection of Figure 3.8 shows that the separation objectives have been fulfilled. The insertion of chemical transformation stages allows the operation on market demand. Dealkylation or transalkylation may convert a surplus of toluene to more valuable benzene and xylenes, while the isomerization improves the yield in p-xylene to 100%, if needed.

3.7
Azeotropic Distillation

Azeotropic distillation deals with the separation of mixtures involving one or several azeotropes. This problem, which in the past was tackled by means of experience and intuition, is today approached by means of systematic methods based on a deeper thermodynamic analysis. Here, we review the indispensable aspects for process synthesis. More details can be found in recent specialized books [8, 14].

3.7.1
Residue Curve Maps

A residue curve map (RCM) consists of a plot of the phase equilibrium of a mixture submitted to distillation in a batch vessel at constant pressure. RCM is advantageous for analyzing ternary mixtures. More exactly, a residue curve shows explicitly the evolution of the residual liquid of a mixture submitted to batch distillation.

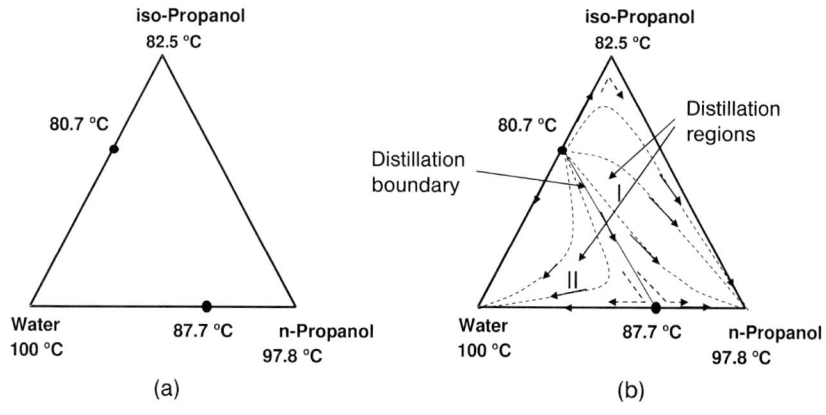

Figure 3.9 Sketch of a residue curve map: isopropanol, n-propanol, water.

Starting with different initial mixtures allows coverage of the entire compositional space. Temperature and vapor composition are computed implicitly.

Figure 3.9 highlights the key feature of a RCM by means of the mixture isopropanol, n-propanol, water, in which two binary azeotropes of water-alcohol occur:

- Fixed points: boiling points of pure components and azeotropes. They can be nodes (stable and unstable) and saddles.
- Direction of residue curves, pointing from lower to higher temperatures. An unstable node is a point from which trajectories emerge: the lowest boiler. A stable node is a point to which trajectories end up: the highest boiler. A saddle is an intermediate transition point (intermediate boiler) to which trajectories go and leave.
- A distillation boundary connects two fixed points: node, stable or unstable, to a saddle. The distillation boundaries divide the separation space into *separation regions*. The shape of the distillation boundary plays an important role in the assessment of separations.

As shown in Figure 3.9, a RCM can be sketched for a rapid evaluation of a separation by knowing only the fixed points. Firstly, the position of the binary azeotropes and of the ternary azeotrope is located. Then, the boiling points for pure components and azeotropes are noted. The behavior of characteristic points, node or saddle, is determined by taking into account the direction of temperatures, from lower to higher values. Finally, the distillation boundaries are drawn as straight lines by connecting the nodes with the corresponding saddles.

Accurately, a RCM is obtained by solving the differential equation describing the evolution in time of the liquid composition in a batch distillation still:

$$\frac{dx_i}{d\theta} = [K_i(x_i, y) - 1]x_i \qquad (3.1)$$

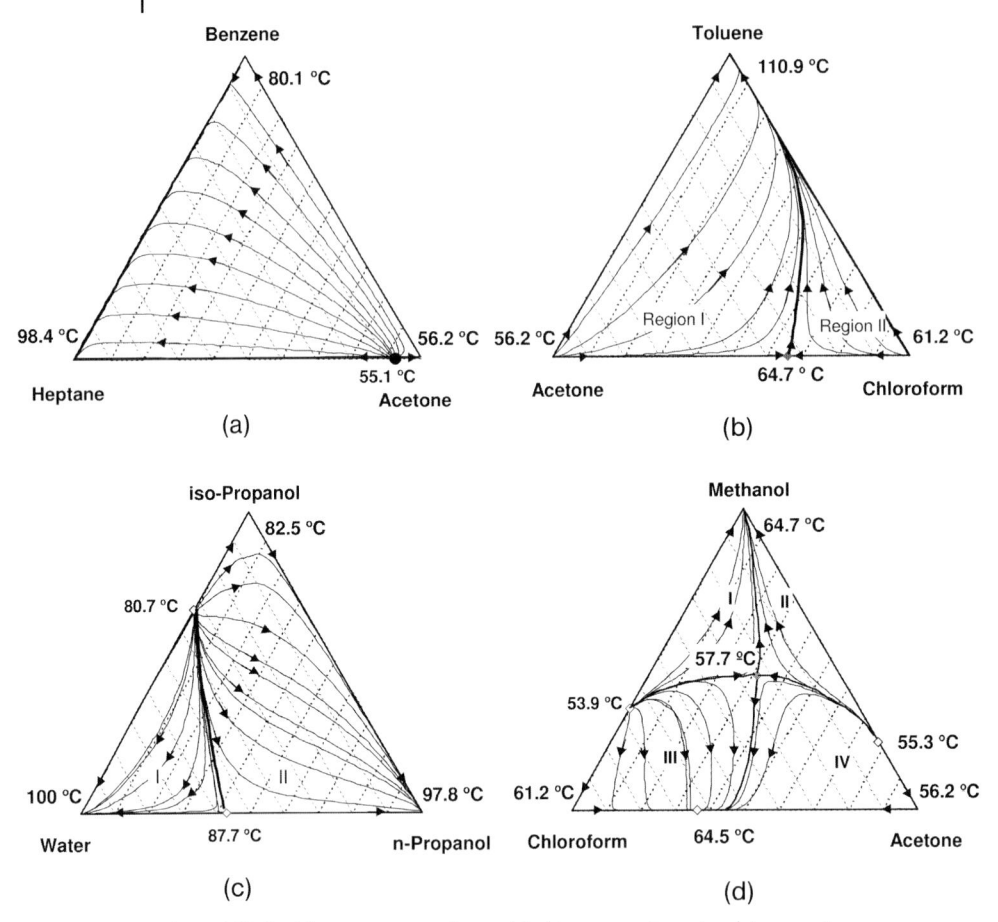

Figure 3.10 Residue curve maps for nonideal ternary systems involving azeotropes.

where $\theta = Vt/L$ is a dimensionless "wrapped" time, V the molar vapor distillation rate and L the molar liquid holdup. The equilibrium constants $K_i(x_i, y_i) = y_i/x_i$ are a function of composition. As an example, Figure 3.10(c) shows the RCM calculated rigorously, to be compared with that sketched in Figure 3.9.

Similarly, a distillation line map (DLM) shows the distribution of liquid composition on the stages of a continuous distillation column at infinite reflux and for infinite number of stages. DLM is obtained even simpler by computing successive dew and bubble points as described by the relation:

$$x_{i,0} \xrightarrow{K_{i,0}} y_{i,0} = x_{i,1} \xrightarrow{K_{i,1}} y_{i,1} = x_{i,2} \xrightarrow{K_{i,2}} y_{i,2} \ldots \quad (3.2)$$

Although RCM and DLM are not identical, the assessment of a distillation process gives similar results in both diagrams. The representation is often called ∞/∞ analysis because it assumes infinite reflux and an infinite number of stages. The predictions are correct for zeotropic mixtures, but only qualitative for azeotropic mixtures, namely when a distillation boundary should be crossed.

The analysis tools available in simulation packages allow rapid tracing of RCM. The only concern remains the confidence in the azeotropic points, as well as the shape of boundaries. The regression of interaction parameters from experimental data should be done methodically [5]. The predictions by UNIFAC should be used only for screening. Checking the L–L solubility against experimental data is of highest importance when heterogeneous azeotropes are formed.

Figure 3.10 shows typical RCM for nonideal mixtures involving azeotropes. For the mixture acetone/heptane/benzene (plot a) there is only one distillation field. The problem seems similar to a zeotropic system, except for the fact that the minimum boiler is a binary azeotrope and not a pure component. With the mixture acetone/chloroform/toluene (plot b) there is one distillation boundary linking the high-boiler with the low-boiler azeotrope. Consequently, there are two distillation regions. Similar behavior shows the plot c, with two azeotropes. The mixture acetone/chloroform/methanol (plot d) has four azeotropes (3 binaries and 1 ternary) displaying a behavior with four distillation regions.

RCM is a powerful graphical tool for assessing the feasibility of separations by distillation. As a simple example, Figure 3.11 presents the generation of alternatives for separating a zeotropic mixture with A, B, C from lowest to highest boiler. The initial feed is the point F. Let us consider the segment d_1b_1 as a representation of a simple distillation process (one feed and two products). This obeys the rules:

1. Feed and products are collinear on the material balance line. The position of points respects the lever rule.
2. Top and bottom products belong to the same residues curves.

Note that the last condition implies infinite reflux and infinite number of stages. The assessment of feasibility of a design for zeotropic mixtures is fully correct, the only problem left being the sizing.

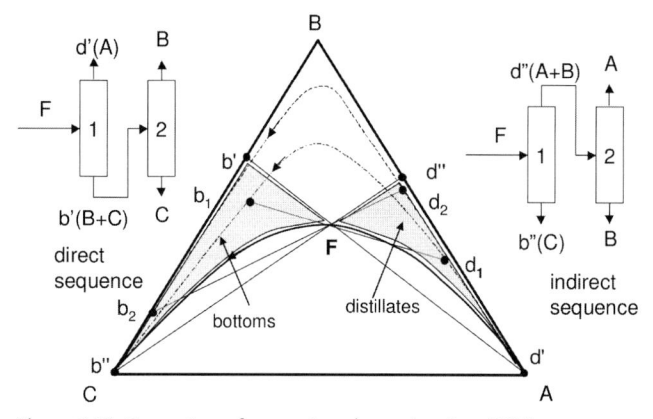

Figure 3.11 Generation of separation alternatives in a RCM.

For example, the separations d_1b_1 and d_2b_2 are sloppy splits with different amounts of A in the distillate, and accordingly with different recoveries. The separation of pure A at the top is represented by the split $d'b'$, which corresponds to a "direct sequence". Accordingly, the separation of B/C in a second column is represented by the edge BC, on which b' is the feed. Similarly, the first split in an "indirect sequence", in which C is separated in bottoms and A/B at the top, is shown by the segment $d''b''$. The locus of all splits between the above limit cases allows the regions of attainable products to be defined.

For azeotropic mixtures the analysis of the situation is more complex, as will be seen from the next section.

3.7.2
Separation by Homogeneous Azeotropic Distillation

The separation problem is defined as finding at least one feasible separation sequence for breaking a binary azeotrope AB by means of an entrainer C using only homogeneous distillation. The solution of this problem depends greatly on the existence of distillation boundaries. There are two possibilities:

1. Both pure products A and B belong to the same distillation field since there is no distillation boundary (one distillation field).
2. The pure products A and B are separated by a distillation boundary and belong to different distillation fields. In this case, the distillation boundary must be sufficiently curved, such that one split can cross it and gives opportunities for separating the other pure component.

A general rule in split assessment can be formulated as follows: pure components only as nodes, while entrainer and azeotropes are saddles since they are recycled.

3.7.2.1 One Distillation Field

If the RCM has only one distillation field the assessment of separations is similar to zeotropic mixtures. Let us examine breaking a minimum azeotrope AB with a medium boiling entrainer C (Figure 3.12). AB azeotrope and B are nodes, while A and C are saddles. In "direct sequence" the pure A can be obtained as top distillate even if this is a saddle! This is in apparent contradiction with the rule requiring pure component distillate. It can be shown that separation is feasible by using two feeds, with the entrainer in a higher position, while the concentration profiles in both stripping and rectification zones point to the same saddle (C). In addition, the residue curves pass through A before reaching $(AB)_{az}$, and the difference in the boiling points is small. The indirect sequence is feasible too, with the merit of avoiding the azeotrope. Both alternatives are similar in number of stages and energy consumption [5].

Table 3.16 gives the criteria for entrainer selection [7]. Medium-boiler entrainer is valid for both minimum and maximum azeotropes, but this condition is difficult

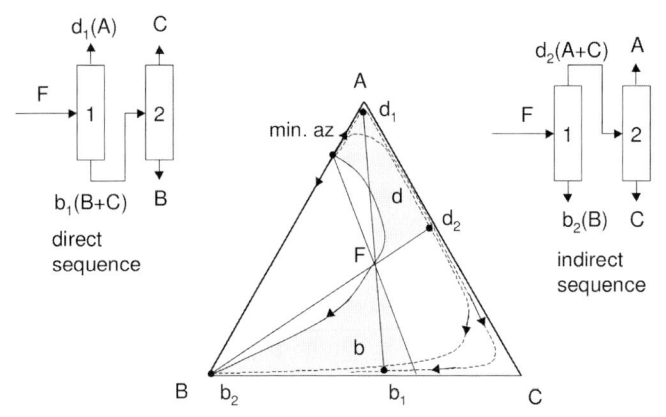

Figure 3.12 Separation of an azeotropic mixture in one distillation region.

Table 3.16 Criteria for entrainer selection for separations in one distillation field.

Entrainer C	Minimum AB azeotrope	Maximum AB azeotrope
Low boiler	Medium-boiler max azeo with A	No
Medium boiler	Yes	Yes
High boiler	Yes (extractive distillation)	Medium-boiler min azeo with B

to meet. For minimum AB azeotrope the entrainer may be: (1) a low boiler giving medium-boiling maximum azeotrope with A, and (2) a maximum boiler. The second case is similar to the extractive distillation. For maximum AB azeotrope, the entrainer may be a high-boiler giving a medium-boiling minimum azeotrope with B, but is difficult to find entrainers giving an opposite azeotrope to the original one. Hence, finding an entrainer to handle separations on one distillation field is quite limited.

Figure 3.13 presents suitable RCMs [5]. A and B but not both must be saddles, except in extractive distillation. Two columns are sufficient, either as a direct or as indirect sequence. Entrainer and mixture can be merged in the feed, except the extractive distillation, where the entrainer goes on the top. As an example we cite the separation of acetone from its azeotrope with heptane by using benzene. Contrary to expectations, the indirect sequence has better indices of investment and energy consumption.

3.7.2.2 Separation in Two Distillation Fields

Figure 3.14 displays a situation when the components A and B to be separated form a maximum boiling azeotrope. The entrainer C is the highest boiler. There

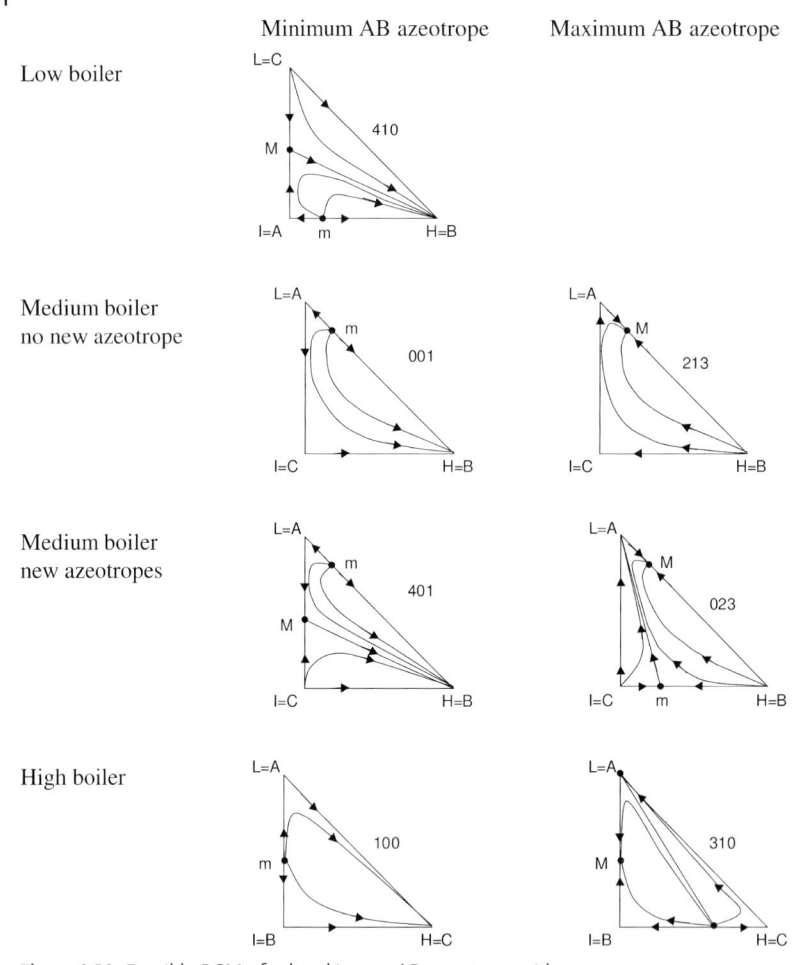

Figure 3.13 Feasible RCMs for breaking an AB azeotrope with an entrainer C by separation in a single distillation field.

is distillation boundary linking C and max(AB), which in turn generated two distillation fields. If the boundary is straight then the separation of A and B follows the rules given before. If the boundary is highly curved new opportunities appear by boundary crossing.

The following rules should be considered [8, 11, 13, 14]:

1. The feed must be sufficiently close to the distillation boundary, but on the concave side.
2. Both products A and B must belong to the same residue curve, either on the same side with the feed, or on the other side of the boundary when this can be crossed.

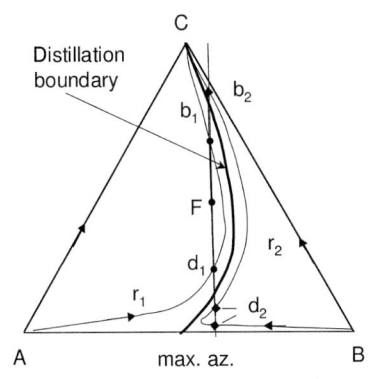

Figure 3.14 Separation regions with a curved distillation boundary.

The selection of an entrainer with boundary crossing is based on the observation that in a RCM both A and B must be nodes, stable or unstable. By consequence the pure components are separated either as overhead or bottom products. Table 3.17 gives a list of recommended heuristics [13, 14]. In all cases, the distillation boundary must be highly curved, although "how curved" cannot be specified theoretically. In this case the simplest entrainer choice is a low-boiler entrainer for minimum AB azeotrope, and a high-boiler entrainer for maximum AB azeotrope, again not easy to meet in practice.

The feasibility of a separation process with boundary crossing is not guaranteed only by RCM analysis. Experimental VLE and an investigation of feasibility in a bench distillation setup are necessary. The following example illustrates the use of computer simulation for investigating alternative separation flowsheets. The simulation indicates that the boundary can cross already in the first split at finite reflux, making easier the next separation steps. Hence, under practical conditions the feasibility region could be larger than that predicted by the ∞/∞ analysis.

Table 3.17 Criteria for entrainer selection for systems with boundary crossing.

Entrainer C	Minimum AB azeotrope	Maximum AB azeotrope
Low boiler	The entrainer has to boil lower than the minimum AB azeotrope.	New maximum azeotropes with both A and B. At least one has to boil higher than the AB azeotrope.
Medium boiler	New minimum azeotrope with A, the low-boiling component.	New maximum azeotrope with B, the high-boiling component.
High boiler	New minimum azeotropes with both A and B. At least one has to boil lower than the AB azeotrope.	The entrainer has to boil higher than the maximum AB azeotrope

Example 3.3: Separation Alternatives of the Mixture Acetone/Chloroform/Toluene [5]

Acetone (A) (nbp 56.2 °C) and chloroform (B) (nbp 61.2 °C) forms a maximum-boiling azeotrope (nbp 54.7 °C). Examine the feasibility of separating high-purity components using as an entrainer toluene (C) (nbp 110.9 °C). The mixture is 50 kmol/h with 30% chloroform and 70% acetone.

The simulation is performed with the Radfrac module in ASPEN Plus™ and the NRTL liquid-activity model [20]. Tracing the RCM (Figure 3.15) shows that acetone and chloroform are unstable nodes, toluene is a stable node and the maximum-boiling azeotrope acetone-chloroform is a saddle. The distillation boundary shows a strong curvature.

Figure 3.15 (left-hand) presents the conceptual flowsheet for the first alternative.. After mixing the initial mixture f_1 with C, the feed becomes m_1. The first $d_1 b_1$ recovers A of high purity. Note that d_1 and b_1 belong to the same residue curve, in the left distillation field. This split has not yet crossed the distillation boundary, but the point b_1 can be pushed very close to it. On the contrary, the second split can cross it, the products d_2 and b_2 jumping on the convex side. The column can be designed to deliver rich C as bottoms and AB mixture as top. Next, B can be obtained at high purity as distillate, while the azeotrope AB is recycled. Hence, the separation sequence has three columns. Pure components are recovered as top distillate, while the entrainer is recycled as bottoms. If the initial feed has an excess of B, the first split will take place in the right-side region. By recycle of the entrainer and mixing with the B-rich mixture from the first split, again a three-column sequence can be generated.

The concentration profiles are displayed in Figure 3.15 (right-hand). The first column has 50 theoretical trays with feed on 20, entrainer/feed ratio 2 and reflux ratio 3.5. A bound in the concentration profile takes place around the feed. In the stripping zone the toluene carries out preferentially the chloroform. The rectification part separates mainly the binary acetone–chloroform, the entrainer concentration being negligible. The profile of the second column

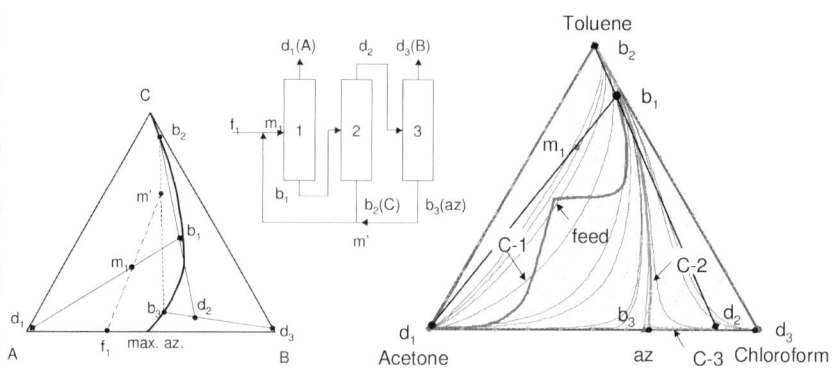

Figure 3.15 Separation of acetone and chloroform with toluene (alternative 1).

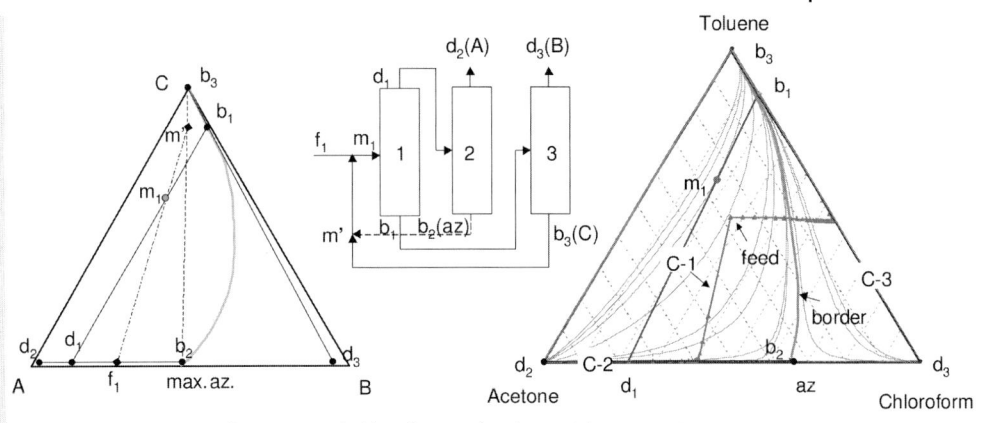

Figure 3.16 Separation of acetone and chloroform with toluene (alternative 2).

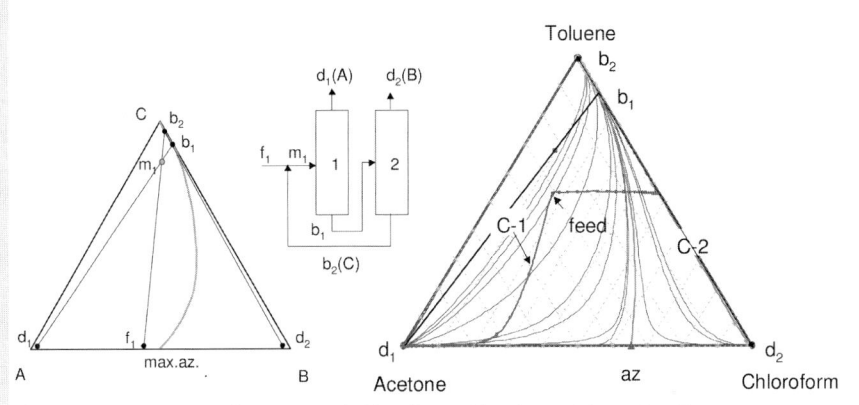

Figure 3.17 Separation of acetone and chloroform with toluene (alternative 3).

shows that this works completely on the right side of the distillation border, even if the feed seems to be on the left side. The concentration profile follows closely the residue curves. The third column is a binary separation, without much interest. Hence, the simulation indicates that the separation sequence is feasible and leads to high-purity products. Then, sizing and column optimization can be done by standard procedures.

Figure 3.16 displays a second alternative, with a "sloppy" split in the first column. The points b_1 and d_1 are on the same residue curve inside the same distillation region. Because d_1 can be pushed on the side AB, the next binary split will give high purity A and AB azeotrope, which is recycled. If the distillation border is high concave the point b_1 could approach closely the side BC. As a result, the third split can deliver good purity B and entrainer to be recycled. This alternative seems interesting but uncertain with respect to achievable purity. Fortunately, the simulation indicates that the distillation border can be deeply crossed.

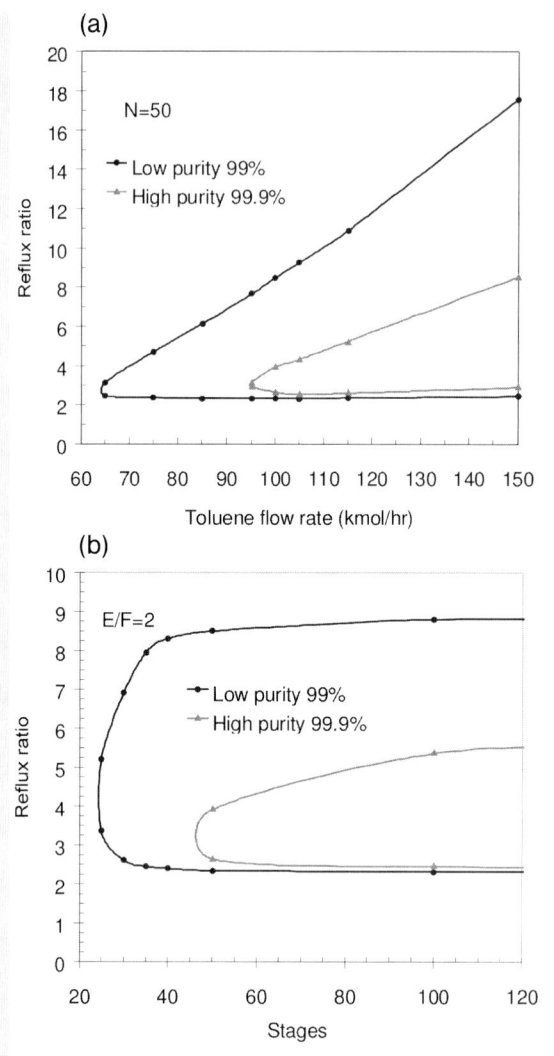

Figure 3.18 Chloroform purity dependence on entrainer, reflux and number of stages.

The above analysis suggests a third alternative (Figure 3.17) with only two columns. The first split could deliver high-purity acetone, while the second split would give chloroform with acetone as impurity. The representation predicts that chloroform purity would not exceed 98% for a reasonable amount of entrainer. Again, computer simulation gives a much better solution. The concentration profile for the first column shows clearly that the distillation border is crossed at finite reflux, and high purity can be obtained in the second split.

An important topic in azeotropic distillation regards the purity. This can be managed by means of both entrainer and reflux ratios. For a given purity there is a minimum entrainer ratio that can be determined from a RCM plot. In

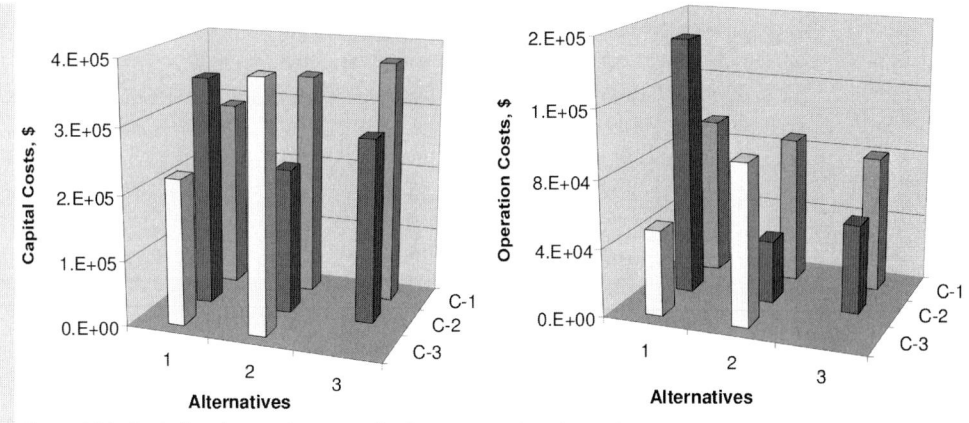

Figure 3.19 Capital and operation costs for three separation alternatives.

contrast with zeotropic systems the reflux cannot be raised infinitely. There is a minimum and a maximum reflux ratio giving the same purity.

Figure 3.18(a) displays curves of constant purity as functions of reflux ratio and entrainer flow rate, when the distillate rate and number of stages are constant. The sequence has two columns. Two situations are examined, low and high purity of chloroform. It can be observed that there is a minimum entrainer ratio to achieve a given purity. For the same entrainer ratio there are minimum and maximum reflux ratios. At low purity there is a wider reflux range, while at high purity the variation is limited. Figure 3.18(b) shows curves at constant purity as functions of reflux and number of stages, keeping the entrainer ratio constant. The existence of a minimum number of stages for the desired purity is clearly indicated. When the column has more stages than the minimum, there are two steady states for the reflux necessary at given purity.

Finally, it is interesting to compare the alternatives as capital and operation costs (Figure 3.19). The most expensive columns are those crossing the boundary, as well as for separating the binary chloroform/toluene. The alternative with two columns is the cheapest. It also has the lowest operation costs. The first alternative is the most expensive in operation. On the basis of total annual cost the best alternative is by far the one with only two columns. The alternative with a sloppy split is ranked in second place, because of its cheaper operation.

3.7.3
Separation by Heterogeneous Azeotropic Distillation

Liquid–liquid separation helps to overcome the constraint of the distillation boundary that penalizes the homogeneous azeotropic distillation. As a result, the heterogeneous azeotropic distillation has much larger applications. The classical example

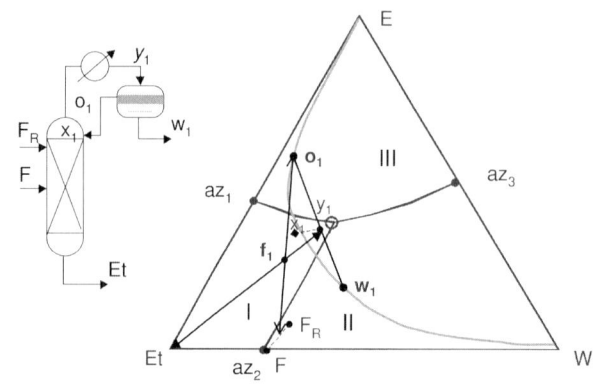

Figure 3.20 Separation of ethanol by azeotropic distillation.

is breaking the azeotrope ethanol/water by employing a hydrocarbon entrainer, as benzene or cyclohexane. Figure 3.20 shows the ternary diagram ethanol (Et), water (W), entrainer (E). The components give three minimum-boiling azeotropes, as well as a ternary heterogeneous azeotrope, which is the lowest boiler. The desired product, ethanol and water, are both nodes, but situated in different distillation regions. The distillation boundary is almost straight and cannot be "crossed" efficiently by homogeneous distillation. Therefore, another strategy should be investigated.

It can be observed that a top vapor mixture having the composition of the ternary azeotrope can split after condensation into two liquid phases: organic phase rich in entrainer o_1, aqueous phase rich in water w_1, with ethanol distributed in both. For feasible separation it is essential that o_1 and w_1 are situated in different distillation regions. By appropriate mixing with recycle streams, compositions are produced for feeding distillation columns that can deliver ethanol and water as high-purity products. Thus, the key feasibility condition for separation is that the point characterizing the ternary azeotrope must fall inside the heterogeneous region.

Let us consider for simplification an initial feed of azeotropic composition (point F). The column receives as reflux the organic phase o_1 and recycled mixture F_R such to produce the overall feed f_1 situated in the distillation region I, in which both ethanol and ternary azeotrope are nodes. The material balance can be designed to give a bottom product of high-purity ethanol and a vapor distillate of composition close to ternary azeotrope y_1 falling into the heterogeneous region. Next, two-phase separation occurs by liquid–liquid split. The organic one is recycled to the azeotropic column. The aqueous stream of composition w_1 is located in region II, where both water and ternary azeotrope are nodes. Two alternatives can be imagined (Figure 3.21).

In the first one (Figure 3.21, left-hand) two columns are needed. The first separates the ternary azeotrope in top, while the second the binary azeotrope az_2 and water. In a second alternative (Figure 3.21, right-hand), only a single column is necessary. Water is obtained directly as bottoms, while the top distillate is enriched

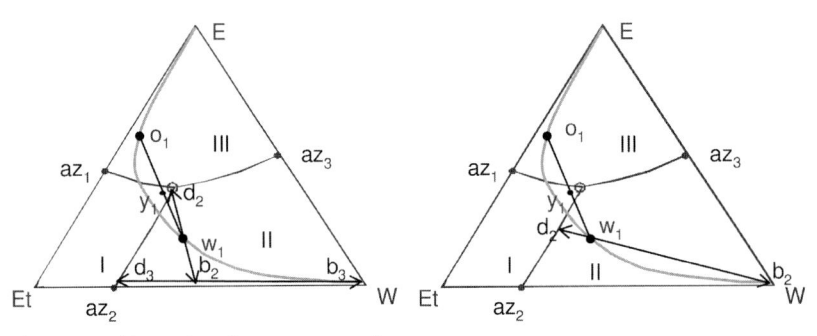

Figure 3.21 Alternatives for water removal.

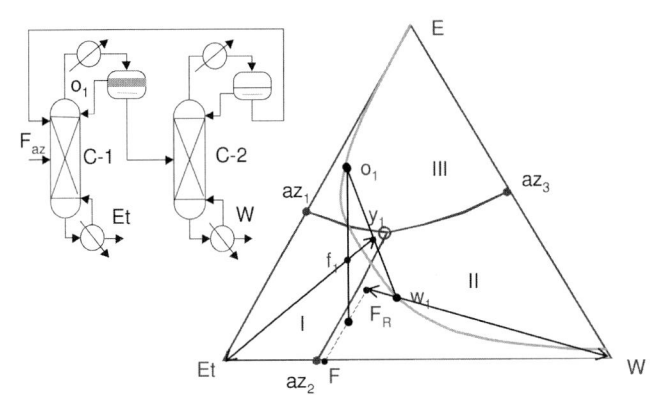

Figure 3.22 Separating ethanol from water by heterogeneous azeotropic distillation.

in ethanol, some water and entrainer is recycled either in decanter or fed into the ethanol column.

On the basis of the above analysis several alternatives with two or three columns can be imagined. Figure 3.22 depicts only the alternative with two columns, the first for ethanol recovery, while the second is for water removal. This flowsheet accepts relatively concentrated solutions. When diluted solutions are treated the enrichment could be done separately, and hence the flowsheet will contain a maximum of three columns. However, the initial diluted feed may be sent to the water column. In consequence, only two columns are necessary.

A more detailed discussion from an economic and operability viewpoint can be found in Doherty and Malone [8]. Despite the apparent advantage, the two-column sequence seems not to be the most economical because of the large entrainer recycle and large number of stages. On the contrary, the two-column sequence plus enrichment column, in total three units, offers the best compromise between investment and solvent-recycle costs.

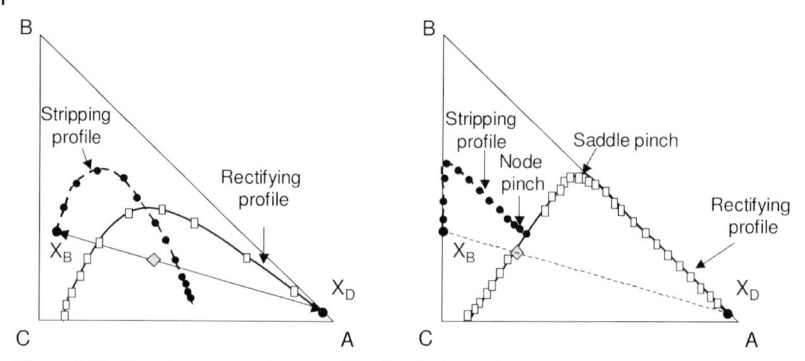

Figure 3.23 Drawing composition profiles by the boundary value method.

3.7.4
Design Methods

For zeotropic systems the standard Fenske–Underwood–Gilliland (FUG) method for predesign followed by rigorous simulation is largely applied [9]. However, this approach is not appropriate for highly nonideal mixtures. The boundary value method developed by Doherty and Malone [8] has received a large acceptance and is now implemented in some software. Figure 3.23 explains the principle for a simple column with one feed and two products. The components A, B and C are in the decreasing order of volatility. The calculation consists of solving tray-by-tray the material and energy balances starting from known specifications of distillate and bottoms. Supplementary specifications are reflux ratio L/D and reboil ratio V/B. In this way, separate profiles for stripping the rectification and stripping sections can be computed (Figure 3.23, left-hand). The design is feasible if the profiles cross themselves. The number of stages and feed location can be obtained by minimizing the "distance" between profiles or by other measure.

At minimum flows, pinch zones can be identified where the composition changes very little on successive stages. The profiles just intersect each other at the feed position. In Figure 3.23 (right-hand) a saddle pinch can be identified in the rectifying profile. This profile corresponds to the minimum reflux in a direct sequence, since the distillate is practically the pure component A. The minimum reflux is practically independent of the bottoms composition. Similarly, a minimum reboil ratio may be identified by simulation. The determination of the minimum number of stages is subtler, since the composition of the top distillate and bottoms are not independent, but the simulation may produce a reasonable estimation.

3.8
Reactive Separations

Combining chemical reaction and separation in a compact device can lead to significant economic advantage in term of investment and operation costs. Reactive distillation (RD) is the most attractive reactive separation. A large variety of reactions could be tackled by RD technology, such as etherifications, esterifications, hydrolysis, acetalization, aldol condensation, hydration/dehydration, alkylation, chlorination, hydrogenation, *etc.* One of the most spectacular applications is the manufacturing of methylacetate, replacing advantageously a process involving about ten distillation and extraction columns.

Heterogeneous catalysis is preferred over homogeneous catalysis. A critical issue is the catalyst design, which should ensure compatibility between the reaction and the separation. The temperature profile dominated by the vapor–liquid equilibrium, as well as the residence-time distribution controlled by the hydrodynamics of internals must comply with the achievable reaction rate and with the desired selectivity pattern.

As displayed in Figure 3.24, a RD device consists, in principle, of a reaction zone surrounded by top and bottom separation sections. Let us consider the reaction A + B ↔ P + R, encountered in many applications. The reaction takes place essentially in the liquid phase. The reactants A and B can be introduced somewhere in the separation section, but usually at the top and the bottom of the reaction zone. The products P and R leave at the top and bottom separation sections, because of the volatility difference, or aided by a mass-separation agent. In the reaction zone the temperature profile is usually almost flat. The reaction rate profile can display a maximum along the reaction zone, whose position depends on the ratio of reactants adjustable by the internal reflux.

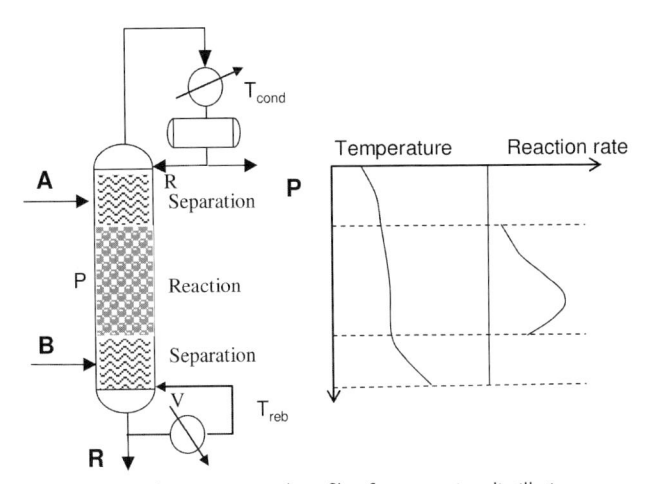

Figure 3.24 Column setup and profiles for a reactive distillation process.

From the reaction-engineering viewpoint a RD device presents definite advantages, as:

(a) Countercurrent flow of reactants giving higher reaction rates than in co-current.
(b) Internal reactants recycle, by reflux and boilup, instead of costly external recycles.
(c) *In-situ* removal of product can shift an equilibrium reaction to completion.
(d) It can remove the constraint set by a distillation boundary. This advantage may be compromised by the occurrence of *reactive azeotropes* (see Appendix A).
(e) Efficient heat integration, namely in the case of exothermic reactions.

The first two items are particularly powerful. The result is that a reactive distillation setup offers the possibility of achieving simultaneously high conversion for both reactants, with stoichiometric consumption of reactants at optimal selectivity. The third item indicates that the reactive distillation is of great interest for equilibrium constrained reactions. Taking advantage of exothermal reactions depends on the temperature level that can be allowed by the phase equilibrium.

There are also disadvantages:

1. When a reactant is gaseous or very light the driving force for reaction is lower compared with a homogeneous liquid-phase reaction at higher pressure.
2. With slow reactions the reaction rate may be not compatible with the liquid residence time and volumetric holdup corresponding to optimal hydraulic operation.

The above limitations can be counteracted by an adequate catalyst design. In general, more active catalysts are necessary for RD processes than with standalone reactors.

3.8.1
Conceptual Design of Reactive Distillation Columns

The conceptual design of reactive distillation systems is challenging. The analysis in residue curve maps offers valuable insight but tracing RCM for reactive systems is more difficult particularly for a finite reaction rate. Appendix I describes the theoretical background. Although a unified theoretical approach is missing, the engineering practice can take great profit from the existence of powerful and robust simulation models inside commercial simulators. However, unlike conventional distillation, the conceptual design of reactive distillation systems requires experimental investigation, both at laboratory and industrial level.

The state-of-the-art can be found in the monograph edited by Sundmacher and Kienle [16], as well as in the books of Doherty and Malone [8], Stichlmair and Fair [14]. The case study referring to the manufacturing of fatty esters presents a practical approach to the design of reactive-distillation columns.

3.9
Summary

The synthesis of separations can be handled by means of a systematic approach. The first separation step is essential, since it allows the decomposition into smaller problems for gas, liquid and solid subsystems. For each subsystem the strategy consists of generating a feasible quasioptimal separation sequence based on the identification of tasks and the assessment of suitable separation techniques. The decomposition in separation tasks can be managed by means of logical selectors. The assignment of tasks is based on the detection of a characteristic property among components. The split generation process is supported by the use of heuristics.

Several separation methods can be available for executing the same task. The task-oriented methodology promotes the application of novel techniques. Combining distillation with other separation methods is appropriate for solving difficult separations. Because the methodology develops alternatives rather a single design, the method can be enhanced by optimization methods, such as those based on mixed-integer liner programming.

The separation of zeotropic mixtures is constrained only by the optimal sequencing of a potentially large number of separators. A rapid solution can be found by producing a shortlist of candidates following the total vapor rate. On the contrary, with azeotropic mixtures the key problem is the feasibility, namely finding a suitable mass-separation agent. A residue curve map is a valuable tool for getting thermodynamic insights. The separation of mixtures exhibiting homogeneous azeotropes and distillation boundaries is particular difficult. Boundary crossing considerably enlarges the alternatives, but it should not be applied unless experimental data or reliable simulation evidence is available. Heterogeneous azeotropic distillation is more versatile.

Combining chemical reaction and separation in a reactive-distillation device, can lead to significant economic advantage in term of investment and operation costs. Because the chemical driving force can be lowered by vapor–liquid equilibrium, this should be compensated by a more active catalyst.

References

1 Barnicki, S.D., J.R. Fair, Separation system synthesis: A knowledge based approach: Liquid mixture separations, *Ind. Eng. Chem. Res.*, 29, 431–439, 1990
2 Barnicki, S.D., J.R. Fair, Separation system synthesis: A knowledge based approach: Gas/Vapor mixtures, *Ind. Eng. Chem. Res.*, 31, 1679–1694, 1992
3 Barnicki S.D., J.J. Siirola, Separation system synthesis, in Kirk-Othmer *Encyclopedia of Chemical Technology*, 4th edn, Vol. 21, 923–962, 1997
4 Biegler, L., I., Grossmann, A. Westerberg, Systematic Methods of Chemical Process Design, Prentice Hall, Upper Saddle River, NJ, USA, 1998
5 Dimian, A.C., Integrated Design and Simulation of Chemical Processes, CAPE series 13, Elsevier, Amsterdam, The Netherlands, 2003

6 Douglas, J.M., Conceptual Design of Chemical Processes, McGraw-Hill, New York, USA, 1988
7 Doherty, M.F., G.A. Caldarola, Design and synthesis of homogeneous azeotropic distillation, *Ind. Eng. Chem. Fundam.*, 24, 474–485, 1985
8 Doherty, M., M. Malone, Synthesis of Distillation Systems, McGraw-Hill, New York, USA, 2001
9 Kister, H.Z., Distillation Design, McGraw-Hill, New York, USA, 1992
10 Seader, J.D., E.J. Henley, Separation Process Principles, John Wiley, New York, USA, 1998
11 Seider, W., D.J.D., Seader, D. R. Lewin, Product and Process Design Principles, Wiley, New York, USA, 2003
12 Smith, R., Chemical Process Design and Integration, Wiley, New York, USA, 2004
13 Stichlmair, J., J.R. Herguijuela, Separation regions and processes of zeotropic and azeotropic ternary distillation, *AIChEJ*, 38, 1523–1535, 1992
14 Stichlmair, J.G., J.R. Fair, Distillation, Principles and Practice, Wiley-VCH, Weinheim, Germany, 1999
15 Strathmann, H., Membrane and Membrane Separation Processes, Ullmann's Encyclopaedia of Industrial Chemistry, Vol. A16, Wiley-VCH, Weinheim, Germany, 1990
16 Sundmacher, K., A. Kienle (eds.), Reactive Distillation, Wiley-VCH, Weinheim, Germany, 2003
17 Westerberg, A.W., O.M. Wahnshaft, Synthesis of distillation based separation processes, in Adv. Chem. Eng., Vol. 23, Process Synthesis, Academic Press, 1996
18 Scottish Environment Protection Agency, Guidance on gas treatment technologies for landfill gas engines, www.environment-agency.gov.uk, 2004
19 Meyers, R.A., Handbook of petroleum refining processes, McGraw-Hill, New York, USA, 1996
20 Aspen Plus, release 12.1, Aspen Technology, Cambridge, Massachusetts, USA

4
Reactor/Separation/Recycle Systems

4.1
Introduction

In Chapter 2, the main steps of the hierarchical approach to process synthesis were presented. In every step, one fundamental problem is tackled and specific techniques are applied for both analysis and synthesis. The hierarchical approach proposed by Douglas [1] and proved successful in a large number of applications does not pay enough attention to one important issue, namely the control of the chemical plant. This is a significant topic because recycle of raw materials, energy integration and reduced size or the lack of buffer vessels are characteristics of modern plants. In these conditions, the interaction between units is so strong that a classical approach to control does not work. Therefore, we talk about the "plant-wide control" problem. The recent book of Luyben et al. [2] emphasizes that "How a process is designed fundamentally determines its inherent controllability ... In an ideal project dynamic and control strategies would be considered during the process synthesis and design activities".

To provide an answer to this challenge, we take a systemic approach. We consider that a process plant consists of several subsystems interconnected through material and energy streams (Figure 4.1). The subsystems will be called *basic flowsheet structures* (BFSs), defined as parts of the plant for which (local) control objectives are assigned and can be achieved using only manipulated variables that are interior to the BFSs. Unit operations are the simplest BFSs. Often, some units interact so strongly that they must be treated as one entity. Examples are heat-integrated reactors, complex distillation arrangements, azeotropic distillation with solvent recycle, *etc.* Delimiting the BFS depends on what design/modeling details are available and how the control objectives are assigned.

The task of plantwide control is to harmonize the BFSs in such a way that the whole system operates in the required manner. This is achieved by assigning control objectives to BFFs. The controllability of the BFSs is a necessary (but not sufficient) condition for the controllability of the entire plant. Consequently, integrating design and control consists of two steps:

Chemical Process Design: Computer-Aided Case Studies. Alexandre C. Dimian and Costin Sorin Bildea
Copyright © 2008 WILEY-VCH Verlag GmbH & Co. KGaA, Weinheim
ISBN: 978-3-527-31403-4

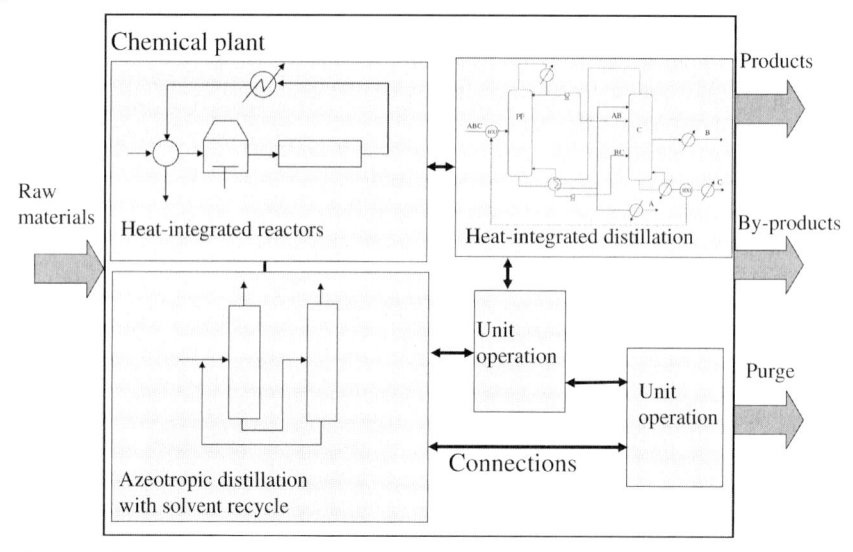

Figure 4.1 Systemic representation of chemical plants.

(a) Design basic flowsheet structures with good controllability properties. This is possible for unit operations, where a lot of industrial experience exists. However, it is an open field of research for more complex subsystems, such as heat-integrated reactors or reactive distillation.

(b) Couple the BFS in such a way that a controllable system is obtained. Interaction between the chemical reactor and the separation section, due to material recycle and plantwide control, is the subject of this chapter.

By including the "reactor/separation/recycle" level in the hierarchical approach, plantwide control can be considered at an early stage of design. In most cases, the separation is considered as a black-box. By "black-box" we mean that some targets are set, for example as species recovery or product purities. The separation is then modeled based on simple input-output component balances. The decisions to be taken and the detail of the results obtained depend on the information about the chemical reactor that is available.

In the simplest case, only the stoichiometry of chemical reactions is known, but no kinetic information is available. This allows designing the control structure, which is choosing the controlled and manipulated variables and their pairing. Note that the knowledge of stoichiometry is also necessary for designing the structure of the separation system, which is determined by the composition and the thermodynamic properties of the reactor-outlet mixture.

When the kinetics (at least for the main reactions) and the type of the reactor are known, the performance of the control structure can be assessed. Firstly, by considering a steady-state model that incorporates kinetic reactor and black-box separation and performing sensitivity studies, it is possible to assess the feasibility of operating procedures, for example changing the raw materials, production rate or product distribution in multiproduct plants. Moreover, dangerous phenomena can be revealed. One example is the high sensitivity to disturbances or uncertain design parameters, known as the "snowball effect" [2, 3]. A plant showing multiple steady states is another potentially unsafe situation. The designer must ensure that the desired steady state is reached during start-up and maintained during operation. Using the steady-state model, it is possible to identify unstable operating points. However, stability is a property characterizing the dynamic behavior and it cannot be guaranteed based on steady-state considerations. Note that in systems where the unreacted raw materials are recycled, steady-state operation is not always possible because the reactor has a limited capacity of converting the reactants into products.

For the designer, understanding the mass balance of the plant is a key requirement that can be fulfilled only when the reactor/separation/recycle structure is analyzed. The main idea is that all chemical species that are introduced in the process (reactants, impurities) or are formed in the reactions (products and byproducts) must find a way to exit the plant or to be transformed into other species [4]. Usually, the separation units take care that the products are removed from the process. This is also valid for byproducts and impurities, although is some cases inclusion of an additional chemical conversion step is necessary [5, 6]. The mass balance of the reactants is more difficult to maintain, because the reactants are not allowed to leave the plant but are recycled to the reaction section. If a certain amount of reactant is fed to the plant but the reactor does not have the capacity of transforming it into products, reactant accumulation occurs and no steady state can be reached. The reaction stoichiometry sets an additional constraint on the mass balance. For example, a reaction of the type A + B → products requires that the reactants A and B are fed in exactly one-to-one ratio. Any imbalance will result in the accumulation of the reactant in excess, while the other reactant will be depleted. In practice, feeding the reactants in the correct stoichiometric ratio is not trivial, because there are always measurement and control implementation errors.

It should be remarked that understanding the mass balance and the way it can be controlled is of great help during flowsheeting. To explain, we note the concept of *design degrees of freedom*, which is the number of variables that must be specified to completely define the process. It can be calculated by subtracting the number of equations from the number of variables. The *control degrees of freedom* is the number of variables that can be controlled. This is equal to the number of manipulated variables, namely the control valves in the process. It turns out [7] that for many processes the number of *design* degrees of freedom is equal to the number of *control* degrees of freedom. While developing the simulation model of the flowsheet, different sets of degrees of freedom can be chosen as specifications. Sometimes, although the correct number of variables has been specified, convergence is very slow or even impossible. From our experience, convergence is easy when the specified variables correspond to the controlled variables of a feasible plantwide control structure. Therefore, taking into account the way plantwide control will be implemented can speed up the convergence of the flowsheet, while lack of convergence usually indicates problems of the corresponding control structure.

Most of the time, the designer will use a detailed reactor model and black-box separation with specified targets. However, there are opposite situations when a close look at the separation shows that this could be very difficult. In these cases, one can set targets for the reactor design that are easier to achieve, and focus on the difficult problem of designing the separation section. For example, acrylonitrile is produced by amoxydation of propylene. The reactor effluent contains small amounts of gaseous byproducts such as water and carbon dioxide. Separating the propylene from these species for the purpose of recycling seems difficult. By consequence, it has to be completely converted in the reaction.

In the rest of the chapter, we will analyze the behavior of several reactor/separation/recycle systems where reactions with different stoichiometries take place. We will start by presenting two different concepts behind plantwide control structure, namely self-regulation and feedback control. In Section 4.3 we will examine a one-reactant, first-order reaction taking place at a fixed temperature in a CSTR. We will exemplify how dimensional and dimensionless mass-balance models can be set up. These models will be used to draw conclusion about the performance of the two different types of control structures. We will also show an unexpected result, namely the very similar behavior of different types of reactors placed in recycle systems. Section 4.4 will introduce more complexity by considering two reactants involved in a second-order reaction. For such systems, we will demonstrate that using plantwide control structures that employ the self-regulating property leads to state multiplicity. The following section will return to the first-order reaction occurring, this time in a tubular reactor. Additionally, we will take into account the heat effect of the reaction, leading to temperature varying along the reactor. This example will illustrate again the occurrence of multiple steady states and will show how information about stability can be obtained based on steady-state arguments. In the last section we will present results concerning a case study, namely the toluene hydro-dealkylation (HDA) plant. The chapter ends with conclusions. A key lesson from this chapter is to avoid relying on self-regulation of the mass balance, and to use feedback for the purpose of controlling species inventory.

4.2
Plantwide Control Structures

In recycle systems, the design of the chemical reactor and the control of the reactants' inventory are interrelated [8]. Figure 4.2 shows two different ways of controlling the inventory in a simple system. The first strategy consists of setting the feed on flow control. Consider the simple example presented in Figure 4.2(a). When more fluid is fed to the vessel, the level increases. The outlet flow rate, which is proportional to the square root of the liquid level, will also increase. After some time, the feed and outlet flows are equal, and a state of equilibrium is reached.

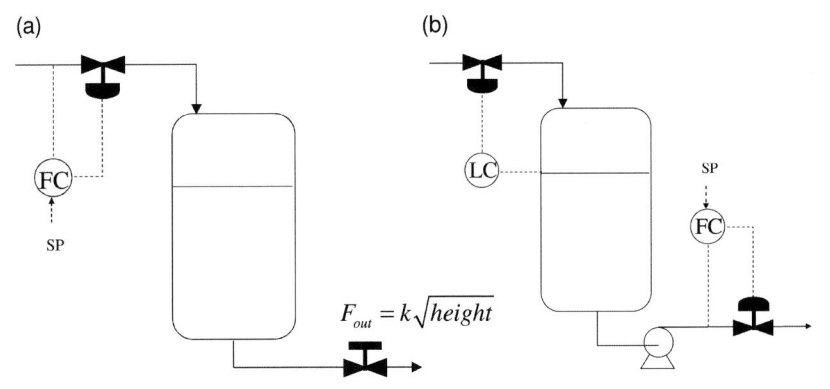

Figure 4.2 Alternatives for inventory control.

We say that the inventory is *self-regulating*. Similarly, the plantwide control can fix the flow rate of reactant at the plant inlet. When the reactant accumulates, the consumption rate increases until it balances the feed rate. This strategy is based on a self-regulation property. The second strategy is based on *feedback control* of the inventory. This consists of measuring the component inventory and implementing a feedback control loop, as in Fig. 4.2(b). Thus, the increase or decrease of the reactant inventory is compensated by less or more reactant being added into the process.

The self-regulation strategy, which could be designated as "classical", is recommended only if the reactor is large enough and the per-pass conversion is high. Nonlinearity of the plant could lead to phenomena, such as state multiplicity and closed-loop instability [9], as well as high sensitivity to production-rate changes, process disturbances or design-parameter uncertainty ("snowball effect"). The approach has the advantage of directly setting the production rate. A supplementary benefit is that product distribution is fixed in the case of some complex reactions.

The feedback-control strategy can be implemented through the following steps [8]:

- fix the reactor-inlet (recycle plus fresh feed) flow rate,
- determine the reactant inventory somewhere in the recycle loop, by measuring the appropriate level, pressure or concentration,
- adjust the fresh feed to keep the inventory at a constant value.

The important advantage of this strategy is that the reactor behaves as "decoupled" from the rest of the plant. The production is manipulated indirectly, by changing the recycle flows, which could be seen as a disadvantage. However, it handles nonlinear phenomena better, such as for example the snowball effect or state multiplicity. Additionally, this strategy guarantees the stability of the whole recycle system if the individual units are stable or stabilized by local control.

4.3
Processes Involving One Reactant

4.3.1
Conventional Control Structure

Let us consider a CSTR/separator/recycle system, where the first-order reaction A → P takes place. Figure 4.3(a) presents the conventional control of the plant. The fresh feed flow rate is kept constant at the value F_0. The reactor holdup V is controlled by the effluent. The reaction takes place at a constant temperature, which is achieved by manipulating the utility streams. Dual-composition control of the distillation column ensures the purities of the recycle and product streams.

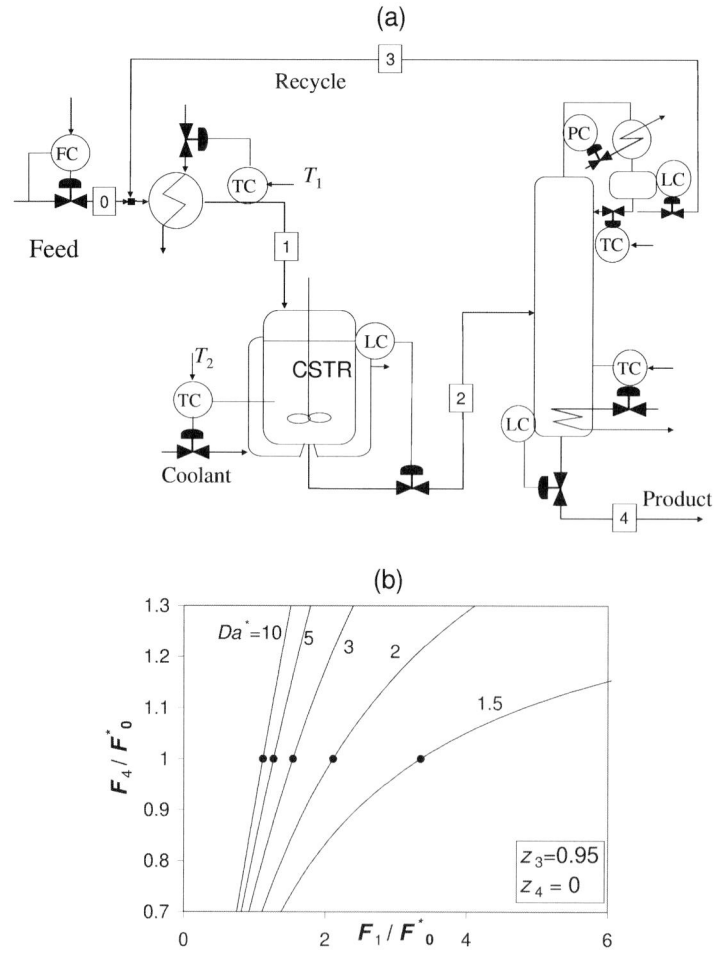

Figure 4.3 First-order reaction in a CSTR/separation/recycle system. (a) Conventional control structure. (b) Production rate F_4 versus the reactor-inlet flow rate F_1.

Assuming a liquid-phase reaction, constant density, perfect control of the reactor holdup and temperature and perfect control of the recycle and product purities, the system is described at steady state by the following equations:

Reactor:

$$F_1 z_1 - F_2 \cdot z_2 - k \cdot V \cdot c_0 \cdot z_2 = 0 \tag{4.1}$$

$$F_2 - F_1 = 0 \tag{4.2}$$

Separation:

$$F_2 \cdot z_2 - (F_3 \cdot z_3 + F_4 \cdot z_4) = 0 \tag{4.3}$$

$$F_2 - (F_3 + F_4) = 0 \tag{4.4}$$

Mixing point:

$$F_0 \cdot z_0 + F_3 \cdot z_3 - F_1 \cdot z_1 = 0 \tag{4.5}$$

$$F_0 + F_3 = F_1 \tag{4.6}$$

In Eqs. (4.1) to (4.6), k is the reaction rate constant. F_k and z_k are the molar flow rate and reactant mole fraction in stream k. Note that the assumption of constant density implies that all streams have the same total molar concentration, which is denoted by c_0.

The control structure fixes the values of F_0, k, V, z_3 and z_4. Therefore, (4.1) to (4.6) can be solved for six unknowns: z_1, z_2, F_1, F_2, F_3, and F_4. The input/output mass balance (4.2) + (4.4) + (4.6) requires $F_4 = F_0$. Then, Eqs. (4.2), (4.3) and (4.5) give:

$$\frac{F_1}{F_0} = \frac{F_2}{F_0} = \frac{z_3 - z_4}{z_3 - z_2} \tag{4.7}$$

$$\frac{F_3}{F_0} = \frac{z_2 - z_4}{z_3 - z_2} \tag{4.8}$$

Substituting these expressions into Eq. (4.1) leads to:

$$k \frac{V}{F_0} c_0 \cdot z_2 = \frac{(z_0 - z_4) \cdot (z_3 - z_2)}{(z_3 - z_2)} = (z_0 - z_4) \tag{4.9}$$

Evidently, the factor $(z_3 - z_2)$ can be simplified if, and only if, it is nonzero.

At this point it is useful to introduce dimensionless variables. Thus, we choose as the reference value the flow rate of the feed stream at the nominal production rate, F_0^* and define the dimensionless flows:

$$f_k = \frac{F_k}{F_0^*} \quad (4.10)$$

Then, the flow rate and composition of the reactor-outlet and the recycle flow rate are given by:

$$(z_2, f_2, f_3)_1 = \left(\frac{z_0 - z_4}{\mathrm{Da}}, \frac{\mathrm{Da} \cdot (z_3 - z_4)}{z_3 \cdot \mathrm{Da} - (z_0 - z_4)}, \frac{z_0 - z_4 - \mathrm{Da} \cdot z_4}{z_3 \cdot \mathrm{Da} - (z_0 - z_4)} \right) \quad (4.11)$$

where the Damkohler number is defined as:

$$\mathrm{Da} = k \frac{V}{F_0^*} c_0 \quad (4.12)$$

Note that the overall mass balance requires the equality of the feed and product flow rates, $F_0 = F_4$. Consequently, the Damköhler number accounts for the production rate (F_0^*), reactor design (V) and reaction kinetics (k).

In Eq. (4.11) all flow rates are positive if, and only if, the following conditions are fulfilled:

$$\frac{z_0 - z_4}{z_3} < \mathrm{Da} < \frac{z_0 - z_4}{z_4} \quad (4.13)$$

The first inequality characterizes recycle systems with reactant inventory control based on self-regulation. It occurs because the separation section does not allow the reactant to leave the process. Consequently, for given reactant feed flow rate F_0, large reactor volume V or fast kinetics k are necessary to consume the whole amount of reactant fed into the process, thus avoiding reactant accumulation. The above variables are grouped in the Damköhler number, which must exceed a critical value. Note that the factor z_3 accounts for the degradation of the reactor's performance due to impure reactant recycle, while the factor $(z_0 - z_4)$ accounts for the reactant leaving the plant with the product stream.

A separation unit is meaningful only if the desired concentration of reactant in the product stream, z_4, is smaller than the concentration at the reactor outlet. This leads to the second inequality in Eq. (4.13).

The system (4.1) to (4.16) admits a trivial solution, which is obtained when $z_3 - z_2 = 0$ and Eq. (4.9) cannot be simplified. This solution is unfeasible, since it implies infinite flow rates,

$$(z_2, f_2, f_3)_2 = (z_3, \infty, \infty) \quad (4.14)$$

The control structure presented in Figure 4.3(a) has the advantage of simplicity. It achieves production-rate changes directly: modifying the fresh feed flow rate leads to new values for all flow rates and concentrations. A new steady state is reached when the consumption rate $kV c_0 z_2$ equals the feed rate. This is possible as long as the condition (4.13) is fulfilled. However, if the reactor volume is too small, the variations of the process flow rates (reactor exit, recycle) induced by changes in the input feed might be very large. This high sensitivity is known as the "snowball effect".

The dependence of the reactor-inlet flow rate F_1 versus the production rate F_4, parameterized by the Damköhler number is presented in Figure 4.3(b). Dots are used to show the nominal operating points, when the dimensionless production rate is unity. Consider the case of a large reactor, for example $Da^* = 5$. When a 30% increase of the production rate is attempted by increasing the feed rate, the flow rates to the reactor and to the separator, $F_1 = F_2$, increase by 40%. This change is acceptable and can be tolerated by the separation. Consider now the design characterized by $Da^* = 2$. In this case, a 30% production-rate increase requires increasing the flow to the reactor and to separation by 100%. Evidently, the hardware cannot cope with such a large disturbance.

The conclusion of this analysis is that plantwide control structures relying on self-regulation can be used, but the snowball effect is avoided only when the reactor volume is large enough. For a first-order reaction involving one reactant, the reactor could be considered "sufficiently large" if $Da > 3$.

4.3.2
Feasibility Condition for the Conventional Control Structure

Let us consider a reactor/separator/recycle process, where n reactants A_j give m products and intermediates P_k. The network of r reactions can be described by:

$$v_A^T \cdot A = v_P^T \cdot P \qquad (4.15)$$

where v_A^T and v_P^T are matrices of stoichiometric coefficients, not necessarily positive.

Let us assume that the separation section does not allow reactants A_j to leave the process. Then, the overall mass balance can be written as:

$$v_A \cdot \xi = F_0 \qquad (4.16)$$

where ξ is the vector of reaction extents and F_0 is the vector of fresh reactant flow rates.

Obviously, the linear system (4.16) has at least one solution ξ for any vector F_0, if the following condition is fulfilled:

$$\text{rank}(v_A) = n \leq r \qquad (4.17)$$

The flow rates of fresh reactants can be set at arbitrary values, but within stoichiometric constraints. Then, the internal flow rates and concentrations adjust themselves in such a way that, for each reactant, the net consumption rate equals the feed flow rate. Therefore, the reactant inventory becomes self-regulating. When $n = r$, Eq. (4.16) has a unique solution. Consequently, the type of reactor, the kinetic expression, the reactor volume, or the value of the rate constants do not influence either the selectivity or the production rate.

In Chapter 9, the butane/butene alkylation for isooctane production is discussed. The reactions can be represented by the following stoichiometry.

$$C_4H_8 + iC_4H_{10} \rightarrow C_8H_{18}$$
$$\text{(A)} \quad \text{(B)} \qquad \text{(P)}$$

$$C_4H_8 + C_8H_{18} \rightarrow C_{12}H_{26}$$
$$\text{(A)} \quad \text{(P)} \qquad \text{(R)}$$

As the rank condition (4.17) is fulfilled, a control structure fixing the flow rates of both reactants is feasible. However, the secondary reaction has only a small contribution to the mass balance of the system. Consequently, the self-regulation strategy leads to a severe snowball effect.

4.3.3
Control Structures Fixing Reactor-Inlet Stream

The control structure discussed in this section is presented in Figure 4.4(a). The reactor-inlet flow rate is fixed at the value F_1. Reactor effluent controls the reactor holdup V, while the coolant flow rate controls the reactor temperature. Dual composition control is used for the distillation column. The reactant is fed on level control. For illustration purposes, a buffer vessel was considered. This increases the equipment cost and might be unacceptable due to safety or environmental concerns. An alternative is to feed the reactant in the condenser drum of the distillation column. This strategy achieves the regulation of reactant inventory, because any imbalance is reflected by a change of the holdup.

The system is still described by Eqs. (4.1) to (4.6). However, the control structure now fixes the values of F_1, V, k, z_3 and z_4. Therefore, the model can be solved for 6 unknowns: F_0, F_2, F_3, F_4, z_1, z_2. Among them, the production rate is given by:

$$F_4 = F_0 = \frac{F_1 \cdot k \cdot V \cdot c_0 \cdot z_3}{F_1 \cdot (z_0 - z_4) + k \cdot V \cdot c_0 (z_3 - z_4)} \quad (4.18)$$

Dividing the numerator and denominator by F_0, the following dimensionless form is obtained:

$$f_4 = \frac{f_1 \cdot Da \cdot z_3}{(z_0 - z_4) \cdot f_1 + Da \cdot (z_3 - z_4)} \quad (4.19)$$

It is worth mentioning that when the control structure presented in Figure 4.4 is used, there are no feasibility constraints similar to Eq. (4.13). Therefore, an operating point always exists.

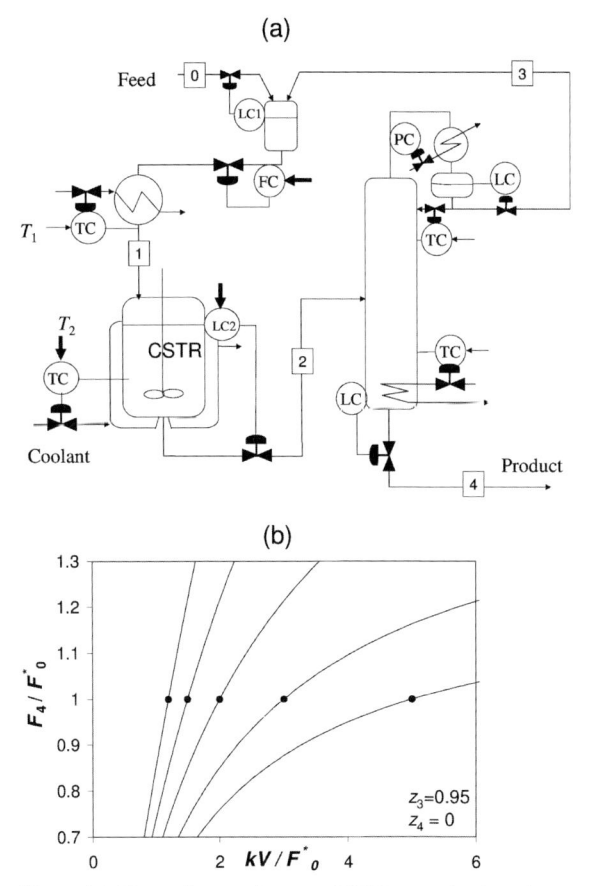

Figure 4.4 First-order reaction in a CSTR/separation/recycle system.
(a) Control structure fixing reactor-inlet flow.
(b) Production rate F_4 versus the reaction conditions kV.

Production-rate changes can be implemented in several ways, depending on the reactor design:

(a) If the reactor volume is large, production-rate changes are achieved by increasing the reactor-inlet flow F_1. The reactor is capable of converting the additional amount of reactant; therefore the increase of the recycle rate is marginal. The net effect is a reduction of the reactant inventory. This is seen by the level controller LC1 (reactant inventory), which increases the feed rate. The results presented in Figure 4.3(b) show that this mode of operation is feasible whenever $Da^* = (k^* V^* c_0)/F^*_0 > 3$, as changes of ±30% in F_4 can be achieved with reasonable changes of F_1. Note that a large reactor is often optimal from an economic viewpoint, because it minimizes the cost of separation.

(b) When the reactor volume is small, production-rate changes are achieved by a) increasing the reactor holdup V through setpoint of reactor level controller

LC2) or b) the reaction temperature T_2 (if allowed), which affects the rate constant k. Figure 4.4(b) shows that, whenever $Da^* < 2$, this strategy requires moderate changes of the manipulated variables. We note that there are many situations when a large reactor cannot be used. For example, in the polyethylene production, the high operating pressure requires a small reactor. In addition, the molecular weight distribution of the final product and difficult heat transfer due to the gel effect favor low conversion, and thus small reactors.

Note that one could combine the two modes of operation, for example changing simultaneously the reactor-inlet flow rate and the reactor holdup or temperature.

The conclusion of this analysis is that plantwide control structures that use feedback to control reactant inventory do not show the snowball effect. These structures can be applied for both large and small reactors, the difference being the variable manipulated for achieving production-rate changes.

4.3.4
Plug-Flow Reactor

In this section we replace the CSTR by a plug-flow reactor and consider the conventional control structure. Section 4.5 presents the model equations. The energy balance equations can be discarded when the heat of reaction is negligible or when a control loop keeps constant reactor temperature manipulating, for example, the coolant flow rate. The model of the reactor/separation/recycle system can be solved analytically to obtain (the reader is encouraged to prove this):

$$\frac{Da \cdot X \cdot z_3}{(z_0 - z_4) - X \cdot (1 - z_3)} = \ln \frac{1}{1-X} \qquad (4.20)$$

where the per-pass conversion is defined as $X = 1 - z_2/z_1$

Similarly to the CSTR, case Eq. (4.14), $X = 0$ is a trivial solution satisfying the model equations irrespective of the Da value. However, it is unfeasible, corresponding to infinite flow rates.

It can be shown that the nontrivial solution is feasible (positive flow rates) if, and only if:

$$\frac{z_0 - z_4}{z_3} < Da < \ln \frac{1}{z_4} \qquad (4.21)$$

Note that first inequality of Eq. (4.21) is identical to the feasibility constraint (4.13) characterizing the reactor/separation/recycle system involving a CSTR.

For both the CSTR and PFR systems, at $Da_T = (z_0 - z_4)/z_3$ two different manifolds of steady states cross each other, in the combined space of state variables and parameters. According to the bifurcation theory, this is a *transcritical* bifurcation point. Here, an exchange of stability takes place: for $Da < Da_T$, the trivial solution

$X = 0$ is stable; this state loses stability at $Da = Da_T$; simultaneously, the nontrivial solution gets physically meaningful values and gains stability.

4.4
Processes Involving Two Reactants

In this section we analyze processes involving the second-order $A + B \rightarrow 2P$ reaction. Such processes have been studied, among others, by Luyben and Tyreus [10]. It has been noticed [11] that certain control structures lead to state multiplicity and instability. Here, we use dimensionless models to derive general feasibility and stability conditions. The reader is encouraged to check carefully the balance equations, writing them first in the dimensional form, and then deriving the dimensionless versions. To solve these equations, software such as Maple or the symbolic toolbox of Matlab can be used.

4.4.1
Two Recycles

If the reactants A and B are lighter and heavier, respectively, than the product P, the flowsheet involves two distillation columns and two recycle streams (Figure 4.5). The control structure includes loops for reactor level and temperature, as well as for distillation columns top and bottom purity. The following dimensionless equations can be derived:

$$f_0^A + f_3 \cdot z_{A,3} - f_2 \cdot z_{A,2} - Da \cdot z_{A,2} \cdot z_{B,2} = 0 \tag{4.22}$$

$$f_0^B + f_5 \cdot z_{B,5} - f_2 \cdot z_{B,2} - Da \cdot z_{A,2} \cdot z_{B,2} = 0 \tag{4.23}$$

$$f_0^A + f_0^B + f_3 + f_5 - f_2 = 0 \tag{4.24}$$

$$f_2 \cdot z_{A,2} - f_3 \cdot z_{A,3} = 0 \tag{4.25}$$

$$f_2 \cdot z_{B,2} - f_5 \cdot z_{B,5} = 0 \tag{4.26}$$

$$f_0^A + f_3 = f_{Rf,A} \tag{4.27}$$

$$f_0^B + f_5 = f_{Rf,B} \tag{4.28}$$

The model contains 7 equations and 12 variables. Fixing the reactor holdup and reaction temperature fixes the Da number. $z_{A,3}$ and $z_{B,5}$ are given by the separation performance. Two additional specifications are needed. The plantwide control that we recommend (Figure 4.5) does not rely on self-regulation. The reactor-inlet flow rate of both reactants $f_{Rf,A}$ and $f_{Rf,B}$ are fixed. Fresh feed rates f_0^A and f_0^B are used to control inventories at some locations in the plant. Note that an arbitrary flow rate can be used as reference in definition of the dimensionless quantities.

If this control structure is used, a unique steady state exists. Therefore, as long as each unit is stable or stabilized by local control, the entire plant is stable. This

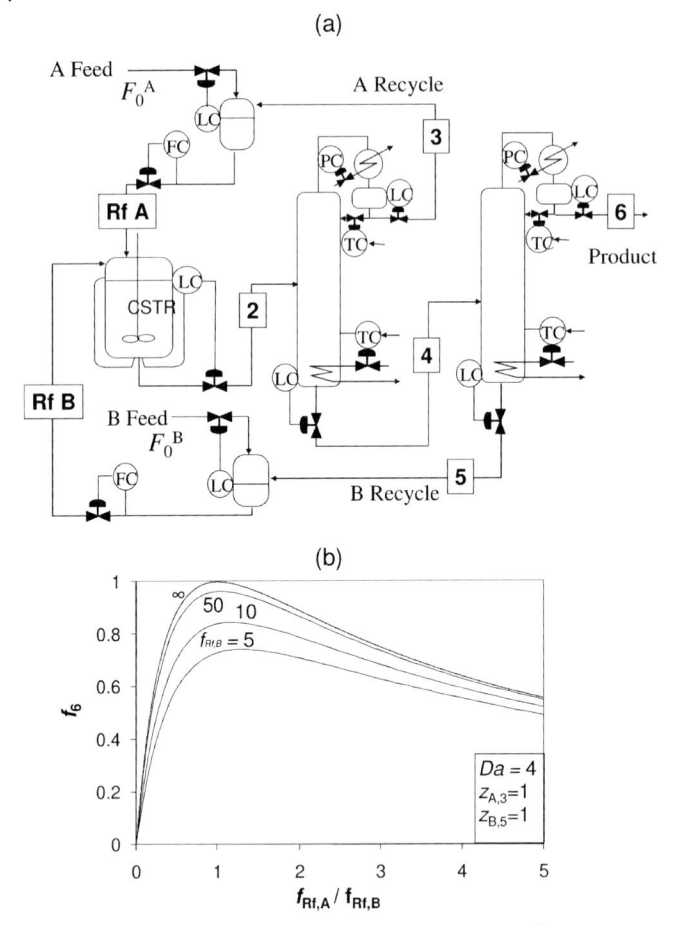

Figure 4.5 Second-order reaction in CSTR/separation/recycle system.
(a) Control structure fixing reactor-inlet flow rates.
(b) Production rate f_6 versus reactor-inlet flow rates $f_{Rf,A}$, $f_{Rf,B}$.

overcomes the disadvantage of setting the production rate in an indirect manner. One reactor-inlet flow rate (for example $f_{Rf,B}$) can be fixed, and production changes can be achieved by manipulating the setpoint of the other flow controller ($f_{Rf,A}$). At high excess of B, the production rate is linear and sensitive, as shown in Figure 4.5. As a result, reactor-inlet flow rate of A can be used to manipulate the production. For fixed, finite $f_{Rf,B}$ and Da = 4, the maximum throughput $f_{6,max} = 1$ is obtained for:

$$f_{Rf,A} = -1 + \sqrt{1 + 6f_{Rf,B} + f_{Rf,B}^2} \tag{4.29}$$

Similarly, if the nominal operating point is at high excess of A, the reactor-inlet flow rate of B can be used to manipulate the production.

4.4.2
One Recycle

If both reactants A and B are lighter (or heavier) than the product P, the flowsheet involves only one distillation column and one recycle stream. Therefore, the strategy presented in Figure 4.5 cannot be applied. If no composition measurements are used, the plantwide control has to rely on self-regulation, as shown in Figure 4.6. One fresh reactant feed (A) is on flow control. This flow rate is used as reference value in the dimensionless model, therefore $f_0^A = 1$. The flow rate of the stream containing the recycled reactants and the fresh B is fixed at f_{Rf}. The second feed, f_0^B, controls the inventory at some location in the plant, for example the reflux drum of the distillation column or an intermediate storage tank. Separation performance is given by $z_{P,3}$.

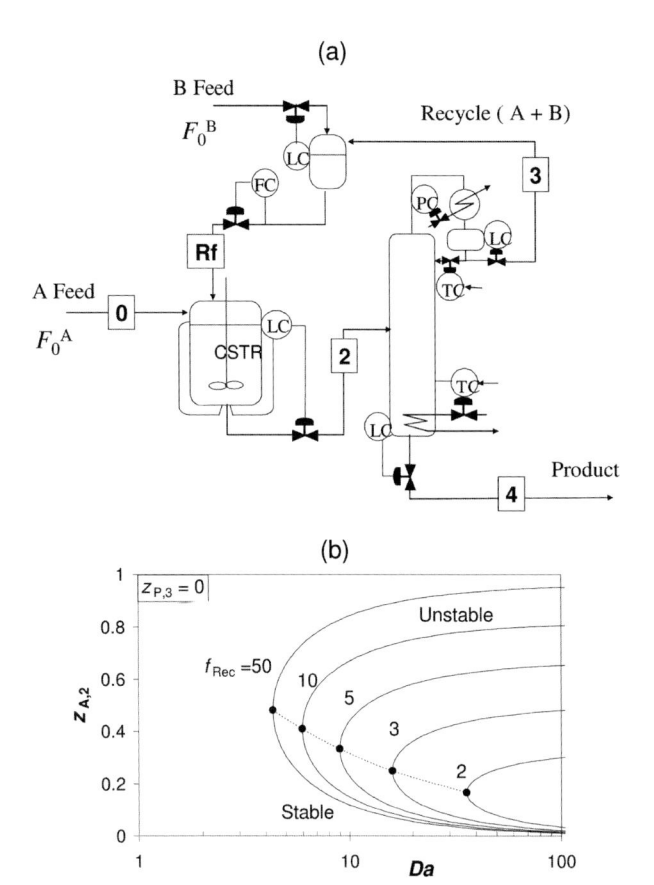

Figure 4.6 Second-order reaction in CSTR/separator/recycle system.
(a) Control structure relying on self-regulation.
(b) Reactor-outlet concentration $z_{A,2}$ versus Damkohler number diagram showing multiple steady states.

The steady-state model consists of Eqs. (4.22) to (4.26) and:

$$z_{A,3} + z_{B,3} + z_{P,3} = 1 \tag{4.30}$$

$$f_0^B + f_3 = f_{Rf} \tag{4.31}$$

The reader is asked to confirm that all degrees of freedom are fulfilled and the model can be solved. The steady-state model has a closed-form solution. Here we only present the $Da - z_{A,2}$ dependence in Eq. (4.32) and Figure 4.6(b):

$$Da = \frac{1 + f_{Rf}}{z_{A,2}} \cdot \frac{1}{(f_{Rf} - 1)(1 - z_{P,3}) + z_{A,2}(1 + f_{Rf})} \tag{4.32}$$

For fixed Da either zero or two solutions exist. When two solutions exist, the state of low-conversion (high-reactant concentration) is unstable. The instability of this branch can be proven by comparing the slopes of the reactant feed and consumption rate. This technique will be exemplified in the next section.

The turning point of the $Da - z_{A2}$ diagram, representing the feasibility and stability boundary, is given by:

$$(Da, z_{A,2})^F = \left(\frac{4}{(1 - z_{P,3})^2} \frac{(f_{Rf} + 1)^2}{(f_{Rf} - 1)^2}, \frac{(1 - z_{P,3})}{2} \frac{f_{Rf} - 1}{f_{Rf} + 1} \right) \tag{4.33}$$

We leave demonstration of relationships (4.32) and (4.33) as an exercise for the reader.

4.5
The Effect of the Heat of Reaction

In this section, we will focus on exothermal reactions taking place in reactor/separation/recycle systems (Figure 4.7). We will choose PFRs, because the temperature is varying along the axial coordinate, which makes isothermal operation difficult. We will consider the conventional control structure and will show that the heat effects can lead to a multiplicity of states. Because these are undesired, the analysis presented here is a strong argument favoring control structures fixing the reactor-inlet flow rate.

4.5.1
One-Reactant, First-Order Reaction in PFR/Separation/Recycle Systems

Assuming pure reactant feed $z_0 = 1$, plug flow, constant physical properties and one-reactant first-order reaction, the following dimensionless model can be derived:

4.5 The Effect of the Heat of Reaction

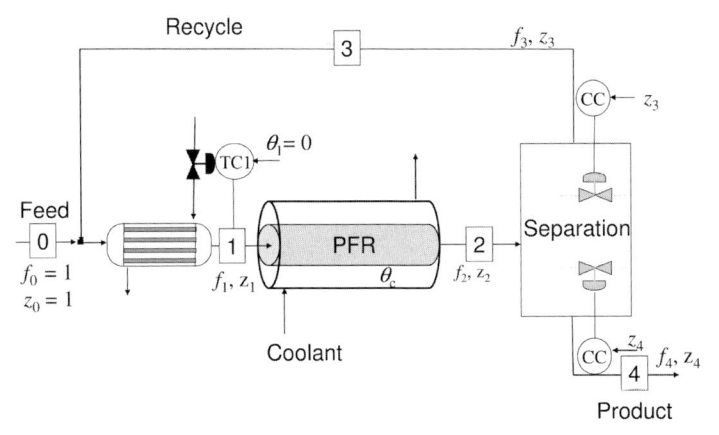

Figure 4.7 PFR/separation/recycle system carrying on a first-order exothermal reaction.

Reactor:

$$\frac{dx}{d\xi} = \frac{Da}{1+f_3} \cdot (1-x) \cdot \exp\left(\frac{\gamma\theta}{1+\theta}\right) \tag{4.34}$$

$$\frac{d\theta}{d\xi} = \frac{Da}{1+f_3}\left(B \cdot z_1 \cdot (1-x) \cdot \exp\left(\frac{\gamma\theta}{1+\theta}\right) - \beta \cdot (\theta - \theta_c)\right) \tag{4.35}$$

$$x(0) = 0; \quad \theta(0) = 0 \tag{4.36}$$

Input-output mass balance:

$$(1 - z_4) = (1 + f_3 z_3) \cdot x(1) \tag{4.37}$$

Mixing point:

$$1 + f_3 \cdot z_3 = z_1 \cdot (1 + f_3) \tag{4.38}$$

The dimensionless variables and parameters are: axial coordinate $0 \leq \xi \leq 1$, conversion $x(\xi)$, temperature $\theta(\xi)$, recycle flow rate f_3 and reactor-inlet concentration z_1; plant Damköhler number Da, activation energy γ, adiabatic temperature rise B, heat-transfer capacity β, coolant temperature θ_c, concentration of recycle and product streams z_3, z_4. For convenience, $X \equiv x(1)$ will stand for conversion at reactor outlet.

The dimensionless variables and parameters are defined as:

$$B = \frac{(-\Delta H)c_0}{\rho c_p T_{\text{ref}}}; \quad Da = k(T_{\text{ref}})\frac{T}{F_0}c_0; \quad \beta = \frac{UA}{V\rho c_p k(T_{\text{ref}})c_0^{n-1}}; \quad \gamma = \frac{E_A}{RT_{\text{ref}}}$$

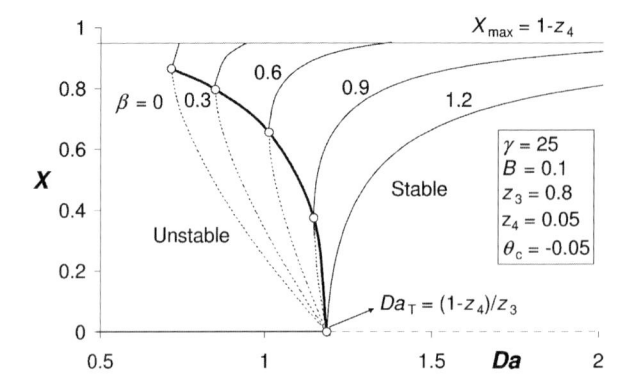

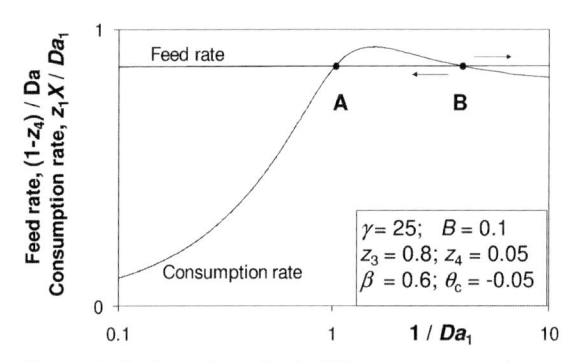

Figure 4.8 Exothermal reaction in PFR–separation recycle.
(a) Bifurcation diagrams showing multiple steady states.
(b) Instability of the low-conversion state can be demonstrated by steady-state arguments.

Because in many practical situations T_1 is kept constant by a heat exchanger placed upstream of the reactor (as in Figure 4.7), the reactor-inlet temperature is used as a reference value, $T_{\text{ref}} = T_1$.

The model equations can be solved by a shooting technique: start with an initial guess X for the conversion $x(1)$, calculate the recycle flow rate f_3 from Eq. (4.37) and reactor-inlet concentration z_1 from Eq. (4.38), integrate the PFR Eqs. (4.34) and (4.35), check and update the guess X. This implies that it is theoretically possible to reduce the model to one equation with one variable:

$$g(X, \text{Da}, \gamma, B, \beta, \theta_c, z_3, z_4) = X - x(1) \tag{4.39}$$

Figure 4.8(a) presents the X versus Da dependence, obtained for different values of the heat-transfer capacity β, and fixed values of the remaining model parameters. The entire diagram (including the unfeasible domain) contains one turning point. For large β, the turning point is located in the unfeasible region $X < 0$ (not shown). As β decreases, the turning point moves to larger conversion and finally enters the feasible region, leading to state multiplicity.

When two steady states exist, the low-conversion one is unstable. This can be demonstrated showing that *a slope condition* is not fulfilled:

- The amount of reactant consumed by the reactor is given by $F_1 z_1 X$, where F_1 is the (dimensional) molar flow at reactor inlet.

- The quantity $F_1 z_1 X$ is represented versus the reactor-inlet flow rate F_1 in Figure 4.8(b). The dimensionless variables $F_1/(kVc_0) = 1/\mathrm{Da}_1$ and $F_1 z_1 X/(kVc_0) = z_1 X/\mathrm{Da}_1$ are used as coordinates, where Da_1 is defined using the reactor-inlet flow rate as reference.

- In the reactor/separation/recycle system, the steady-state values of the reactor-inlet flow rate F_1 are the intersections of this curve with the horizontal line representing the net amount of reactant fed in the process. This is given, in a dimensionless form, by $F_0(1 - z_4)/(c_0 kV) = (1 - z_4)/\mathrm{Da}$.

Let us consider a small, positive deviation of the reactor-inlet flow rate, from the steady state B. At the right of point B, the amount of reactant fed in the process is larger than the amount of reactant consumed. Reactant accumulation occurs, leading to a further increase of the recycle and reactor-inlet flow rates; hence the steady state B is unstable. This is independent of the dynamics, because the proof is based only on steady-state considerations.

The turning point of the Da–X diagram represents a feasibility boundary. Knowledge of this point is important for two reasons. Firstly, an optimization procedure might suggest a small reactor, corresponding to an operating point close to the unfeasible region. Such a system might suffer from serious operability problems. If the reaction kinetics was overestimated, or the feed flow rate deviates from the nominal design value, the operating point falls at the left of the turning point of the Da–X map, in the region where no feasible steady state exists. In this case, reactant accumulation occurs, and the plant has to be shut down. Secondly, stability is a requirement that must be fulfilled by any design. Because the upper branch of the Da–X diagram is the stable one, the coolant temperature θ_c and heat-transfer capacity β must be chosen such that the required conversion is larger than the conversion at the turning point. The reader is referred to [12] for the equation defining the turning points and a more detailed discussion of the results.

Here, we only note several general conclusions. Firstly, an unstable operating point is more likely when low conversion is required. Instability can be removed by lowering the coolant temperature θ_c, and increasing the heat-transfer capacity β. If this is not achievable, for example due to restricted heat-transfer area or heat-transfer coefficient, increasing the reactor-inlet temperature T_1 is an option. This has the favorable effect of decreasing the dimensionless parameters γ and B. These results explain the conclusions of previous case studies. For example, high reactor-inlet temperature was found to be beneficial from a dynamic point of view [13, 14]. Luyben [15] concluded that controllability worsens as specific reaction rates, activation energies, and tube diameters increased. Designs with large reactor

4.6
Example – Toluene Hydrodealkylation Process

The hydrodelakylation (HDA) plant [1] will demonstrate that including the reactor/separation/recycle step in the hierarchical design procedure allows considerations of plantwide control issues. We will show that for a reasonable control structure, the plant shows multiple steady states. By analyzing the reactor/separation/recycle structure, the designer is able to choose a stable operating point, which is robust with respect to disturbances, throughput changes, raw materials variability or uncertain design parameters. Moreover, we will illustrate the dangerous nonlinear phenomena that can occur in recycle systems.

In the HDA process, benzene is produced by toluene hydrodealkylation, a mildly exothermic reaction ($\Delta^r H_{950\,K} = -50450\,J/mol$) with moderate activation energy ($E_a = 217\,600\,J/mol$):

$$C_6H_5-CH_3 + H_2 \rightarrow C_6H_6 + CH_4$$

In this study, the following reaction rate expression is used:

$$r = k \cdot c_T \cdot c_H^{0.5}$$
$$k = 5.943 \times 10^{14} \cdot \exp\left(-\frac{34138\,K}{T}\right) \quad (m^3/kmol)^{1/2}\,s^{-1} \tag{4.40}$$

The reactor/separation/recycle structure of the plant is presented in Figure 4.9. Hydrogen and toluene, fresh and recycled, are mixed and fed to a tubular reactor. The reaction effluent, containing mainly toluene, hydrogen, benzene and methane is sent to a gas–liquid separator. From the liquid stream, benzene and eventual

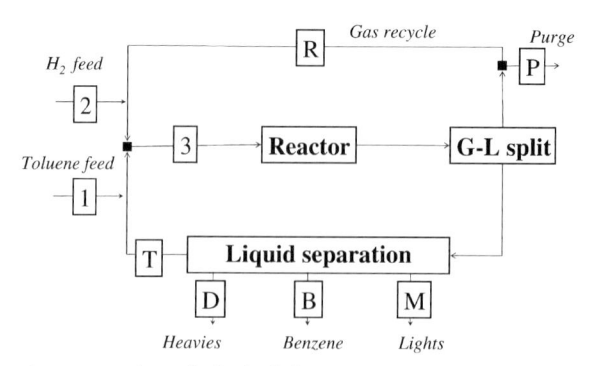

Figure 4.9 Toluene hydrodealkylation: reactor/separation/recycle structure of the plant.

light and heavy byproducts are removed, while toluene is recycled. The vapor stream contains both hydrogen and methane. Here, we choose not to separate these components, but to use a purge for methane removal.

By neglecting the secondary reactions and assuming perfect separation, the following equations can be derived:

$$F_B = F_1 \tag{4.41}$$

$$F_P = F_2 \tag{4.42}$$

$$F_1 = F_2 \cdot (y_{H,2} - y_{H,P}) \tag{4.43}$$

$$\frac{y_{T,3}}{y_{H,3}} = \frac{F_1/X}{F_R \cdot y_{H,P} + F_2 \cdot y_{H,2}} \tag{4.44}$$

$$F_T = F_1 \cdot \frac{1-X}{X} \tag{4.45}$$

F_j and $y_{k,j}$ denote the mole flow rate of stream j and the molar fraction of the k component (H–hydrogen, T–toluene) in stream j, respectively; X is the reaction conversion.

Equation (4.41) says that the production of benzene is equal to the amount of toluene fed to the process. Equation (4.42) expresses the fact that the number of moles of "gas" components does not change due to the reaction: for each mole of hydrogen that is consumed, one mole of methane is formed. Equation (4.43) reflects the reaction stoichiometry: in moles, the reacted toluene must be equal to the feed hydrogen minus hydrogen leaving the plant with the purge. Equation (4.44) gives the ratio toluene/hydrogen at reactor inlet, while Eq. (4.45) gives the amount of recycled toluene.

In addition, the conversion X depends on the reactor volume according to the characteristic equations of the adiabatic tubular reactor.

$$X = X(V) \tag{4.46}$$

Equations (4.37) to (4.42) involve eleven variables; therefore five degrees of freedom must be specified. We assume constant purity of the fresh hydrogen, $y_{H,2} = 0.95$. The control structure fixes the fresh toluene flow rate $F_1 = 120$ kmol/h and hydrogen/toluene ratio at reactor inlet, $y_{T,3}/y_{H,3} = 1/5$. Specifying two additional variables, for example reactor volume V and gas recycle flow rate F_R, the mass-balance equations can be solved for six unknowns: F_2, F_B, F_T, F_P, X and $y_{H,P}$. This is left as an exercise for the reader.

Alternatively, one can use flowsheeting software, such as for example Aspen Plus. Setting the reactor/separation/recycle structure is straightforward by using an adiabatic plug-flow reactor and black-box separation models. Figure 4.10 presents the conversion X vs. reactor volume V, obtained for different values of the gas recycle flow rate F_R. From Figure 4.10 is obvious that the operating point A is a good candidate: it is located on the stable branch, far from the turning point. In

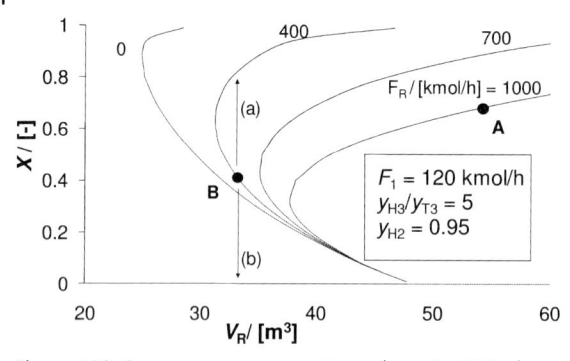

Figure 4.10 Conversion versus reaction volume in HDA plant.

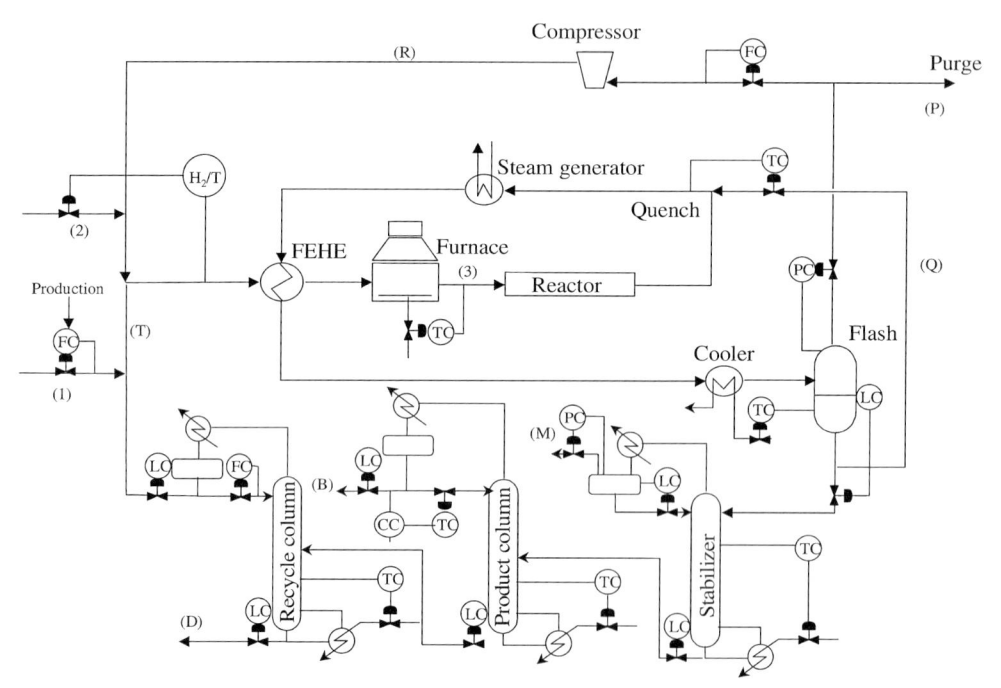

Figure 4.11 Plantwide control of the toluene hydrodealkylation plant.

contrast, design B is located on the unstable branch, and close to the unfeasibility region.

Figure 4.11 present the complete flowsheet together with the control structure. The reaction takes place in an adiabatic tubular reactor. To avoid fouling, the temperature of the reactor-outlet stream is reduced by quenching. A feed-effluent heat exchanger (FEHE) recovers part of the reaction heat. For control purposes, a furnace is included in the loop. The heat-integrated reaction system is stabilized

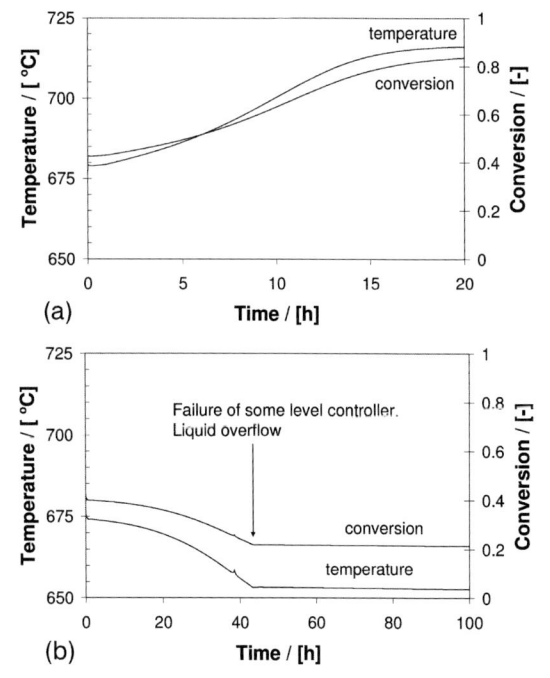

Figure 4.12 HDA plant: Instability of the low-conversion branch.

by controlling the reactor-inlet temperature and the temperature of the quenched stream. After cooling, the flash separation of the reactor effluent leads to two recycles. From the liquid recycle, lights, product and heavies are separated by distillation. Purging a fraction of the gas recycle is necessary to avoid methane accumulation. The units in the plant were designed and rigorous simulation was performed using Aspen Plus. The simulation was then exported to Aspen Dynamics, where control loops were provided and tuned.

For design A, no operating problems are encountered, as expected. The plant is robust in the face of various disturbances (e.g. ±25% production change) or uncertain design parameters.

In contrast, design B suffers from serious operability problems. Starting from steady state, reaction ignition occurs (Figure 4.12a). Reaction extinction takes place when the reactor-inlet temperature deviates by only −2 °C from the design value (Figure 4.12b). The plant moves towards the trivial ($X = 0$) steady state. All flow rates increase and one of the levels becomes uncontrollable.

An alternative control strategy fixes the reactor-inlet toluene flow rate [16]. Fresh toluene is fed into the condenser drum of the last distillation column, on level control. Production-rate changes can be achieved by changing the setpoint of the toluene reactor-inlet flow, or the setpoint of the reactor-inlet temperature controller. When this control structure is used, the whole range of conversion becomes stable. Drawing of this control structure is left as an exercise to the reader.

4.7
Conclusions

The "reactor/separation/recycle" level allows plantwide control issues to be included in the hierarchical approach at an early level of design. In most cases, the Separation is considered as a black-box for which targets are set, for example as species recovery or product purities. A stoichiometric or a kinetic reactor can be used. In the first case, plantwide control structures can only be proposed, while in the second case these can be also evaluated.

The plantwide control deals, mainly, with the mass balance of the species involved in the process. The species inventory can be maintained based on two different principles, namely self-regulation and feedback control. Control structures based on self-regulation set the flow rates of fresh reactants at values determined by the production rate and stoichiometry. Control of inventory by feedback consists of fixing one flow rate in each recycle loop, evaluating the inventory by means of concentration or level measurements, and reducing the deviations from the setpoint by change of the feed rate of fresh reactants.

Equation (4.17) gives the condition for feasibility of control structures based on self-regulation. The physical explanation is that sufficient reactions should be available in order to balance, for each component, the fresh feed with the overall consumption rate. A second feasibility condition requires that the reactor should be large enough. High sensitivity of a recycle with respect to fresh feed, known as the "snowball effect", can be avoided by designing the system for high per-pass conversion. For multireactant/multireaction systems or when heat effects are involved, state multiplicity occurs. The instability of the low-conversion branch restricts the selection of the operating point.

When feedback control is applied, the recommended location for fixing flow rates is the reactor inlet. This strategy decouples the reactor from the rest of the plant. Changing the production rate can be achieved by modifying:

(a) The setpoint of the reactor temperature, pressure or level controller. The method is appropriate for small reactors, operating at low per-pass conversion.

(b) The setpoint of the reactor-inlet flow controller. The method is recommended for large reactors, when reactant conversion is high.

When several reactants are involved, the two strategies may be combined. The suggested approach may be summarized as follows: design the plant for high conversion of the reference reactant, set its feed on flow control, fix the recycle flows, and adjust the makeup of the other reactants to keep the reactor-inlet flows or concentrations at constant values.

From all these results, we conclude that plantwide control strategies could take advantage from self-regulation. However, undesired nonlinear phenomena may occur. Therefore, we recommend thorough steady-state sensitivity and dynamic stability studies, before attempting a practical implementation. The reactor/separation/recycle level is the first moment during the design process when these studies can be performed.

References

1. Douglas, J.M., Conceptual Design of Chemical Processes, McGraw-Hill, New York (1988)
2. Luyben, W.L., B. Tyreus, M.L. Luyben, Plantwide Process Control, McGraw-Hill, New York, USA (1999)
3. Luyben, W.L., Snowball effects in reactor/separator processes with recycle, Ind. Eng. Chem. Res. 33, 299–305 (1994)
4. Downs, J., Distillation control in a plantwide control environment, in Practical Distillation Control, van Nostrand Rheinhold, New York (1992)
5. Groenendijk, A.J., A.C. Dimian, P. Iedema, Systems approach for evaluating dynamics and plantwide control of complex plants, AIChE J., 41, 133 (2000)
6. Dimian, A.C., A.J. Groenendijk, P. Iedema, Recycle interaction effects on the control of impurities in a complex plant, Ind. Eng. Chem. Res., 40, 5784 (2001)
7. Luyben, W.L., Design and control degrees of freedom, Ind. Eng. Chem. Res. 35, 2204–2212 (1996)
8. Bildea, C.S., A.C. Dimian, Fixing flow rates in recycle systems: Luyben's rule revisited, Ind. Eng. Chem. Res., 42, 4578–4585 (2003)
9. Bildea, C.S., A.C. Dimian, P.D. Iedema, Nonlinear behavior of reactor/separator/recycle systems, Comput. Chem. Eng., 24, 209–215 (2000)
10. Luyben, M.L., B.D. Tyreus, W.L. Luyben, Analysis of control structures for reaction/separation/recycle processes with second-order reactions, Ind. Eng. Chem. Res., 35, 758 (1996)
11. Luyben, W.L., M.L. Luyben, Essentials of Process Control. McGraw-Hill, New York (1997)
12. Bildea, C.S., S. Cruz, A.C. Dimian, P. Iedema, Design of tubular reactors in recycle systems, Comput. Chem. Eng., 28, 63–72 (2004)
13. Reyes, F., W.L. Luyben, Design and control of a gas-phase adiabatic tubular reactor process with liquid recycle, Ind. Eng. Chem. Res., 40, 3762–3774 (2001)
14. Reyes, F., W.L. Luyben, Extensions of the simultaneous design of gas-phase adiabatic tubular reactor systems with gas recycle, Ind. Eng. Chem. Res., 40, 635–647 (2001)
15. Luyben, W.L., Effect of design and kinetic parameters on the control of cooled tubular reactor systems, Ind. Eng. Chem. Res., 40, 3623–3633 (2001)
16. Luyben, M.L., B.D. Tyreus, W.L. Luyben, Plantwide control design procedure. AIChE J., 43, 3161–3174 (1997)

5
Phenol Hydrogenation to Cyclohexanone

5.1
Basis of Design

This introductory case study presents the key features of a conceptual process design by using the systematic methods presented in the previous chapters. The selected process is the manufacture of cyclohexanone, a key intermediate in the production of ε-caprolactam and adipic acid, which are basic materials for nylon-type polymers.

5.1.1
Project Definition

The nominal plant capacity is 120 000 metric tonnes per year cyclohexanone of oxime purity. In addition, the plant should be capable of switching on the production of KA-oil, a cyclohexanone/cyclohexanol mixture, used as an intermediate for the manufacturing of adipic acid. The plant will be located on an integrated industrial site with moderately continental climate. Optimal energy consumption is aimed by taking appropriate heat-integration measures. The amount of organic waste should be kept below 0.5% from the production rate. No discharge of wastewater to the environment is allowed, namely containing phenol and other organics, as well as no release of toxic emissions.

Cyclohexanone is a colorless high-boiler liquid with the normal boiling point at 156.7 °C. Some quality specifications are given in the Table 5.1.

Cyclohexanone can be characterized as a volatile, combustible and toxic liquid. Health and safety precautions should be observed, as stipulated in US EPA and OSHA recommendations, as well as in the ESIS-ECB European database. The storage and manipulation of both phenol and hydrogen cause hazard and safety problems. Phenol is highly corrosive, alone or in water solution. Hydrogen is particularly dangerous by the risk of fire and explosion. For these reasons, it is highly desirable to avoid large storage capacities, long transport pipes, as well as to minimize the inventory of units handling hazardous materials.

For more generality, this case study will develop a process capable of producing cyclohexanone and cyclohexanol in any desired proportion. We will voluntarily

Chemical Process Design: Computer-Aided Case Studies. Alexandre C. Dimian and Costin Sorin Bildea
Copyright © 2008 WILEY-VCH Verlag GmbH & Co. KGaA, Weinheim
ISBN: 978-3-527-31403-4

Table 5.1 Quality specifications for two typical products [1, 2].

	High purity	KA-oil
Color	Colorless	Colorless
Ketone content, % min	99.5	89
Cyclohexanol, %	400 ppm	10
Phenol	None	Acidity 0.03% max
Water, max	200 ppm	200 ppm
Total organic impurities[a]	100 ppm	
Distillation range, 95% at 101.3 kPa	152–157 °C	

a) 2-heptanone 2 ppm, cyclohexenyl-cyclohexanone 1 ppm.

complicate the approach in order to illustrate more generic features, by enlarging the number of design decisions and generating more alternatives.

5.1.2
Chemical Routes

The main industrial routes for cyclohexanone manufacture have as starting points cyclohexane and phenol, by oxidation and hydrogenation, respectively. Another interesting method is based on the hydration of cyclohexene obtained by selective hydrogenation of benzene. The intermediate cyclohexanol is further dehydrogenated or separated if desired.

The process based on phenol hydrogenation considered here can be described by the following overall stoichiometric equation:

$$C_6H_5\text{-}OH + {}_xH_2 = {}_yC_6H_{10}O + {}_zC_6H_{11}\text{-}OH \tag{5.1}$$

with $y + z = 1$ and $x = 2y + 3z$. If the byproducts are neglected, the molar ratio x gives directly the yield. For example, producing KA-oil with 90% mol. cyclohexanone requires an effective molar ratio hydrogen/phenol $x = 2.1$.

The hydrogenation of phenol can take place either in vapor or in liquid phase. Both processes today employ palladium-based catalyst, but with different supports and activators.

In vapor-phase phenol hydrogenation the operating conditions are usually temperatures of 140–170 °C and a pressure slightly above atmospheric [1, 11]. Older processes based on nickel-type catalyst have two distinct reaction steps, namely full hydrogenation to cyclohexanol followed by dehydrogenation. Modern processes based on palladium-type catalysts can achieve over 90% yield in cyclohexanone in a single reactor. The product is mainly the KA-oil, but more cyclohexanol may be supplied on demand. The suppression of the dehydrogenation reactor including expensive equipment for achieving high temperatures represents a significant economic advantage.

Liquid-phase hydrogenation of phenol operates at temperatures below the atmospheric boiling point, around 140–150 °C. High selectivity is claimed, over 99% at 90% conversion. In addition, the process needs less catalyst inventory and is inherently safe [2].

Today the main route for cyclohexanone manufacturing is liquid-phase oxidation of cyclohexane. The synthesis involves the formation of cyclohexyl-hydroperoxide, further converted to cyclohexanone, cyclohexanol and byproducts, as illustrated by the following scheme:

$$C_6H_{12} + O_2 = C_6H_{11}OOH \tag{5.2}$$

$$3C_6H_{11}OOH \rightarrow 2C_6H_{11}OH + C_6H_{10}O + H_2O + O_2 + \text{byproducts} \tag{5.3}$$

The reaction is conducted at 140–180 °C and 0.80–2 MPa in a series of CSTRs or in a single tower oxidizer. The reaction selectivity depends highly on the catalyst. For example, a cobalt-based soluble catalyst gives a ratio ketone/alcohol of about 3.5. In order to maximize the yield the conversion is kept low. Note that a more selective process based on boric acid was developed, but the oxidation agent is expensive and the technology rather complicated.

Although cyclohexane oxidation dominates the market, because of cheaper raw materials, the hydrogenation of phenol remains competitive, offering better selectivity with fewer environmental and safety problems. In addition, this process allows efficient valorization of phenol-rich wastes from coal industries. Recently built plants make use of this technology, as reported by the engineering group Aker-Kvaerner (www.kvaerner.com, 2004). The availability of low-price phenol is the most important element for profitability. Besides the well-known cumene process, a promising route is the selective oxidation of benzene with N_2O on iron-modified ZSM-5 catalyst [12]. In this way, the price of phenol may become independent of the market of acetone.

5.1.3
Physical Properties

Some physical properties of the main species are listed in Table 5.2. Bringing phenol in reaction conditions implies vaporization at low partial pressure. Vacuum is necessary for carrying out separations by distillation. Phenol forms azeotropes with both cyclohexanol and cyclohexanone. If unconverted phenol should be recycled this could affect the global yield by recycling desired product too. If water appears as a byproduct, it gives azeotropes with both cyclohexanone and cyclohexanol. Because these azeotropes are low boilers they can be removed easily by distillation.

5 Phenol Hydrogenation to Cyclohexanone

Table 5.2 Physical properties of main species [1–3].

	Phenol	C-hexanone	C-hexanol	Hydrogen
Mol. Weight	94.11	98.15	11.16	2
Melting point, °C	40.9	−47	25.15	–
Normal boiling point, °C	181.9	156.4	161.1	−253
Liquid density, g/l	0.933	0.9493	0.9455	–
Solubility in water g/100°g	Large	3.6	9.0	None
Water solubility, g/100 g		12.6	5.7	–
Heat of vaporization, kJ/mol	46.18	45.51	44.92	
Azeotropes with water °C, mol. fr. water	None	96.3/0.56	97.8/0.80	
Azeotropes with phenol °C, mol. fr. phenol	None	185.8/0.75	183.0/0.80	

5.2
Chemical Reaction Analysis

5.2.1
Chemical Reaction Network

Figure 5.1 illustrates the key reactions implied in the manufacturing of cyclohexanone by phenol hydrogenation. The reactions are of consecutive type, in which the desired product is an intermediate. Small amounts of cyclohexene might appear at higher temperature by cyclohexanol dehydration. Additional reactions can lead to heavies by polymerization or benzene and cyclohexane by disproportionation.

Table 5.3 shows the enthalpy and Gibbs free energy of formation of species at 289.13 K and 1 atm [3]. On this basis standard ΔH_f^0 and ΔG_f^0 can be determined, as shown in Table 5.4. The hydrogenation of phenol is highly exothermic, while the dehydrogenation of cyclohexanol is moderately endothermic. The conversion

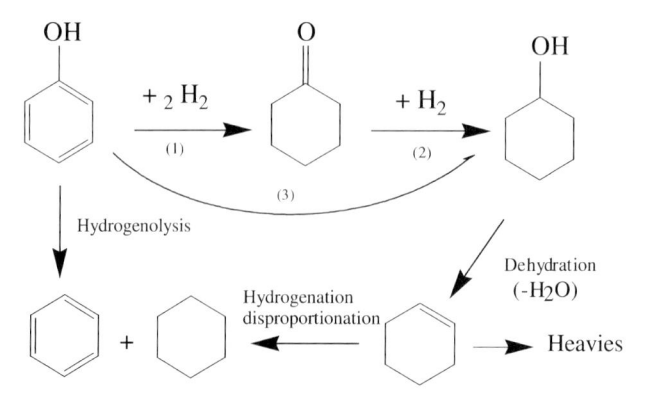

Figure 5.1 Reaction network by phenol hydrogenation.

Table 5.3 Standard enthalpy and Gibbs free energy of formation (kJ/mol).

	Phenol	C-hexanone	C-hexanol	C-hexene	C-hexane	Benzene	Water
ΔH_f^0	−96.4	−230.12	−294.55	−4.32	−123.1	82.88	−241.81
ΔG_f^0	−32.55	−90.87	−118.05	106.90	32.26	129.75	−228.42

Table 5.4 Enthalpy and Gibbs free energy of reactions in standard conditions (kJ/mol).

	Reaction	ΔH_R^0	ΔG_R^0	Remark
1	$C_6H_5\text{-}OH + 2H_2 \rightarrow C_6H_{10}=O$	−133.72	−58.32	Exothermic, reversible, favored by low temperature
2	$C_6H_{11}\text{-}OH \rightarrow C_6H_{10}=O + H_2$	64.43	27.18	Endothermic, reversible, favored by high temperature
3	$C_6H_{11}\text{-}OH \rightarrow C_6H_{10} + H_2O$	48.42	−3.47	Endothermic, reversible, favored by high temperature

in both reactions is constrained by the chemical equilibrium. The dehydration of cyclohexanol to cyclohexene has a small negative ΔG_f^0, suggesting that this reaction is very likely thermodynamically.

An important aspect is the formation of impurities. Light impurities may originate predominantly by cyclohexanol dehydrogenation to cyclohexene and water, but at higher temperatures cyclohexane and benzene could appear. The formation of cyclohexene is thermodynamically favored, although selective catalyst may solve this problem. With respect to heavy impurities, the most probable is the formation of cyclohexyl-cyclohexanone from cyclohexene and cyclohexanone. The amount could be kept below 0.5 wt% by operating below 160 °C [11]. Note also that the formation of heavies is more likely in the dehydrogenation reaction than in the hydrogenation step. The compilation of different sources shows that an overall yield in products of a minimum of 98% could be achieved in industrial conditions.

5.2.2
Chemical Equilibrium

5.2.2.1 Hydrogenation of Phenol

The study of chemical equilibrium can detect thermodynamic constraints on the achievable conversion and selectivity. In this section we make use of the Gibbs free-energy minimization method available in Aspen Plus™ [9]. We assume that both cyclohexanone and cyclohexanol are products. The curves in Figure 5.2 show the evolution of the phenol equilibrium conversion, yield and selectivity with the ratio hydrogen/phenol at temperatures of 180, 200, 220 °C and a pressure of 3 bar.

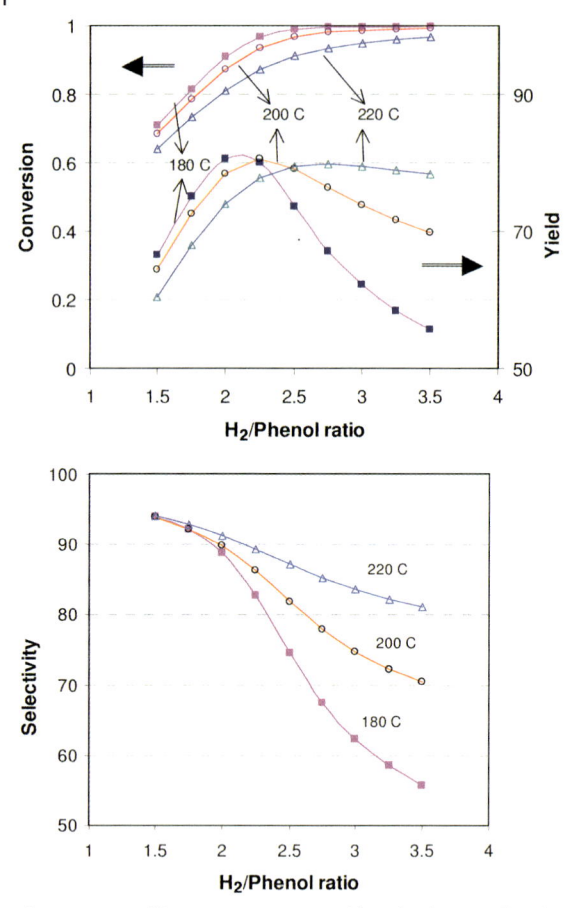

Figure 5.2 Equilibrium conversion, yield and selectivity for phenol hydrogenation.

It may be observed that lower temperature favors higher equilibrium conversion. This could exceed 99% for temperatures below 180 °C. A larger H_2/phenol ratio leads to higher equilibrium conversion, but to lower selectivity. Keeping the ratio of reactants low is more favorable for selectivity, although a ratio below two is technically not interesting. A maximum in selectivity is observed, which is sharper at lower temperature. As the temperature increases, this maximum becomes less sensitive because of the compensation effect of the dehydrogenation reaction. In conditions of thermodynamic equilibrium a maximum yield of 80% may be obtained at 180 °C and a H_2/phenol ratio slightly over 2, with both selectivity and conversion about 90%.

Supplementary constraints in the operating parameters arise from the VLE. At the reactor inlet the H_2/phenol ratio and temperature should be high enough to keep the mixture in the gaseous state. For example, at atmospheric pressure these should be above 3 and 150 °C, respectively. For this reason the dilution of the reaction mixture with an inert, such as methane or nitrogen, should lead to better selectivity by requiring a lower H_2/phenol ratio, but this operation mode would imply higher recycling costs.

In practice, because of kinetic effects, the product distribution offered by different catalysts may exhibit deviations from the chemical-equilibrium analysis. However, the examination of numerous papers dealing with catalysis issues confirms the main trends: operate at lower temperature and keep the H_2/phenol ratio as low as possible to promote the formation of cyclohexanone.

5.2.2.2 Dehydrogenation of Cyclohexanol

The computation of the chemical equilibrium at cyclohexanol dehydrogenation is apparently trivial. The following computation could change this viewpoint and illustrate the key role of the accuracy in thermochemical data. Table 5.5 gives the enthalpy and Gibbs free energy of formation in standard state at 1 atm and 298 K from the database of Aspen Plus and estimated values by the methods of Benson, Gani and Joback [9], as well as data retrieved from the monograph of Poling et al. [3] abbreviated here PPC. Note that Gani's method is the most used in Aspen Plus. In some cases the differences are larger than 5 kJ/mol, considered as the minimum acceptable, namely for cyclohexanone and cyclohexanol. As a consequence, there are large discrepancies in equilibrium composition, as can be seen in Figure 5.3. For example, at 600 K and 2 atm the equilibrium conversion by Benson, PPC and Gani methods are of 60%, 90% and 100%, respectively. To clarify this dilemma we should search for experimental data. Figure 5.3 also shows data calculated by Frenkel et al. [13] at 1 atm from thermodynamic functions obtained by statistical methods, the authors claiming good agreement with experiments. These data verify PPC at 2 bar, and not far from predictions at 1 bar. The conclusion is that in this case the data from PPC are better, although not very accurate. As a result,

Table 5.5 Enthalpy and Gibbs free energy of formation of chemical species.

	Phenol	C-hexanone	C-hexanol	C-hexene
ΔH_{f0} (kJ/mol)				
PPC	−96.4	−230.12	−294.55	−4.32
Aspen Plus	−96.48	−236.63	−304.16	−2.28
Gani	−97.83	−236.63	−304.16	−2.28
Benson	−93.35	−225.6	−288.47	−3.52
Joback	−96.48	−230.21	−265.08	−3.73
ΔG_{f0} (kJ/mol)				
Prausnitz et al.	−32.55	−90.87	−118.05	106.9
Aspen Plus	−34.26	−114.29	−126.08	111.00
Gani	−34.26	−114.29	−126.08	111.00
Benson	−31.16	−89.19	−121.04	108.96
Joback	−32.94	−90.79	−112.73	61.76

For H_2O the data are 241.81 and 2228.42 kJ/mol.

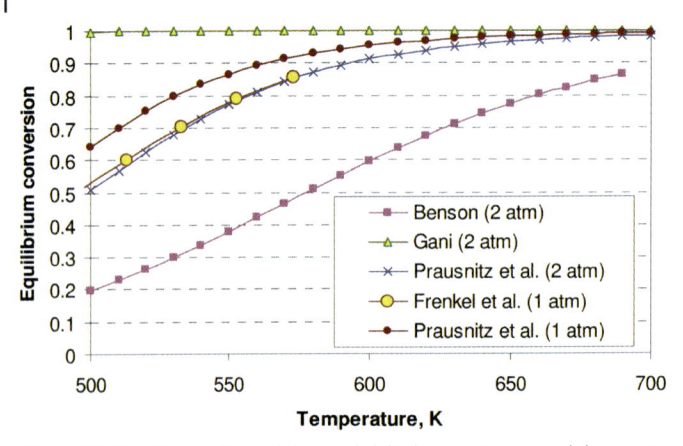

Figure 5.3 Equilibrium for cyclohexanol dehydrogenation to cyclohexanone.

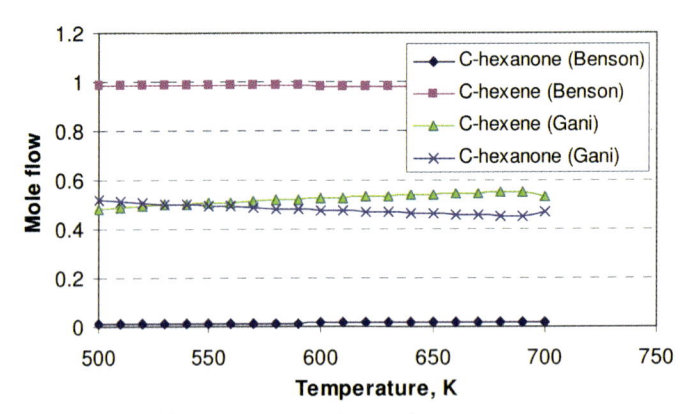

Figure 5.4 Equilibrium composition by simultaneous dehydrogenation and dehydration of cyclohexanol, at 2 atm and 1 mol initial reactant.

in order to achieve high conversion relatively high temperatures should be used, above 600 K.

In a second attempt we consider as a secondary reaction the dehydration of cyclohexanol to cyclohexene (Figure 5.4). The results are even more confusing. Benson's method predicts that the dehydration would be dominant at about 98% conversion, while following the Gani's method both dehydration and dehydrogenation are in competition each at around 50% conversion. The occurrence of dehydration seems intuitively reasonable, since water formation should give a lower Gibbs free energy. Moreover, it seems to confirm some technological reports indicating that traces of cyclohexene and water form even without a catalyst by simply passing cyclohexanol vapor through a heated tube at 400–450 °C [1]. Thus, even if the dehydration is favored thermodynamically, in practice the problem is solved by using a more selective catalyst for dehydrogenation, a lower temperature and a shorter contact time.

5.2.3
Kinetics

5.2.3.1 Phenol Hydrogenation to Cyclohexanone

The catalysts for hydrogenating phenol to cyclohexanone are based in general on group VIII metals, such as platinum or palladium, impregnated on different supports, such as alumina, zeolites, silica gel, active carbon, or more recently on carbon fibers. It is agreed that the palladium-based catalysts are more selective than those involving platinum [18]. The selectivity depends strongly on chemical formulation and physical morphology of the support, as illustrated by some performance data in terms of phenol conversion/selectivity: Pd/Al_2O_3 40%/45%; $Pd/CaO\text{-}Al_2O_3$ 98%/20%; Pd/zeolite 33%/44%; Pd membrane 75%/90% [16]. The cited reference claims a selectivity of 96% with a special Pd/MgO-type catalyst at 423 K, but for a phenol conversion less than 75%. Attaining high selectivity (over 95%) at high conversions (over 80%) remains a challenge.

In general, a reaction kinetics following a LHHW model is suitable, but the identification of parameters remains demanding. For some catalysts power-law models may be appropriate, for others not. For example, reaction orders identical with stoichiometric coefficients were suitable for Pd/Al_2O_3 doped with different metals. On the contrary, for Pd/MgO reaction orders with respect to phenol ranging from −0.5 to 0.5 were observed [17]. However, the bibliographic search was not able to find a quantitative kinetic model for Pd-type catalysts suitable for reactor design.

In order to get a qualitative idea, Table 5.6 presents kinetic constants for the consecutive/parallel reaction scheme given in Figure 5.1 obtained with a Pd-type catalyst (Park et al. [15]). Hydrogen was in large excess so that first-order kinetics may be assumed. Note that kinetic constants are reported as the mass load W/F_{phenol}, the phenol being produced by the evaporation of aqueous solutions. The nature of the support is the determinant for selectivity, but the activity is also affected. The most selective catalyst is Pd deposited on activated carbon (AC), but

Table 5.6 Rate constants for phenol hydrogenation following the parallel (k_1, k_3) and series reactions (k_1, k_2) at 423 K over Pd catalyst (after Park et al. [15]).

Catalyst	k_1 (mol g^{-1} h^{-1})	k_2 (mol g^{-1} h^{-1})	k_3 (mol g^{-1} h^{-1})	$k_1/(k_2 + k_3)$
Pd/SiO_2	2.5×10^{-3}	3.1×10^{-3}	1.1×10^{-3}	0.59
Pd/Ta_2O_5	5.3×10^{-3}	1.4×10^{-3}	6.9×10^{-4}	2.53
Pd/AC	2.1×10^{-3}	6.0×10^{-5}	6.1×10^{-5}	17.5
Pd/graphite	2.4×10^{-3}	2.0×10^{-3}	6.9×10^{-4}	0.89

this is also the slowest. As discussed, lower temperature favors the formation of cyclohexanone. Following these data, phenol conversion of about 75% can be achieved with selectivity better than 92% at $W/F = 128\,\text{g\,mol}^{-1}\text{h}$.

In this project, we make use of platinum-type catalyst on silica gel. Although this is less selective than more modern palladium-based catalysts, kinetic data are available in the literature as an LHHW model [2], better suited for flexible reactor design. The reaction rate equations are:

Hydrogenation of phenol to cyclohexanone:

$$-r_2 = k_1 K_A p_A (K_B p_B)^2 / (1 + K_A p_A + K_B p_B + K_C p_C + K_D p_D)^3 \quad (5.4)$$

Hydrogenation of cyclohexanone to cyclohexanol:

$$-r_2 = k_2 K_C p_C (K_B p_B) / (1 + K_A p_A + K_B p_B + K_C p_C + K_D p_D)^2 \quad (5.5)$$

In the above equations the symbols A, B, C, D designate phenol, hydrogen, cyclohexanone and cyclohexanol. Table 5.7 presents the model parameters at 423 K and 1 atm. The model takes into account the effect of the products on the reaction rate in the region of higher conversion. This feature is particularly useful for describing the product distribution in consecutive catalytic-type reactions. Note that the adsorption coefficients are different in the two reactions. Following the authors, this assumption, physically unlikely, was considered only to increase the accuracy of modeling.

The above kinetics is valid for small particles when the process rate is controlled by the chemical reaction at the surface. Diffusion effects should be accounted for large-size particles. Table 5.8 presents the calculation of the effectiveness factor [24] for spherical particles of 6 mm diameter and a mixture 1:3 phenol/hydrogen at 2 bar and 423 K. Other data are: BET internal surface $S = 40\,\text{m}^2/\text{g}$, mean pore radius 150 Å, catalyst density $\rho_p = 1000\,\text{kg/m}^3$, particle void fraction $\xi = 0.3$,

Table 5.7 Rate constants for LHHW kinetics by the hydrogenation of phenol [20].

Quantity	Reaction (1)	Reaction (2)
k_i (mol kg^{-1} h^{-1})[a]	877	28.7
K_A (atm^{-1})[a]	9.3	15
K_B (atm^{-1})[a]	1.1	0.91
K_C (atm^{-1})[a]	19	7.7
K_D (atm^{-1})[a]	8	2.6
A_i (mol kg^{-1} s^{-1})[b]	1234.28	40.40
E_i (kJ kmol^{-1})[b]	30 000	30 000

Observations: [a] kinetic data at 423 K and 1 bar; [b] Arrhenius-type expression for k_i.

Table 5.8 Calculation of catalyst effectiveness.

Quantity	Value	Observation
Reaction rate	$r_1 = 28.540$ mol/g cat h $= 0.00792$ kmol/m^3·s	Eq. (5.4)
Phenol concentration	$c_A = 0.0144$ kmol/m^3	$c_A = Px_A/RT$
First-order constant	$k_1 = 0.55$ s^{-1}	$k_1 = r_1/c_A$
Internal diffusivity	$D_e = (\xi/\tau) \times D_K$ $= (0.3/4) \times 9.26 \times 10^{-7}$ $= 6.945 \times 10^{-8}$ m^2/s	Knudsen diffusivity $D_K = 19400 \dfrac{\xi^2}{\tau S \rho} \sqrt{\dfrac{T}{M_A}}$
Thiele modulus	$\phi = 8.4$	$\phi = \dfrac{d_p}{2} \sqrt{\dfrac{k_1}{D_{eff}}}$
Beta factor	$\beta - 0.007$	$\beta = (-\Delta H_R) D_{eff} c_{A,s}/\lambda T_s$
Internal effectiveness	$\eta = 0.313$	$\eta = \dfrac{3}{\phi}\left(\dfrac{1}{\tanh\phi} - \dfrac{1}{\phi}\right)$
Reynolds particle	$Re_p = 229$	$U = 0.25$ m/s
Sherwood particle	$Sh = 11.5$	$Sh = Re_p^{0.5} Sc^{0.33}$
External mass-transfer coefficient	$k_c = 6.6 \times 10^{-2}$ m/s	
Mears factor	$M_f = 0.025 \ll 0.15$ mass-transfer resistance negligible	$M_f = r\rho_c(1-\varepsilon_b)/k_c c_{Ab}$
Overall effectiveness	$\Omega = 0.312$	$\Omega = \dfrac{\eta}{1+\eta k_1(1-\varepsilon_b)/k_c a_c}$

tortuosity $\tau = 6$. In addition, the following properties of the reaction mixture are known: density 1.91 kg/m^3, kinematic viscosity 1.31×10^{-5} m^2/s, mass diffusivity (phenol/hydrogen) 2.8×10^{-5} m^2/s, Prandtl number 0.475, thermal conductivity 7.40×10^{-5} kW/m/K.

The computation of the effectiveness factor indicates that the internal diffusion is of significance, but the external mass-transfer resistance is negligible. In the reactor design we will consider a global effectiveness of 0.3. The original data are reported at a temperature of 423 K. To describe the effect of temperature over a larger range we adopt the value 30 kJ/mol for the apparent activation energy. This value is suggested by the rule "one-half true activation energy" that is valid for strong pore resistance, knowing that the activation energy for the true surface reaction is about 60–70 kJ/mol [17]. Accordingly, pre-exponential factors A_i were calculated for the two reactions as given in Table 5.7.

5.2.3.2 Cyclohexanol Dehydrogenation

By making use of an appropriate catalyst the conversion of cyclohexanol can be directed to cyclohexanone or to cyclohexene, by dehydrogenation or dehydration, respectively. It is agreed that the dehydrogenation is associated with basic redox sites, while the dehydration is favored by acid sites [19]. For example, a catalyst based on CuCrO$_4$2CuO.2H$_2$O may produce over 85% cyclohexanone, but more

Table 5.9 Parameters in LHHW kinetic model for cyclohexanol dehydrogenation.

A, frequency factor	E, activation energy	K_C	K_B	K_D	Adsorption heat of species
mol/g h	kJ/mol	–	–	–	kJ/mol
9.40×10^{-8}	108.0	2.68×10^{-9}	3.43×10^{-9}	34.7×10^{-9}	−96.9

than 90% cyclohexene when based on AlPO$_4$. In industry the dehydrogenation can be conducted with good selectivity by using catalysts based on chromium oxide-copper, nickel, zinc-iron, *etc.* [1]. In this project we adopt a ZnO-based catalyst, for which kinetic data were given as LHHW model [21]:

$$-r_3 = kK_D(\gamma_D - \gamma_B\gamma_C/K_y)/(1 + K_B\gamma_B + K_C\gamma_C + K_D\gamma_D)^2 \tag{5.6}$$

Table 5.9 shows the values of a six-parameter model. The overall kinetic constant is formulated by the Arrhenius law $k = A \times \exp(-E/RT)$. The adsorption constants are given as $K = K_{j0} \times \exp(-\Delta H_a/RT)$ with the heat of adsorption equal for all species. Note that in the cited study the equilibrium constant is $K_y = 6.0416 \times 10^6 \times \exp(-7953.5/T)$.

5.3
Thermodynamic Analysis

Handling the separation problem involves essentially the ternary mixture phenol/cyclohexanone/cyclohexanol. As Table 5.10 indicates, phenol is the highest boiler (181.9 °C), followed by cyclohexanol (160.8 °C) and cyclohexanone (155.4 °C). Phenol forms positive azeotropes with cyclohexanone and cyclohexanol, of similar composition (roughly 75% mol phenol) and very similar boiling points.

Firstly, we will examine the VLE of binary mixtures. The Wilson model is selected for liquid activity with Redlich–Kwong EOS for vapor phase. The predictions offered by Aspen Plus [9] are in agreement with experimental data [5], except

Table 5.10 Boiling points for the mixture phenol/cyclohexanone/cyclohexanol [3, 5].

Pressure	1 atm	0.1 atm
Cyclohexanone	155.4	83.8
Cyclohexanol	160.8	99.2
Phenol	181.9	114.1
Azeotrope C-hexanone/phenol	185.0 (0.245 mol)	117.7 (0.250 mol)
Azeotrope C-hexanol/phenol	184.8 (0.230 mol)	116.5 (0.245 mol)

Table 5.11 Binary interaction parameters of the Wilson/Redlich–Kwong model.

Binary	a_{12}	b_{12}	a_{21}	b_{21}
Phenol/c-hexanone	−0.89313	−5.8447	749.8611	2555.162
Phenol/c-hexanol	0	0	297.848	149.241

for the binary cyclohexanone/phenol. Consequently, we proceed to the regression of experimental data [25]. Table 5.11 presents the binary interaction parameters of the Wilson model as the form $A_{ij} = a_{ij} + b_{ij}/T$.

Figure 5.5 indicates that the pressure does not affect significantly the composition of azeotropes, although noticeable increase in volatility of cyclohexanone takes place at lower pressure. The y–x diagram of the mixture cyclohexanone/cyclohexanol indicates quasi-ideal behavior (Figure 5.6). At atmospheric pressure the relative volatility is so low that the distillation is not feasible. Fortunately, this increases significantly by lowering the pressure. Below 150 mm Hg the distillation becomes technically feasible, but still requiring a large number of stages.

Figure 5.7 displays the RCM of the ternary mixture cyclohexanone/cyclohexanol/phenol at 0.1 bar. Note the existence of a distillation boundary that prevents recycling pure phenol, as well as the fact that the difference between the boiling points of azeotropes is only 1 °C. The RCM suggests that cyclohexanone will separate easily by distillation, while phenol will be recycled as an azeotrope, most probably with cyclohexanol.

5.4
Input/Output Structure

For more generality we will consider a process involving two reaction stages:

1. Hydrogenation of phenol to a cyclohexanone/cyclohexanol mixture in a ratio given by the catalyst selectivity.
2. Dehydrogenation of cyclohexanol to cyclohexanone to improve the overall yield.

As the reactions involve the same components, these may be handled by a common separation system. Figure 5.8 presents the input/output structure and the main recycles.

At the input/output level the key design decision regards the conversion of reactants. This should obtain the optimal selectivity in useful products, whilst minimizing the occurrence of harmful species for safety and the environment. When several reactants are involved, the first heuristic recommends the complete conversion of the reactant that is the most expensive or the most difficult to recycle (see Chapter 2). Thus, the complete conversion of phenol would be desirable, but

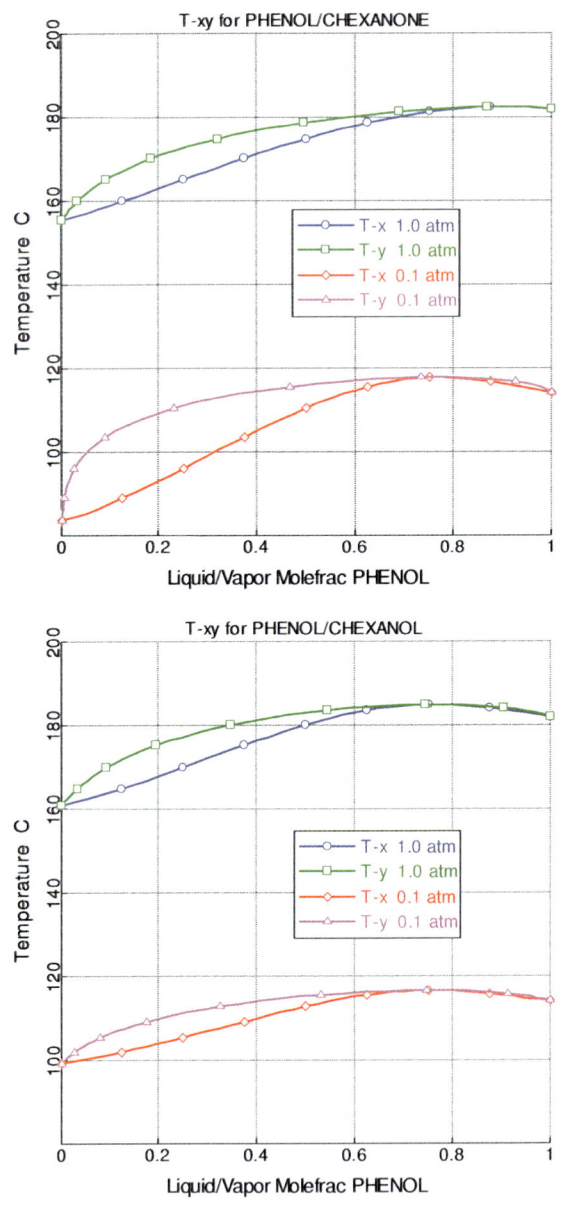

Figure 5.5 T-x-y diagrams for phenol/cyclohexanone and phenol/cyclohexanol.

poor selectivity in cyclohexanone is expected when using a platinum-type catalyst on silica gel.

Thus, as the first design decision we assume partial conversion of phenol to both cyclohexanone and cyclohexanol. As a result, the unreacted phenol must be recycled. Then cyclohexanol is converted to cyclohexanone in a separate reactor,

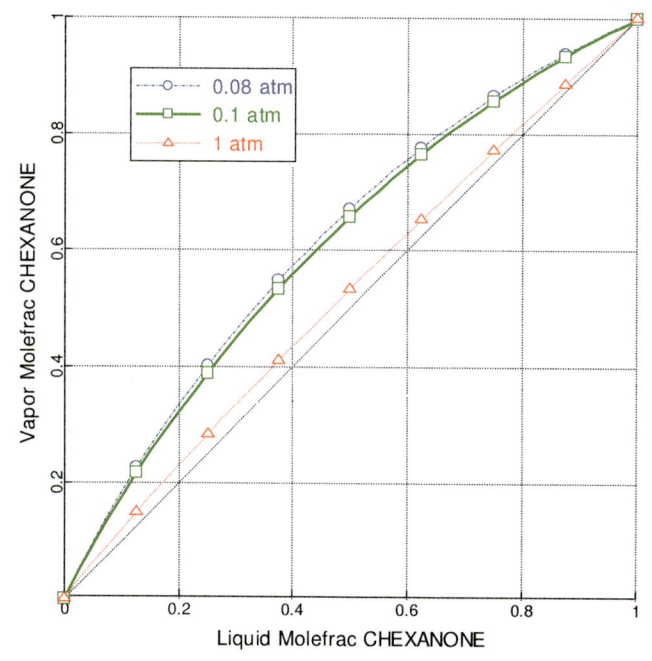

Figure 5.6 Effect of pressure on VLE for the binary cyclohexanone/cyclohexanol.

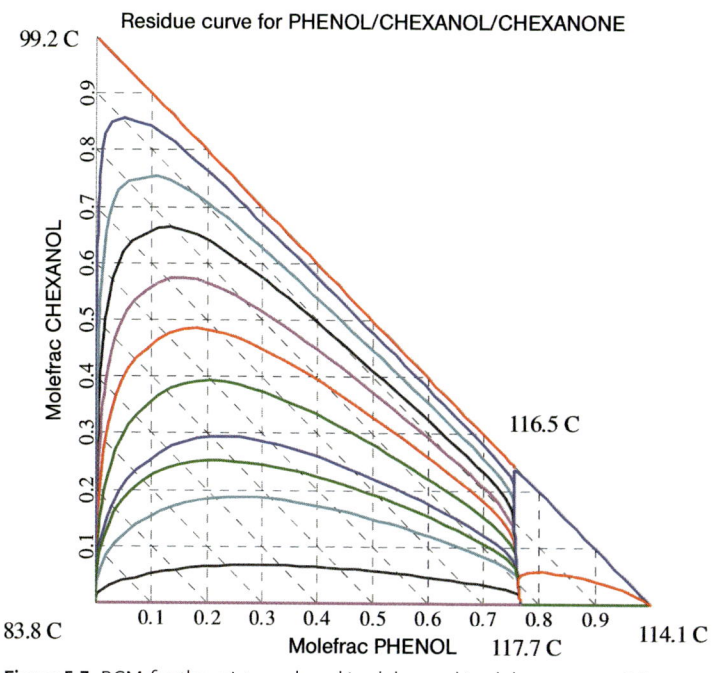

Figure 5.7 RCM for the mixture phenol/cyclohexanol/cyclohexanone at 0.1 atm.

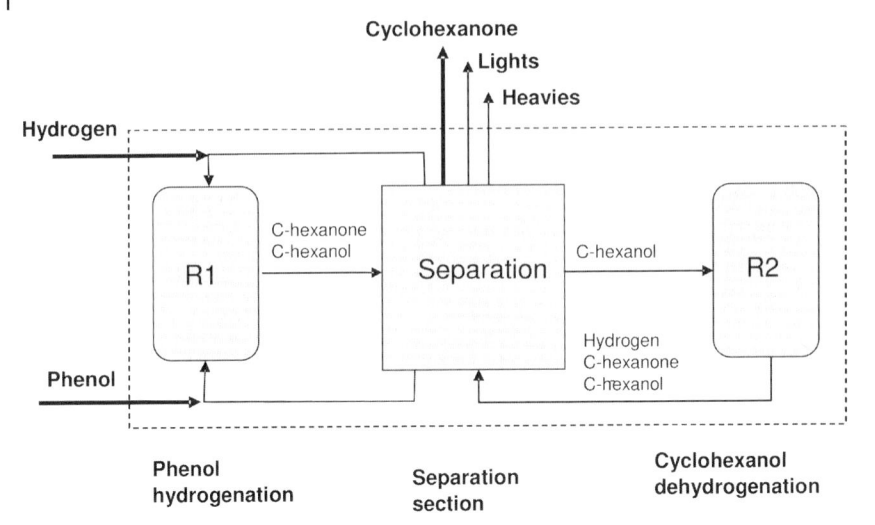

Figure 5.8 Overall material-balance flowsheet.

again with incomplete conversion and recycling. In order to maintain an optimum mass balance the hydrogen produced by dehydrogenation is reused in the hydrogenation step.

As material inputs we consider phenol and hydrogen of 100% purity. Alternatively, a hydrogen feed containing an inert, usually methane or nitrogen, could be used. The presence of an inert is favorable, because it decreases the temperature for phenol evaporation and improves the selectivity by allowing a lower hydrogen/phenol ratio. On the other hand, larger gas recycle implies a more expensive reactor and higher costs for compression. In addition, providing a gas purge becomes necessary to prevent inert accumulation, but with loss in useful reactant and environmental penalty. Evidently, the use of an inert will imply an economic and environmental analysis.

As plant outputs, we may notice cyclohexanone or KA-oil, with small liquid bleeds in lights and heavies, but no hazardous or greenhouse gas emissions.

5.5
Reactor/Separation/Recycle Structure

5.5.1
Phenol Hydrogenation

Figure 5.9 presents the flowsheet prior to heat integration. Fresh and recycled phenol is evaporated and mixed with hydrogen in the evaporator (Ev-1) at about 2 bar. The gas mixture enters the catalytic hydrogenation reactor (R-1). The inlet temperature should be kept strictly constant, in this case at 150 °C, to avoid the

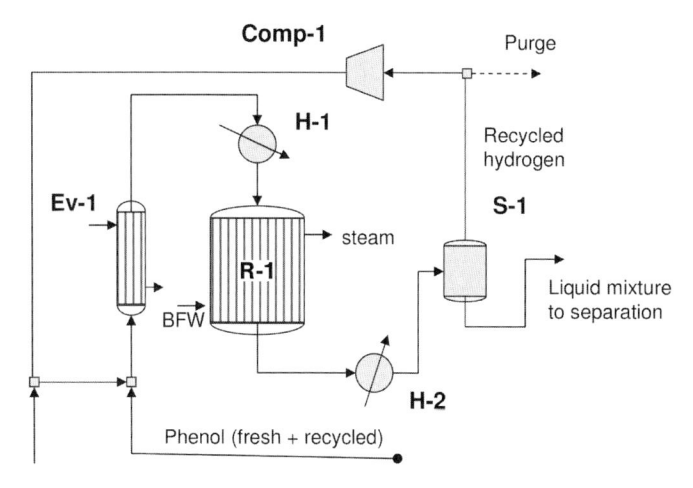

Figure 5.9 Flowsheet for the phenol hydrogenation step.

occurrence of a "hot spot". Therefore, the placement of a small heat exchanger (H-1) in front of the reactor is recommended for control reasons. Reactor cooling is ensured by steam generation at 140–150 °C (3.6–4.7 bar). The reaction mixture is separated in gas and liquid streams in the flash (S-1) after cooling at 33 °C by passing through the exchanger (H-2). The gas containing hydrogen is recycled to the reactor via the compressor (Comp-1), while the liquid phase is sent to separation.

5.5.1.1 Reactor-Design Issues

The reaction rate referred to the mass of catalyst is converted to the reaction volume by the relation $-r_A$ (kmol/m^{3}s) = $-r_A$ (kmol/kg.s) × ρ_b (kg/m^3). The bed density $\rho_b = \rho_c(1-\varepsilon_b)$ is obtained from the particle density ρ_c and the bed void fraction ε_b. A reasonable approximation for spherical particles is $\varepsilon_b = 0.4$. Assuming $\rho_c = 1000$ kg/m^3 gives $\rho_b = 600$ kg/m^3. Accordingly, the pre-exponential factor has to be recalculated taking into account correction factors for both the effectiveness Ω and the bed density $(1-\varepsilon_b)$. One gets $A_1 = 1234.28 \times 0.312 \times (1-0.4) = 231$ kmol/m^3.s and $A_2 = 40.46 \times 0.312 = (1-0.4) = 7.56$ kmol/m^{3}s.

With respect to thermal design, high sensitivity and hot-spot occurrence are expected since the reaction is very exothermic. Several alternatives could be imagined. The simplest is adiabatic PFR with inert recycling. The disadvantage is a drastic increase in reactor size and higher cost for compression. A second solution is a shell and tubes heat-exchanger reactor provided with intensive heat transfer to the cooling agent, such as by liquid thermal fluid or steam generation. The last solution is adopted here. Therefore, the resistance from the gas mixture to the tube wall controls the overall heat transfer.

For the partial heat-transfer coefficients the following equations can be applied [6]:

Cooling of a gas from a tube filled with spheres:

$$\mathrm{Nu} = 3.5\,\mathrm{Re}_p^{0.7} \exp(-4.6 d_p/d_t) \qquad (5.7)$$

Heat transfer from packed beds:

$$\mathrm{Nu} = 2.26\,\mathrm{Re}_p^{0.8}\,Pr^{0.33} \exp(-6 d_p/d_t) \qquad (5.8)$$

$\mathrm{Re}_p = u_s d_p \rho_g/\eta_g$ is the Reynolds number in which d_p refers to particle diameter, u_s superficial velocity, ρ_g gas density and η_g dynamic viscosity. The exponential term accounts for the enhancement effect due to the catalytic bed. In the Nusselt number the characteristic length is the tube diameter, such that $\mathrm{Nu} = \alpha_w d_t/\lambda_g$. For hydrogen-rich mixtures the thermal conductivity is remarkably high.

As a typical design, we consider a superficial gas velocity of 0.25 m/s, 6-mm particles, a tube diameter of 50 mm, as well as a density of 1.39 kg/m³, a viscosity of 1.44×10^{-5} N s/m³, a thermal conductivity of 7.40×10^{-5} kW/m/K, and a heat capacity of 2.32 kJ/kg/K. One gets $\mathrm{Re}_p = 145$, which gives heat-transfer coefficients between 70–100 W/m² K (Table 5.12). The conclusion is that the gas side indeed controls the overall heat transfer.

Table 5.12 Heat-transfer coefficient catalyst–gas phase.

	Re_p	Nu	α_w (W/m² K)
Eq. (5.7)	145	65.6	97.1
Eq. (5.8)	145	53.6	79.3

Preliminary calculations indicate that the hydrogenation reactor exhibits high *parametric sensitivity* with some design elements, such as inlet temperature, cooling agent temperature and residence time. As an example, we present calculations for 140 kmol/h phenol and 700 kmol/h hydrogen. Other elements are: 8000 tubes/50 mm/3 m, boiling water temperature 145 °C, inlet gas temperature 160 °C, inlet pressure 2.0 bar and bed pressure drop 0.2 bar. Figure 5.10 displays temperature profiles for the overall heat-transfer coefficient of 60, 70 and 80 W/m² K. Clearly, a hot spot occurs in the first reactor zone, at about 25% of the total length, with a temperature rise of 35 to 50 °C. This phenomenon is not desirable either for catalyst integrity or for reactor productivity. Indeed, an almost flat temperature profile gives higher conversion. In practice, preventing hot-spot occurrence could be done by filling the entry zone with less active catalyst, by using a higher temperature difference between the gas and thermal agent, and by using an inert. It is good to keep in mind that to avoid high sensitivity the inlet reactor temperature has to be kept rigorously constant at a convenient value. The occurrence of hot spots may upset the dynamic simulation. This problem may be tackled by considering two PFRs in series, the first containing less active catalyst.

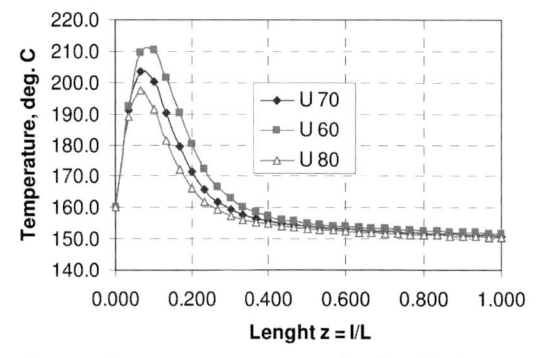

Figure 5.10 Hot-spot occurrence in the phenol hydrogenation reactor.

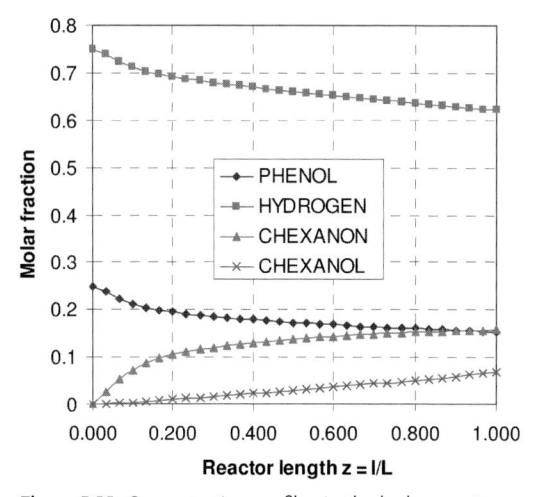

Figure 5.11 Concentration profiles in the hydrogenation reactor.

Figure 5.11 presents typical concentration profiles in a PFR with cooling. As known for consecutive reactions $A \xrightarrow{k_1} P \xrightarrow{k_2} R$, the maximum achievable yield in the intermediate P depends on the ratio k_1/k_2. There is an optimal per-pass conversion and residence time. Good selectivity may be obtained either at lower conversion with less selective catalyst, but spending more for recycles, or with more selective catalyst at higher conversion. Thus, operating the reaction system at variable per-pass conversion may be used to manipulate the selectivity pattern.

Design Alternative with Partial Conversion of Phenol In this section we demonstrate the advantage of performing reactor analysis and design in a recycle structure. As explained in Chapter 2, in contrast with a standalone viewpoint this approach allows the designer to examine systemic issues, the most important being the flexibility with respect to production rate and target selectivity, before

5 Phenol Hydrogenation to Cyclohexanone

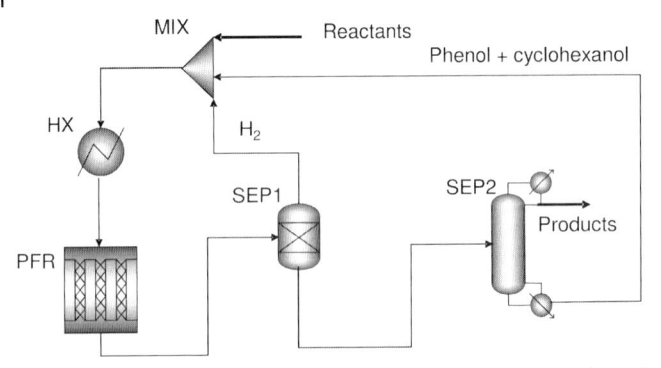

Figure 5.12 Flowsheet for the reactor design with separations and recycles.

the assessment of separations. In addition, strategic decisions regarding the plant-wide control of component inventory may be examined. Figure 5.12 displays the flowsheet for reactor simulation with black-box separators and recycle. Here we consider two separators to account for distinct recycles of hydrogen and phenol, As the thermodynamic study indicates, the recycle will be actually the azeotrope phenol/cyclohexanol with $x_{Ph} = 0.76$. The plant receives fresh feeds in phenol and hydrogen that have to be set proportional to the targeted production rate and product yield. The following relation is obvious:

$$Y = 3 - R_0 \tag{5.9}$$

$Y = [\text{ketone}]/[\text{phenol}]_0$ defines the global yield, $R_0 = [H_2]_0/[\text{phenol}]_0$ being the ratio of fresh reactants. Theoretically, the ratio of reactants in the fresh feed should be between 2 and 3, leading to a global yield in cyclohexanone between 1 and 0.

A key parameter in reactor design is the ratio of reactants at the inlet. Usually, this constraint, imposed by the chemist or by the technologist, should be respected by the designer within a tight tolerance. Because phenol must enter the reactor only as vapor at 150 °C, where the phenol partial pressure is 0.42 atm, the minimum H_2/phenol ratio should be about 4:1 if the total gas pressure is 2 atm. This value could drop to 3:1 if the temperature might be raised to 160 °C, or by lowering the pressure, or by dilution with an inert. Hence, it is good to keep in mind that the ratio of reactants at the reactor inlet is in many situations very different than the ratio of fresh reactants.

On the basis of a plant capacity of 120 kton/yr cyclohexanone and assuming 8400 h/yr working time gives a fresh fed of 145.77 kmol/h phenol. For simpler assessment we consider a nominal plant throughput of 150 kmol/h. The PFR is a standard shell and tubes heat exchanger. We consider tubes of 1.5 inches with OD 1.9 in (48.26 mm), ID 1.77 in (44.95 mm) and wall thickness 0.065 in (1.65 mm). By using the above kinetic equations the simulation shows that a reactor with 12 000 tubes of 4000 mm length is capable of ensuring the nominal production rate. Then, the number of shells can be determined by the following relation [4]:

$$N_t = 1298 + 74.86C + 1.283C^2 - 0.0078C^3 - 0.0006C^4 \qquad (5.10)$$

with $C = 0.75(D/d) - 36$, where D and d are the diameters of shell and tube, respectively. One gets six shells of 2000 tubes, each of 2850 mm.

The simulation gives a catalyst volume of $11.2\,m^3$ per shell, in total $67.2\,m^3$, corresponding to 40.3 tonnes. The residence time is variable, depending on the operation conditions, in the most cases it is close to 13 s. The global productivity F/W is $150 \times 94/40300 = 0.350$ kg-phenol/kg-catalyst, or 0.563 kg-phenol/l-catalyst/h. The above performance may be seen as satisfactory when compared with industrial data in some older patents, which reports 0.1 to 0.6 kg-phenol/l-catalyst [11]. We remember that the simulation made use of a platinum catalyst, for which kinetic data were available, which is indeed less active than modern palladium-based catalysts. More recent processes claim productivity larger than 1.6 kg phenol/l-catalyst/h [14] operating at higher temperature (180 °C) with a low hydrogen/phenol ratio.

The reactor should ensure the desired productivity in the condition of a stable and nonsensitive operation point that avoids the snowball effect. Table 5.13 illustrates two operation points: (1) mixed regime cyclohexanone/cyclohexanol with a ratio of 2, and (2) KA-oil regime, when this ratio rises to 9. The kinetics is given by Eqs. (5.4) and (5.5). The change can be realized by manipulating the ratio of fresh feed reactants hydrogen to phenol, from 2.33 to 2.1, respectively. It can be observed that switching between the two regimes considerably affects the per-pass conversion, and in consequence the recycle streams. Particularly sensitive is the flow of the mixture phenol/cyclohexanone/cyclohexanol. This implies that the distillation columns should be designed for much larger flexibility. The hydrogen flow at the reactor inlet is much less affected, while at the reactor outlet it is practically constant. These observations have plantwide control implications, as will be seen later in this chapter.

Table 5.13 The behavior of hydrogenation reactor for two different operation points.

Flows, kmol/h	Mixed regime		KA-oil regime	
	In	Out	In	Out
Phenol	172.96	22.96	265.59	115.59
Hydrogen	767.96	417.96	731.40	416.40
Cyclohexanone	0	100.00	0	135.0
Cyclohexanol	5.47	55.47	27.69	42.69
Flow to distillation, kmol/h		178.43		293.27
Performance				
H_2/Ph plant/reactor inlet	2.33/4.44		2.1/2.75	
Per-pass conversion/selectivity	0.867/0.667		0.565/0.9	
Yield overall	0.667		0.9	
Selectivity ratio (c-one/c-ol)	2		9	

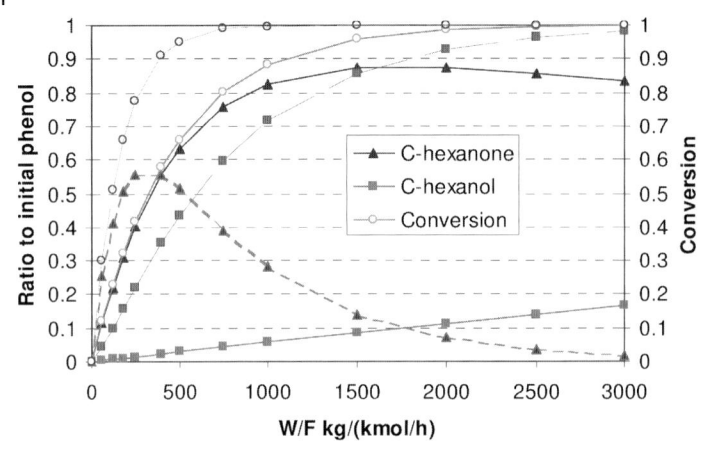

Figure 5.13 Comparative performance of two Pd-type catalysts: Pd/AC – full line, Pd/TaO$_2$ – dashed line; kinetics from Table 5.6.

Design Alternatives with Complete Conversion of Phenol An ideal reactor design would achieve complete per-pass conversion of phenol so as to avoid recycling. Clearly, a high-performance Pd-type catalyst is needed (Table 5.6). For evaluation we select a "fast" catalyst (Pd/TaO$_2$) against a "slow" but more selective catalyst (Pd/AC). Figure 5.13 displays conversion and product distribution at different mass catalyst to phenol flow rate ratios. The calculation is done at a constant temperature of 150 °C by means of equations similar to the analytical solution for first-order kinetics, replacing the space-time V/Q_{v0} by the mass load ratio W(kg catalyst)/F(kmol phenol/h). It may be observed that the Pd/TaO$_2$ (fast) catalyst gives a maximum yield of 55%, but for only 85% conversion (dashed line). If the conversion is pushed to 99% the yield drops dramatically to 40%. For these figures the W/F ratio is 750 kg/kmol-phenol/h equivalent to a productivity of 0.125 kg-phenol/kg-catalyst/h. On the contrary, Pd/AC (slow) catalyst gives a maximum yield of 87% at 96% conversion and W/F of 1500. To achieve a conversion of 99% the W/F ratio should be increased to more than 2200, with the total yield being kept above 85%. The global productivity would be three time less than before.

The above results are semiqualitative. It is clear that more selective catalyst will bring advantages, by increasing the conversion and diminishing the recycle flow. However, if the design decision is "recycle phenol" then achieving good yield is feasible even with less selective catalyst. Moreover, the dehydrogenation step is no longer necessary. The energy consumption for separations should increase only moderately, since the phenol comes as bottoms from the recycle distillation column. As an overall benefit, the process becomes less sensitive to catalyst deactivation and more flexible in operation.

On the contrary, pushing the reaction to full conversion of phenol seems not to be profitable. Because the selectivity drops considerably a second dehydrogenation reactor becomes necessary, for which the investment is justified only at significant

5.5 Reactor/Separation/Recycle Structure | 151

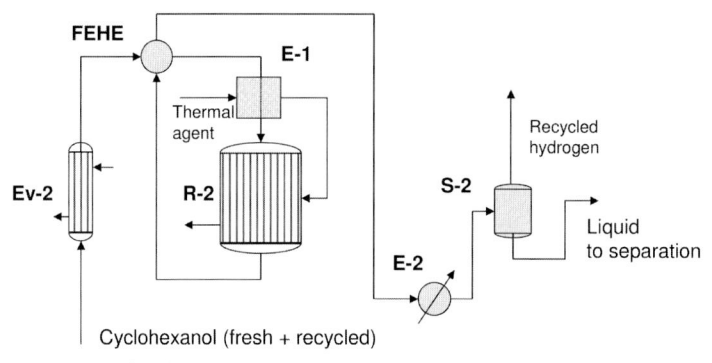

Figure 5.14 Flowsheet for the cyclohexanol dehydrogenation step.

throughput. However, to get a more precise idea about the two alternatives we will proceed with the design of the cyclohexanol dehydrogenation section.

5.5.2
Dehydrogenation of Cyclohexanol

Figure 5.14 presents the flowsheet for cyclohexanol dehydrogenation. After evaporation in (Ev-2), the reactant is preheated by exchange with the product in the feed-effluent-heat-exchanger FEHE, and heated further in the heat exchanger (E-1). This unit can be a furnace or a shell and tube heat exchanger driven by the same thermal agent supplying the heat of reaction. The dehydrogenation reactor itself is of heat-exchanger type with catalyst inside and thermal agent outside.

The impurities can be grouped into two categories: lights (water, cyclohexene, cyclohexadiene) and heavies (phenol, dicyclohexyl-ether, cyclohexenyl- cyclohexanone). To limit their amount, the conversion is kept around 80% with a selectivity of about 98%. The hot reactor effluent is cooled in countercurrent with the feed in FEHE, and finally for phase separation in the heat exchanger (E-2) at 33 °C. The simple flash (S-2) can ensure a sharp split between hydrogen, recycled to hydrogenation reactor, and a liquid phase sent to separation.

5.5.2.1 Reactor Design
We consider a selectivity of 0.5 in the hydrogenation step, a worse scenario, which gives a throughput of 75 kmol/h cyclohexanol. Kinetic data valid for a ZnO-type catalyst is given by Eq. (5.6) and Table 5.9. To keep the level of impurities low the per-pass conversion should be below 85%. We consider a multitubular heat-exchanger reactor operated at 2 atm and 330 °C. The heating agent is a thermal fluid with a constant temperature of 350 °C. The overall heat-transfer coefficient is taken at 70 W/m² K. The predesign calculation, confirmed later, leads to the following configuration: 2000 tubes × 50 mm diameter × 4 m length. The temperature profile shows initially a sharp decrease to a minimum of 313 °C, when the reaction rate is high, followed by a steady increase to about 340 °C. The residence time is

about 7 s. Thus, the dehydrogenation reactor is identical with one shell of the hydrogenation section, but three times more productive.

5.6
Separation System

The separation section receives liquid streams from both reactors. For assessment the residue curve map in Figure 5.7 is of help. The first separation step is the removal of lights. This operation can take place in a distillation column operated under vacuum (200 mmHg) with a partial condenser. Next, the separation of the ternary mixture cyclohexanone/cyclohexanol/phenol follows. Two columns are necessary. In a *direct sequence* (Figure 5.15) both cyclohexanone and cyclohexanol are separated as top products. The azeotrope phenol/cyclohexanol to be recycled is the bottoms from the second split. In an *indirect sequence* (Figure 5.16) the azeotropic phenol mixture is a bottom product already from the first split. Then, in the second split cyclohexanone is obtained as the top distillate, while cyclohexanol is taken off as the bottom product. The final column separates the phenol from the heavies.

Because the volatility of cyclohexanone increases considerably at low pressure, the separations in both (C-2) and (C-3) columns should take place under vacuum, between 50 and 100 mmHg. The column pressure should give a bottom temperature in the range 125–135 °C, in agreement with the available steam for heating. The most difficult step in separation would be (C-2) in direct sequence, and (C-3) in indirect sequence, since both deal with the binary cyclohexanone/cyclohexanol. Because of higher amount of cyclohexanone in the combined feed, the direct sequence should require less energy. By combining the reaction and separation sections the final flowsheet is obtained, as pictured in the Figure 5.17. This is submitted to simulation, as described in the next section.

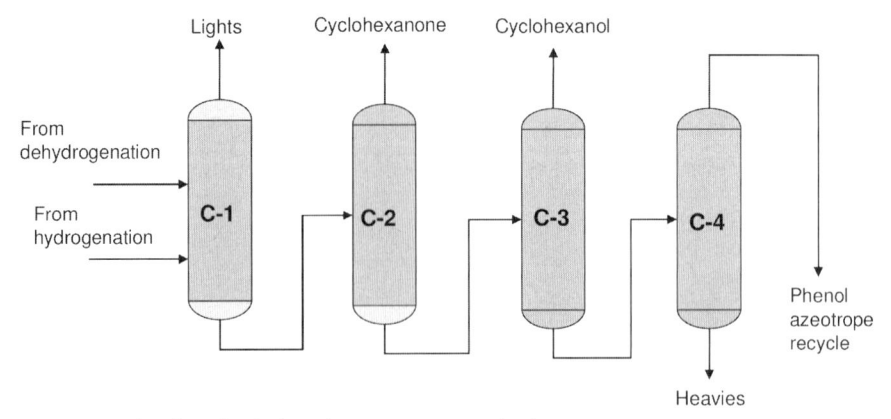

Figure 5.15 Flowsheet for the liquid-separation system by direct sequence.

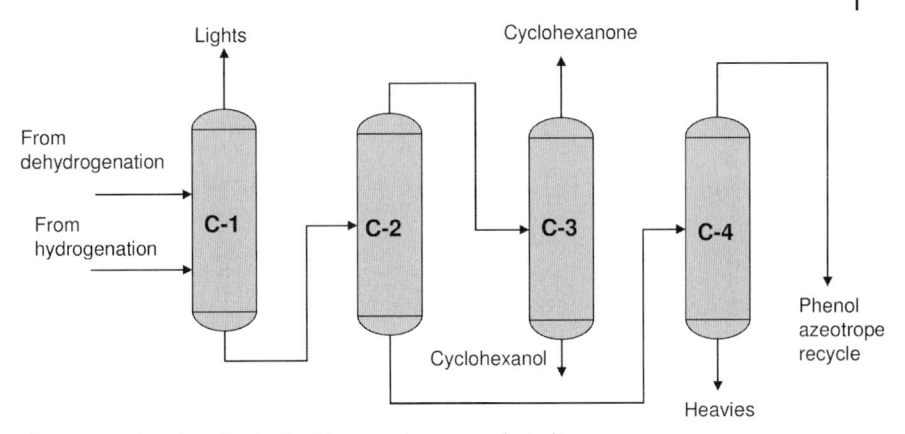

Figure 5.16 Flowsheet for the liquid-separation system by indirect sequence.

5.7
Material-Balance Flowsheet

5.7.1
Simulation

The first objective of flowsheeting is to obtain a consistent description of the material-balance envelope. The reactor model should be of the kinetic type, at least for the main reactions, in order to account for the effect of variable flow rates and composition of recycles coming from separations. Stoichiometric or yield reactor models can be employed for describing secondary reactions and formation of impurities. In a first attempt the separators may be black-boxes provided with appropriate specifications.

The flowsheet displayed in Figure 5.17 has three recycles. The first two loops consider the hydrogenation section with recycles of hydrogen from SEP1 and phenol from SEP2, respectively. Phenol in recycle was allowed to contain 24% cyclohexanol. The third recycle consists of unconverted cyclohexanol after dehydrogenation issued from SEP3. The last two recycles are interconnected through the liquid-separation section, simulated by the black-box SEP2. The stoichiometric reactor RSTOIC describes representative secondary reactions. In this case, the lights are formed via the dehydration of cyclohexanol to cyclohexene and water. Regarding the heavies, only the formation of cyclohexyl-cyclohexanone was considered. In addition, this component is available in the standard database of Aspen. Other impurities lighter or heavier than those considered will have no affect either on the separation sequence or on the design of the separators. In this case, the selectivity by cyclohexanol dehydrogenation was considered as 98%, the rest being converted to cyclohexene, from which 20% goes to cyclohexyl-cyclohexanone.

Getting convergence with the simplified flowsheet shown in Figure 5.18 does not raise particular problems. On this basis, a realistic description of streams is

154 | 5 Phenol Hydrogenation to Cyclohexanone

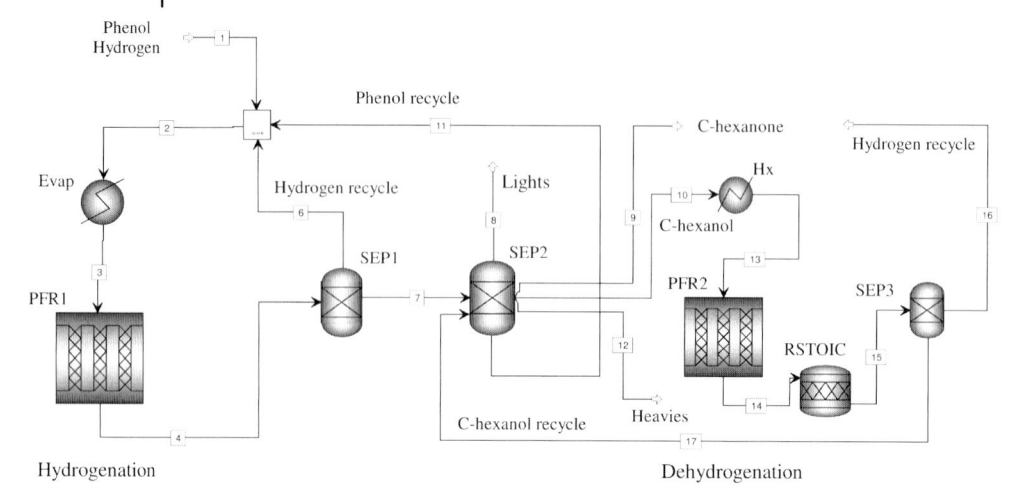

Figure 5.17 Preliminary simulation of the material-balance envelope.

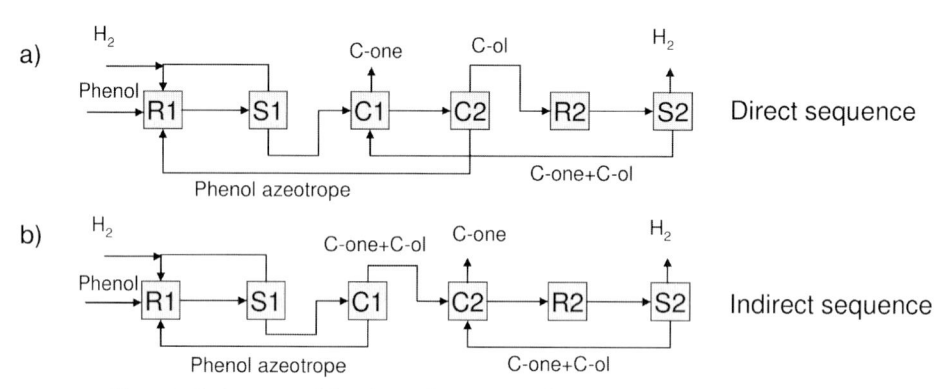

Figure 5.18 Structure of flowsheet after inserting the separation sequence.

obtained, which can serve further for designing the units. After the insertion of the separation section with direct and indirect sequences the flowsheet alternatives displayed in Figure 5.18 are obtained. This time, the simulation could be confronted with serious convergence problems. The user might be tempted to overdesign the units, increasing the number of iterations or switching to more sophisticated algorithms. Actually, he/she should be aware that some unit specifications, correct in standalone mode, may be in conflict when considering the interactions over the whole flowsheet.

The treatment of conflicting specifications leading to convergence problems has been developed elsewhere [7]. For example, this situation arrives when the distillation columns are specified by fixed product flow rates. These specifications, correct for standalone columns, lead to nonconvergence when the units are placed in recycles. The explanation is that during the iterative solution it is impossible to find a perfect match of the plant material balance of components if all the outputs are "blocked". On the contrary, if some outputs are relaxed, or if some specifications are of ratio type, the chance of getting convergence is much better because the flow rates of recycles and components may adapt themselves to fulfil the material balance. In some cases, infinite solutions might exist, if components could accumulate in a recycle at any values. Setting the flow rate of recycle at an acceptable level from the technology viewpoint may fix the problem.

As an example, let us consider the flowsheet based on the direct sequence (Figure 5.18a). The first recycle loop, including a hydrogenation reactor and a phenol recycle column, can be converged by specifying the bottoms to feed ratio of C2. Alternatively, setting the total recycle flow rate of bottoms at 80 kmol/h gives robust convergence. This approach is consistent with the reactor design done previously with black-box separators. In addition, it anticipates a good control strategy. For the cyclohexanol dehydrogenation recycle loop, a normal specification would be the top product of C1, in fact the total cyclohexanone production. Indeed, convergence is obtained for several operating points, but with the result of a considerable amount of cyclohexanone in the top of C2 as well as in recycle. Setting the distillate/feed ratio as a specification leads to a solution that is physically consistent, with only trace of cyclohexanone in the top product of C2. On the contrary, the indirect sequence leads to easier convergence (Figure 5.18b) by specifying the recycle flow from C1 and top distillate from C2.

The reader will observe that the direct sequence of separations leads to nested recycle loops, while the indirect sequence gives two independent recycle loops. Contrary to expectation, more favorable for the direct sequence, both schemes are equivalent in energy consumption. The explanation is that the largest energy consumer is the separation between the key components, cyclohexanone and cyclohexanol, which is the first split in the direct sequence, and the second split in the indirect sequence.

The conclusion is that the indirect separation sequence should be more attractive, because it is equivalent in energy, but with fewer interactions in operation. In addition, the split in C2 is phenol-free, offering better catalyst protection in dehydrogenation.

5.7.2
Sizing and Optimization

After obtaining convergence of the material balance the sizing and optimization of the main units may follow. These are the distillation columns. Tray and packing sizing can be done directly in Aspen Plus [9] on the basis of L/V traffic. Table 5.14 presents the results of the optimized sizing in terms of energy consumption. For the first split, the reflux is reduced to the lowest limit obtaining robust convergence when the recycle is fixed at 80 kmol/h. For the second split the D/F and reflux ratio are optimized specifying cyclohexanone purity of a minimum of 98 mol%, while the cyclohexanol should contain no more than 4 mol% cyclohexanone.

Table 5.14 Optimal sizing of the distillation columns in indirect sequence.

Columns	Sizing
First split, composition (mol. fr.) Top: c-hexanone 0.5 c-hexanol 0.5 Bottom: c-hexanone 0.022 c-hexanol 0.241 phenol 0.737	30 theoretical stages (15 feed) 40 real trays Pressure: 50–200 torr (0.066–0.266 bar) Temperature: 80.6–141.6 °C $Q_c = 4.139$ MW; $Q_{reb} = 4.160$ MW Reflux ratio 1.0 Koch flexitray 4 passes: D max 2.5 m
Second split, composition (mol. fr.) Top: c-hexanone 0.98 c-hexanol 0.02 Bottom: c-hexanone 0.04 c-hexanol 0.96	40 theoretical stages (16 feed) 50 real trays Pressure: 50–237 torr (0.066–0.316 bar) Temperature: 73.9–125.3 °C $Q_c = 5.81$ MW, $Q_{reb} = 5.63$ MW Reflux ratio 2.18 Koch flexitray 4 passes: D max 3.3 m

The columns' pressure deserves special attention. The top of both columns is set at 0.066 bar with a pressure drop per tray at 500 Pa, a conservative value for vacuum operation. The total pressure drop is accounted for in real stages with 75% overall efficiency. Koch flexitrays with 4 passes are adopted as internals, since these give smaller diameter. However, relatively large values are obtained, of 2.5 and 3.2 m for the bottom and top trays respectively, due to the high throughput and deep vacuum.

5.8
Energy Integration

In this subsection we focus only on the main aspects of heat saving. We leave the development of a heat-exchanger network by pinch analysis as an exercise. It is obvious that the strategy should maximize the use of heat developed by the exothermic hydrogenation reaction and minimize the heat consumption in the endothermic dehydrogenation step. The insertion of simple heat exchangers into the mass-balance flowsheet gives useful physical insights before a more sophisticated approach. For example, coupling the duties of heaters and coolers of the inlet and outlet flows around the reactors offers a simpler simulation of the feed-effluent-heat-exchanger (FEHE) units, typically used in energy saving.

Figure 5.19 presents the main features of the heat integration for the two-reactor flowsheet. The most relevant fact is that the heat developed by the phenol hydrogenation, about 6.8 MW, could completely cover the need of the end-product column C-3. The heat of the reactor effluent can be used only marginally for feed preheating, about 0.7 MW. On the contrary, saving energy by coupling effluent

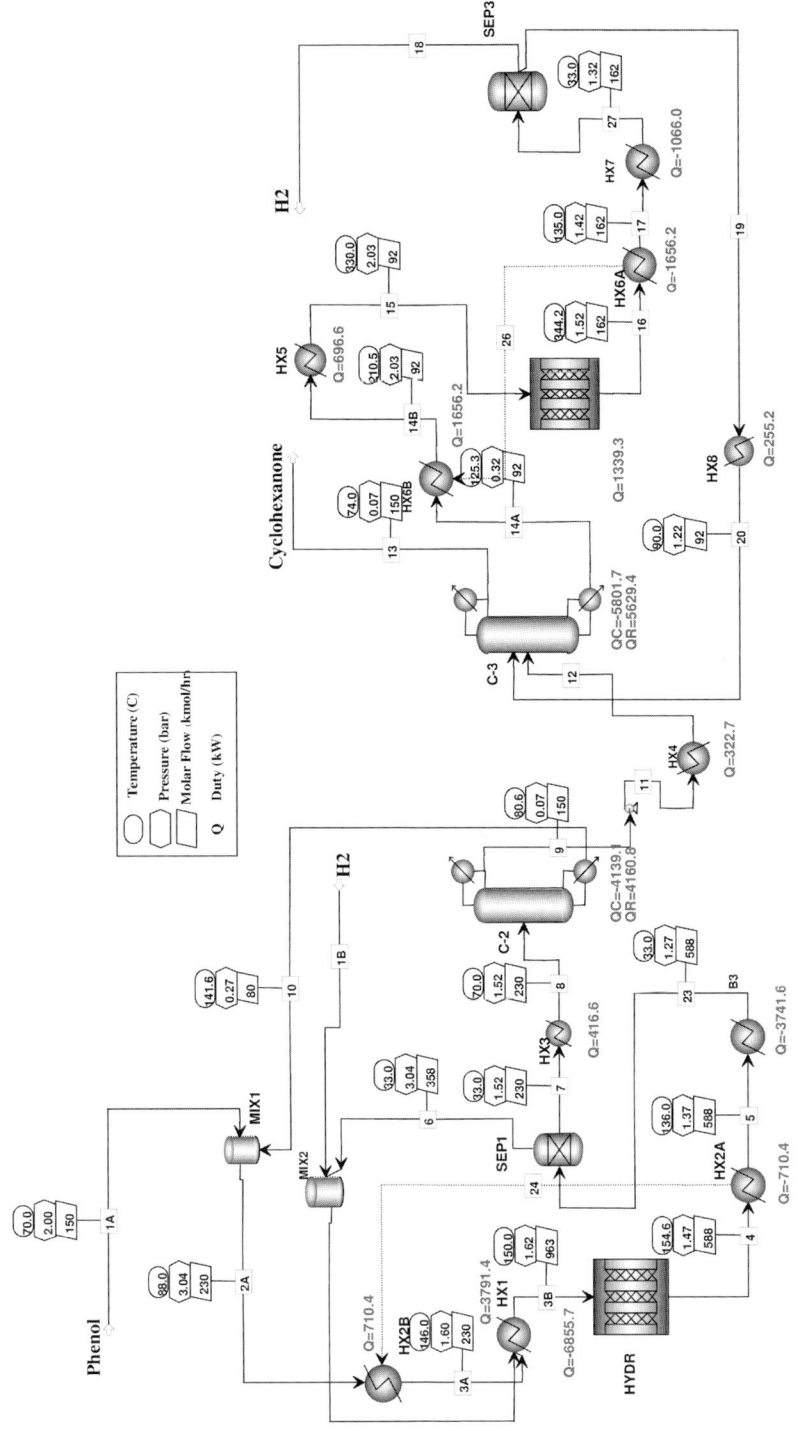

Figure 5.19 Flowsheet with recycles, separation and heat integration.

and feed preheating is more favorable around the dehydrogenation reactor, saving about 1.6 MW.

In terms of energy consumption the total MP steam needed for evaporation and distillation is Q_{t1} = 3.79 + 4.16 + 5.63 MW = 13.58 MW. Taking into account a production of 4.1 kg/s cyclohexanone, the process needs a minimum of 3.312 MJ/kg* (symbol * for product). To account for the columns for lights and heavies separation, an extra of 30% might be added, giving an energy consumption for the hydrogenation plus separation section of about 4.305 MJ/kg*. For steam at 160 °C (6.18 bar) the enthalpy of vaporization is 2.083 MJ/kg steam, which gives a total consumption of 2.06 kg steam/kg*. If generated steam is used to drive the cyclohexanol column a saving of 0.66 kg steam/kg* is achieved, dropping the energy consumption to 1.40 kg steam/kg*. Note that this value is slightly lower than the 1.6 kg steam/kg* reported for a competitive process that makes use only of a hydrogenation step [14]. Considering a hydrogenation temperature around 180 °C HP steam can be raised, which is more effective for driving the distillation columns and/or producing electricity. On the other hand, a lower pressure drop and a more efficient use of energy may be achieved by using structured packing instead of trays for cyclohexanol distillation.

To the above we should add the energy for the dehydrogenation reaction stage. It may be observed that about 1600 kW may be saved by inserting a simple FEHE between the incoming cyclohexanol stream and the outlet reactor mixture. It remains that about 2010 kW should be supplied to the dehydrogenation reaction, including the preheating. Although this amount is not excessive, its cost is much higher than the equivalent steam because of the high temperature, above 340 °C. Expensive thermal fluid, Dowtherm A (400 °C and 10 bar) or $NaNO_2$ eutectic salt (454 °C) should be used.

It is now clear that the suppression of the dehydrogenation step is the most radical step in saving energy. This is possible only if a suitable catalyst may be found to meet the specifications of KA-oil in just one reaction step. This option will be discussed in the next section.

5.9
One-Reactor Process

From the analysis in the previous sections it may be concluded that a process for manufacturing cyclohexanone with one reaction section is feasible by employing a Pd-type catalyst. As discussed in Section 5.2.3, this catalyst may give selectivity over 90%, although not complete phenol conversion. This approach is preferred in more recent projects, as disclosed by Aker Kvaerner (www.kvaerner.com, 2004). Figure 5.20 presents a conceptual flowsheet. In this case a direct separation sequence is more suited because it recovers the selling products as top distillate and minimizes the heat consumption. In addition, nonreacted phenol is recycled as an azeotrope with cyclohexanol and not with cyclohexanone. Note that the separation of heavies is performed during the evaporation of phenol, saving one vacuum-distillation column.

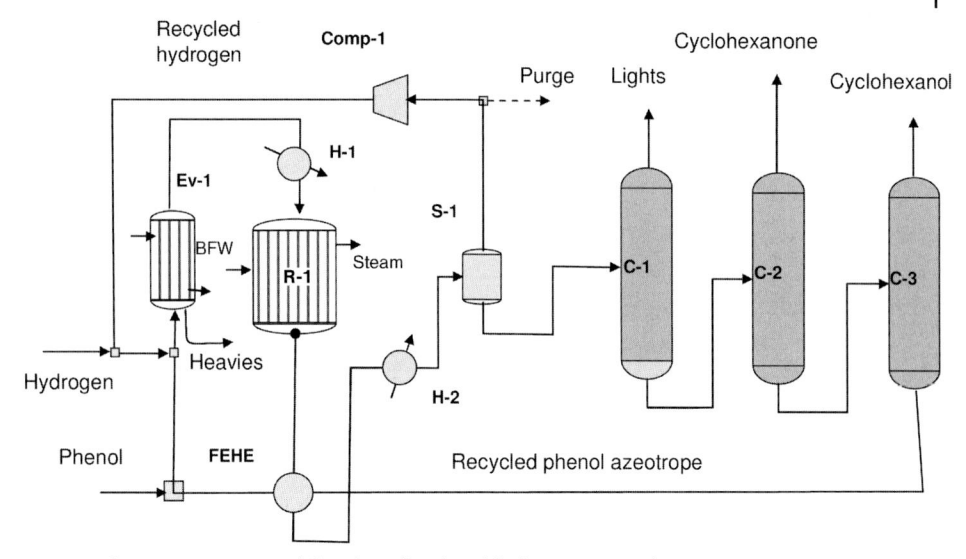

Figure 5.20 Conceptual flowsheet for phenol hydrogenation with one reactor.

The results of simulation are shown in Figure 5.21, prior to energy integration but with key data for streams and duties. The shell-and-tube chemical reactor ensures the phenol hydrogenation to the conversion required by the target selectivity. A stoichiometric reactor placed after the hydrogenation reactor accounts for secondary reactions: 1% conversion of cyclohexanol to water and cyclohexene (lights), followed by 30% conversion to cyclohexyl-cyclohexanone (heavies). To keep the simulation robust SEP1 is an ideal black-box and not a flash unit. The distillation columns operate under vacuum, at top pressures of 0.2, 0.1 and 0.08 bar, respectively. The internals should be selected so as to minimize the pressure drop.

Some comments regarding the design of the distillation columns are useful. For the tower (C-1) the cyclohexene concentration in bottoms has to be pushed down to 1 ppm to meet the purity specifications. About 20 theoretical stages are necessary, with high-position feed location (stage 7). Although the top-product flow rate is very small, sufficient internal reflux should be ensured for separation and hydraulics. Small losses of cyclohexanone in the top distillate are unavoidable, but may be minimized by suitable control. For example, a column equipped with Norton IMTP packing of 38 mm has a 1.1-m diameter operating at a pressure drop of about 0.14 bar.

The column (C-2) supplies pure cyclohexanone. About 40 theoretical stages, correct feed location (17) and sufficient high reflux ratio (4 to 5:1) are necessary to achieve higher purities. Hydraulic calculations at 80% flooding and four passes give max/min diameters of 3.25/2.35 m for Koch flexitrays and 3.90/2.5 m for Glitch ballast trays, with a total pressure drop below 0.25 bar at 40 mm weir height. Combining plates and packing could reduce the pressure drop further offering good separation.

160 | *5 Phenol Hydrogenation to Cyclohexanone*

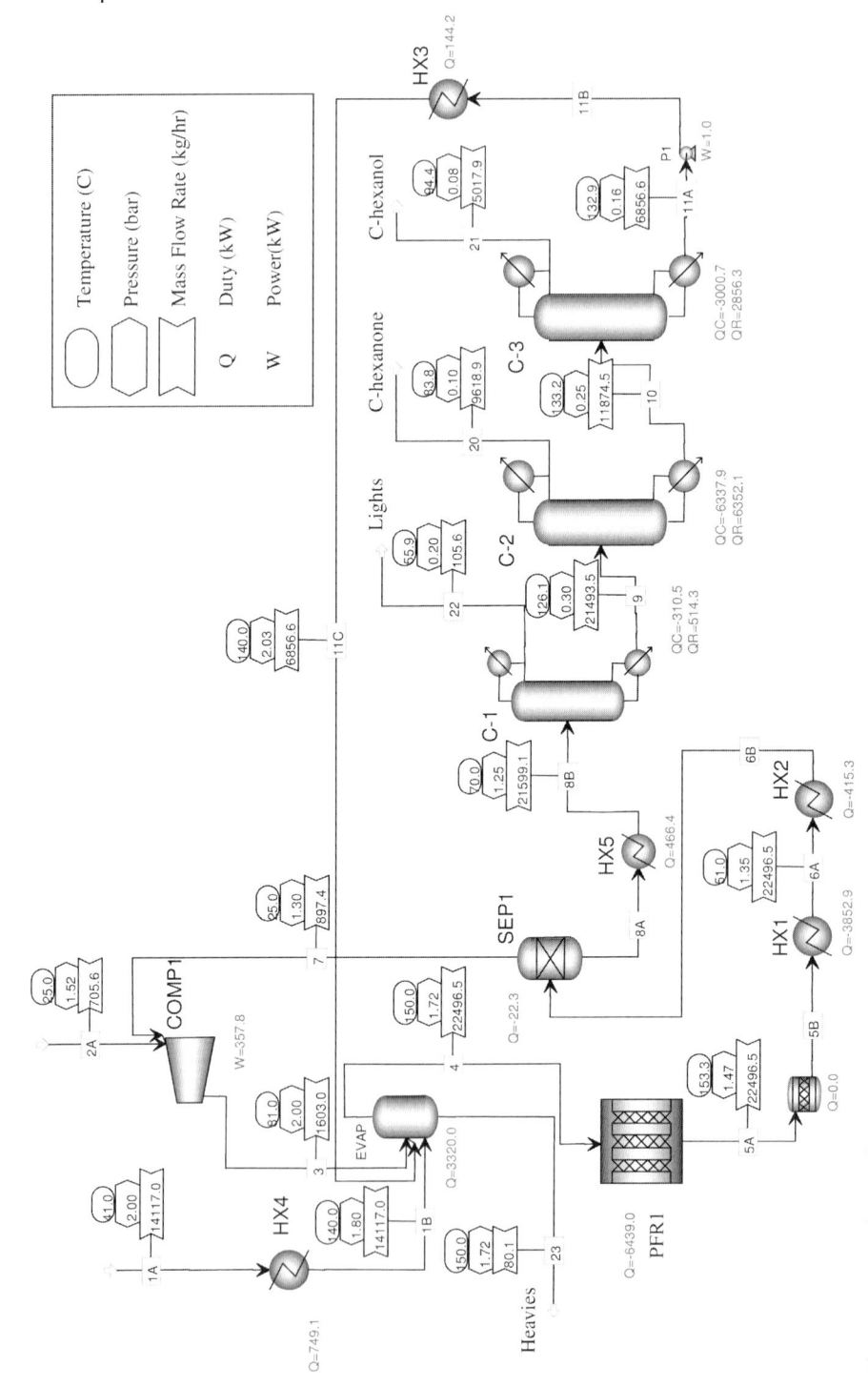

Figure 5.21 Simulation of one-reactor phenol hydrogenation process.

The column (C-3) delivers cyclohexanol as top and a phenol/cyclohexanol mixture to be recycled as bottoms. The column is designed to ensure bottoms as close as possible to the azeotrope phenol/cyclohexanol, while minimizing losses in cyclohexanone. This column has 30 theoretical stages and operates at moderate reflux, below 3:1, which leads to diameters of 2.5/1.8 m.

The examination of heat duties in Figure 5.21 suggests heat-integration opportunities. A major impact has used the energy released in reactor for driving the distillation column (C-2), the largest consumer. In addition, the reactor effluent can preheat the fresh and recycled phenol, as well as ensure a suitable feed condition for (C-1).

5.10
Process Dynamics and Control

5.10.1
Control Objectives

From a plantwide viewpoint the process control should ensure:

1. safe operation, including start-up and shut-down,
2. production rate at nominal capacity with flexibility of $\pm 10\%$, as well as target selectivity with respect to cyclohexanone/cyclohexanol ratio,
3. good purity product, of minimum 98% cyclohexanone,
4. minimum losses in organic waste and gas emissions,
5. minimum time of transitions between operation points.

The first issue is fulfilled if the plant does not operate in unstable points or dangerous conditions. In this process the most risky place is the chemical reactor, where hot spots may occur. There are fire and explosion risks too, but these should be normally managed by standard safety measures.

The second issue regards the optimal plantwide material balance. It is clear that the raw materials must be fed only in amounts required by the target production and selectivity. The control structure of fresh feeds should allow flexibility, within predefined limits, both in production rate and selectivity, but avoiding large variation of recycles that might upset some units (snowball effect).

The control of product purity should cope with flexibility requirements. Excessive purity turns into higher operation costs, but lower purity is unacceptable. The problem may be solved in principle by adequate local control loops. The same is valid for waste minimization. However, when the separators are involved in the recycle loops, both the design and control must take into account the effect of interactions.

Finally, responsiveness and fast dynamics of transitions may be achieved by suppressing unnecessary storage tanks and by optimizing the inventories of units.

Dynamic simulation starts by proper sizing of units and by the evaluation of the key plantwide control structures. The manipulated variables should have sufficient power to reject various disturbances and to track the controlled variables on new setpoints. The selection of controlled variables should comply with the requirements of sensitivity and linearity with respect to manipulations.

The dynamic simulation file prepared in Aspen Plus is exported in Aspen Dynamics [10]. We select the flow-driven simulation mode. Aspen Dynamics files have already implemented the basic control loops for levels and pressures. Units with fast dynamics, such as the evaporator or some heat exchangers, may be handled as steady state. The implementation of control loops for the key operational units, chemical reactor and distillation columns, take into account some specific issues from the plantwide perspective, which are developed in detail in Luyben et al. [8].

5.10.2
Plantwide Control

In this section we will examine several plantwide control structures. The control objective is to change the production rate by ±10% while achieving the selectivity and purity targets. As discussed in Chapter 4, plantwide control structures can be classified with respect to the strategy employed for controlling reactants inventory [22]. Four different alternatives are presented in Figure 5.22.

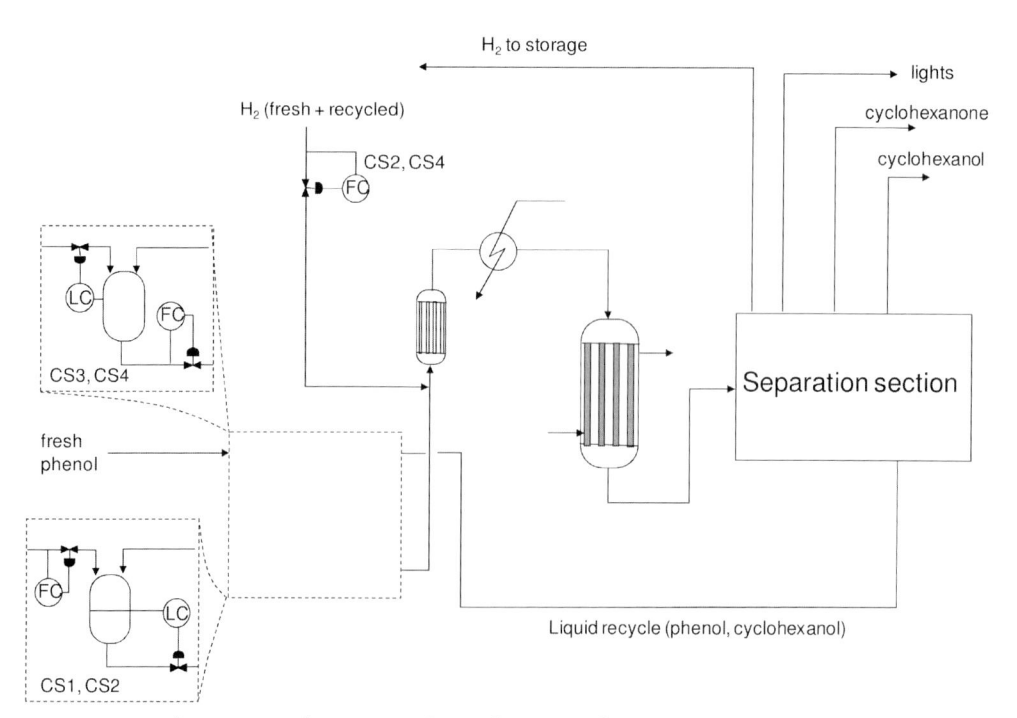

Figure 5.22 Implementation of control structures for one-reactor process.

5.10 Process Dynamics and Control

Control structure CS1 fixes the flows of both reactants at plant input. Therefore, the principle of self-regulation is employed. When the reactant feeds are fixed, the internal flow rates reach values for which the reaction conversion and selectivity reflect the amount of fresh reactants, according to Eq. (5.9). Change of production rate or product distribution is achieved directly, by changing the flows of fresh reactants.

In contrast, control structure CS4 uses feedback to control the reactant inventory. The reactor-inlet flow rates are fixed and the fresh reactants are brought in the process on level control (phenol) or on pressure control (hydrogen). Production changes are implemented by manipulating the reactant-inlet flows. Three different modes are denoted by CS4a (flows of both reactants are changed), CS4b (phenol flow changed) and CS4c (hydrogen flow changed).

The two strategies are combined in control structures CS2 and CS3, when one fresh reactant is on flow control (phenol or hydrogen, respectively). This stream is manipulated when production changes are attempted. In each case, the flow of the other reactant is fixed at the reactor.

In the following, we compare the performance of these control structures. The results of dynamic simulation are presented in Figures 5.23 to 5.28. A summary is displayed in Table 5.15, where the nominal point is given in the first line, the streams on flow control are in the grey cells, while *italic* values show changes from the nominal operating point.

The plant was designed for nominal fresh feeds of 150 kmol/h phenol and 350 kmol/h hydrogen that corresponds to a theoretical yield of 100 kmol/h cyclohexanone and 50 kmol/h cyclohexanol.

In the structure CS1 the flow rates of both fresh reactants are fixed. As shown in Chapter 4, this strategy can be used for complex reactions, when the number of independent reactions matches the number of products for which the self-regulation principle is applied. Figure 5.23 indicates the response at a step change of

Table 5.15 Plantwide control structures of the material balance.

CS	Feed streams		Reactor inlet		Recycles		C-hexanone		C-hexanol	
	Ph-OH	H_2	Ph-OH	H_2	Ph-OH	H_2	$C_6H_{10}O$		$C_6H_{11}OH$	
	kmol/h	kmol/h	kmol/h	kmol/h	kmol/h	kmol/h	kmol/h	purity	kmol/h	purity
Nominal	149.44	348.5	220	388	70.51	39.45	99.81	0.98	48.63	0.977
CS1	*165*	*383*	241.6	433.5	76.6	50.5	111	0.99	52.88	0.98
CS2	*165*	340.8	305.6	388	140.6!	47.1	100.7	0.99	51.5	0.77!
CS3	162.8	*383*	220	433.8	57.2	50.8	105.4	0.98	56.1	0.98
CS4a	162.7	377.2	*242*	*426*	79.3	48.8	110	0.99	51.5	0.98
CS4b	150.6	346.8	*242*	388	91.3	41.1	104.3	0.99	45.3	0.98
CS4c	160.6	377.4	220	*426*	59.4	48.6	104.6	0.98	55	0.98

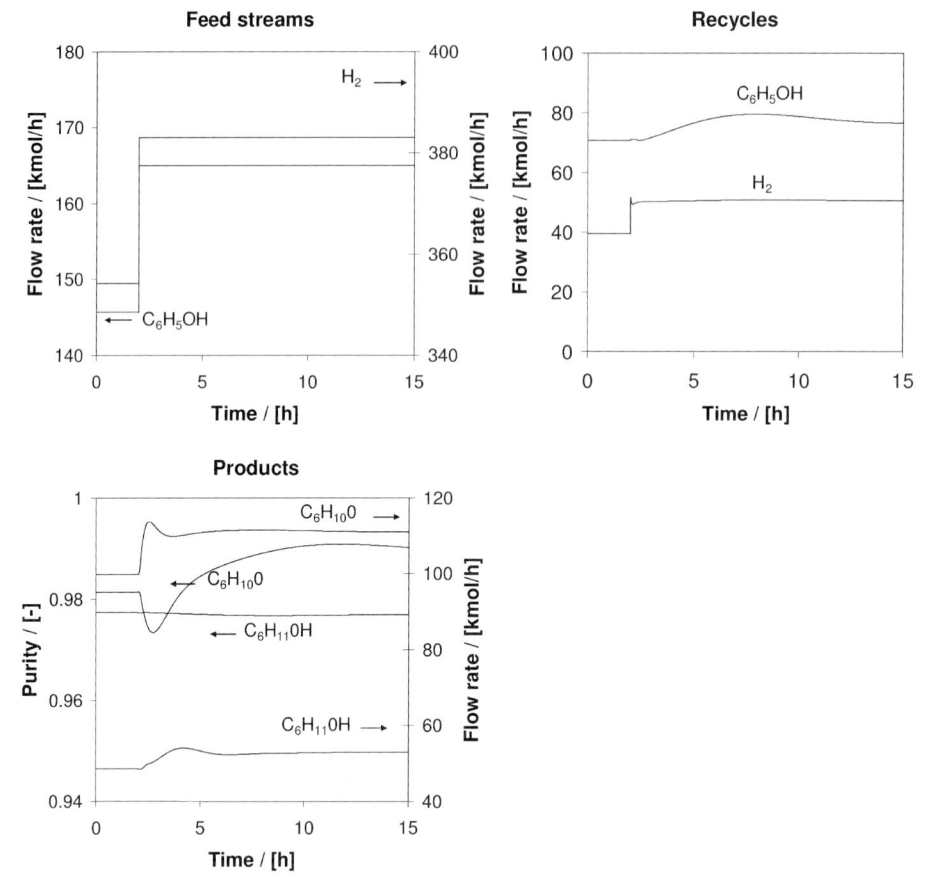

Figure 5.23 CS1: Production rate by simultaneously change of both fresh feeds. At $t = 2$ h, phenol fresh feed increases from 149.5 to 165 kmol/h; hydrogen fresh feed from 348.5 to 383 kmol/h.

10% in phenol and 9% in hydrogen. As both reactions contribute significantly to the reactants consumption, the snowball effect does not occur. For the new feeds, the theoretical yield is 112 kmol/h cyclohexanone and 53 kmol/h cyclohexanol, very close to the simulation results. The concentration of product remains at the nominal values. The simulation proves the advantage of control structures based on self-regulation: the product distribution can be varied simply by changing the ratio of reactants. However, the dynamics is quite slow, about 10 h being necessary to reach the new steady state.

In the structure CS2 (Figure 5.24) the phenol fresh feed is raised to 165 kmol/h, while the hydrogen flow to reactor is kept constant at the value from the nominal steady state at 388 kmol/h. This control structure does not work. Phenol accumulates in recycle because the hydrogen is insufficient for reaction, leading to plant upset.

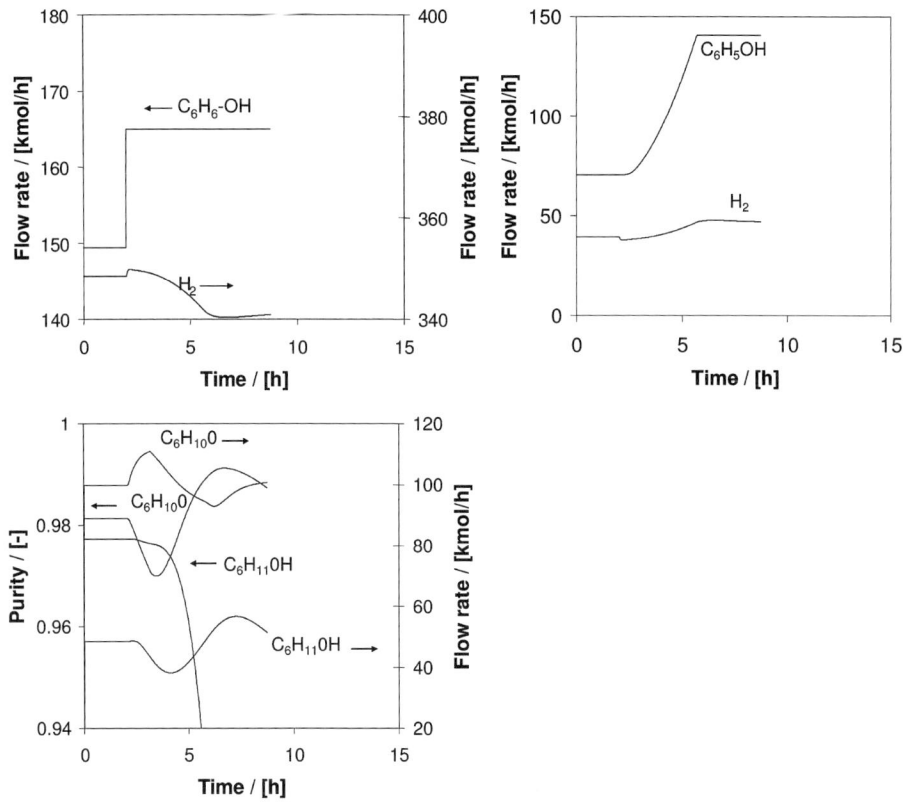

Figure 5.24 CS2: Production rate by phenol fresh feed and fixed hydrogen recycle. At $t = 2\,h$, fresh phenol increases from 149.5 to 165 kmol/h; hydrogen on makeup.

In CS3 (Figure 5.25), the hydrogen fresh feed is increased by about 9% from 348.5 to 380, while the recycle flow of phenol remains fixed to 220 kmol/h. This control structure works well. Both the production of cyclohexanone and cyclohexanol is increased by about 4%, while phenol makeup increases with 8%. The purity of both products remains above 98%. A somewhat shorter transition time is obtained. The fact that hydrogen "pushes" the plant better than phenol is quite surprising, but it can be explained by the fact that there is no snowball effect on the gas-recycle side.

In CS4a (Figure 5.26) the reactor-inlet flow rates of both reactants are manipulated. This control structure works very well and it seems to be the best in terms of high purity and short transition times.

Two additional simulations are presented in Figures 5.27 and 5.28, where reactor-inlet phenol and reactor-inlet hydrogen are, respectively, varied, while keeping constant the flow rate at the reactor inlet for the other reactant. These alternatives work satisfactorily from the control viewpoint but lead to variation in selectivity, which probably is not desired.

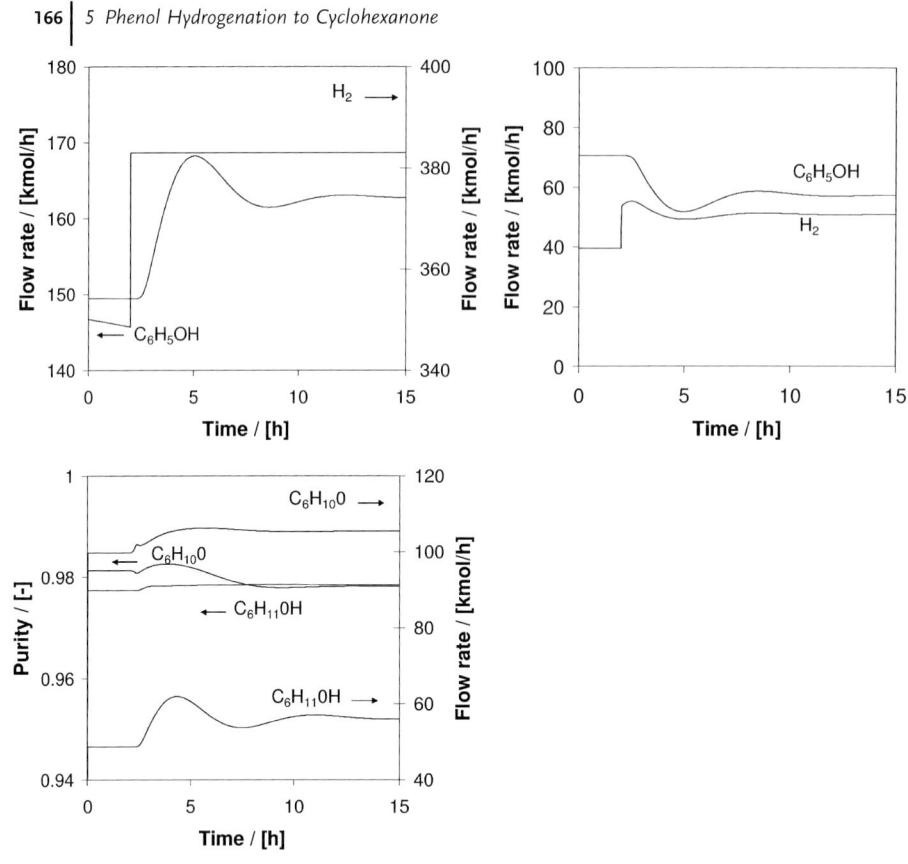

Figure 5.25 CS3: Production rate by hydrogen fresh feed and fixed phenol recycle. At $t = 2\,\text{h}$, fresh hydrogen increases from 348.5 to 383 kmol/h; phenol on makeup.

5.11
Environmental Impact

After closing the material and heat balances, we will examine the potential environmental impact (PEI) of the design. The basic information is the stream report. Table 5.16 shows material- and heat-balance data for a fresh feed of 150 kmol/h phenol and 350 kmol/h hydrogen, in total 14 822.5 kg/h. The products are cyclohexanone 9618.9 and 5017.9 cyclohexanol in the molar ratio 2:1. After simulation it is found that the amount of waste is 150.6 kg/h lights and 80 kg/h heavies. These data lead to a global yield of raw materials of 98.75%.

For analyzing the potential impact of a conceptual design we make use of the package WAR (Waste Reduction Algorithm) developed by the Environment Protection Agency in the USA [23]. The graphical interface allows the user to enter information for different alternatives, or to import it from a process simulator. Input data includes the chemicals species, the flow rates of the charac-

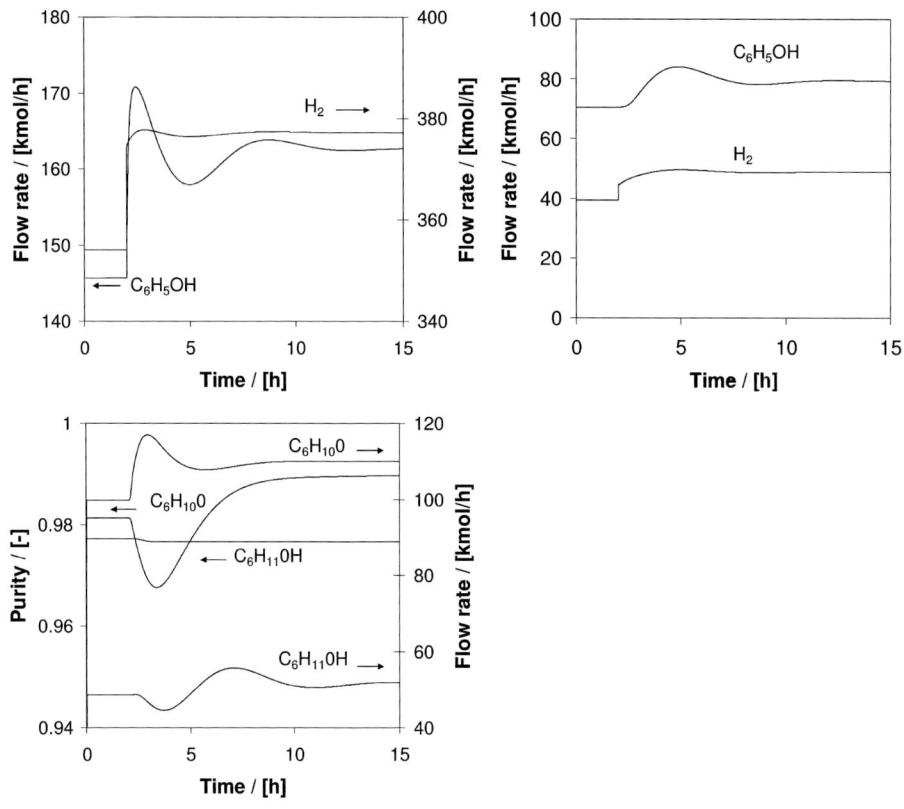

Figure 5.26 CS4a: Production rate by manipulating both reactor-inlet reactants. At $t = 2\,\text{h}$, hydrogen increases from 388 to 426 kmol/h; phenol from 220 to 242 kmol/h.

teristic streams, as feeds, products, waste, and optionally the amount of energy implied.

The WAR algorithm involves the concept of a potential environmental impact (PEI) balance. The PEI indices provide a relative indication of the environmental friendliness or unfriendliness of a chemical process. These are:

HTPI – human toxicity potential by ingestion,
HTPE – human toxicity potential by inhalation or dermal exposure,
TTP – terrestrial toxicity potential,
ATP – aquatic toxicity potential,
GWP – global-warming potential,
AP   – acidification potential,
ODP – ozone-depletion potential,
PCOP – photochemical oxidation potential.

HTPI and HTPE characterize the toxicity of materials with respect to ingestion and inhalation. TTP and ATP point out harmful effects of pollution on soil and

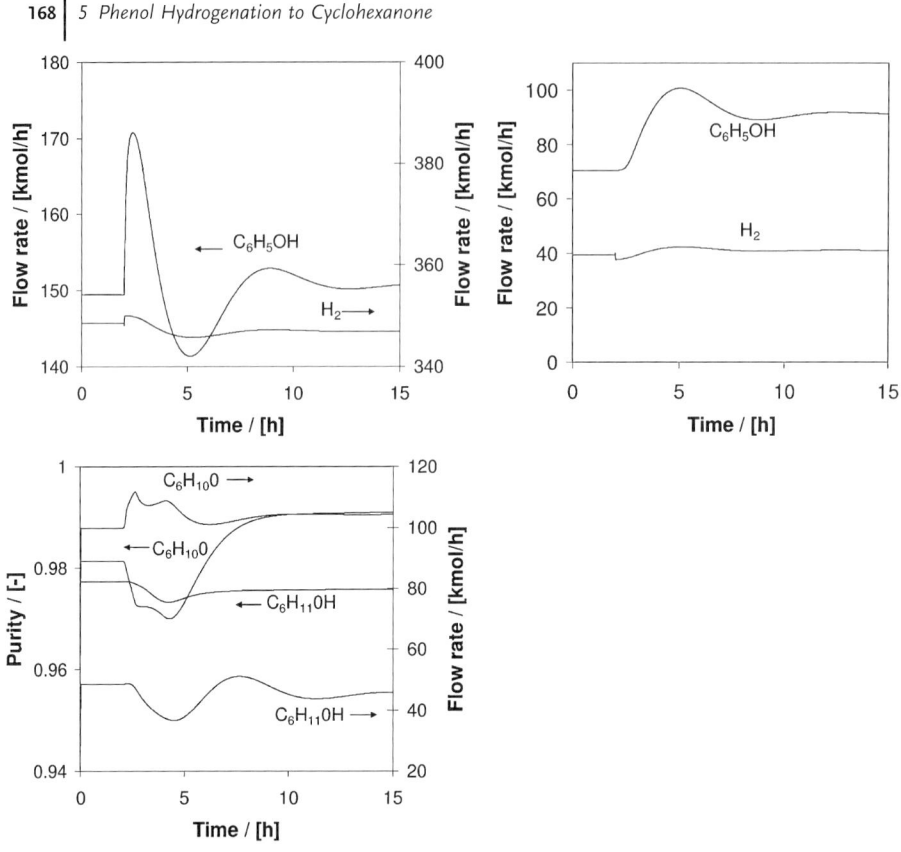

Figure 5.27 CS4b: Production rate by manipulating one reactor-inlet reactant. At $t = 2\,h$, phenol at reactor inlet increases from 220 to 242 kmol/h; hydrogen kept constant at 388 kmol/h.

water. GWP, AP, ODP and PCOP concern the impact of pollutants on the atmosphere. The database of the package WAR contains PEI for a large number of chemicals, either originated from different compilations or estimated by simulation. Higher values mean stronger impact. Weighting factors may be used to describe a particular industrial site.

In this case, the stream data can be imported directly from the Aspen Plus report file. The program WAR calculates automatically PEI values for each category and for the whole process on an hourly basis (PEI/h), or with respect to product unit (PEI/kg). The effect on the environment of the energy used for driving the process can be taken into consideration too. Figure 5.29 illustrates typical data for three cases:

1. including in analysis both light and heavy waste (base case);
2. excluding the phenol from heavies;
3. neglecting the effect of lights but keeping phenol in heavies.

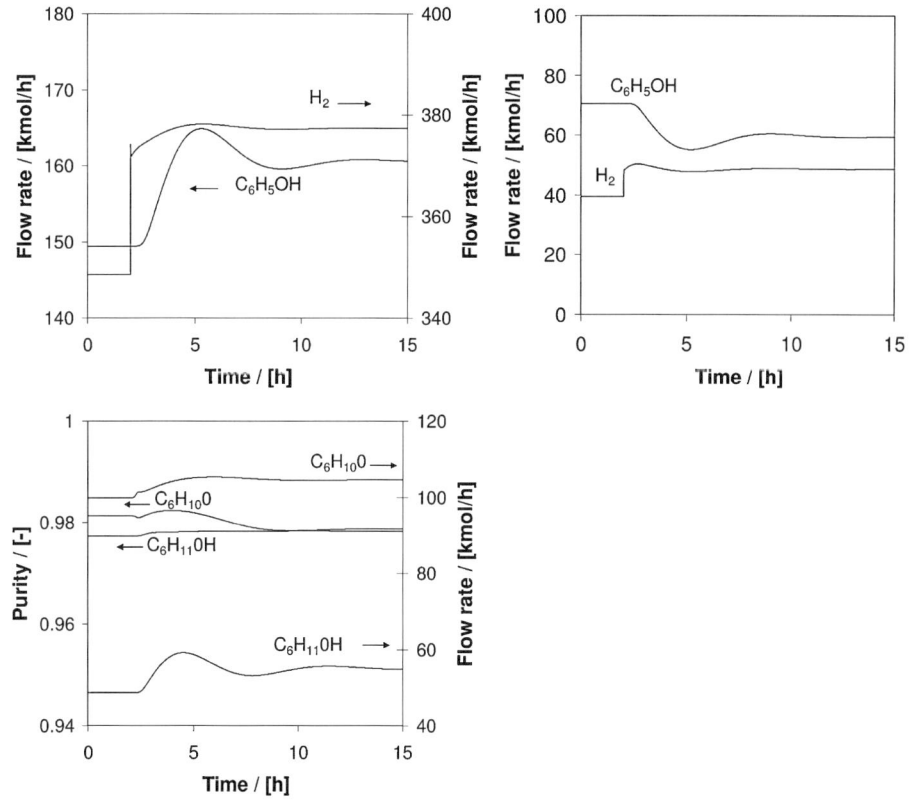

Figure 5.28 CS4c: Production rate by manipulating one reactor-inlet reactant. At $t = 2\,\text{h}$, hydrogen at reactor inlet increases from 388 to 426 kmol/h; phenol kept constant at 242 kmol/h.

It may be seen that the presence of phenol in heavy waste has the most significant environmental impact, chiefly because of its toxicity to humans and potential soil contamination. On the contrary, the environmental impact of lights is reduced, but still not negligible. The other PEIs with impact on the atmosphere are zero. Indeed, this is not surprising because there are no gaseous emissions.

The conclusion of this short analysis is that particular attention should be paid to the selection and design of the evaporator, in order to reduce the amount of phenol in the residual waste. In general, this process should not raise critical environmental and health problems. The amount of waste to environment may be reduced even more. Thus, by evaporation/distillation, about 40 kg/h phenol and 50 kg/h cyclohexanol can be recovered from waste. In this way the overall material yield can be increased to 99.35%. Additionally, cyclohexene can be reconverted in cyclohexanol and recycled into the process. In this way the amount of waste may be decreases to below 0.5%, consistent with the initial target.

Table 5.16 Material balance for one-reactor process.

Stream	Feed	C-hexanone	C-hexanol	Lights	Heavies
Mole flow kmol/h					
Phenol	150.0	0	0	0	0.4405
Hydrogen	350.0	0	0	0	0
C-hexanone	0	97.624	0.311	0.543	5 ppm
C-hexanol	0	0.376	49.793	0.016	0.0229
C-hexene	0	1 ppm	0	0.470	0
Water	0	0	0	0.671	0
Heavies	0	0	0	0	0.2014
Total flow kmol/h	500.0	98.0	50.105	1.70	0.665
Total flow kg/h	14822.51	9618.95	5017.93	105.57	80.06

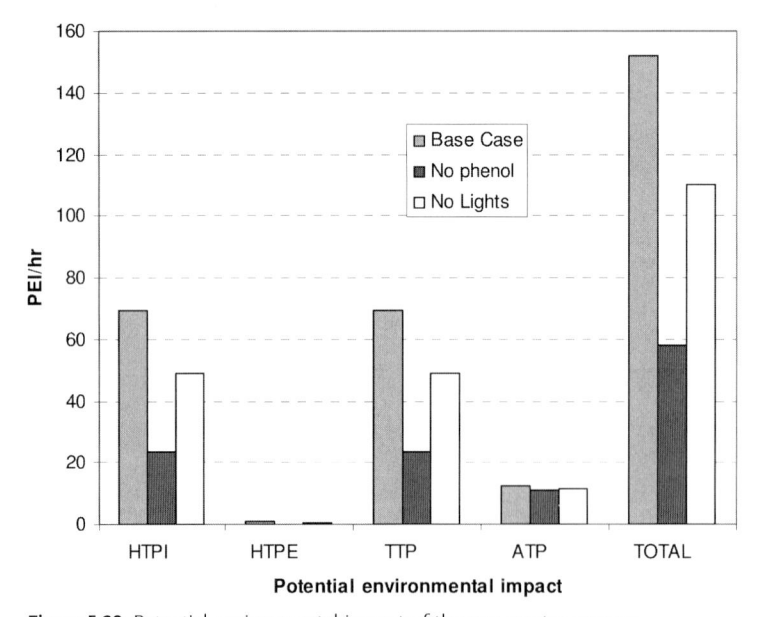

Figure 5.29 Potential environmental impact of the one-reactor process.

5.12
Conclusions

The case study of cyclohexanone manufacturing by phenol hydrogenation illustrates the basic principles of the conceptual development of a process flowsheet. For more complexity we consider a two-step process. Firstly, phenol is submitted to hydrogenation, in which both cyclohexanone and cyclohexanol are formed. After separation, the dehydrogenation of alcohol to ketone takes place in a second reactor. The goal is to raise the overall yield to the specified target, which can be only cyclohexanone or a mixture of ketone/alcohol (KA-oil). Both reaction steps are conducted in gaseous phase and make use of a solid catalyst. In the hydrogenation step, modern palladium-based catalyst can ensure per-pass selectivity better than 90% at phenol conversions between 60–70%. However, in this project we consider a less-selective platinum-based catalyst for which kinetics data are available as LHHW model. The results show that even with such a catalyst an economic process can be designed.

Both reactions are constrained by chemical equilibrium. The calculation emphasizes the need for accurate thermodynamic data. The operational parameters can be chosen such that kinetics not equilibrium is the determinant for selectivity.

This example illustrates the usefulness of designing the chemical reactor by taking into consideration the structure of recycles. In this way, the flexibility of the reactor design to changes in plant production rate or in the performance of separation units can be early assessed. For phenol hydrogenation a heat-exchanger-type PFR is used with a catalyst in tubes and raising steam outside. Since the reaction is highly exothermic high parametric sensitivity occurs. However, a robust and flexible design can be found.

In the two-step process the two reactors are coupled by the same separation system. Phenol gives azeotropes with both cyclohexanone and cyclohexanol. The relative volatility of cyclohexanone to cyclohexanol is very low at normal pressure, but it rises significantly under high vacuum. Alternative separation schemes are evaluated based on direct and indirect sequences. Both are equivalent in energy consumption, although the indirect sequence is more suitable by a decoupling effect.

Energy integration shows that important saving, up to 40%, can be obtained by using the steam generated in the hydrogenation reactor for driving at least one distillation column, namely the most difficult cyclohexanol/cyclohexanone split. Even more energy saving can be achieved by the complete suppression of the dehydrogenation section. Thus, the key factor for getting high efficiency is the availability of a selective catalyst. Even working at low conversion with one reactor is far more economical than employing two reactors. Moreover, producing only KA-oil removes the most energetic step, the separation of cyclohexanone and cyclohexanol.

The proposed one-reactor flowsheet is similar to recent technologies, both in structure and economic performance. Further flowsheet simplification would be difficult to obtain. Liquid-phase hydrogenation could suppress the evaporator, but it requires a more complicated reactor technology. As a result, the availability of low-cost phenol can make this technology highly competitive with the cyclohexane oxidation.

References

1. Ullmann's Encyclopaedia of Industrial Chemistry, Wiley-VCH, Weinheim, Germany, 2002
2. Kirk-Othmer Encyclopedia of Chemical Technology, Wiley-Interscience, 2006
3. Poling, B.E., Prausnitz, J.M., O'Connell J.P., The Properties of Gases and Liquids, 5th edn, McGraw-Hill, New York, USA, 2001
4. Perry, J.H., Chemical Engineer's Handbook, 7th edn, McGraw-Hill, New York, USA, 1997
5. Ghmeling, J., Menke, J., Krafczyk, J., Fischer, K., Azeotropic Data, Wiley-VCH, Weinheim, Germany, 1994
6. Froment, G.F., Hofmann, H., Design of fixed-bed gas-solid catalytic reactors, in Chemical Reaction and Reaction Engineering, Marcel Dekker, New York, USA, 1987
7. Dimian, A.C. Integrated Design and Simulation of Chemical Processes, CACE 13, Elsevier, Amsterdam, The Netherlands, 2003
8. Luyben, W.L., Tyreus, B.D., Luyben, M.L., Plantwide Process Control, McGraw-Hill, New York, USA, 1999
9. Aspen Plus, release 12, Aspen Technology, Cambridge, MA, USA
10. Aspen Dynamics, release 12, Aspen Technology, Cambridge, MA, USA
11. Stamicarbon, US Pat. 3,3305,586, 1967
12. Solutia Inc., US Pat. 5.808.167, 1998
13. Frenkel M.L., Yursha I.A., Kabo G.Y., Thermodynamic parameters of cyclohexanol dehydrogenation, *J. Appl. Chem. (Russia)* 62(5), 1173, 1989
14. Dodgson, I., Griffin, K., Barberis, G., Pignatari, F., Tauszik, G., A low cost phenol to cyclohexanone process, *Chem. Industry (London)*, 24, 830–834, 1989
15. Park, C., Keane, M.A., Catalyst support effects: gas-phase hydrogenation of phenol over palladium, *J. Colloid Interf. Sci.* 266, 183–194, 2003
16. Claus, P., Berndt, H., Mohr, C., Radnik, J., Shin, E., Keane, M.A., Pd/MgO catalysts characterization and phenol hydrogenation activity, *J. Catal.*, 192, 88–97, 2000
17. Mahata, N., Vishwanathan, V., Kinetics of phenol hydrogenation over supported palladium catalysts, *Catal. Today*, 49, 65–69, 1999
18. Talukdar, A.K., Bhattacharyya, K.G., Hydrogenation of phenol over platinum and palladium catalysts, *Appl. Catal. A: General*, 96, 229–239, 1993
19. Bautista, F.M., Campelo, J.M., Garcia, A., Luna, D., Marinas, J.M., Quirós, R.A., Romero, A.A., Influence of acid-base properties of catalysts in the gas-phase dehydration-dehydrogenation of cyclohexanol, *Appl. Catal.*, 243, 92–107, 2003
20. Hancil, V., Beranek, L., Kinetics of consecutive catalytic reactions: hydrogenation of phenol on platinum catalyst, *Chem. Eng. Sci.*, 25, 1121–1126, 1970
21. Gut, G., Jaeger, R., Kinetics of the catalytic dehydrogenation of cyclohexanol to cyclohexanone on a zinc oxide catalyst, *Chem. Eng. Sci.*, 37(2), 319–326, 1982
22. Bildea, C.S., Dimian, A.C., Fixing Flow Rates in Recycle Systems: Luyben's Rule Revisited. *Ind. Eng. Chem. Res.*, 42, 4578–4585, 2003
23. Young, M.D., Cabezas, H., Designing sustainable processes with simulation: WAR algorithm, *Comp. Chem. Eng.*, 23, 1477–1491, 1999
24. Fogler, R.H., Elements of Chemical Reaction Engineering, Prentice-Hall, Upper Saddle River, NJ, USA, 4th edn, 2005
25. Murogova, R.A. et al., *Zh. Prikl. Khim.*, 45, 824, 1972; Karaeva, et al., *ibid*, 48, 2659, 1975

6
Alkylation of Benzene by Propylene to Cumene

6.1
Basis of Design

6.1.1
Project Definition

Isopropylbenzene, also known as cumene, is among the top commodity chemicals, taking about 7–8% from the total worldwide propylene consumption. Today, the cumene is used almost exclusively for manufacturing phenol and acetone.

This case study deals with the design and simulation of a medium size plant of 100 kton cumene per year. The goal is performing the design by two essentially different methods. The first one is a classical approach, which handles the process synthesis and energy saving with distinct reaction and separation sections. In the second alternative a more innovative technology is applied based on reactive distillation.

Table 6.1 presents the purity specifications. The target of design is achieving over 99.9% purity. It may be seen that higher alkylbenzenes impurities are undesired. Ethyl- and butylbenzene can be prevented by avoiding olefins and butylenes in the propylene feed. N-propylbenzene appears by equilibrium between isomers and can be controlled by catalyst selectivity.

In this project we consider as raw materials benzene of high purity and propylene with only 10% propane. As an exercise, the reader can examine the impact of higher propane ratios on design.

6.1.2
Manufacturing Routes

General information about chemistry, technology and economics can be found in the standard encyclopaedic material [1, 2], as well as in more specialized books [3]. The manufacturing process is based on the addition of propylene to benzene:

$$C_6H_6 \text{ (Bz)} + C_3H_6 \text{ (P)} = C_6H_5-C_3H_7 \text{ (IPB)} \tag{6.1}$$

Chemical Process Design: Computer-Aided Case Studies. Alexandre C. Dimian and Costin Sorin Bildea
Copyright © 2008 WILEY-VCH Verlag GmbH & Co. KGaA, Weinheim
ISBN: 978-3-527-31403-4

Table 6.1 Specifications for cumene [1, 3].

Cumene purity	99.94 wt% min.
Bromine index	<5
Ethylbenzene	<100 ppm
n-propylbenzene	<200 ppm
Butylbenzene	<100 ppm

Beside isopropyl benzene (IPB) a substantial amount of polyalkylates is formed by consecutive reactions, mostly as $C_6H_5\text{-}(C_3H_7)_2$ (DIPB) with some $C_6H_5\text{-}(C_3H_7)_3$ (TPB). The main reaction has a large exothermal effect, of $-113\,kJ/mol$ in standard conditions. The alkylation reaction is promoted by acid-type catalysts. The synthesis can be performed in gas or liquid phase. Before 1990 gas-phase alkylation processes dominated, but today liquid-phase processes with zeolite catalysts prevail. Recent developments make use of reactive distillation.

Cumene processes based on zeolites are environmentally friendly, offering high productivity and selectivity. The most important are listed in Table 6.2 [4]. The catalyst performance determines the type and operational parameters of the reactor and, accordingly the flowsheet configuration. The technology should find an efficient solution for using the reaction heat inside the process and and/or making it available to export. By converting the polyalkylbenzenes into cumene an overall yield of nearly 100% may be achieved.

Figure 6.1 illustrates a typical conceptual flowsheet [3]. Propylene is dissolved in a large excess of benzene (more than 5:1 molar ratio) at sufficiently high pressure that ensures only one liquid phase at the reaction temperature, usually between 160 and 240 °C. The alkylation reactor is a column filled with fixed-bed catalyst, designed to ensure complete conversion of propylene. The reactor effluent is sent to the separation section, in this case a series of four distillation columns: propane (LPG) recovery, recycled benzene, cumene product and separation of polyisopropylbenzenes. The flowsheet involves two recycles: nonreacted benzene to alkylation and polyalkylbenzenes to transalkylation. The minimization of recycle flows and of energy consumption in distillation are the key objectives of the design. These can be achieved by employing a highly active and selective catalyst, as well as by implementing advanced heat integration.

Table 6.2 Technologies for cumene manufacturing based on zeoltes.

Process		Mobil-Raytheon	CD-Tech	Q-max/UOP	Enichem
Zeolite	3-DDM	MCM-22	Y	Beta	Beta
Reactor	Fixed bed	Fixed bed	Catalytic distillation	Fixed bed	Fixed bed

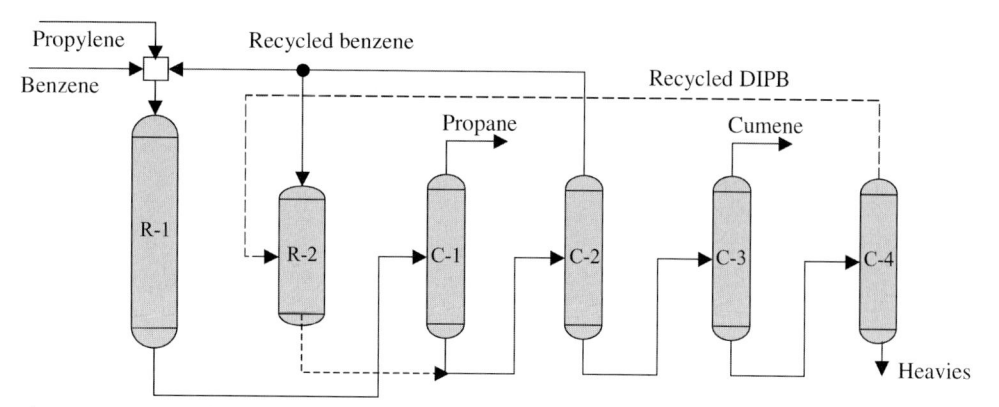

Figure 6.1 Conceptual flowsheet for manufacturing cumene by Dow-Kellogg process [3]: (R-1) alkykation, (R-2) transalkylation, (C-1) propane column, (C-2) benzene recycle column, (C-3) cumene column, (C-4) polyisopropylbenzenes column.

Table 6.3 Overall process material balance after Dow-Kellog technology [3].

Material	kg/ton cumene	Utilities	Per metric ton cumene
Feed:			
Benzene (100%)	653	Fuel for heating purposes (hot oil, steam, fired reboiler)	0.60 Gcal
Propylene (100%)	352	Steam (export) as LP at 3.5 atm	525 kg
Product	1000	Cooling water at maximum cooling air	1.08 m^3
Cumene			
Heavies	6	Electricity	17.0 kWh
LPG	variable		

Table 6.3 illustrates a typical material balance of a cumene plant using Dow-Kellog technology [3]. The propylene may contain up to 40% propane, but without ethylene and butylene. Beside cumene, variable amounts of LPG can be obtained as subproducts. Energy is also exported as LP steam, although it is consumed as well as other utilities (fuel, cooling water, electricity).

6.1.3
Physical Properties

Table 6.4 presents some fundamental physical constants. Critical pressures of propane and propylene are above 40 bar, but in practice 20 to 30 bar are sufficient

6 Alkylation of Benzene by Propylene to Cumene

Table 6.4 Basic physical properties of components in the outlet reactor mixture.

Components	Formula	M_w	T_f (°C)	T_b (°C)	P_c (bar)	T_c (°C)
Propylene	C_3H_6	42	−186.3	−47.6	46.0	91.75
Propane	C_3H_8	44	−181.7	−42.17	42.5	96.67
Benzene	C_6H_6	78	6.6	80.1	48.8	289.0
1-hexene	C_6H_{12}	84	−139.7	63.5	31.4	231.0
Cumene	$C_6H_5\text{-}C_3H_7$	120	−96.9	152.5	32.1	357.85
Di-isopropylbenzene	$C_6H_4\text{-}(C_3H_7)_2$	162	–	210.5 (para)	–	–
Tri-isopropylbenzene	$C_6H_4\text{-}(C_3H_7)_3$	204	–	(232)	–	–

to ensure a high concentration of propylene in the coreactant benzene. From the separation viewpoint one may note large differences in the boiling points of components and no azeotrope formation. In consequence, the design of the separation train should not raise particular problems. Since the liquid mixtures behave almost ideally a deeper thermodynamic analysis is not necessary. The use of vacuum distillation is expected because of the high boiling points of the polyalkylated benzenes.

6.2
Reaction-Engineering Analysis

6.2.1
Chemical-Reaction Network

The mechanism of benzene alkylation with propylene involves the protonation of the catalyst acidic sites [5, 6] leading to isopropylbenzene, and further di-isopropylbenzenes and tri-isopropylbenzenes. By the isomerization some n-propylbenzene appears, which is highly undesirable as an impurity. The presence of stronger acid sites favors the formation of propylene oligomers and other hydrocarbon species. Therefore, high selectivity of the catalyst is as important as high activity. It is remarkable that the polyalkylates byproducts can be reconverted to cumene by reaction with benzene. Below, the chemical reactions of significance are listed:

Alkylation

Benzene + $CH_2=CH-CH_3$ (Propylene) $\longrightarrow$ Cumene/isopropylbenzene (IPB) with $CH_3-CH-CH_3$ substituent (i)

Transalkylation

$$\text{Diisopropylbenzene (DIPB)} + \text{Benzene} \longrightarrow 2\ \text{(IPB)} \qquad (ii)$$

$$\text{Triisopropylbenzene (TIPB)} + \text{Benzene} \longrightarrow \text{(DIPB)} + \text{(IPB)} \qquad (iii)$$

Polyalkylation

$$\text{(IPB)} + C_3H_6 \longrightarrow \text{(DIPB)} + C_3H_6 \longrightarrow \text{(TIPB)} \qquad (iv)$$

Secondary isomerization

$$C_3H_6 + \text{Benzene} \longrightarrow \text{n-Propylbenzene} \qquad (v)$$

Dimerization

$$2\ C_3H_6 \longrightarrow H_3C-CH=CH-(CH_2)_2-CH_3 \qquad (vi)$$

Hexene

Other side reactions

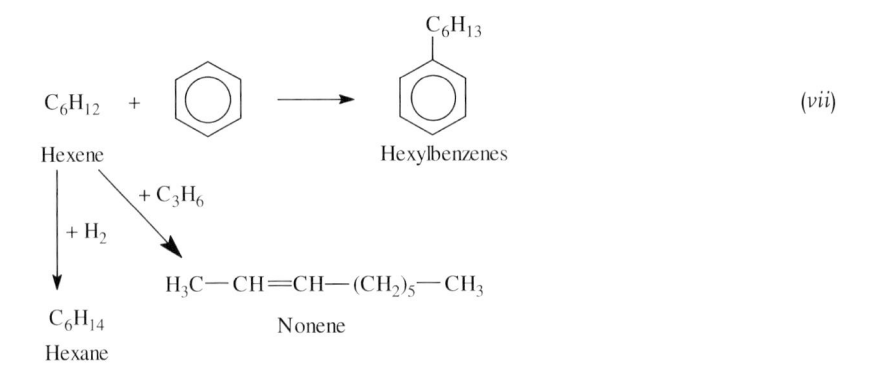

$$C_6H_{12} \text{ (Hexene)} + C_6H_6 \longrightarrow C_6H_5\text{-}C_6H_{13} \text{ (Hexylbenzenes)} \qquad (vii)$$

Hexene $\xrightarrow{+H_2}$ C_6H_{14} (Hexane)

Hexene $\xrightarrow{+C_3H_6}$ $H_3C\text{—}CH\text{=}CH\text{—}(CH_2)_5\text{—}CH_3$ (Nonene)

6.2.2
Catalysts for the Alkylation of Aromatics

In general, the alkylation of aromatics is dominated today by liquid-phase processes based on zeolites. The term "zeolitic" refers to molecular sieves whose framework consists essentially of silica and alumina tetrahedra. The complexity of tetrahedral groups may be linked in polynuclear structures. Five types of zeolites are the most applied: beta, Y, ZSM-12, MCM-22 and mordenite [5, 6]. These catalysts are characterized by large pore opening necessary for achieving high selectivity. As an illustration, Figure 6.2 shows the spatial structure of a beta-zeolite. Both the specific three-dimensional structure and the chemical composition, including the presence of doping elements, lead to substantial differences in activity and selectivity. The catalytic properties depend strongly on the surface treatment and on the activation procedure.

Since industrial catalysts are employed as pellets, the mass- and heat-transfer effects can play an important role. The internal diffusion is often the critical step controlling the overall process rate. The use of an efficient catalyst is the decisive element in designing a competitive process.

Table 6.5 presents some global yield data, including transalkylation. Zeolite-beta is often mentioned among the best suited for fixed-bed operation, with selectivity

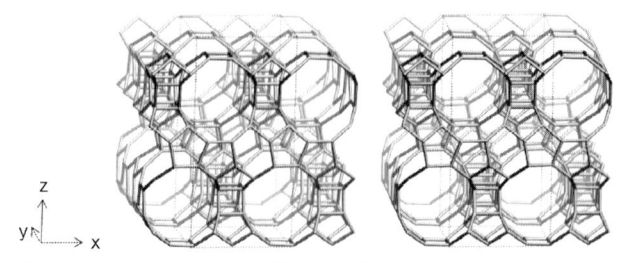

Figure 6.2 Spatial structure of beta-zeolite.

Table 6.5 Selectivity obtained with different zeolite catalysts in cumene synthesis [4].

	Zeolite-beta	Mordenite	MCM-22	Zeolite-Y
Overall selectivity on propylene (%)	99.87	98.61	98.74	98.30

Table 6.6 Selectivity and DIPB distribution at different temperature and propylene conversions [6].

Catalyst	T (°C)	$X_{propylene}$[a] (%)	Selectivity[b] (%)			Iso/n ratio	DIPBs distribution (%)		
			Cumene	DIPB	Oligo		ortho	meta	para
MCM–22	180	76.05	92.12	7.34	0.32	1650	10	30	60
		97.97	90.56	9.03	0.27	830	8	32	60
	220	91.70	90.78	8.84	0.18	790	7	33	60
		96.28	89.54	9.60	0.11	460	5	38	57
Beta	180	76.25	92.16	6.96	0.41	920	6	42	52
		97.34	90.76	8.33	0.25	900	5	44	51
	220	89.90	89.34	10.07	0.21	720	5	46	49
		98.34	88.67	10.58	0.15	460	3	51	46

Reactions conditions: total pressure, 3.5 MPa; B/P mole ratio, 7.2. Catalyst with Si/Al ratio about 16.
a) Different propylene conversion were achieved by changing the WHSV.
b) Selectivity referred to propylene.

in cumene around 90%. Other studies prefer MCM-22 because of better stability against deactivation [7]. As Table 6.6 shows, the selectivities of zeolite-beta and MCM-22 are similar in the range of temperature of 180–220 °C and benzene/propylene ratios of 3.5–7.2. Modified Y-type zeolites were found capable of selectivity over 97% at lower temperature [8], and are therefore recommended for catalytic distillation. Recent patents show that the new superactive zeolite catalysts are suitable for both alkylation and transalkylation reactions.

At quasiequal selectivity the differences in performance of catalysts can be justified by the amount of trace impurities produced. Because operating at lower temperature is more favorable, material efficiency seems to be in contradiction with heat integration.

Table 6.7 displays some physical properties of zeolites. A study issued from industry [8] demonstrates the significant role of mass-transfer resistances, even for small particles below 1 mm, reporting that pore diffusion may decrease the effectiveness from low to very low values (0.4 to 0.06). The external mass-transfer resistance is much less important. In consequence, in commercial operation only a small part of the catalyst is effectively used, typically less than 10%. Since the

Table 6.7 Physical properties of a zeolite catalyst.

Surface area, m^2/g	500–800	Particle size	Extrudates 1.6–2.4 × 4–10 mm
Particle porosity	0.5	Void fraction	0.35–0.4
Particle density, kg/m^3	1000	Tortuosity	5

reduction of particle size is restricted in practice for technological reasons, another alternative is the use of a surface-coated monolith catalyst. However, high exothermicity raises problems with respect to temperature control. In conclusion, the subtle combination of chemical and physical factors leads to a large variability in the behavior of the commercial catalysts with respect to reaction rate and selectivity.

6.2.3
Thermal Effects

A critical issue in reactor design is exploiting at best the high exothermicity of the alkylation reaction. Note that the thermal effect corresponds roughly to the evaporation of 3.67 moles benzene. A measure of exothermicity is the *adiabatic temperature rise* illustrated in Figure 6.3 as a function of the molar ratio benzene/propylene with the inlet temperature as a parameter. Higher dilution with benzene can make it fall significantly, from 120 °C to less than 60 °C; the inlet temperature plays a minor role. On the other hand, higher benzene/propylene ratio gives better selectivity, but increases the cost of separations. As a result, the ratio benzene/propylene is a key optimization variable.

Other measures for better temperature control could be employed, such as a series of reactors with intermediate cooling, or injection of a cold inert. The simulation shows that these methods have no significant effects on the overall yields, although they may offer a better protection of the catalyst in long-time operation.

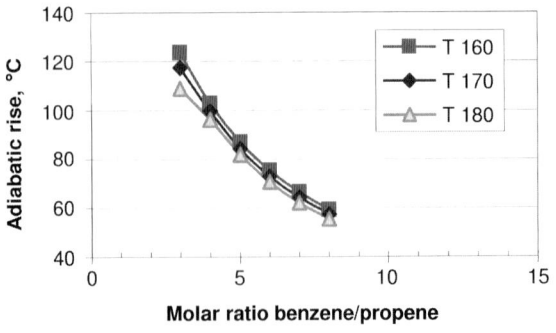

Figure 6.3 Adiabatic temperature rise as a function of the reactant ratio and inlet reactor temperature.

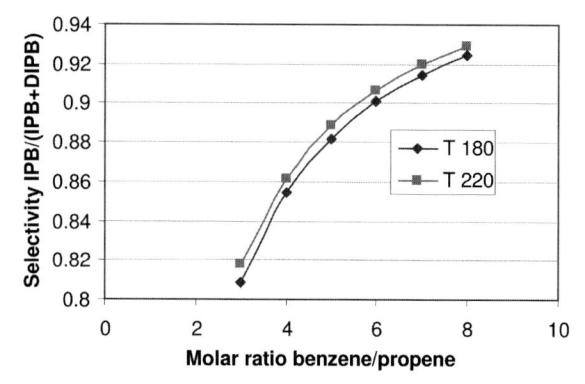

Figure 6.4 The variation of selectivity with the molar ratio benzene/propylene.

6.2.4
Chemical Equilibrium

Chemical equilibrium indicates that more than 99% conversion of propylene may be achieved for benzene/propylene ratios larger than three. However, the selectivity remains a problem. Figure 6.4 shows the variation of selectivity defined as cumene formed per mole of propylene, when only di-isopropylbenzene is the byproduct. Increasing the ratio from 3 to 9 moles gives a significant selectivity improvement from 82% to over 92%. From this point of view the performance of beta-zeolites reported in Table 6.6 seems to achieve its thermodynamic limit. Higher temperature is beneficial for getting higher yield, but the effect is limited.

6.2.5
Kinetics

The examination of patents reveals that the operation conditions for the alkylation of benzene with propylene are temperatures between 150 and 230 °C and pressures between 25 and 35 bar. The catalyst productivity expressed as WHSV is in the range 1–10 (based on the reaction mixture) at benzene/propylene molar ratios ranging from 5 to 8.

As mentioned, from the reaction kinetics viewpoint the behavior of zeolite catalysts shows large variability. In addition, the apparent kinetics can be affected by pore diffusion. The compilation of literature revealed some kinetic equations, but their applicability in a realistic design was questionable. In this section we illustrate an approach that combines purely chemical reaction data with the evaluation of mass-transfer resistances. The source of kinetic data is a paper published by Corma et al. [7] dealing with MCM-22 and beta-zeolites. The alkylation takes place in a down-flow liquid-phase microreactor charged with catalyst diluted with carborundum. The particles are small (0.25–0.40 mm) and as a result there are no diffusion and mass-transfer limitations.

6 Alkylation of Benzene by Propylene to Cumene

The surface chemical reaction seems to follow the Eley–Rideal (ER) mechanism, in which the adsorption of propylene is predominant over benzene, as represented by the equation:

$$r_A = \frac{k_0 K_P c_P}{1 + K_P c_P} \tag{6.2}$$

in whick k_0 is the surface reaction constant, K_P the adsorption constant of propylene, and c_P its the bulk propylene concentration. The mentioned reference supplies experimental values for $k_1 = k_0 K_P$ but not for K_P. Because on the interval of interest $K_P c_P \ll 1$ we may assume a pseudo-first-order reaction.

The overall process can be affected by pore diffusion and external mass transfer. Molecular diffusion coefficients D_{PB} may be calculated by Aspen Plus. Effective pore diffusion may be estimated by the relation $D_{P,\text{eff}} = D_{PB}(\varepsilon_p/\tau_p) = 0.1 D_{PB}$, in which ε_p is the particle porosity and τ_p the tortuosity. Furthermore, the Thiele modulus and internal effectiveness can be calculated as:

$$\phi = \frac{d_p}{2}\sqrt{\frac{k_1}{D_{P,\text{eff}}}} \quad \text{and} \quad \eta = \frac{3}{\phi^2}\left(\frac{\phi}{\tanh \phi} - 1\right) \tag{6.3}$$

For calculating the external mass transfer k_l a value of $Sh = 2$ can be safely assumed. The specific mass-transfer area per unit of bed volume is $a_p - 6(1-\varepsilon_b)/d_p$, in which ε_b is the bed void fraction. The combination of resistances leads to the following expression for the apparent kinetic constant:

$$k_{1,\text{app}} = [1/k_1\eta + 1/k_1 a_p] \tag{6.4}$$

Table 6.8 presents the details of calculations for spherical particles with an equivalent diameter of 2.4 mm. It may be observed that the pore diffusion considerably affects the process rate, particularly at higher temperatures. The external mass transfer plays a minor role. Their combination leads to a global effectiveness that drops from 75% to 35% when the temperature varies from 160 to 220 °C. Based on the above elements the apparent reaction constant may be expressed by the following Arrhenius law:

Table 6.8 First-order apparent reaction constant for benzene alkylation with propylene.

T °C	k_7 1/s	D_{PB} m²/s	D_{eff} m²/s	Thiele η	k_{eff} 1/s	$k_l a$ 1/s	$k_{1,\text{app}}$ 1/s	η_{overall}
160	0.0040	1.45×10^{-8}	1.45×10^{-9}	1.99	0.807	0.0032	0.0363	0.00296	0.74
180	0.0087	1.73×10^{-8}	1.73×10^{-9}	2.69	0.711	0.0062	0.0433	0.00541	0.62
200	0.0272	2.19×10^{-8}	2.19×10^{-9}	4.23	0.542	0.0147	0.0548	0.01161	0.43
220	0.0463	2.60×10^{-8}	2.60×10^{-9}	6.06	0.475	0.0220	0.0650	0.01645	0.35

$$k_1 = 6510\exp(-52564/RT) \tag{6.5}$$

in which the reaction rate is given in kmol/m^3.s and the activation energy in kJ/kmol. Table 6.6 enables an estimation of a first-order reaction constant for the DIPB formation as:

$$k_2 = 450\exp(-55000/RT) \tag{6.6}$$

The above kinetic equations have been tested by the simulation of an adiabatic PFR. For an inlet temperature of 160°C, a benzene/propylene ratio of 7 and a spatial time WHSV of 10 a total conversion of propylene may be reached with selectivity around 90%. In conclusion, the kinetic data corresponds to a fast industrial catalyst and may be reasonably used in design.

6.3
Reactor/Separator/Recycle Structure

The following reactor performance in recycle is the aim: over 99.9% per/pass propylene conversion over 88% cumene selectivity, adiabatic temperature rise below 70°C, but a maximum catalyst temperature of 250°C. The inlet pressure should be sufficiently high to ensure only one liquid phase. Thermodynamic calculations at 35 bar indicate bubble temperatures of 198 and 213°C for propylene/benzene ratios of 1/4 and 1/7, respectively (Figure 6.5). The reactive mixture can be maintained as liquid up to about 250°C, since the concentration of propylene diminishes gradually by reaction.

In a first attempt, we simulate the reactor as an adiabatic PFR. We consider a diameter of 1.3 m and a total length of 7 m, which ensure propylene conversion over 99.9%. The feed consists of 100 kmol/h propylene at molar benzene/

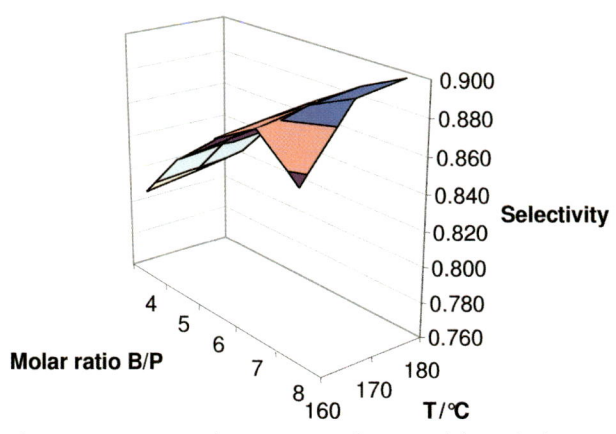

Figure 6.5 Reactor residence time as a function of the molar benzene/propylene ratio.

Table 6.9 The performance of an adiabatic PFR function of inlet temperature and excess of benzene.

B/P ratio	160 °C		170 °C		180 °C	
	T_{out}	S	T_{out}	S	T_{out}	S
4	267.7	0.8117	276.0	0.8167	280.9	0.8215
5	250.8	0.8403	258.7	0.8451	266.5	0.8496
6	238.4	0.8608	246.7	0.8654	254.9	0.8697
7	228.8	0.8755	237.5	0.8807	246.9	0.8847
8	218.9	0.8542	230.2	0.8926	238.9	0.8966

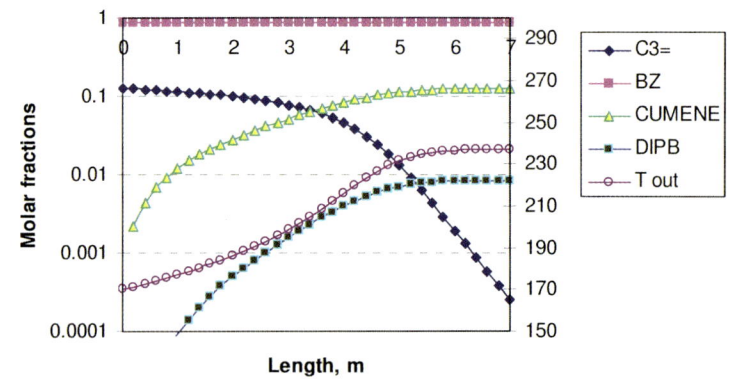

Figure 6.6 Profile of concentrations and temperature in an adiabatic PFR for cumene synthesis.

propylene ratios between 4 and 8. The inlet temperature varies between 160 and 180 °C. The reaction mixture may circulate downwards or upwards.

The results of reactor simulation are presented in Table 6.9 and in Figure 6.6. It may be observed that the larger the benzene excess the better the selectivity. Higher inlet temperature is favorable too, but the effect is limited. A larger excess of benzene can limit the adiabatic temperature rise from 100 °C to less than 60 °C. Other options are possible, such as a series of reactors with intercooling, partial cold feed of reactants, or internal heat exchangers, but we did not find a substantial advantage over a single-bed adiabatic reactor. For example, a series of two adiabatic reactors with intercooling gives a minimum amount of DIPB by splitting the propylene in the ratio 3:1, but the effect is small with an excessive temperature rise in the first reactor.

To simplify the analysis, we make use of a single adiabatic reactor. The sizing elements given before ensure the desired production rate. Figure 6.6 displays concentration and temperature profiles for an inlet temperature of 170 °C and a benzene/propylene ratio of 7. The above kinetic model gives per-pass selectivity of propylene to cumene of about 88%, in good agreement with the experimental data for MCM-22 and beta-zeolite [7], as well as a reasonable amount of DIPBs. The formation of secondary products reaches a plateau toward the exit of the reactor. The DIPB's distribution is about 5% ortho, 40% meta and 55% para, with less than 0.3% oligomers. Dividing the total throughput by the amount of catalyst gives a weight hourly space velocity (WHSV) of 11, or a residence time of 5 min. These values are in good agreement with the industrial practice.

To increase the overall yield, the DIPB is reconverted to cumene in a transalkylation reactor in the presence of a large excess of benzene. The same zeolite catalyst may be used. Practical data for the design are temperatures of 140–150 °C, benzene/DIPB ratios between 20–30 and a WHSV of 2 to 3.

6.4
Mass Balance and Simulation

At the reactor outlet the reaction mixture has a temperature of 230 °C and a pressure of 34 bar, the molar composition being 86.6% benzene, 12.6% cumene and 0.8% DIPB. Other components are *lights*, in this case the propane entered with the feed, and *heavies*, lumped as tri-propylbenzene.

By examining a list of physical properties in Table 6.4, it can be seen that the freezing point cannot be exploited for separations. The relatively wide boiling points show good opportunities for separations by distillation. Note that if hexene forms by propane dimerization it will accumulate in the benzene recycle loop because its removal is very difficult.

Because of the quasi-ideal VLE of binary mixtures, the boiling points of components suggests the sequencing of separations. Following the heuristics in Chapter 3 the removal of lights has to be done in the first place. The first separation (C-1) is the depropanizer column. The pressure is selected so as to ensure the condensation of the top product by air cooling. Next, follows the separation of the ternary mixture benzene/IPB/DIPB. Because of the large benzene amount the "direct sequence" is the best choice. The recovery of benzene takes place in the column (C-2). If follows the separation cumene/DIPB in the column (C-3), this time operated under vacuum and constrained by the reboiler temperature. DIPB recovered from heavies in the vacuum distillation column (C-4) is sent to the transalkylation, together with an appropriate amount of recycled benzene. It may be seen that the above flowsheet is the same as the technology displayed in Figure 6.1. The simulation follows the flowsheet in Figure 6.7.

The design of the distillation columns deserves some comment. A pressure of 12 bar is convenient for (C-1) because it gives a bottoms temperature below 200 °C and a condenser temperature of 34 °C. The design should prevent loss of benzene in the top product, which is used as GPL fuel. The feed temperature is kept at optimum, by trading the hot utility consumption against recovery. Note that in this case the shortcut predesign by Underwood–Gilliland is very different from the rating design by Radfrac. The following configuration ensures a high recovery

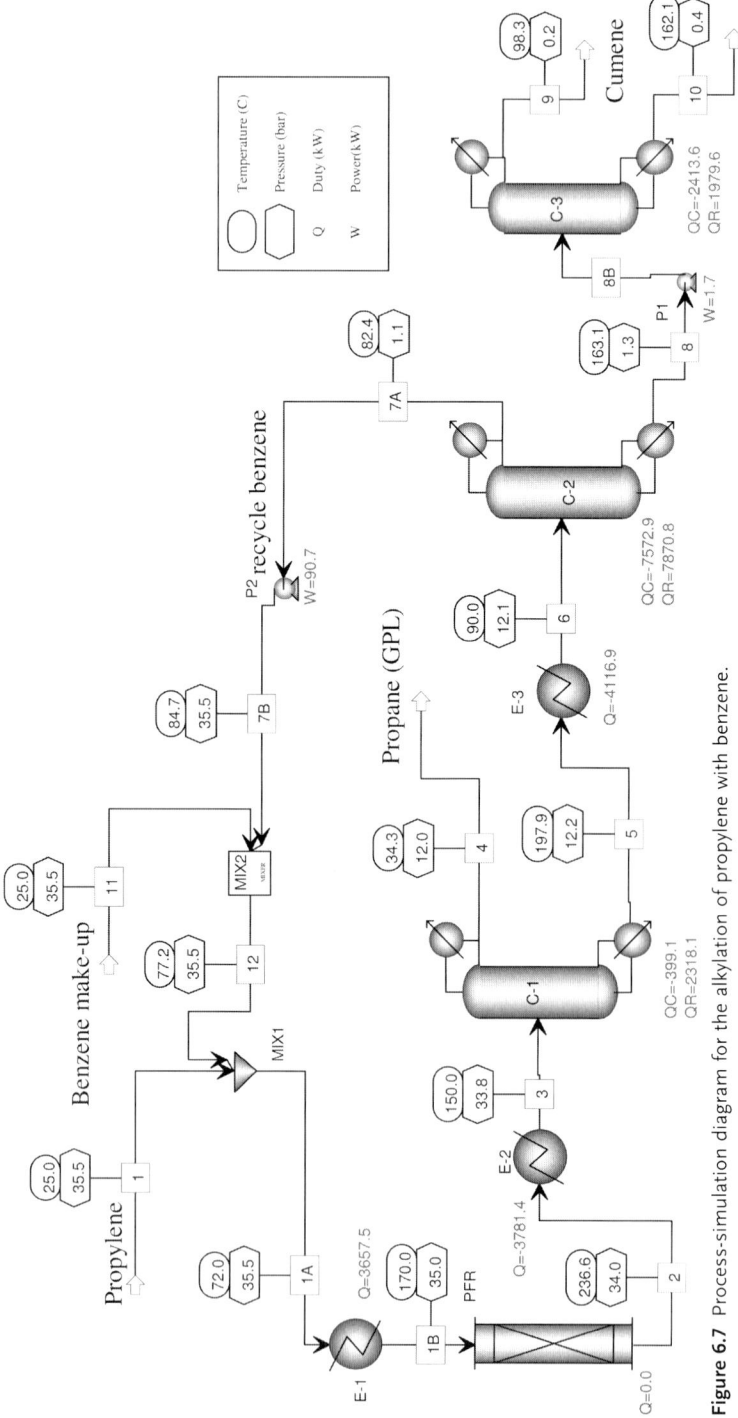

186 | *6 Alkylation of Benzene by Propylene to Cumene*

Figure 6.7 Process-simulation diagram for the alkylation of propylene with benzene.

of propane with less than 100 ppm benzene: 16 theoretical stages with feed on 5 and a temperature of 150 °C, and a reflux of 4300 kg/h.

The recycle column (C-2) is characterized by a very large distillate/feed ratio. Therefore, the design should minimize the reboiler duty. In addition, the pressure should be low enough to avoid excessive reboiler temperature. High recovery of benzene in top is desirable (over 99.9%) but small amounts of cumene are tolerated. In a first attempt we consider a column of 25 stages operated at atmospheric pressure, with feed location in the middle and a reflux ratio of 0.26.

The column (C-3) for cumene distillation operates under vacuum to avoid an excessive bottom temperature. A number of 30 stages and a reflux ratio of 1.2 are sufficient to ensure good-purity cumene with less than 100 ppm benzene.

In order to focus on the main issues of process integration, we disregard the distillation column for heavies, as well as the transalkylation section. A preliminary simulated flowsheet in Aspen Plus [9] is shown in Figure 6.8, with values of temperatures, pressures and heat duties. The fresh feed of propylene is 110 kmol/h. Note that *design specifications* are used for the fine tuning of the simulation blocks. The fresh benzene is added in the recycle loop as makeup stream so as to keep the recycle flow rate constant. This approach makes the convergence easier.

6.5
Energy Integration

Based on the preliminary flowsheet a table of streams for heat integration are built (Table 6.10). On this basis a pinch-point analysis can be done by using a specialized software, such as SuperTarget [10]. In this way, targets for energy by can be determined for ΔT_{min} of 10 °C, as shown in Figure 6.8. The minimum energy requirements are Q_h = 9143.4 kW and Q_c = 11 063.9 kW. By taking advantage only from process/process heat exchange a saving in heat up to 43% in hot utility and 40% in cold utility can be achieved. Since the reaction is highly exothermal, we expect possible export of energy too.

Table 6.10 Hot and cold stream table.

	Cold streams				Hot streams		
Name	T_s (°C)	T_t (°C)	Duty (kW)	Name	T_s (°C)	T_t (°C)	Duty (kW)
Reactor_in	72	170	3 657.5	Reactor_out	236.6	150.0	−3 781.4
				Feed_C-2	197.9	90.0	−4 116.9
Reb_C-1	198.9	197.9	2 318.1	Cond_C-1	36.3	35.3	−399.1
Reb_C-2	162.1	163.1	7 870.8	Cond_C-2	83.4	82.4	−7 572.9
Reb_C-3	161.1	162.1	1 979.6	Cond_C-3	99.3	98.3	−2 413.6
Total			15 826.6				−18 283.9

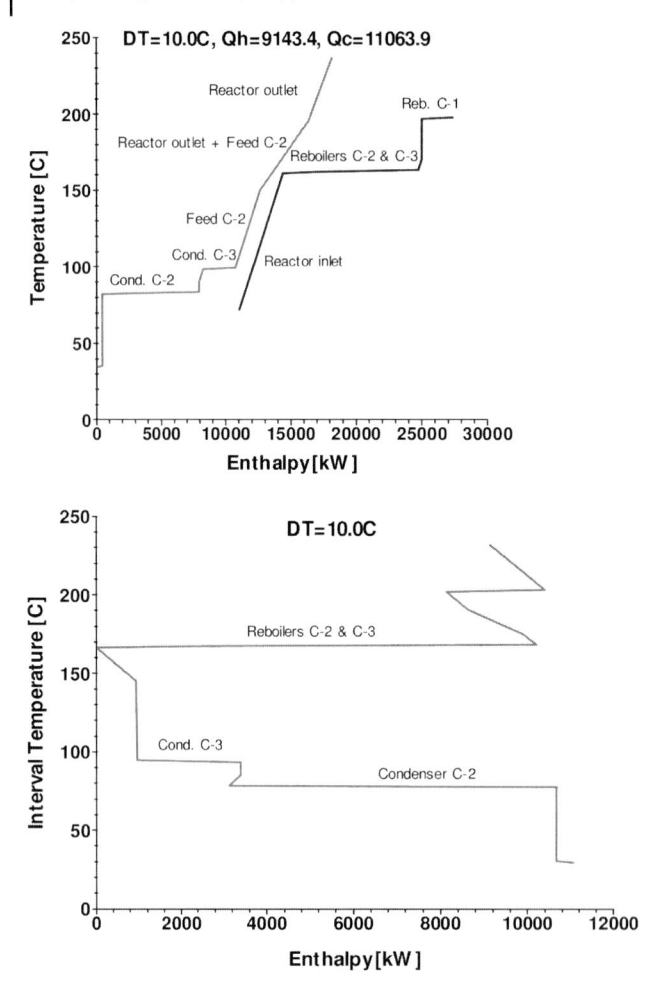

Figure 6.8 Composite curves (left) and grand composite curves (right) in the preliminary analysis.

The examination of composite curves indicates that the pinch is situated between the reactor and the benzene column. The largest energy consumer is by far the benzene column, with reboiler and condenser duties of more than 7.5 MW. By exchange between the reactor outlet and inlet streams considerable energy is saved, but only a modest amount of about 1 MW can be used for steam generation. The target temperatures of the hot streams (condensers) of the columns C-2 and C-3 show that most of the energy is rejected in the environment, at a temperature slightly below 100 °C. In consequence, a first design modification is suggested: raise the pressure in columns so as to recover the energy of condensation in a more useful form, such as low-pressure steam. On the other hand, since the temperature difference top–bottom is large (about 80 °C) the hot-utility temperature

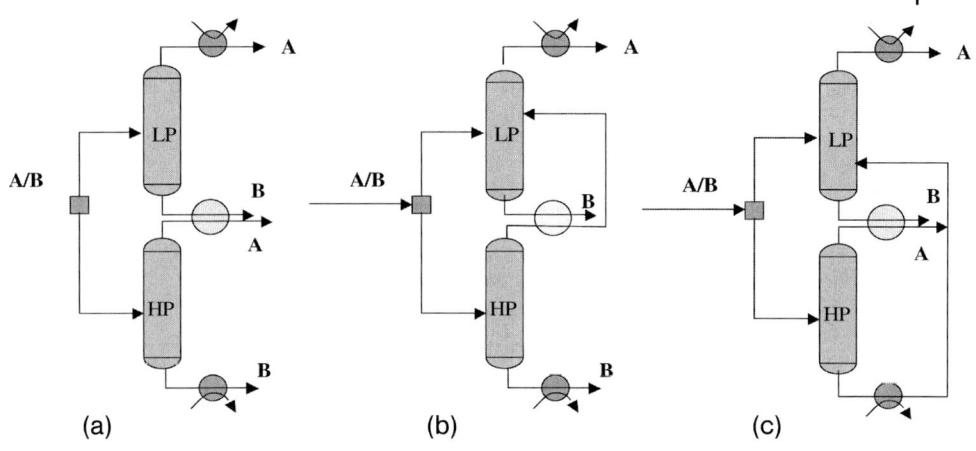

Figure 6.9 Energy saving by multieffect distillation:
(a) standard arrangement, (b) sloppy split of heavy component,
(c) sloppy split of light component.

for driving the reboiler will set a hard constraint on the maximum column pressure. Obviously, the reflux and feed temperature of the columns have to be optimized.

In the following, we focus the attention on energy saving for the benzene recycle column. Since the separation regards essentially a binary mixture with very large distillate/bottoms ratio, the analogy with multieffect evaporation seems rational. Accordingly, we split the original feed into two parts for two columns in parallel, operated at high and low pressure, respectively. In addition, the temperature of top and bottom may be manipulated by the product composition. By this approach, called multieffect distillation, three possibilities of heat integration could be imagined, as illustrated in Figure 6.9.

In alternative (a) pure products are obtained in each column. Since the relative volatility diminishes with the pressure, higher reflux is needed in the HP column. The balance of duties can be obtained by adjusting the split of the feed. Roughly speaking, by double-effect distillation the energy consumption is divided by two. In alternative (b) there is a large temperature difference between top and bottom that may be exploited by a sloppy split in the HP column with the heavy component, while in alternative (c) this is done with the light component. Alternative (c) is the best for the present case study since it allows a lower temperature of the hot utility.

The modified flowsheet is presented in Figure 6.10: the low-pressure column (C-2A) operates at 3 bar, while the high-pressure column (C-2B) operates at 8 bar. Note that the bottom stream of (C-2B) may contain an important amount of benzene. The examination of the profiles shows an imbalance in vapor flow in the lower part of the column. In consequence, the duty of the reboiler of (C-2B) may be ensured to a large extent by a side-stream heater. In this way, the heat can be transferred at a more convenient temperature level and higher driving force. In

6 Alkylation of Benzene by Propylene to Cumene

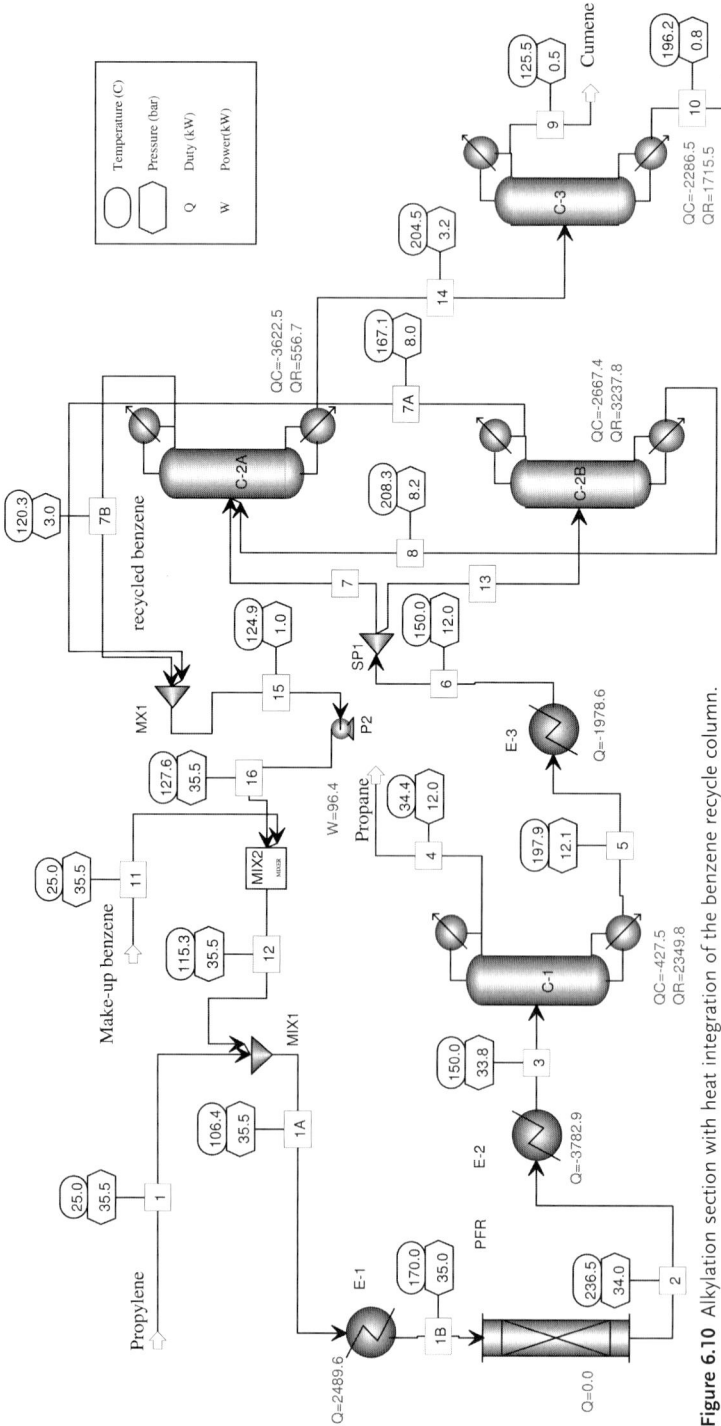

Figure 6.10 Alkylation section with heat integration of the benzene recycle column.

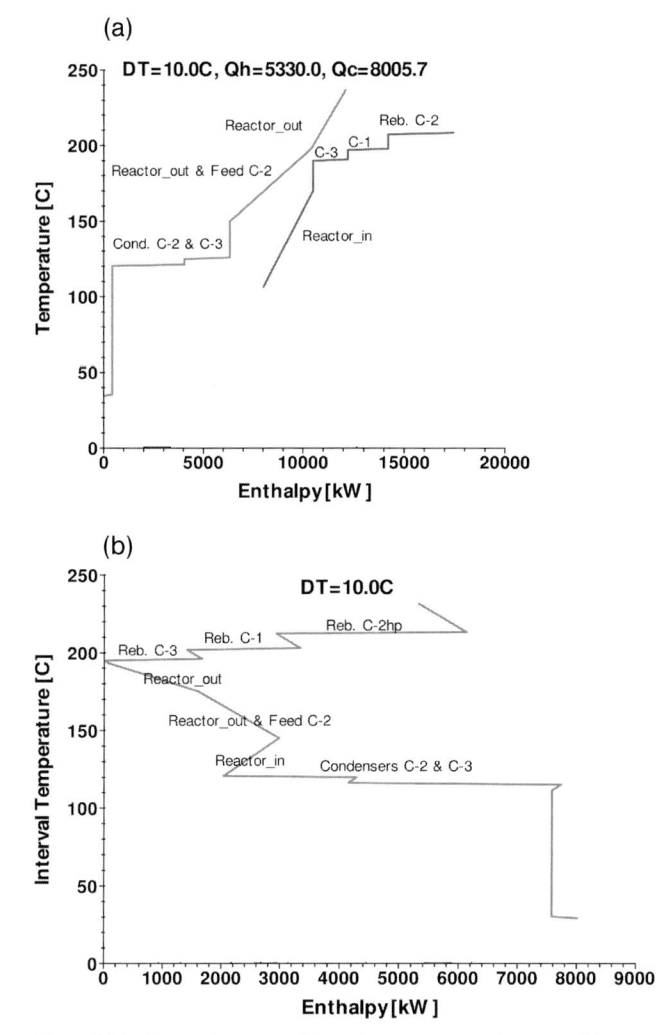

Figure 6.11 Composite curves (a) and grand composite curve (b) after the heat integration of the benzene recycle column.

the simulated flowsheet the reflux is identical in both columns at 0.26. The simulation shows that about 2660 kW may be saved by using the condenser of (C-2B) as the reboiler for (C-2A). Since the net distillate flow of C-2A is larger, a supplementary reboiler duty of about 560 kW is necessary. The net hot-utility consumption of benzene distillation drops from 7870 to 3794 kW, representing a saving of 51.8%. Furthermore, by slightly increasing the pressure in the column (C-3) allows the generation of low-pressure steam. For both columns (C-2) and (C-3) the hot utility can be ensured by Dowtherm A or another similar thermal fluid.

The impact of the above approach on the overall energy management can be examined again by plotting the composite curves (Figure 6.11). The minimum energy requirements are now $Q_h = 5330$ kW and $Q_c = 8005$ kW, much lower than before. But the salient element is that an amount of 2000 kW can be exported as process steam with a pressure of about 5 bar, as indicated by Figure 6.12 (right-hand). The reactor inlet may be matched against the reactor outlet and the feed to the column (C-2). The heat available at higher temperature may used to generate medium-pressure steam. Figure 6.12 presents the flowsheet with heat integration around the alkylation reactor that includes the preheater PREH1, feed-effluent heat exchanger FEHE, and the steam generator SG.

6.6
Complete Process Flowsheet

At this stage we introduce the second reaction step, the transalkylation of DIPBs in cumene. The final process flowsheet is presented in Figure 6.13. The benzene recycle is split to achieve a benzene/DIPB ratio larger than 10. The same catalyst as for alkylation can be used at about 160 °C. The transalkylation may be modeled as an equilibrium reactor. The equilibrium conversion is 90%, so that the overall yield in cumene can rise to over 99.4%. After the separation of heavies in the distillation column (C-4) the mixture is directed to the column (C-2LP). Note the presence of a trim heater FHEAT for controllability reasons. Because of the two recycles of benzene, the feed preheating to reactors is split accordingly. The major part is taken by the heat exchanger for the alkylation section, the other for preheating the benzene excess from transalkylation to the column (C-2LP). In this way the duty of the reboiler of (C-2LP) is lowered. The only stream available for MP steam generation is the reactor outlet. Because the net benzene recycle to alkylation diminishes, the adiabatic temperature rise is higher. To keep a maximum of 240 °C at the exit, the inlet reactor temperature has been reduced slightly to 165 °C.

Some observations regarding the flowsheet convergence should be mentioned. The convergence of the flowsheet in Figure 6.13 with only benzene recycle is difficult. To get convergence, the material balance around the distillation column (C-2) should be finely adjusted. If the recycle of benzene is not of high purity, accumulation of cumene occurs, which in turn increases the amount of DIPB. When the transalkylation reactor is introduced, the flowsheet convergence becomes easy and robust. In fact, we deal with consecutive reactions $A \xrightarrow{B} P \xrightarrow{B} R$ for which multiple steady states may occur. The presence of back reaction $R + A \rightarrow 2P$ has a stabilizing effect on the plantwide material balance, as shown in Chapter 4.

At the end of the project we can compare the energy values obtained in this project with the performance of commercial processes, as reported in Table 6.3. The targeting procedure gives about 7600 kW hot utility for a production of 3.66 kg/s cumene, equivalent with $7600/3.66 = 2076$ kJ/kg $= 0.497$ Gcal/ton. This value is close to the 0.6-Gcal/ton reported for a modern process. The potential exported energy is given by the steam generated in SG1, condensers (C-2LP) and (C-3), in total $2796+3971+1743 = 8600$ kW, or 2352 MJ/ton-cumene, which is

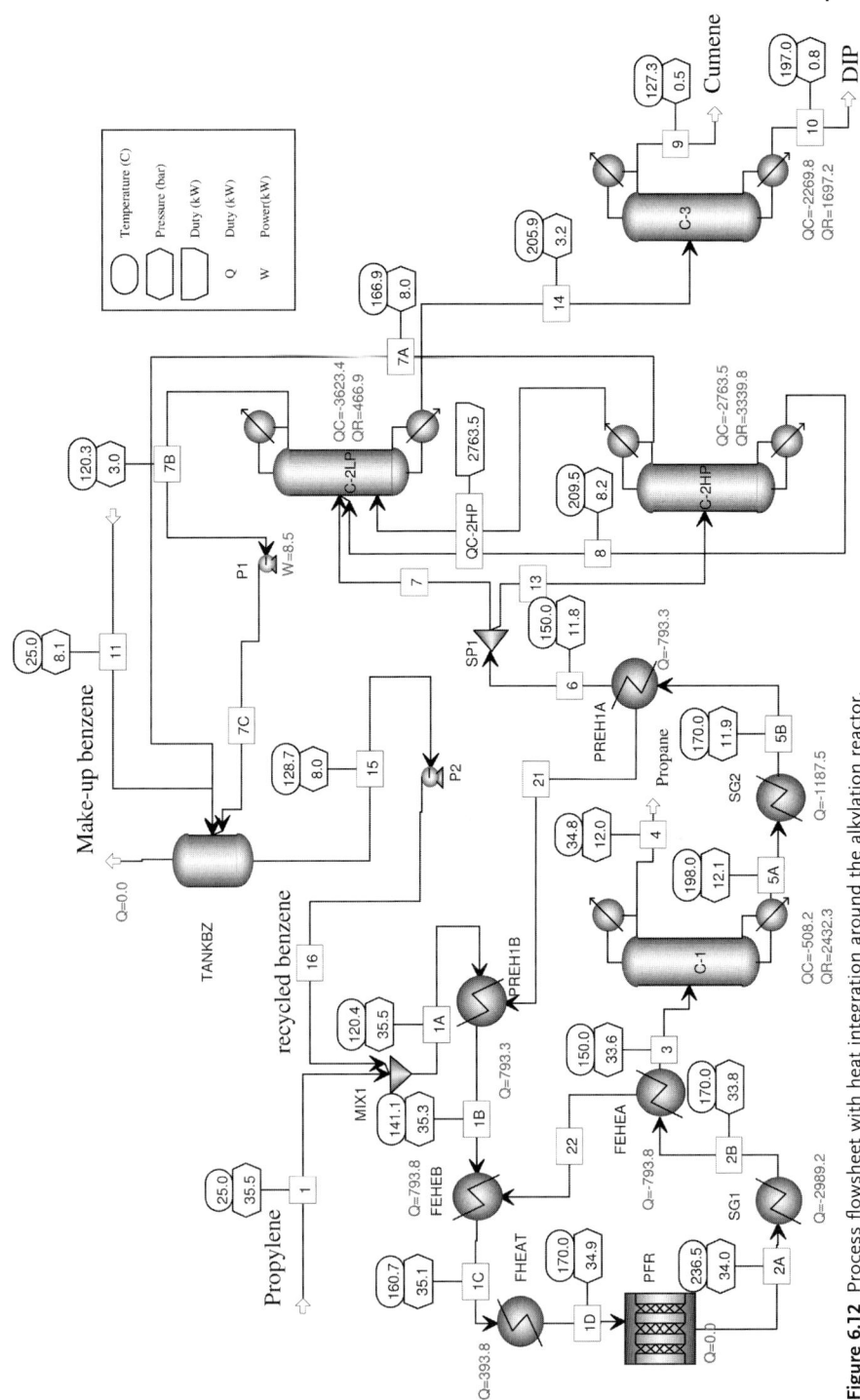

Figure 6.12 Process flowsheet with heat integration around the alkylation reactor.

6 Alkylation of Benzene by Propylene to Cumene

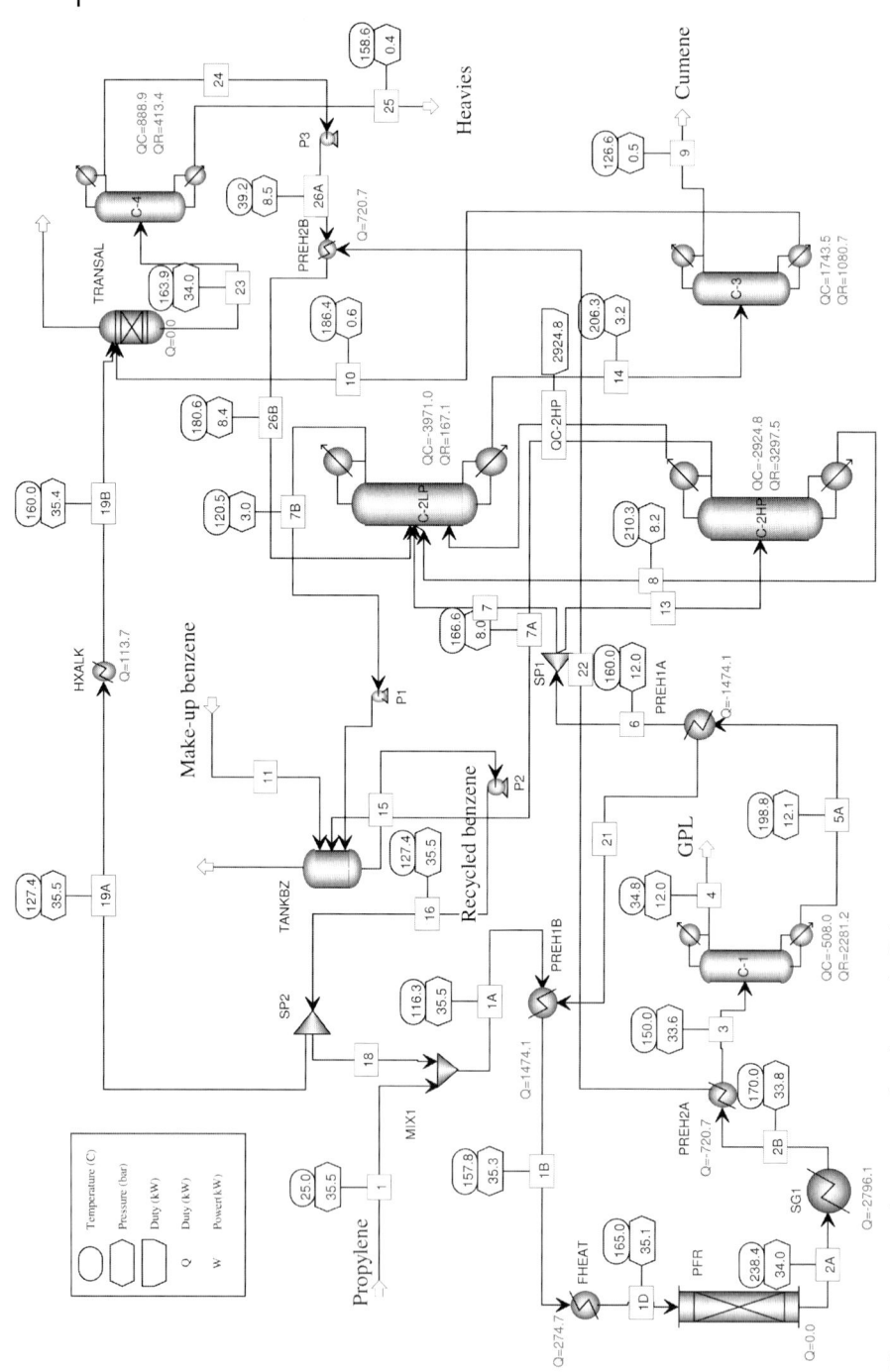

Figure 6.13 Complete process flowsheet with alkylation, transalkylation and energy integration.

equivalent to 1090 kg steam/ton-cumene (steam of 3.5 bar has an enthalpy of vaporization 2150 kJ/kg). This value is double compared with the 525 kg/h steam at 3.5 bar in Table 6.3.

It is useful to re-examine the position of the optimum benzene amount in recycle after performing the heat integration. We can define as an objective the minimization of a "loss function" that includes the cost of DIPB (reconverted to cumene), plus the cost of hot utility in the recycle column, minus the value of the raised steam:

$$\text{Loss} = \text{DIPB} \times 2 \times \text{Price\_of\_cumene} + Q_h \times \text{Price\_hot\_utility} - Q_c \times \text{Price\_LPsteam}$$

The optimization variable is the flow rate of the recycled benzene. As a constraint, the outlet reactor temperature is limited at 250°C. The first term decreases, while the second and third terms increase with higher B/P ratio. As a numerical example, we consider the following prices: 72 $/kmol (600 $/ton) cumene, 0.150 $/kWh hot utility (high-temperature thermal fluid), as well as 0.015 $/kWh for the generated LP steam. The Aspen Plus optimization routine finds an optimum at a B/P ratio around 7. Note that the optimum is rather flat, but also very sensitive to prices. For lower values of the hot utility (probable) the optimum shifts to the high B/P bound, in this case 10. This analysis demonstrates that the reaction selectivity toward the main product is the key optimization variable.

6.7
Reactive Distillation Process

The design developed so far handles reaction and separation as distinctive stages. A simpler design integrating reaction and separation in the same unit would be preferable in a reactive distillation (RD) setup. The alkylation of organic aromatics was one of the first applications of RD technology [11, 12]. A standard configuration of a RD column consists of three sections: rectification, reaction and stripping. The reaction zone can be realized by means of reactive trays or reactive packing. The company CD-Tech, one of the pioneers in field, proposed catalytic bales, as illustrated by Figure 6.14, but other possibilities for catalyst implementation are available [13].

Here, the product IPB is much heavier than the reactants, benzene and propylene, making possible easy separation in bottoms. On the other hand, propylene is much lighter than benzene, which should be used in large excess for better selectivity. It is desirable that the propylene reacts completely to avoid a new separation problem. Therefore, the column should keep only two sections, reaction and stripping. Benzene and propylene are fed at the top and at the bottom of the reaction section creating a countercurrent flow of reactants. The reaction takes place in the liquid phase in the presence of a zeolite catalyst. This time, the minimum reflux rate is dictated not by the separation but by the amount of benzene that must be condensed to remove the heat of reaction. From the heat-

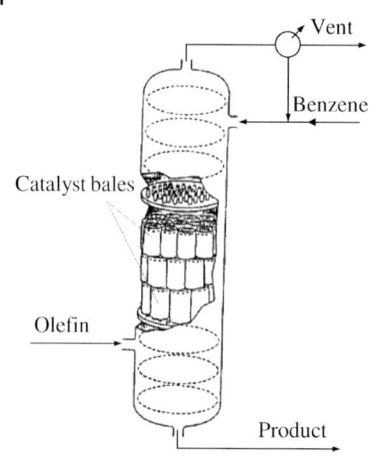

Figure 6.14 Catalytic distillation column for cumene synthesis.

recovery viewpoint the pressure should be selected to get the highest possible temperature in top. From the reaction viewpoint the temperature should favor both the reaction rate and the desired selectivity. Better selectivity is obtained at lower temperature (see Table 6.6). The above aspects are contradictory and a compromise has to be found.

If, from the heat-saving viewpoint the advantage of reactive distillation over fixed-bed reactor technology seems obvious, from reaction-engineering viewpoint there are some physical constraints that may be seen as disadvantages. Firstly, when a reactant is highly volatile the driving force for reaction in the liquid phase is diminished by the vapor–liquid equilibrium. This is the case with the propylene. The reaction rate is about $r_{IPB} = kx_B P y_P / K_P$, in which P is the column pressure and K_P is the VLE constant of propylene, As a result, the concentration of the propylene in liquid at the operating pressure (11 bar) is about an order of magnitude lower with respect to the fully dissolved gas at higher pressure (25 bar). Secondly, the flow of reactants and products in the countercurrent favors the formation of secondary polyalkylation species with respect to a cocurrent PFR. This negative effect on selectivity is amplified by the fact that in a RD setup the amount of benzene is limited by the optimum reflux rate needed for heat integration.

Therefore, adopting the solution of reactive distillation instead of separate reaction and separation units does not lead automatically to a more efficient process. Matching the conditions of separation and reaction in the same device requires careful design. The element with the highest impact is the chemical reaction. The key condition for an efficient and competitive process by reactive distillation is the availability of a superactive catalyst capable to compensate the loss in the driving force by phase equilibrium, but at the same time ensuring a good selectivity pattern.

Figure 6.15 presents a compact flowsheet based on catalytic distillation, as simulated with Aspen Plus^{TM} [9]. Benzene and propylene are fed in countercurrent in

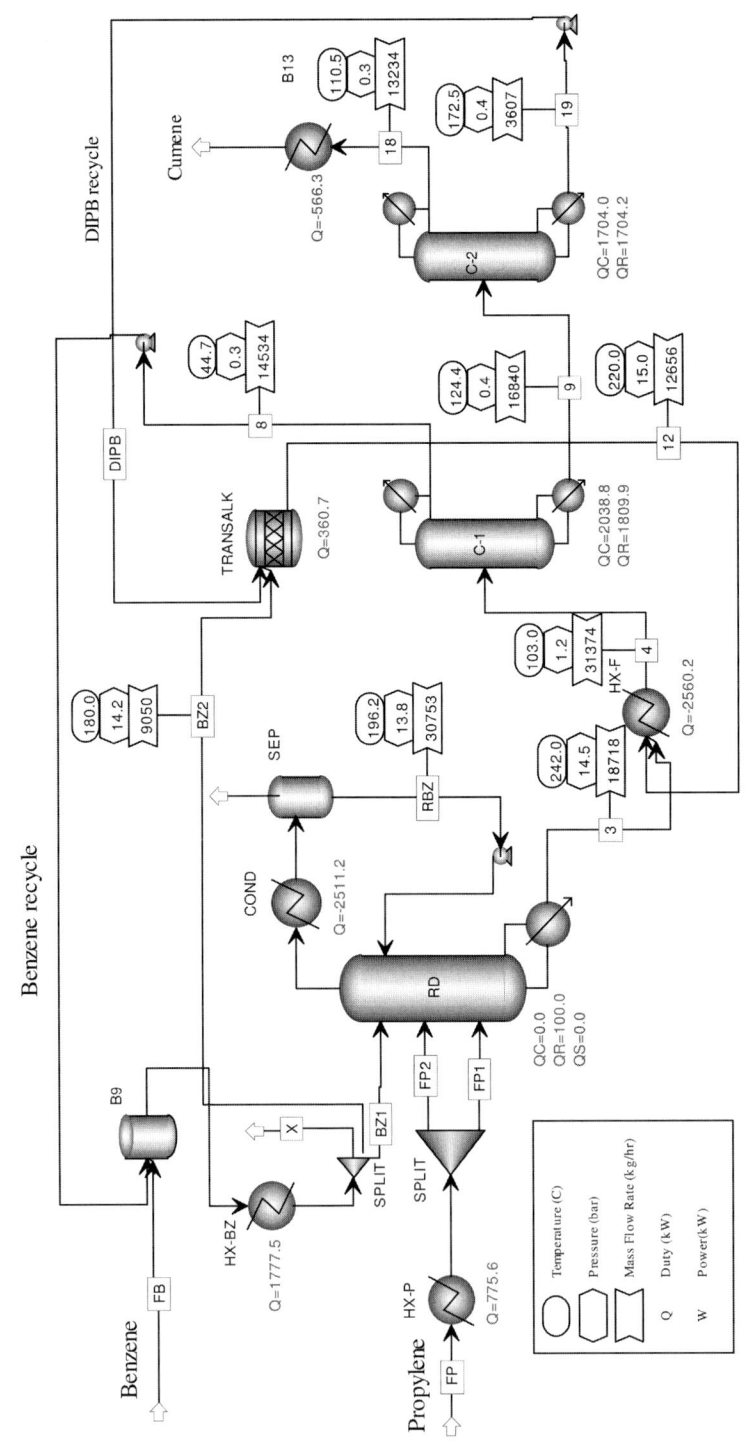

Figure 6.15 Flowsheet for cumene synthesis making use of catalytic distillation.

a RD column, the propylene feed being split into two parts. The column is designed for total conversion of propylene. The top vapor consisting essentially of nonreacted benzene is condensed and sent back as reflux. The mixture after reaction is further treated in the columns (C-1) for separating the excess of benzene and in (C-2) for separating the cumene and DIPB. Both columns operate under vacuum at about 0.3 bar. DIPB is sent to transalkylation and to be reconverted to cumene.

As a numerical example we consider a column with 39 reactive stages each of $1\,m^3$ holdup. The pure propylene feed of 110 kmol/h split into two equal parts enters the column on plates 20 and 35. This operation takes better advantage of the propylene concentration and offers better temperature control. Benzene is fed on the top stage in excess of 60% above the stoichiometry. This excess is necessary mainly to limit the temperature in bottom, but helps the selectivity to IPB too. Taking into account the reflux around the column, an overall benzene/propylene ratio larger than five may be realized. The following kinetic data are used in simulation [14, 15]:

Cumene formation: $r_1 = k_1 C_P C_B^{0.7}$ with $k_1 = 2.6 \times 10^6 \exp(-77000/RT)$ (6.7)

DIPB formation: $r_2 = k_2 C_P C_{IPB}$ with $k_2 = 2.0 \times 10^6 \exp(-80000/RT)$ (6.8)

with the reaction rate in kmol/m³s, concentrations in mol/l and energy of activation in kJ/kmol.

A pressure of 14 bar gives a good compromise between the above aspects. The RD column is simulated as reboiled stripper with reactive stages. Although the highly exothermic reaction should make unnecessary the use of a heat source, we consider just a small reboiler to prevent residual propylene entrained in the bottom. For this reason, few reactive stages below the low feed of propylene are useful.

Figure 6.16 displays the temperature profile and liquid-phase molar fractions for cumene and DIPB. It may be observed that the temperature is practically constant over the reactive sections with a first plateau at 200 °C and a second one at 210 °C. The top temperature is at 198 °C while the bottom temperature climbs to 242 °C. The explanation may be found in the variation of concentrations for cumene and DIPB in the liquid phase. The maximum reaction rate takes place on the stages where propylene is injected. The cumene concentration increases rapidly and reaches a flat trend corresponding to the exhaustion of the propylene in liquid phase. It may be seen that the amount of DIPB increases considerably in the second reaction zone. This variation is very different from that with a cocurrent PFR. The above variations suggest that the productivity could be improved by providing several side-stream injections and/or optimizing the distribution of catalyst activity.

It is interesting to examine the energetic performance of the new design. A simple examination of the duties from Figure 6.16 may highlight the possibilities for energy saving. The heat content of the RD outlet may cover the needs of feed preheating. The steam generated in the condenser, about 2.5 MW, may be exported

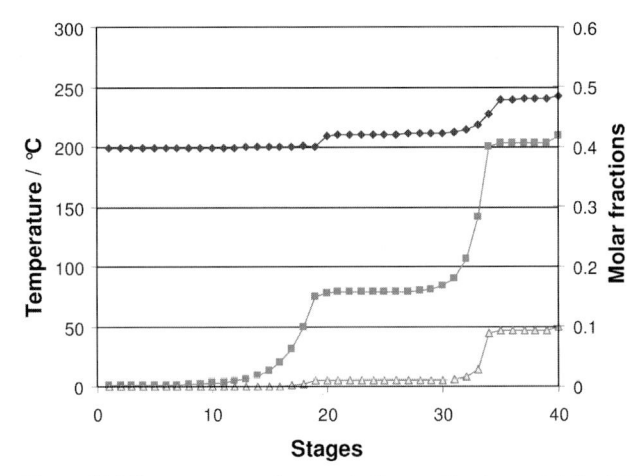

Figure 6.16 Temperature and concentration profiles in a catalytic distillation column.

as steam of 6 bar. Very low pressure steam may be raised in (C-2) and upgraded by compression to ensure the heat for the column (C-1). The high-temperature hot utility is needed only for the vacuum distillation columns.

In conclusion, compared with the previous design the flowsheet based on reactive distillation appears to be more economical as hardware and more efficient from an energy-integration viewpoint. However, the premise of feasibility is the availability of a catalyst with superior properties in terms of activity and selectivity compared with those used in a liquid-phase reactor.

6.8 Conclusions

An efficient process can be designed for the manufacturing of cumene by the alkylation of benzene by making use of zeolite catalysts available today. Simple adiabatic reactor technology is appropriate, but the operating pressure should be sufficiently high to ensure only liquid-phase reaction. To limit the formation of byproducts by consecutive polyalkylation a large ratio benzene/propylene is used, which in turn implies large benzene recycle and considerable energy consumption. The energy spent for benzene recycling can be reduced considerably by heat integration, namely by double-effect distillation. In addition, the heat developed by reaction can be advantageously recovered as medium-pressure steam. The performance indices of the conceptual design based on literature data are in agreement with the best technologies.

A modern alternative is the use of reactive distillation. At first sight appealing, this raises a number of problems. The reaction rate is considerably reduced with respect to a homogeneous liquid process because of the lower propylene concentration due to phase equilibrium. In addition, the countercurrent flow of reactants and products favors the formation of secondary polyalkylation species. Therefore, catalytic distillation becomes economically interesting only if a suitable catalyst is available. This should ensure much higher activity and better selectivity compared to a liquid-phase process. If these conditions are fulfilled the catalytic distillation is superior by more compact equipment and better use of energy.

References

1 Ullmans Encycloepedia of Chemical Technology, Wiley-VCH, Weinheim, Germany, 2003
2 Weisermel, K., H.J. Arpel, Industrial Organic Chemistry, Wiley-VCH, Weinheim, Germany, 2003
3 Wallace, J.W., Gimpel, H.E.; The Dow-Kellogg Cumene Process, in Meyer's Handbook of Petroleum Refining Processes, McGraw-Hill, New York, USA, 2nd edn, 1997
4 Perego, C., Ingallina, P., Recent advances in the industrial alkylation of aromatics: new catalysts and new processes, Catal. Today, 3–22, 2002
5 Degnan, T. F., Smith, C. M., Venkat, C. R., Alkylation of aromatics with ethylene and propylene: recent developments in commercial processes, Appl. Catal. A; General, 221, 283–294, 2001
6 Perego, C. et al., Experimental and computational study of beta, ZSM-12, Y, mordenite and ERB-1 in cumene synthesis, Microporous Materials 6, 395–404, 1996
7 Corma, A., Martinez-Soria, V., Schnoeveld, E.; Alkylation of benzene with short-chain olefines over MCM-22 zeolite, J. Catal., 192, 163–173, 2000
8 Ercan, C., Dautzenberg, C. Y., Barner, H. E., Mass-transfer effects in liquid-phase alkylation of benzene with zeolite catalysts, Ind. Eng. Chem. Res., 37, 1724–1728, 1998
9 Aspen Plus, release 10, Aspen Technology, Cambridge, Massachusetts, USA
10 Super Target, release 5, Linnhoff-March, Northwich, UK
11 Smith, L.A., US Pat, 4,849,569, 1989; US Pat, 5.446.223, 1995
12 Hsieh et al., US Pat, 5.082.990, 1992
13 Krishna, R., Hardware selection and design aspects for reactive-distillation columns, in Reactive Distillation, status and future directions, Sundmacher, K. and Kiele, A. (eds.), Wiley, New York, USA, 2003
14 Han, M., Li, X., Lin, S., Intrinsic kinetics of the alkylation of benzene with propylene over beta zeolite catalyst, Kinet. Catal., 42 (4), 533–538, 2001
15 Han, M., X. Li, S. Lin, Theor. Fund. Chem. Eng., 36 (3), 259–263, 2000

7
Vinyl Chloride Monomer Process

7.1
Basis of Design

7.1.1
Problem Statement

Vinyl chloride monomer (VCM) is one of the leading chemicals used mainly for manufacturing polyvinyl chloride (PVC). The PVC worldwide production capacity in 2005 was of about 35 million tons per year, with an annual growth of about 3%, placed after polyolefines but before styrene polymers. In the 1990s the largest plant in the USA had a capacity of about 635 ktons [1], but today there are several plants over one million tons. At this scale even incremental improvements in technology have a significant economic impact. Computer simulation, process optimization and advanced computer-control techniques play a determinant role.

In the past the manufacture of VCM raised concerns regarding hazard, safety and pollution. Therefore, the VCM technology was among the first to profit from improvements suggested by process simulation. As result, the modern VCM plants are today among the cleanest and safest in the chemical process industries.

In this project the raw materials are ethylene and chlorine of high purity. The design refers to a capacity of 300 kton/year VCM polymerization grade. Information about technology used in this project can be found in encyclopaedic works [1–3], review papers [4–6], patents, as well as on the websites of leading producers. Table 7.1 shows quality specifications, namely regarding key impurities [2].

The goal of this case study is to illustrate some generic issues raised by the conceptual design of a large-scale process involving several reaction and separation sections interconnected in a complex plant with recycles. The study emphasizes the intricate effects of handling the removal of numerous impurities generated in different reactors by a common separation system, with implications on process dynamics and plantwide control.

Chemical Process Design: Computer-Aided Case Studies. Alexandre C. Dimian and Costin Sorin Bildea
Copyright © 2008 WILEY-VCH Verlag GmbH & Co. KGaA, Weinheim
ISBN: 978-3-527-31403-4

Table 7.1 Purity specifications for vinyl chloride [2].

Impurity	Maximum level (ppm)
Acetylene	2
Acidity, as HCl by wt	0.5
Acetaldehyde	0
Alkalinity, as NaoH by wt.	0.3
Butadiene	6
1-butene, 2-butene	3, 0.5
Ethylene, propylene	4, 8
Ethylene, dichloride (EDC)	10
Nonvolatiles	150
Water	200
Iron, by wt.	0.25

7.1.2
Health and Safety

Health and safety regulations (OSHA) require the monitoring of concentrations of harmful species in all facilities where VCM is produced or used. The employees should not be exposed to more than 1 ppm over 8 h, or no more than 5 ppm for periods exceeding 15 min. Chronic exposures to more than 100 ppm may lead to serious diseases. On the other hand, VCM is flammable by heat, flame and oxidizing agents. Large fires are difficult to extinguish. Because of possible peroxide formation, the VCM must be stored or transported under an inert atmosphere. The use of stabilizers prevents polymerization during processing and storage. VCM is generally transported in railroad tank cars and tank trucks.

7.1.3
Economic Indices

Table 7.2 presents some economic indices of a modern VCM technology [6], in which stoichiometric values are indicated in parentheses. The yield for chlorine may rise over 98%, the yield for ethylene being slightly lower because of losses by combustion.

7.2
Reactions and Thermodynamics

7.2.1
Process Steps

Most of the VCM technologies are based on "balanced" processes. By this is meant that all intermediates and byproducts are recycled in a way that ensures a tight

Table 7.2 Economic indices of VCM processes [6].

Raw material (kg/t VCM)	
• ethylene	462 (448)
• chlorine	578–585 (568)
• oxygen	131–139 (128)
Utility/t VCM	
• electric power (kWh)	105–110
• steam (t)	0.05–0.20
• cooling water (m³)	150–175
• fuel (Mcal)	660–700
Operators (men/shift)	4

closure of the material balance to only VCM as the final product, starting from ethylene, chlorine and oxygen. The main chemical steps are explained briefly below.

1. Direct chlorination of ethylene to 1,2-ethylenedichloride (EDC):

$$C_2H_4 + Cl_2 \rightarrow C_2H_4Cl_2 + 218\,kJ/mol$$

2. Thermal cracking (pyrolysis) of EDC to VCM:

$$C_2H_4Cl_2 \rightarrow C_2H_3Cl + HCl - 71\,kJ/mol$$

3. Recovery of HCl and oxychlorination of ethylene to EDC:

$$C_2H_4 + 2HCl + 0.5O_2 \rightarrow C_2H_4Cl_2 + H_2O + 238\,kJ/mol$$

Hence, an ideal balanced process can be described by the overall equation:

$$C_2H_4 + 0.5Cl_2 + 0.25O_2 \rightarrow C_2H_3Cl + 0.5H_2O + 192.5\,kJ/mol$$

Therefore, half of the ethylene and the whole of the chlorine are fed to direct chlorination, while the other half of the ethylene goes to oxychlorination. The process block diagram consists of three plants, as shown in Figure 7.1. Note that the purification of all EDC streams produced by synthesis, as well as recycled, can be done in a common separation section. In order to increase the overall yield, the chlorine contained in subproducts, light or heavies, can be recovered by oxidation to HCl.

Since the overall reaction is very exothermal, the VCM plant should be able to cover a large part of its energy needs.

Most of the chlorinated waste is produced in the oxychlorination step. Therefore, employing only direct chlorination of ethylene is more beneficial from the envi-

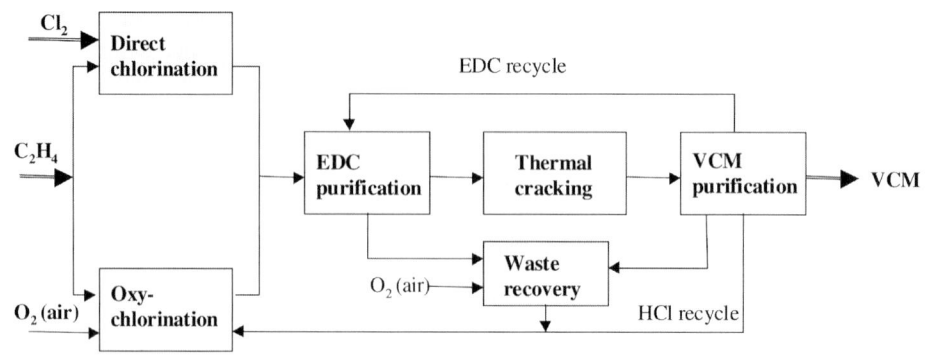

Figure 7.1 VCM manufacturing by a balanced process using oxyclorination.

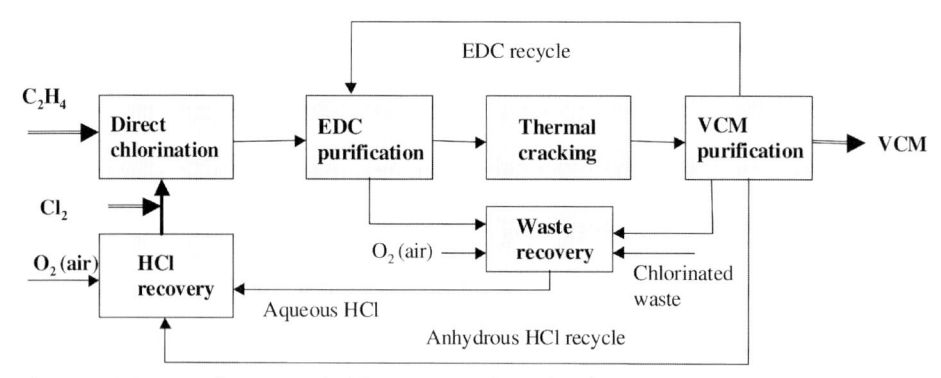

Figure 7.2 VCM manufacturing with chlorine recovery by HCl oxidation.

ronmental viewpoint (Figure 7.2), but chlorine has to be recovered from the byproduct HCl, as by means of the classical Deacon reaction:

$$2HCl + 0.5O_2 \rightarrow Cl_2 + H_2O$$

In a new process proposed by Kellogg the oxidation of HCl makes use of nitrosylsulfuric acid (HNSO$_5$) at 4 bar and 260–320 °C [9]. The amount of byproducts diminishes drastically raising the overall yield over 98%. The gaseous emissions are reduced practically to zero. Chlorinated waste from other processes may be incinerated to HCl, and in this way recycled to the manufacturing of VCM. The process is safer because the contact of hydrocarbon mixtures with oxygen is eliminated. A capital cost reduction of 15% may be estimated compared with the oxychlorination process.

Finally, it is worthy to mention the advent of a new generation of technologies based on ethane oxychlorination, as commercialized by EVC International [10]. The process is described by the following global equation:

$$2C_2H_6 + Cl_2 + 3/2O_2 \rightarrow C_2H_3Cl + 3H_2O$$

The technology profits from the extensive experience gained by the oxychlorination of ethylene, but is based on a completely different catalyst. The new process claims a cost reduction of about 30%, because the ethane price is about one third that of ethylene.

7.2.2
Physical Properties

Table 7.3 displays key physical properties of the main components. Large differences in the physical state may be observed. Ethylene and HCl are gases, chlorine and VCM may be handled as liquids at lower temperatures and adequate pressure, while ethylene-dichloride is a liquid. Therefore, the use of higher pressures and lower temperatures is expected in various processing steps.

At this stage, it is interesting to examine the properties of all chemical species, products, byproducts, intermediates and impurities with impact on the process design. Table 7.4 gives a list of normal boiling points, as well as origin and exit points. This issue will be clearer after presenting the chemistry of each reaction step. It may be seen that traces of acetylene will follow HCl in recycle. Since some chlorinated species have boiling points close to EDC, its purification will be demanding.

Table 7.3 Physical properties of the main components.

Name	Ethylene	Chlorine	HCl	VCM	EDC
Molecular formula	C_2H_4	Cl_2	HCl	C_2H_3Cl	$C_2H_2Cl_2$
Molecular weight	28.05	70.906	36.46	62.5	98.96
Normal boiling point, °C	−103.8	−34.5	−85.1	−13.4	83.4
Critical temperature, K	282.4	417	324.6	429.7	561
Critical pressure, bar	50.4	77	83.1	56	53.7
Critical volume, cm^3/mol	129	124	81	169	220
Liquid density, kg/m^3 (°C)	577 (−110)	1563 (−34)	1193 (−85)	969 (−14)	1250 (16)
ΔH_{vap} at nbp, kJ/mol	13.553	20.432	16.161	20.641	32.029
Explosion limit in air, %	2–36	None	None	4–22	6–16

7.3
Chemical-Reaction Analysis

A key aspect in the VCM technology is the removal of numerous byproducts and impurities formed in the three reaction systems. By passing through interconnected recycles some impurities produced in one reactor may induce the formation of impurities in other reactors. For example, impurities generated by the EDC

Table 7.4 Chemical species in VCM process.

Component	Formula	Nbp, (°C)	Origin	Exit
C2H2	C_2H_2	−84.0	CK	VCM separation, sent to selective hydrogenation
HCl	HCl	−66.5	CK	VCM separation, to Oxy
MECL	CH_3Cl	−24.0	CK, Oxy	VCM purification
VCM	C_2H_3Cl	−13.8	CK	VCM separation, product
BUTD	C_4H_6	−4.5	CK	EDC purification, lights
ETCL	C_2H_5Cl	12.3	DC, CK, Oxy	EDC purification, lights
CLP	C_4H_5Cl	37.0	CK	EDC purification, lights
DCM	CH_2Cl_2	40.7	CK, Oxy	EDC purification, lights
DCE =	$C_2H_2Cl_2$	47.7	CK, Oxy	EDC, purification lights
TCM	$CHCl_3$	61.3	CK, Oxy	EDC purification, lights
CCl4	CCl_4	76.7	CK, Oxy	EDC purification, lights
EDC	$1,2\text{-}C_2H_4Cl_2$	82.4	DC, Oxy	Intermediate, to CK
TRI	C_2HCl_3	86.7	CK, Oxy	EDC purification
CLAL	$C_2H_3Cl_3$	97.7	Oxy	Separating Oxy
TCE	$C_2H_3Cl_3$	113.9	DC, CK, Oxy	EDC purification, heavies
Heavies	$C_4H_8Cl_2$	155.1	CK, Oxy	EDC purification

DC: direct chlorination; CK: cracking; Oxy: oxychlorination

cracking but not removed from HCl will generate other impurities in oxychlorination, which in turn, if not removed by EDC purification, will deteriorate the selectivity of the cracking and direct chlorination steps, accumulating even more impurities in recycles, *etc.* Hence, the handling of impurities is the key issue in managing the overall plant material balance by VCM manufacturing. Because of simulation constraints, namely in dynamic mode, the number of components involved in the chemistry of different steps should be limited. However, in order to get consistent material balance it is preferable to use stoichiometric equations and avoid mass-yield expressions.

7.3.1
Direct Chlorination

The direct chlorination of ethylene consists of reaction between dissolved gaseous reactants in the liquid EDC following the equation:

$$C_2H_4 + Cl_2 \rightarrow C_2H_4Cl_2 \quad \Delta H^0_{298} = -218 \text{ kJ/mol} \tag{1}$$

The catalyst is of Lewis-acid type, in most cases $FeCl_3$, in concentrations of 0.1 to 0.5 wt%. Secondary reactions take place. The most important byproduct is the 1,1,2-trichlorethane (TCE), as described by:

$$C_2H_4 + 2Cl_2 \rightarrow C_2H_3Cl_3 + HCl \tag{2}$$

Very likely is the chlorination of EDC to trichloroethane (TRE):

$$C_2H_4Cl_2 + Cl_2 \rightarrow C_2H_3Cl_3 \text{ (TCE)} + HCl \tag{3}$$

The formation of impurities involves the occurrence of radicals. For this reason small amounts of oxygen, often present in the chlorine produced by electrolysis, may increase the selectivity to EDC by inhibiting the secondary reactions.

The use of high-purity reactants is recommended to avoid the formation of a larger spectrum of impurities that might complicate even more the EDC purification. Slight excess of chlorine is preferred in order to ensure complete ethylene conversion.

The direct chlorination of ethylene may be conducted following two techniques:

1. Low-temperature chlorination (LTC).
2. High-temperature chlorination (HTC).

In LTC the reactor is a simple gas–liquid contact column operating at temperatures between 50–70 °C, below the mixture boiling point (Figure 7.3a). Lower temperature is advantageous for achieving higher selectivity, over 99%. Two methods can be used for removing the heat of reaction, such as an internal heat-transfer device, for example a cooling coil, or external heat exchanger with recycling of cold EDC. As a disadvantage we note that rejecting the heat of reaction to the environment at low temperature is highly inefficient. Another major drawback is the catalyst removal from EDC by costly operations and sources of pollution.

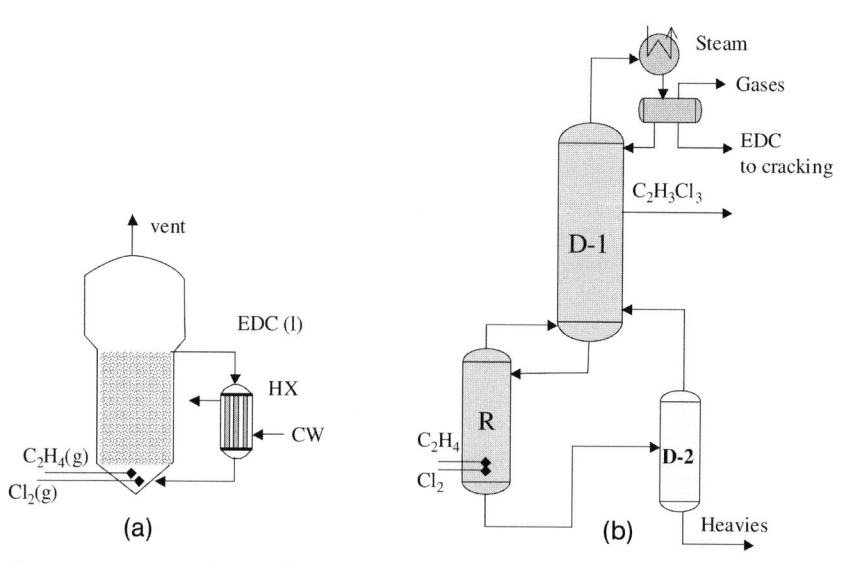

Figure 7.3 Reaction techniques for the direct chlorination of ethylene.

In HTC, the reaction is conducted at the boiling point of EDC at 1.5 to 5 bar and 90 to 150 °C. In this manner the heat of reaction, which is seven times the heat of EDC vaporization, can be used advantageously for purification. The chemical reactor may be integrated as a reboiler of a distillation column, or designed as independent equipment (Figure 7.3b). A disadvantage of HTC would be lower selectivity, but some patents claim that modified catalysts and/or the use of inhibitors make possible operation up to 150 °C.

The overall process rate involves the addition of resistances for mass transfer and chemical reaction. With respect to chemical rate, Orejas [11] proposes the following equations obtained by the regression of industrial data:

$$-r_1 = k_{D1} c_{C_2H_2} c_{Cl_2} \quad \text{with} \quad k_{D1} = 11493 \exp(-2156.58/T) \, \text{m}^3 \, \text{mol}^{-1} \, \text{s}^{-1} \quad (7.1)$$

$$-r_1 = k_{D2} c_{C_2H_2} c_{Cl_2}^2 \quad \text{with} \quad k_{D2} = 8.517 \times 10^9 \exp(-7282.21/T) \, \text{m}^6 \, \text{kmol}^{-2} \, \text{s}^{-1} \quad (7.2)$$

At 60 °C the reaction rate constants are 17.7 and 2.71, in the above units. These data are in good agreement with experimental values of 13.2 and 2.39 [12].

7.3.2
Oxychlorination

The oxychlorination step is described by the following global reaction:

$$C_2H_4 + 2HCl + 0.5O_2 = C_2H_4Cl_2 + H_2O \quad \Delta H^0_{298} = -295 \, \text{kJ/mol} \quad (4)$$

The catalyst widely used is based on copper(II) chloride impregnated on alumina. The highly exothermal reaction is carried out at temperatures around 200 °C and pressures of 1.5–5 bar, in fixed-bed or fluid-bed reactors. The fluid-bed technique offers more intensive heat transfer, prevents the occurrence of hot spots and allows more efficient catalyst regeneration. Ethylene conversion of 93–97% can be achieved with selectivity in EDC of 91–96%.

As mentioned, the spectrum and amount of impurities formed during oxychlorination is much larger compared with direct chlorination. Some key impurities are listed below: 1,1,2-trichloroethane (TCE), chloral (CCl_3-CHO), trichloroethylene (TRI), 1,1- and 1,2-dichloroethylenes, ethyl chloride, chloro-methanes (methyl-chloride, methylen-chloride, chloroform), as well as polychlorinated high-boiling components. In particular, chloral needs to be removed immediately after reaction by washing because of its tendency to polymerization.

The formation of TRI is undesired, because its removal by distillation is very difficult. In fact, TRI and EDC form a low-boiling point azeotrope very close to EDC. The formation of TRI in the oxychlorination reactor is due to the acetylene entrained with the HCl byproduct from cracking. Two solutions can be adopted:

- selective catalytic hydrogenation of acetylene in the HCl recycle stream;
- chlorination of TRI to heavies after concentration by EDC distillation.

7.3 Chemical-Reaction Analysis

Some processes make use of the first method. The second one has been highlighted by studies in the field of process dynamics and plantwide control [7, 8]. More generally, the chemical conversion of impurities is a powerful method for diminishing positive feedback effects through recycles with negative influence on operation and control.

Secondary reactions manifest, as explained next. An amount of ethylene is lost by combustion at higher temperature:

$$C_2H_4 + O_2 = Cl_2 + H_2O \tag{5}$$

Small amounts of chlorine appear since the copper catalyst is active for oxidation:

$$HCl +_{0.5}O_2 = Cl_2 + H_2O \tag{6}$$

Chlorine is involved further in radical reactions producing many chlorinated species. At higher temperature even the C–C bond in chlorinated products can break, forming chloromethanes. All these reactions lead finally to a wide spectrum of impurities. Stoichiometric equations are given below and may be used for material-balance purposes, although the true reaction mechanism is much more complex:

$$C_2H_4Cl_2 + Cl_2 = C_2H_3Cl_3 \text{ (TCE)} + HCl \tag{7}$$

$$C_2H_2 + 2Cl_2 = C_2HCl_3 \text{ (TRI)} + HCl \tag{8}$$

$$C_2H_4 + 2Cl_2 + O_2 = Cl_3C\text{–}CHO \text{ (CLAL)} + HCl + H_2O \tag{9}$$

$$C_2H_6 + Cl_2 = C_2H_5Cl \text{ (ETCL)} + HCl \tag{10}$$

$$C_2H_4 + 2Cl_2 = C_2H_3Cl_3 \text{ (TEC)} + HCl \tag{11}$$

$$C_2H_2 + 3Cl_2 = 2CHCl_3 \text{ (TCM)} \tag{12}$$

$$C_2H_4 + 4Cl_2 + O_2 = 2CCl_4 + 2H_2O \tag{13}$$

$$C_2H_4 + Cl_2 = C_2H_2Cl_2 \text{ (DCE=)} +_2HCl \tag{14}$$

$$2C_2H_4Cl_2 = C_4H_8Cl_2 \text{ (heavies)} + HCl \tag{15}$$

Despite large-scale industrial application, kinetic data for the catalytic oxychlorination are missing. For academic purposes the following kinetic equation may be applied [15]:

$$r = \frac{K_1 k_2 p_{HCl}^2 p_{C_2H_2}}{1 + K_1 p_{HCl}^2 (1 + k_2 p_{C_2H_2}/k_3 p_{O_2})} \tag{7.3}$$

The reaction rate is given in Nm^3 EDC/(m^3-catalyst h), the pressure in atm. The temperature dependence of the constants is as follows:

$$\ln k_1 = -7.27 + 19300/RT; \quad \ln k_2 = 24.87 - 21400/RT;$$
$$\ln k_3 = 12.8 - 6200/RT$$

The design of a fluid-bed reactor is difficult with conventional tools. For this reason, we will consider in this project the EDC from oxychlorination as an external source entering the purification system.

7.3.3
Thermal Cracking

At high temperature, the EDC decomposes into VCM and HCl by a complex reaction mechanism discussed further in this section. The endothermic reaction takes place at temperatures between 480–550 °C and pressures from 3 to 30 bar. The reaction device consists of a long tubular coil placed in a furnace (Figure 7.4). The first part, hosted in the convection zone, preheats the reactant up to the temperature where the pyrolysis reaction rate becomes significant. The second part, the reaction zone, is placed in the radiation chamber. The tube diameter is selected so as to give a superficial gas velocity between 10–20 m/s. The coil length should ensure a space-time of 5 to 30 s.

About 20–30 chemical components have been identified by EDC pyrolysis [1]. Many originate from species carried with the EDC issued from the synthesis steps. For this reason, advanced purification of EDC sent to cracking is required. Some impurities, such as methylchloride (<60 ppm) and butadiene (<100 ppm), are undesired because they are difficult to remove during VCM purification. Other unsaturated components, such as chloroprene or trichloroethylene lead to polymeric residues that can cause the fouling of reactors. It is worth mentioning that the process itself is capable of producing "good impurities". These species may act as a source of radicals with the effect of increasing the reaction rate and selectivity at substantially lower reaction temperature. Chloromethanes, such as $CHCl_3$ and CCl_4, are examples of good impurities. If these are not in present in EDC in sufficient amounts, they are added deliberately.

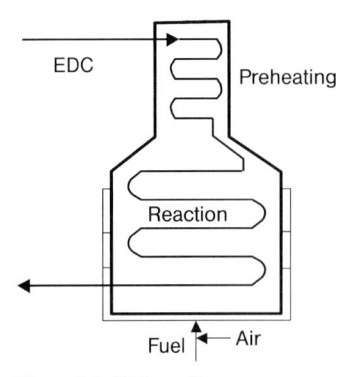

Figure 7.4 EDC cracking reactor.

The reaction mechanism by the EDC cracking in industrial conditions is extremely complex. Ranzi et al. [13, 14] proposed a scheme involving more than 200 elementary reactions as well as 40 molecular and radical species. The software SPYRO is available for the detailed design of the reaction system, including the reaction coil and the furnace (www.spyro.com). The package can also be used for monitoring the performance in operation and prevent problems, such as fouling of tubes by coke formation.

Some reactions with implications for the reactor design are given below:

$$1,2\text{-}C_2H_4Cl_2 = C_2H_3Cl + HCl \tag{16}$$

$$C_2H_3Cl = C_2H_2 + HCl \tag{17}$$

$$1,2\text{-}C_2H_4Cl_2 = C_2H_4 + Cl_2 \tag{18}$$

$$C_2H_5Cl = C_2H_4 + HCl \tag{19}$$

$$C_2H_4 + C_2H_2 = C_4H_6 \text{ (BUTD)} \tag{20}$$

$$C_2H_2 + C_2H_3Cl = C_4H_5Cl \text{ (CLP)} \tag{21}$$

The reactions (21) and (22) are particularly important. The first one leads to the main products. The second one emphasizes the formation of acetylene, further involved in the formation of other impurities, such as benzene and vinyl-acetylene. The third and the fourth reactions explain the formation of light unsaturated hydrocarbons. In the presence of free radicals they may produce a variety of higher molecular species, some unsaturated as butadiene (reaction (25)) and chloroprene (reaction (26)). The chloroprene, formed by the addition of acetylene to vinyl-chloride, is highly undesired.

Next, the reactions leading to light and heavy chlorinated hydrocarbons are worth mentioning:

$$1,2C_2H_4Cl_2 + Cl_2 = C_2H_3Cl_3 \text{ (TCE)} + HCl \tag{22}$$

$$C_2H_3Cl_3 = C_2H_2Cl_2 \text{ (DCE)} + HCl \tag{23}$$

$$C_2H_3Cl +_2Cl_2 = C_2HCl_3 \text{ (TRI)} + HCl \tag{24}$$

$$C_2H_4 + C_2H_2 = C_4H_6 \text{ (BUTD)} \tag{25}$$

$$2C_2H_4Cl_2 = C_4H_8Cl_3 \text{ (heavies)} + Cl_2 \tag{26}$$

Highly undesired is the formation of carbon deposit. This phenomenon is favored by higher wall temperature, by the presence of heavy chlorinated hydrocarbons, as well as by some heavy impurities, namely trichloroethylene (TRI). Preventing the coke formation is a major problem in operating the furnace for EDC cracking. Keeping the reaction temperature below 500 °C prevents the coke formation but decreases the reaction rate. Therefore, as already mentioned, it is rational to use "initiators", such as nitromethane, chloroform or carbon tetrachloride.

Table 7.5 Kinetic parameters for EDC pyrolysis (after Ranzi et al. [14]).

Reaction	A (s^{-1})	E (cal/mol)	k (480 °C)	k (550 °C)
1,2-C$_2$H$_4$Cl$_2$ = C$_2$H$_3$Cl + HCl	0.36 × 10^{+14}	58 000	5.30 × 10^{-4}	1.43 × 10^{-2}
C$_2$H$_3$Cl = C$_2$H$_2$ + HCl	0.5 × 10^{+14}	69 000	4.73 × 10^{-7}	2.39 × 10^{-5}
1,2-C$_2$H$_4$Cl$_2$ = C$_2$H$_4$ + Cl$_2$	1.0 × 10^{+13}	72 000	1.274 × 10^{-8}	7.62 × 10^{-7}

Ranzi et al. [13] demonstrated that the radical generation by initiation is by far more important than by chlorine elimination or C–C bond breaking.

Table 7.5 presents kinetic parameters of first-order rate constants for the key reactions by EDC pyrolysis by assuming purely molecular mechanism [14]. Following this data the main reaction just begins at 480 °C, but much higher temperature is necessary to reach an efficient reaction rate. On the other hand, because of larger activation energies the secondary reactions are promoted by increasing temperature. For these reasons, the pyrolysis temperature profile is an optimization issue. This depends on a large number of factors, such as for example pressure, heat-load distribution, nature and amount of impurities in EDC, as well as the use of initiators.

7.4
Reactor Simulation

7.4.1
Ethylene Chlorination

A detailed model for a gas–liquid column with external recirculation loop has been published by Orejas [11]. The model takes into account the axial dispersion and mass transfer from bubbles. An important conclusion is that the mass-transfer rate is fast compared with the chemical reaction. As a result, a pseudohomogeneous model for liquid-phase reaction may be applied for design purposes.

The simulation of the ethylene chlorination can be done as an adiabatic PFR with a liquid superficial velocity between 0.1 and 0.3 m/s. The results obtained by Aspen Plus [19] by using a scheme such as that shown in Figure 7.3(a) with the kinetic equations (7.1) and (7.2) is in good agreement with the industrial practice.

7.4.2
Pyrolysis of EDC

The pyrolysis reactor can be simulated in Aspen Plus as PFR with power-law kinetics and temperature profile or heat duty. To validate the kinetic data, we consider an initial flow rate of 73 000 kg/h EDC at a reaction temperature of 530 °C and 18 bar. The reactor consists of 16 tubes in parallel with an internal diameter of

Table 7.6 Reactions input in the stoichiometric reactor.

	Reaction	Conversion	Reference reactant
1	$C_2H_2 + VCM \rightarrow CLP$	0.5	C_2H_2
2	$2EDC \rightarrow HEAVIES + Cl_2$	1-S	EDC
3	$VCM + Cl_2 \rightarrow TCE$	0.5	Cl_2
4	$EDC + Cl_2 \rightarrow TCE + HCl$	0.2	Cl_2
5	$C_2H_2 + 2Cl_2 \rightarrow TRI + HCl$	1	Cl_2

100 mm. The linear velocity at the reactor inlet/outlet is of about 6 m/s and 10 m/s, respectively, sufficient to ensure a highly turbulent regime but a moderate pressure drop. The length is 250 m, corresponding to a spatial time of about 30 s.

A comparison with industrial data shows that the kinetic data from Table 7.5 gives somewhat conservative results. The temperature should be raised to more than 550 °C to achieve conversions of about 50%. It is known that modern processes operate at much lower temperatures and make use of initiators. To obtain more realistic results the pre-exponential factor for the pyrolysis reaction was modified to 1.14×10^{14}, while the pre-exponential factor of the acetylene production increased to $5 \times 10^{+14}$. The reactions (23) to (25) were neglected, while the reactions (26) to (31) were accounted for by a stoichiometric approach.

Table 7.6 gives details of the stoichiometric reactor in Aspen Plus. Note that the reaction 2 is expressed in term of a selectivity variable S, representing the degradation of EDC in heavies. This reaction is responsible also for chlorine production, further involved in other byproducts. The selectivity S may be related with the conversion X of EDC as follows:

$$S = 0.989 + 0.0506X - 0.0652X^2 \quad (0.4 < X < 0.65) \tag{7.4}$$

S expresses, in a synthetic manner, the influence of the reaction conditions on the overall selectivity. This information is essential for studying the flexibility of design. Note that the number one in the last reaction is just convention: we assume that the whole amount of free chlorine must be consumed in the last reaction, for which the real selectivity is of $1-(0.5 + 0.2) = 0.3$.

7.5
Separation System

7.5.1
First Separation Step

The synthesis of separations begins by examining the partial flow rates of components at the reactor outlet. A typical mixture at 530 °C and 18 bar is:

Component	EDC	VCM	HCL	C2H2	CLP	TCE	HEAVIES	TRI
Flow kg/h	35714	23440	13733	5	23	29	46	10

After pyrolysis, the mixture is submitted to rapid cooling by quench. This operation prevents further decomposition of VCM, but also removes the coke and other high-molecular impurities. Several alternatives are possible, as illustrated in Figure 7.5. The alternative (a), described in many schemes, makes use of flash separation with liquid recycle and external cooling. The disadvantage is energy loss because of the low temperature for heat recovery. The alternatives (b) and (c) employ a drastic cooling of the outlet gas by a liquid EDC stream coming from a downstream unit, a flash or distillation column. The exit gas temperature from quench can be such that it preserves a good driving force for the heat exchanger HX used for energy recovery by steam generation [16]. Next, the mixture is cooled and submitted to separation by a flash.

Next, the behavior of the reaction mixture submitted to a simple flash will be investigated. Because of the presence of polar species we select a thermodynamic

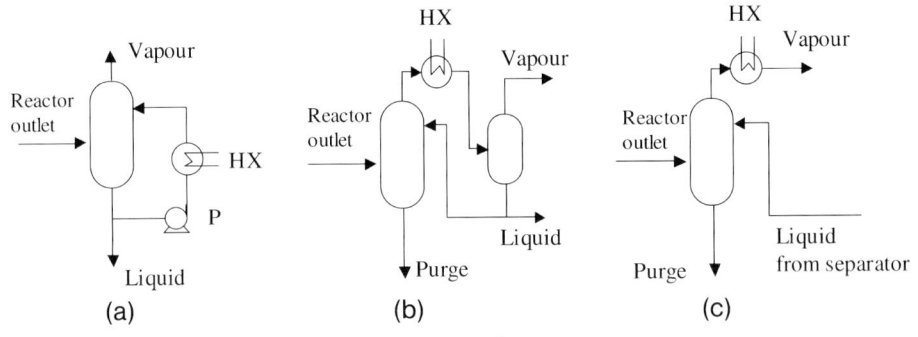

Figure 7.5 Alternatives for the quench of the reactor outlet.

Table 7.7 Phase equilibrium of the reactor-outlet mixture at 135 °C and 15 bar.

Component	F	SR-Polar EOS			NRTL-RK		
		X	Y	K	X	Y	K
EDC	0.3240	0.7809	0.2477	0.317	0.7487	0.2320	0.310
VCM	0.3367	0.1747	0.3638	2.082	0.1630	0.3744	2.297
HCL	0.3382	0.0413	0.3877	9.375	0.0856	0.3930	4.590
C2H2	0.0002	1.87E–05	0.0002	10.499	4.70E–05	0.0002	4.211
CLP	0.0002	0.0004	0.0002	0.517	0.0004	0.0002	0.480
TCE	0.0002	0.0007	0.0001	0.158	0.0006	0.0001	0.157
HEAVIES	0.0003	0.0017	0.0001	0.055	0.0014	7.8E–05	0.053
TRI	6.5E–05	0.0001	5.2E–05	0.366	0.0001	4.9E–05	0.358

option SR-Polar EOS. The second choice is the NRTL/Redlich–Kwong model, for which binary interaction parameters are known. Table 7.7 presents the results at 135 °C and 15 bar. Both models give similar results, although with some differences for C_2H_2 and HCl, the most volatile species. Hence, using equation-of-state models may be seen as sufficiently accurate for the first split by flash, but in assessing the distillations we will make use of the NRTL/RK model.

7.5.2
Liquid-Separation System

The mixture is shared approximately in equal parts between HCl, VCM and EDC. Applying the heuristics in Chapter 3, the following "direct sequence" can be developed:

Split 1: top HCl (+C_2H_2)/bottoms (VCM + EDC)
Split 2: top crude VCM/bottoms crude EDC
Split 3: top Lights (CLP)/bottoms crude EDC
Split 4: top pure EDC/bottoms TCE and heavies

The pressures in the distillation columns are selected for heat-integration reasons, namely the availability of a low-temperature utility for HCl condensation and cooling water for VCM distillation. Figure 7.6 presents the flowsheet. The distillation column (C-1) separating HCl operates at 10–12 bar, corresponding to top temperatures of −29 to −20 °C. The separation of VCM takes place in the column (C-2) at lower pressures, usually 4 to 6 bar, for which normal or slightly refrigerated cooling water can be used. The crude VCM obtained at the top of (C-2) is submitted to further purification in order meet the specifications listed in Table 7.1.

The purification of EDC from lights and heavies is performed in the towers (T-1) and (T-2). It is interesting to note that in practice two additional small columns

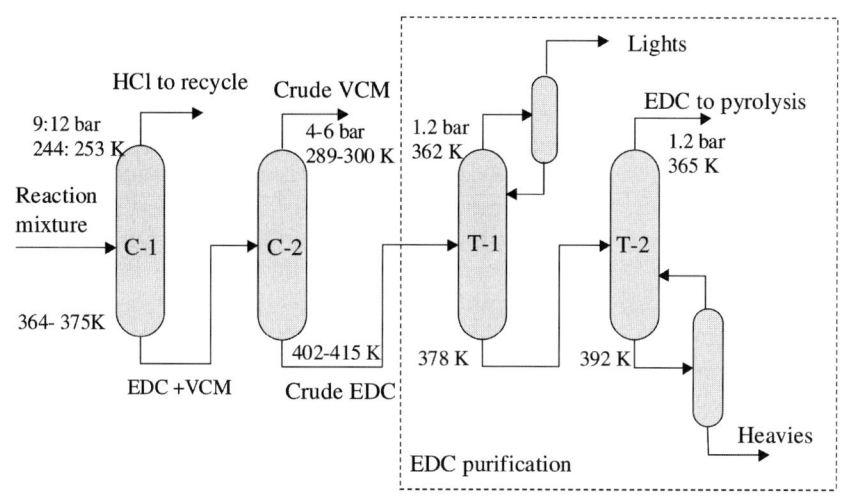

Figure 7.6 The structure of separations by a "direct sequence".

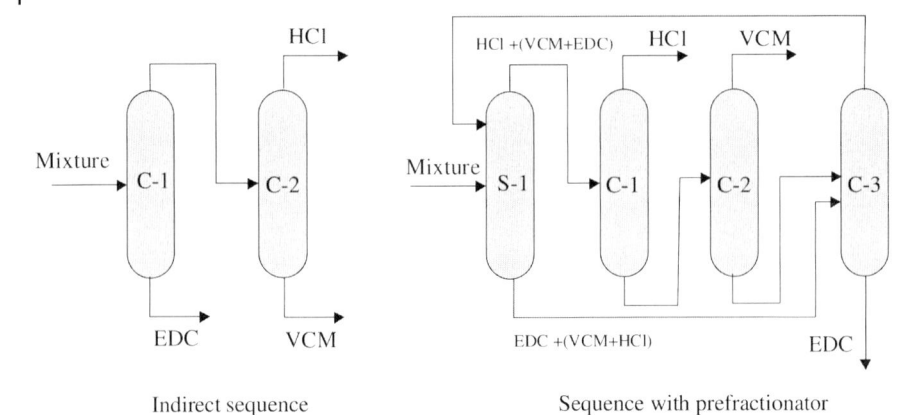

Figure 7.7 Alternatives for the separation of the HCl/VCM/EDC mixture.

are provided to minimize the losses in EDC passed in distillate and bottoms, respectively. These columns also have an important role in process dynamics and control. In addition, the chemical conversion of the nonsaturated impurities to heavies may take place by injecting chlorine either in a separate reactor or directly in the column (T-1).

The above separation sequence, although the most used in industry, is not unique. Another possibility would consist of adopting the "indirect sequence". In this case, EDC separates in the first split as bottoms, followed by VCM/HCl distillation, as pictured in Figure 7.7 (left-hand). This alternative is penalized by excessive bottom temperature in the first split at pressures above 5 bar. In addition, in the second step an intermediate compression would be necessary for an efficient separation HCl/VCM.

However, another interesting solution can be imagined, as in Figure 7.7 (right-hand). Indeed, a sloppy split in a prefractionator (S-1) can make the reboiler temperature compatible to heat integration by process/process exchange. The prefractionator could be simple flash or reboiled stripping. It follows two distillation columns for HCl and VCM separations. The last reboiled stripper column (C-3) delivers EDC for final purification. This solution ensures higher quality for VCM, obtained as top distillate and free of HCl. In addition, the scheme can better accommodate the heat integration between different columns, by varying the pressure levels. This scheme was acknowledged in an industrial paper [17].

7.6
Material-Balance Simulation

The reactor and separation systems described above can be assembled in a flowsheet. In a preliminary simulation, the goal is closing the material balance with recycles. In a first approach shortcut models may be used to simulate the distilla-

7.6 Material-Balance Simulation

tion columns. This approach guarantees the consistency of the material balance. Later, rigorous models are inserted for finishing the sizing.

Figure 7.8 presents the flowsheet. Fresh and recycled EDC enters the distillation column of lights C101. The next column C102 separates the heavies. After evaporation and temperature rise the vapor enters the cracking reactor PYRO. A second stoichiometric reactor PYROS describes the secondary reactions. The quench of the outlet mixture consists of cooling with recycled EDC. After cooling and flash separation the mixture is sent to HCl separation in the distillation column C201. The last column C-2 supplies crude VCM and EDC to be recycled. Sampled values for temperature, pressure and duties are displayed for information.

When simulating this flowsheet convergence problems appear. These may be attributed to the following reasons:

1. thermodynamic behavior of some impurities,
2. conflicts between specification of units and convergence requirements.

Firstly, the removal of TRI in the bottom of C101 is hindered by the azeotrope that this impurity forms with the EDC. Another constraint is that the concentration of TRI in the purified EDC should be kept below 1000 ppm to avoid coke formation during pyrolysis. The recovery of CLP in the top distillate must be quantitative, but with less than 8 wt% concentration to prevent polymerization. As a consequence, a substantial amount of EDC should be entrained in the top distillate. For separation reasons, both the number of trays and the L/V traffic in the column C101 should be sufficiently high. Good convergence is obtained with 35 theoretical stages and a boil-up ratio of 0.6. In principle, the removal of TRI could be enhanced by a side stream. Indeed, the liquid concentration profile indicates a maximum of TRI near the top, but the simulation shows that this effect is rather limited. Only by allowing enough EDC in the top distillate can the removal of both TRI and CLP be quantitative.

Similarly, the column C102 should be designed for advanced removal of TCE and of other heavies by allowing some EDC to be drawn in the bottom product. As a result, the excess of EDC in the top of C101 and in the bottom of C102 should be recycled by means of supplementary small columns, as shown in Figure 7.6. It is interesting to note that if the design of these columns is not appropriate, snowball effects occur by accumulation of impurities [9, 10]. We recall that a snowball consists of large variations in the recycle streams generated by small variations in the input or output streams because of excessive sensitivity of the system. Snowball may be an indication about the occurrence of possible multiple steady states.

The simulation of the columns C201 and C202 needs careful analysis of specifications. The design of HCl column must ensure both high recovery and purity of HCl in order to prevent accumulation in recycle. About 30 stages and reflux flow rate at 1000 kg/h are convenient. The distillate flow rate should ensure complete HCl recovery, but with minimum losses in VCM. From the simulation viewpoint, making use of a "design specification" is the best way to finely adjust

218 | 7 Vinyl Chloride Monomer Process

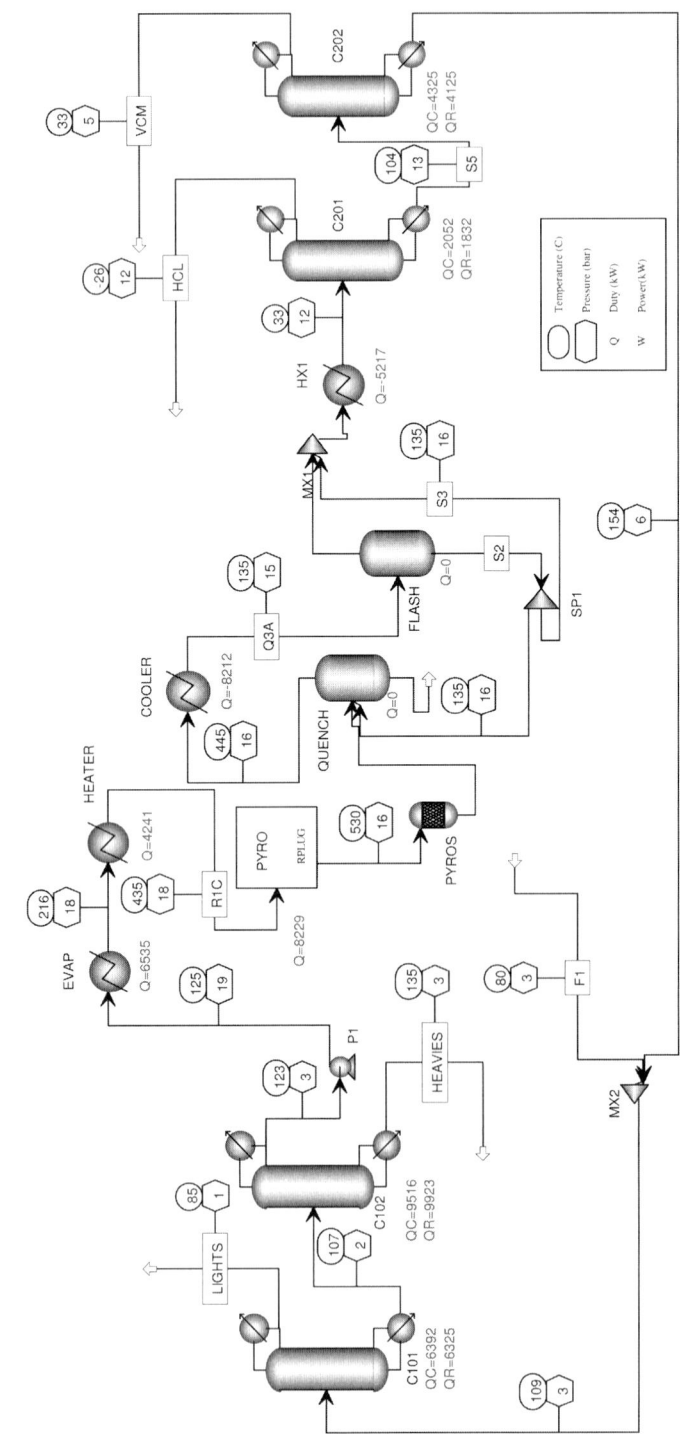

Figure 7.8 Preliminary flowsheet of the VCM plant.

the distillate rate. The next column C202 offers the choice between specifying the top distillate and bottoms flow rate, or the relative mode distillate/feed. We found that the most reliable manner is fixing the rate of the bottom product. This means keeping constant the recycle flow rate of EDC at a value compatible with the performance of the chemical reactor. This approach is equivalent to keeping constant the inlet reactor flow, also a good strategy from the plantwide control viewpoint (see Chapter 4).

7.7 Energy Integration

Table 7.8 presents the pressures, temperatures and duties of the candidate streams for heat integration. The reactor outlet has substantial energetic potential. Note that the dew point is 152 °C at 15 bar. The H–T curve shows that the heat content of the vapor part is sufficient for covering the duty of the EDC evaporator. This solution has the advantage of feeding the reactor with gaseous EDC and preventing the coke formation. If the evaporation takes place inside the furnace an upsetting effect can take place [16]. Note that the position of the heat exchanger can be before or after the quench. The second alternative offers better protection against fouling, while the temperature driving force remains high.

The preheater of the gaseous EDC is integrated in the convection zone of the furnace. The pyrolysis reactor is placed in the radiation zone. The design of the furnace is a complex aspect not treated here, in which heat integration plays an

Table 7.8 Streams for heat integration.

Hot streams	P, T, duty bar, deg. C, MW	Cold streams	P, T, duty bar, deg. C, MW
Flue gases furnace	1.013; 1100→400 9.48	Reactor inlet	22.0; 216.0→435.0 5.53
		Reaction mixture	21.0; 435.0→540.0
Reactor outlet Q1-Q2-Q3-Q4	16.0; 441.8→135.0; −10.43	EDC vaporizer F3-F4	19; 121.0→215.6; 7.50
Separation inlet S4-S4A-S4B	14.0; 135.0→30.0; −7.28	EDC preheater F4-R1	18.0; 215.6→435.0; 530
Condenser C101	1.2; 93.5→90.2; −3.58	Reboiler C101	1.5; 103.8→104.2; 3.18
Condenser C102	3.0; 124.5→123.1; −3.42	Reboiler C102	3.5; 130.4→135.6; 3.73
Condenser C201	11.0; −27.6→−28.9; −0.72	Reboiler C201	11.5; 80.8→91.0; 2.45
Condenser C202	3; 20.8→16.9; −2.98	Reboiler C202	3.5; 128.5→128.6; 2.45

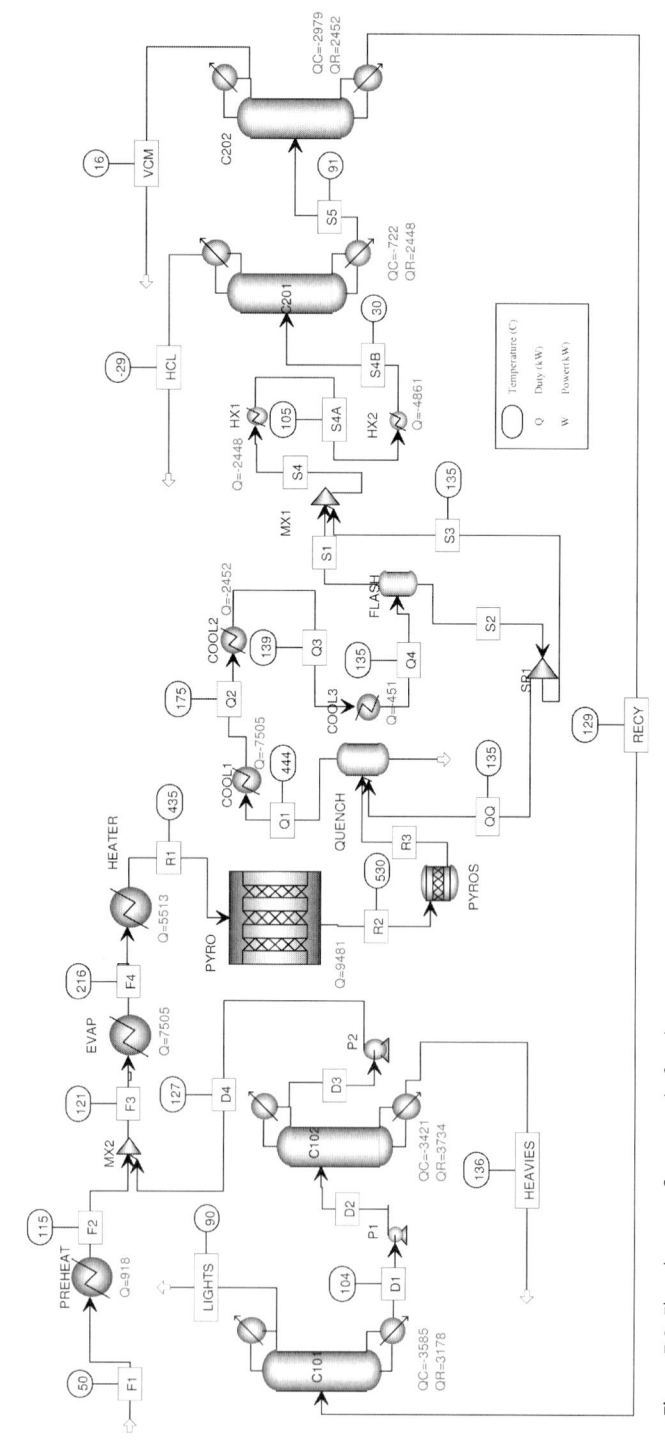

Figure 7.9 Flowsheet configuration before heat integration.

important role too. For example, the heat content of the flue gases can be used for both feed preheating and steam generation.

The enthalpy of the reactor outlet stream can be used for ensuring the reboiler duty of the column C202 (2.45 MW). This operation is simulated by the heat exchanger COOL2. The hot-outlet temperature at 139 °C ensures a minimum temperature difference of 10 °C with respect to the reboiler. Condensation starts during cooling, the final vapor fraction being about 81%. The residual heat may be rejected against air or water cooling. Placing a small heat exchanger before flash separation is recommended for better temperature control. Splitting the liquid after flash is necessary for the quench.

The resulting vapor and liquid streams are merged again, but cooled low enough to ensure a convenient thermal condition of the feed for C201. The optimum temperature is around 30 °C, which minimizes the load of the condenser, driven by an expensive refrigeration agent. A part of the heat can be used to cover the reboiler duty of C201 by the unit HX1. An outlet temperature of 105 °C gives a driving force at the cold end of about 14 °C. The remaining stream to C201 still has a significant potential of about 4.86 MW, but the temperature level is too low for heat integration.

An interesting energy saving arises by the integration of columns C101 and C102 by playing on pressures at 1.1 and 3 bar, respectively. The condenser of C102 has a duty of 3.42 MW, high enough to cover the reboiler duty of C101 of 3.18 MW. Note that the reboiler of C102 has a relatively moderate temperature (136 °C), so that MP steam can be used for heating, in turn generated by upgrading the LP steam produced by direct hot chlorination, or imported from the other sections of the plant.

After this exercise, we may conclude that by adequate heat integration the energy requirements of the pyrolysis/distillation section can be reduced considerably. The hot utility can be covered by internal means, namely by the HP steam raised in the oxychlorination stage. Refrigeration is necessary for low-temperature separation of HCl, but the amount can be minimized by accurate column design. Important saving of energy in driving the distillation columns can be achieved. The hot reactor effluent can be used to drive the reboilers of the columns for separating HCl and VCM. Supplementary saving can be obtained by the direct integration of the columns for the EDC purification. However, care should be paid to controlability aspects.

The above heat-integration scheme is obviously not unique. Figure 7.10 shows the scheme proposed by Lurgi [18] on its website. No temperature indications are given, but after this exercise it can be easily understood. It is interested to observe that the scheme makes use of two FEHE units to recover the energy of the outlet gases to preheat the inlet reactor mixture both for entering the preheater and the reaction zone to the reaction temperature. No unstable behavior take place since the pyrolysis reaction is endothermic. The quencher is in fact a multiflash separator that delivers separate gas and liquid feeds for the HCl column. The selection of pressures can lead to several options that might be explored by the reader.

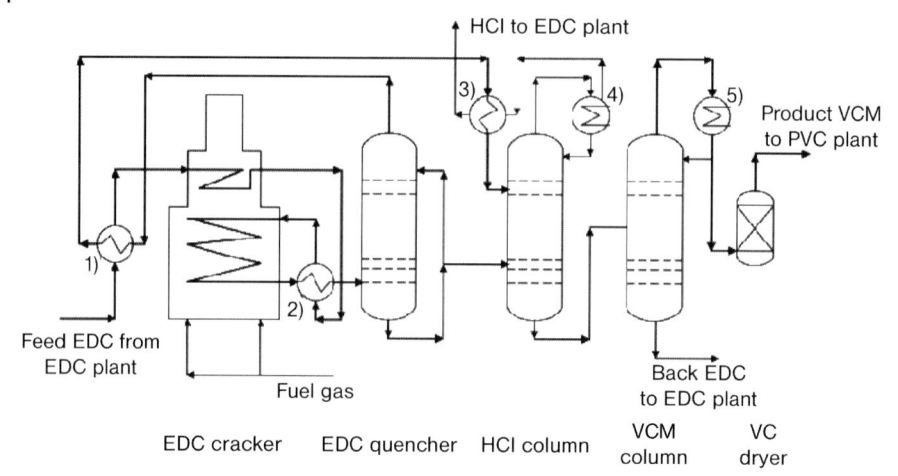

Figure 7.10 Simplified PFD of the EDC cracking/VCM separation section with heat-integration scheme (after Lurgi Company). (1) EDC preheater, (2) Feed effluent heat exchanger (FEHE), (3) EDC quencher condenser, (4) HCl column condenser, (5) VCM column condenser.

7.8
Dynamic Simulation and Plantwide Control

Dynamic simulation is prepared by sizing the units. For each unit with significant dynamics, the holdup is calculated by assuming a residence time, typically in the range of 5–10 min, and finding the corresponding volume. For the distillation columns tray sizing is necessary. Fast units, such as heat exchangers, pumps or mixers can be considered as instantaneous. Note that simulating a model of type "reactor-with-specified-temperature" is not possible, because of a high-index problem. As a result, both the temperature of the heating medium (burn gases) and the heat-transfer coefficient have to be specified. This approach is very approximate as the heat is transferred mainly by radiation. Although the performance of the pyrolysis reactor described in Section 7.4.2 could be reproduced with small adjustments of the reactor-inlet temperature and reactor length, convergence of the steady-state simulation with closed recycle is very difficult. As alternative strategy the open-recycle simulation is exported to Aspen Dynamics, which provides the basic inventory controllers. After adding temperature controllers around heat exchangers, the recycle was closed and simulation was run until a steady state was obtained. Then, control of the distillation columns was implemented.

The plantwide control structure is presented in Figure 7.11. The most interesting points are discussed below. Firstly, we fix the reactor-inlet flow rate and feed the fresh EDC on level control. Control of reactor and cooling section does not raise any problem. Since the HCl column operates mainly as a stripper, the temperature in the bottom is controlled by manipulating the steam rate, so as to ensure

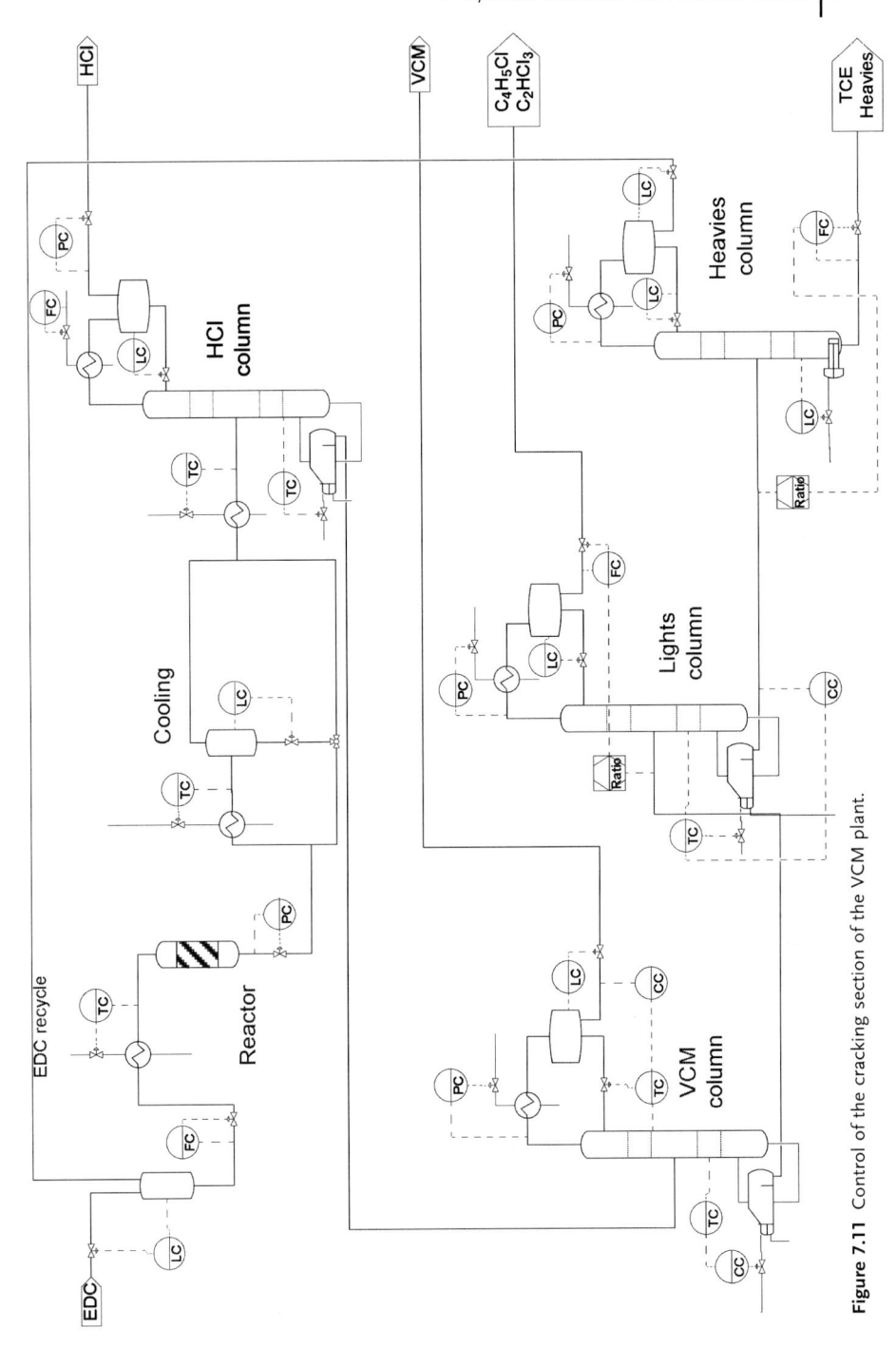

Figure 7.11 Control of the cracking section of the VCM plant.

that HCl is completely removed from the liquid product. At the top of the column the cooling duty is constant, the reflux flow controls level in the condenser drum, while the flow of the vapor product controls the pressure.

At the VCM column, the purity of the product stream (chloroprene less than 10 ppm), is controlled using a concentration/temperature cascade scheme, with reflux as the manipulated variable. Feed-rate changes could be accounted for by keeping constant the ratio reboiler duty/feed through feedforward control. However, when the plant was disturbed, a relatively large amount of VCM escaped in the bottom and ended in the distillate of the lights column. Therefore, a second concentration–temperature cascade controlling the bottoms composition was implemented.

The task of the lights column is to remove the light components from the recycled EDC, with chloroprene and tri-chloroethylene being the most important impurities. Therefore, a concentration-cascade scheme was implemented, with chloroprene concentration and reboiler duty as controlled and manipulated variables, respectively. The distillate to feed ratio was kept constant using feedforward control. This ratio can be used to adjust the level of tri-chloroethylene in the bottom product. The level in the condenser drum was controlled by the reflux. Note that fixing the reflux and controlling the level by distillate does not work, because the distillate rate is very small.

Control of the heavies column is simpler. A fixed fraction of the feed is taken as bottom product, in a feedforward manner. The reboiler duty controls the level in the column sump. Note that this arrangement, which is required because of the small bottoms stream, cannot be implemented if a kettle reboiler is used. The column is operated at constant reflux, while the distillate rate controls the condenser level.

Figure 7.12(a) shows results of the dynamic simulation, obtained when a disturbance of +25 °C in the temperature of the burn gases was introduced at time $t = 5$ h. The disturbance increases the reaction rate, which results in more VCM and HCl. As the recycle decreases, more fresh EDC is necessary to keep constant the reactor-inlet stream. Small variations in the flow rates of lights and heavies are observed. In the same time, the level of impurities at the reactor inlet decreases.

In the second simulation (Figure 7.12b), the reactor-inlet flow was increased from 74 000 kg/h to 81 500 kg/h. Initially, the amounts of VCM produced and of fresh EDC fed increase. However, these flows soon decrease to the initial values. This means that, when reaction conditions are fixed, production-rate changes can be achieved only at the expense of large variations of the recycle flow. Moreover, all the flow rates are very sensitive to disturbances if a control structure fixing the flow rate at the plant inlet is used.

7.9
Plantwide Control of Impurities

The inventory of impurities is a plantwide control problem acknowledged in industry for a long time. This problem can be handled systematically by means of

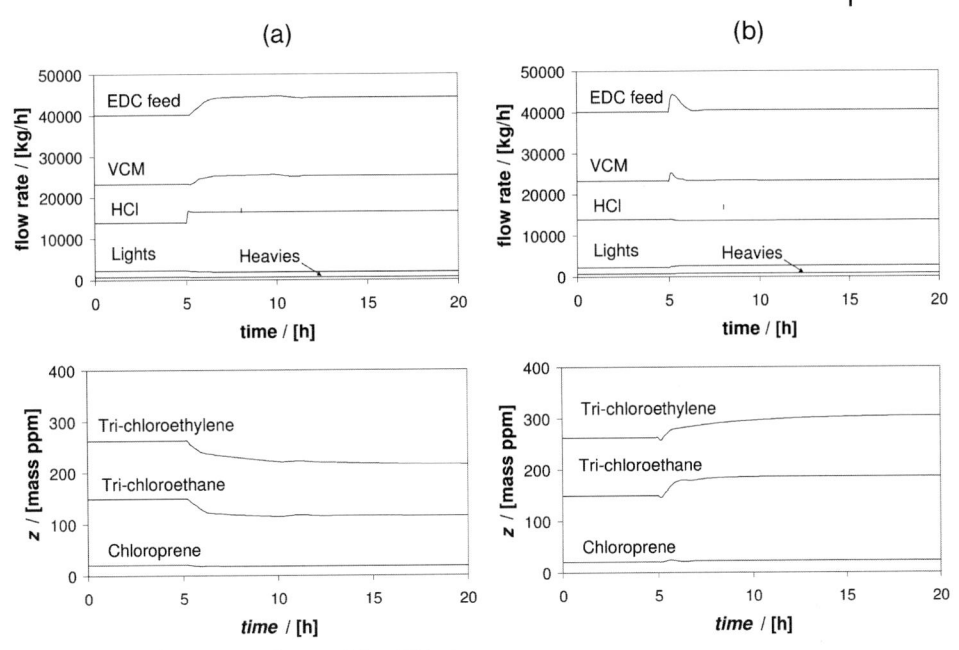

Figure 7.12 Production-rate change of the VCM plant.

dynamic simulation and controllability analysis. The inventory of main components and impurities are interdependent, because they are coupled through recycles. The interactions can hinder or help keeping an acceptable level of impurities, depending on the balance between positive- and negative-feedback effects. The implementation of control structures based on the viewpoint of standalone units can lead to severe conflicts. Hence, a systemic approach based on the quantitative evaluation of the recycle effects is needed.

Figure 7.13 presents a simplified flowsheet, which concentrates the essential features the "balanced" VCM technology, as conceptually developed in the previous sections, but this time with the three plants and recycles in place: chlorination of ethylene (R1), thermal cracking of EDC (R2) and oxychlorination of ethylene (R3). As mentioned in Section 7.3, from plantwide control three impurities are of particular interest: (I_1) chloroprene (nbp 332.5 K), (I_2) trichloroethylene (nbp 359.9 K), and (I_3) tetrachloromethane (nbp 349.8). I_1 and I_2 are "bad", since the first can polymerize and plug the equipment, while the second favors the coke formation by EDC pyrolysis. On the contrary, I_3 has a catalytic effect on the VCM formation, in some patents being introduced deliberately.

The source of I_1 is the pyrolysis reactor, I_2 appears mainly by the oxychlorination, while I_3 is produced in both chlorination and oxychlorination steps. These impurities must be removed selectively from EDC. I_1 and I_2 should be kept at low ppm levels, while I_3 is close to an optimal value. This is the incentive of the plantwide control problem.

7 Vinyl Chloride Monomer Process

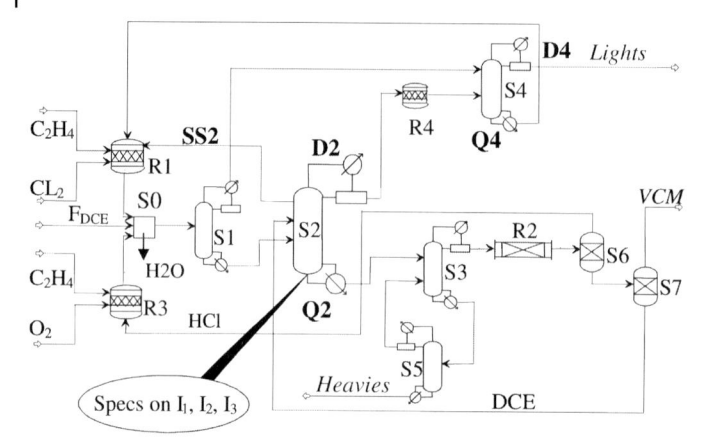

Figure 7.13 Base-case flowsheet of a balanced VCM process for handling impurities.

The plant simulation considers only a reduced number of units, but dynamically representative, as follows. Crude EDC from R1 and R3 are sent to washing/drying in the unit S0. Dissolved gases and very light impurities are removed in S1, and further in the distillation column S4, which is the exit point of the light impurities. After pretreatment, the crude EDC is sent to purification in the distillation column S2, which is the key unit of the separation system. This column receives crude EDC from three reactors. It is also the place where three large recycle loops cross. The top distillate of S2 should remove the light impurities mentioned above, while the purification of EDC from heavies is continued in the distillation columns S3 and S5.

The separation of impurities in S2 is affected by volatility constraints. At 350 K, the top temperature, the volatility of I_1, I_2, and I_3 relative to EDC is about 1.9, 0.94 and 1.6. Therefore, the top distillate of S2 can easily remove I_1 and I_3, but not I_2. Note also that the top distillate of S2 cannot contain more than 8% I_1.

To prevent the accumulation of I_2, a side stream drawn from S2 is sent to the reactor R1, where chlorination to heavies takes place. Because of the constraint on I_1, the top distillate of S2 carries with it a significant amount of EDC, which has to be recovered and recycled to S2 by the column S4. By returning the bottom of S4 to the reactor R1, some amounts of impurities I_1 and I_2 can eliminated, but this operation complicates the flowsheet. Therefore, it is rational to introduce a specialized reactor for the conversion of nonsaturated impurities in heavies by liquid-phase chlorination. This new reactor, designated by R4, placed between S2 and S4, gives the opportunity to develop new flowsheet alternatives by rerouting the streams, as depicted in Figure 7.14. Thus, the heavies created by the conversion of lights can leave the plant through S5 (alternative A), return to S2 (alternative B) or pass directly to S3 (alternative C).

Note that all heavy impurities are removed by the columns S3 and S5, which work in tandem in order to limit the losses in EDC. With respect to design, S2 is

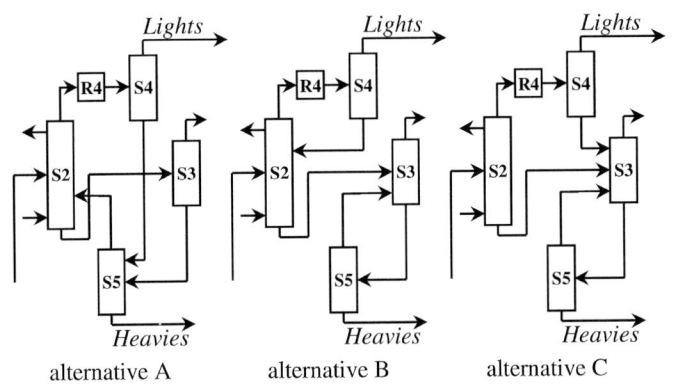

Figure 7.14 Flowsheet alternatives for the removal of impurities.

a big column with a large number of theoretical stages operating at high reflux. The separation in S3 requires only a few stages. On the contrary, S4 and S5 are small units but of particular importance, being the only exit of light and heavy impurities, respectively.

After thermal cracking the reaction mixture is quenched and cooled (nonpresented). The recovery of HCl and the separation of VCM from unreacted DCE take place in units S6 and S7, respectively.

From a simulation viewpoint units S0, S6 and S7 may be considered blackboxes. On the contrary, S1 to S5 are simulated by rigorous distillation columns, as sieve trays. In the steady state all the reactors can be described by a stoichiometric approach, but kinetic models are useful for R1, R2 and R4 in dynamic simulation [7, 8]. As shown before, the reaction network should be formulated so as to use a minimum of representative chemical species, but respecting the atomic balance. This approach is necessary because yield reactors can misrepresent the process.

The quality of the EDC sent to pyrolysis must fulfil strict purity specifications, but too low an impurity level implies high energy consumption. The concentrations of I_1 and I_2 in the bottom of S2 must not exceed 100 and 600 ppm, respectively, while the concentration of I_3 must be kept around its optimal value at 2000 ppm. It is worth mentioning that these contradictory requirements cannot be fulfilled by any standalone design of S2. The effective control of impurities becomes possible only by exploiting the positive-feedback effects of the recycle loops that are balanced by the negative-feedback effects of chemical conversion and exit streams.

The three quality specifications regarding the impurities in EDC, available by direct concentration measurements, such as by IR spectroscopy or online chromatography, are the *outputs* of the plantwide control problem. The degrees of freedom indicate as first choice manipulated variables belonging to the large column S2:D2–distillate flow rate, SS2–side-stream flow rate, and Q2–reboiler duty. We may also consider manipulated variables belonging to the small column

S4, connected to S2 by a recycle, but dynamically faster. Thus, supplementary inputs are: D4 – distillate flow rate, and Q4 – reboiler duty. A major disturbance of the material balance can be simulated by a step variation in an external EDC feed. A second significant disturbance is the amount of impurity I_3 introduced in the plant.

The base case and three alternatives were evaluated by controllability analysis [7, 8], firstly at steady-state. The conclusion is that the loops Q2 (reboiler duty) – I_1 and SS2 (side-stream flow) – I_2 are more interactive than the loop controlling I_3 with D2, D4 or Q4. The use of D4 offers the best decoupling of loops. In the base case and alternative B the effect of the variables belonging to S4 on I_3 is enhanced by closing the other loops, while in alternatives A and C this effect is hindered. However, at this point there is not a clear distinction between the base case and alternatives. A dynamic controllability analysis is needed.

When the interactions between controllers are taken into account, only two controllers, Q2 – I_1 and D4 – I_3, are sufficient to keep I_2 between bounds, because the disturbances affect the outputs in the same direction. The reboiler duty Q2 and the side draw SS2 also affect the impurities I_1 and I_2 in the same direction. So both controllers are supporting each other in rejecting the disturbances. Because the power of Q2 is much larger than SS2, the controller Q2 – I_1 is dominating, and the loop SS2 – I_2 is not needed. Hence, leaving I_2 free, but having the guarantee of a bounded variation, is a rational compromise that preserves the robustness of the control system. Hence, the plantwide control objective could be achieved with only two control loops, Q2 – I_1 and D4 – I_3, which over a practical range of frequency are almost decoupled. Fully dynamic simulation confirmed the analysis by implementing the controllers Q2 – I_1 and D4 – I_3 as P-type only. Note that using manipulated variables from different units is not a current control practice. However, the principle of proximity is preserved, because the columns S2 and S4 are dynamically adjacent.

The above analysis emphasizes that the most significant improvement came from the chemical conversion of impurities by diminishing the positive feedback of recycles. Controllability study and closed-loop simulation indicate that the base case and alternative B have the best dynamic properties. The last modification offers the shortest path of impurities and faster dynamics, together with better protection against plugging. The revamp of the plant would consist in adding the small reactor R4 for chlorination of impurities, repiping the connection from S4 to S2 and replacing the internals in the lower part of the column S2 with fouling-resistant trays or packing.

Summing up, if the inventory of the main components can be handled by local control loops, the inventory of impurities has essentially a plantwide character. The rates of generation, mainly in chemical reactors, and of depletion (exit streams and chemical conversion), as well as the accumulation (liquid-phase reactors, distillation columns and reservoirs) can be balanced by the effect of recycles in order to achieve an acceptable equilibrium state. Interactions through recycles can be exploited to create plantwide control structures that are not possible from a stand-alone unit viewpoint.

7.10
Conclusions

The VCM case study emphasizes the complexity of designing a large chemical plant with an intricate structure due to several reactors and separation sections. A balanced process is designed such that only VCM leaves the plant, with raw materials efficiency close to stoichiometry.

Characteristic of this process is the formation of a large spectrum of chlorohydrocarbon impurities. These originate in the three reaction systems, chlorination, oxychlorination and pyrolysis. The first two are sources of the intermediate EDC, which is a reactant in the third. Some impurities are circulating between the three reactors due to recycle flows. In this way the purification of EDC becomes an intricate problem with plantwide character. Therefore, the handling of impurities implies not only their removal by distillation, but also the minimization at source, namely by improving the reaction conditions. In particular, the yield of pyrolysis can be enhanced by making use of initiators, some being produced and recycled in the process itself. The handling of impurities is at the origin of control problems regarding the plantwide material balance. Chemical conversion of impurities accumulating in recycle prevents the occurrence of snowball effects that otherwise will affect the operation of reactors and separators.

The VCM process offers good opportunities for energy saving by heat integration, namely in distillations. Direct chlorination and particularly oxychlorination are important sources of energy.

References

1 Ullmann's Encyclopaedia, Chlorinated Hydrocarbons, vol. A6, 263–384, 1993
2 Weissermel, K., Arpel, H.J., Industrial Organic Chemistry, Wiley-VCH, 2003
3 Cowfer, J.A., Magistro, A.J., Vinyl Chloride, in Kirk-Othmer Encyclopedia of Chemical Technology, 23, 865–883, 1992
4 McPherson, R., Starks, C.M., Fryar, G.J., Vinyl chloride monomer . . . what you should know, Hydrocarbon Process., pp. 75–92, March 1979
5 Wong, E.W., Ambler, C.P., Baker, W.J., Parks, J.C., Producing high purity VCM product, Hydrocarbon Processing, 129–134, August 1992
6 Saeki, Y., Emura, T., Technical progresses for PVC production, Prog. Polym. Sci., 2055–2131, 2002
7 Groenendijk, A.J., Dimian, A.C., Iedema, P.D., System approach for evaluating dynamics and plantwide control of complex plants, AIChE J., 46, 119–132, 2000
8 Dimian, A.C., Groenendijk, A., Iedema, P.D., Recycle interaction effects on the control of impurities in a complex plant, Ind. Eng. Chem. Res., 40, 5784–5794, 2001
9 Bostwick, L.E., Recovering chlorine from HCl, Chem. Eng., 10, 1986
10 Clegg, I.M., Hardman, R., Vinyl chloride production process, US Pat 5728905, 1998
11 Orejas, J.A., Model evaluation for an industrial process of direct chlorination of ethylene in a bubble-column reactor, Chem. Eng. Sci., 56, 513–522, 2001
12 Wachi, S., Morikawa, H., J. Chem. Eng. Japan, 19, 437–443, 1986
13 Ranzi, E., Dente, M., Rovaglio, M., Pyrolysis and chlorination of small hydrocarbons, Chem. Eng. Commun., 117, 17–39, 1992

14 Ranzi, E., Dente, M., Faravelli, T., Mullick, S., Bussani, G., Mechanistic modeling of chlorinated reacting systems, *La Chimica & L'Industria*, Vol. 72, 905–914, 1990

15 Zhernosek, V.M., Vasilieve, I.B., Avertsov, A.K., Gelbsthtein, A.I., Kinetic and mechanism of the catalytic chlorination of ethylene, *Kinetika i Kataliz*, 12, No. 2, 407–413, 1971

16 Teshima, Y., Onishi, S., TOSOH Corp., European Patent 0270007, 26.11.1987

17 Bruzzi, V., Colaianni, M., Zanderighi, L., *Energy Convers. Mgmt*, 39, No. 16–18, 1853–1862, 1998

18 Lurgi Oel-Gas Chemie A.G., Vinylchloride plant flowsheet (www.lurgi_oel.de)

19 Aspen Plus®, Aspen Dynamics™, Aspen Technology Inc., Cambridge, MA, USA, release 11

8
Fatty-Ester Synthesis by Catalytic Distillation

8.1
Introduction

Fatty esters are important fine chemicals used in manufacturing cosmetics, detergents and surfactants. With the emergence of biodiesel the manufacturing of fatty esters enters in a new period, of large-scale production, for which more efficient technologies are needed. Currently, the fatty esters are produced in batch processes by homogeneous catalysis, namely with sulfuric or p-toluensulfonic acids. There are serious drawbacks, such as costly post-treatment for purifying the product and recycling the excess of alcohol, catalyst lost after reaction, pollution by neutralization with alkali and low productivity. Therefore, a technology based on continuous operation and employing solid catalyst is highly desirable. Organic ion-exchange resin, such as Amberlyst or Nafion, are usually used in esterification of low molecular acids – namely acetic acid – with different alcohols, but at temperatures below 100 °C. For long-chain fatty esters, much higher reaction temperatures are expected. Recent studies [1–4] have revealed that superacidic catalyst based on sulfated zirconia is suitable for the esterification of fatty acids with various alcohols at temperatures between 150 and 180 °C.

A reactive distillation (RD) process would bring evident technological and ecological advantages. An important feature is that the reactants can be fed in the stoichiometric ratio ensuring in this way the maximum efficiency of raw materials. Unlike a batch process, where the excess of alcohol is recovered by costly distillation, higher reaction rate can be achieved by internal alcohol recycle. However, the presence of water as a byproduct makes this wish much more difficult than it appears.

This case study investigates the possibility of applying reactive distillation to the synthesis of fatty-acid esters as a generic multiproduct process. As representative species we consider the lauric (dodecanoic) acid and some alcohols the series C_1–C_8, such as methanol, n-propanol and 2-ethyl-hexanol (isooctanol). The generic reversible chemical reaction is:

$$R_1\text{-COOH} + R_2\text{-OH} \leftrightarrow R_1\text{-COO-}R_2 + H_2O \qquad (1)$$

Chemical Process Design: Computer-Aided Case Studies. Alexandre C. Dimian and Costin Sorin Bildea
Copyright © 2008 WILEY-VCH Verlag GmbH & Co. KGaA, Weinheim
ISBN: 978-3-527-31403-4

As undesired secondary reaction one may note the etherification of alcohols:

$$R_2\text{-OH} + \text{HO-}R_2 \leftrightarrow R_2\text{-O-}R_2 + H_2O \tag{2}$$

A relevant characteristic of the technology should be the ability to remove the water selectively and continuously in order to shift the chemical equilibrium to full conversion. Because the presence of a liquid water phase will lead to rapid deactivation of the solid catalyst, operating conditions for water-free organic liquid should be found. In addition, the thermodynamic behavior of the reaction mixture is nonideal, particularly with respect to the couple alcohol–water.

8.2
Methodology

The strategy of design, illustrated in Figure 8.1, consists of an evolutionary search of the feasible design space by means of a systematic combination of thermodynamic analysis, computer simulation and only limited experiments. The approach is generic for developing a RD process, at least for similar systems. The first element of similarity is the existence of an equilibrium reaction with water as product. This raises the problem of possible aqueous-phase segregation. The second element is the similarity of thermodynamics properties over a class of substrates. However, while the fatty acids and fatty esters manifest a certain

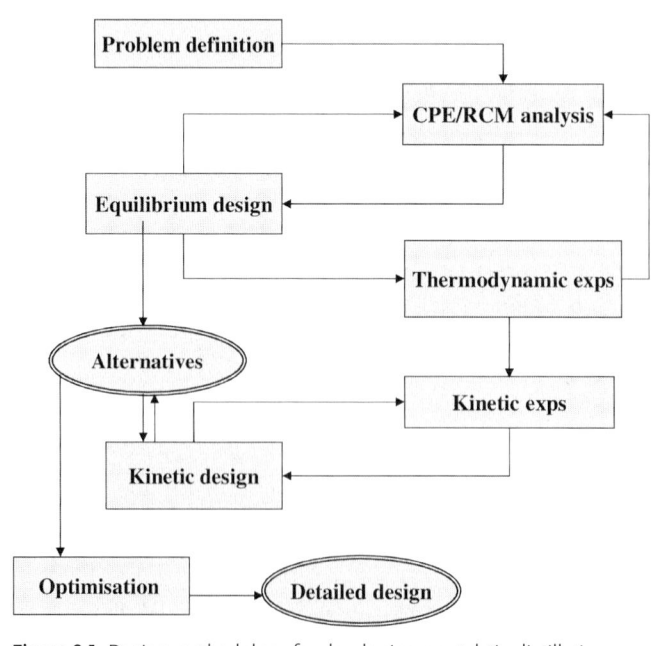

Figure 8.1 Design methodology for developing a catalytic distillation process.

analogy in physical properties, the alcohol–water couple may exhibit significant differences with respect to azeotropy and immiscibility. The third element of similarity is the use of a heterogeneous catalyst. The catalyst helps the designer to manipulate the reaction rate so as to match the separation requirements. At the same time it brings hard constraints with respect to hydraulics of internals (catalyst load, liquid holdup and pressure drop) and residence-time distribution. The catalyst should be applicable for several substrates, alcohols and acids.

Table 8.1 describes the steps of the methodology in more detail. The procedure starts with the *Problem definition*: production rate, chemistry, product specifications, safety, health and environmental constraints, physical properties, available technologies. Then, a first evaluation of feasibility is performed by an *equilibrium design*. This is based on a thermodynamic analysis that includes simultaneous chemical and physical equilibrium (CPE). The investigation can be done directly by computer simulation, or in a more systematic way by building a residue curve map (RCM), as explained in the Appendix A. This step will identify additional *thermodynamic experiments* necessary to consolidate the design decisions, mainly phase-equilibrium measurements. Limitations set by chemical equilibrium or by thermodynamic boundaries should be analyzed here.

As demonstrated in many practical cases the equilibrium-based design is the simplest way to investigate the feasibility, but not for solving the design problem. A more realistic approach makes use of the kinetics of the chemical process. Reliable kinetic data are required. In their absence, the necessity of performing *kinetic experiments* is evident. Computer-aided design may set targets for the synthesis of an effective catalyst, as well as for the optimal planning of kinetic experiments.

The *kinetic design* is the central step in the approach. Two models can be employed: (1) reaction kinetics plus instantaneous phase equilibrium, or (2) reaction kinetics plus rate-based modeling for mass and heat transfer. The first approach is simpler and was found to be sufficient in a large number of industrial cases, including MTBE and methyl acetate manufacturing. The second approach is much more demanding implying the knowledge of specific correlations for the hydrodynamics and mass transfer involving the internals and materials used, which is more difficult to obtain. The evaluation of catalyst effectiveness is needed too.

A central problem is the selection of the appropriate internals. Structured catalytic packing is mostly employed. The most critical problem at this stage is the availability of correlations for describing the hydrodynamics conditions corresponding to the effective use of the catalyst.

A number of *alternatives* can emerge, both as arrangement of internals and flowsheet configuration, such as for example, distribution of reaction and separation zones, feed policy, use of pre- or postreactors. The best alternative is selected taking into account economic and operational aspects, such as for example the minimization of byproducts and the preservation of catalyst activity over a longer period.

The *optimization* step addresses the final tuning of the design. Issues of interest are the distribution of catalyst activity, pressure, temperature profiles, as well as the energy consumptions.

Table 8.1 The steps of the design methodology presented in Figure 8.1.

1. Problem definition and data analysis
   (a) Feedstock and product specifications with respect to purity.
   (b) Constraints: toxicity, safety, environment, technology.
   (c) Catalyst selection: activity, properties, regeneration.
   (d) Data analysis: physical properties of components and mixtures, estimation of missing properties.

2. Chemical and phase equilibria (CPE)
   (a) Phase equilibrium: azeotropes, immiscibility regions, phase diagrams.
   (b) Chemical equilibrium constant and achievable conversion in a batch system.
   (c) Plot of residue or distillation curve map.
   (d) Identification of reactive azeotropes.

3. Equilibrium design
   (a) Specifications of product composition and limiting operating temperatures.
   (b) Selection of column pressure.
   (c) Feasibility analysis by representation in the RCM or by simulation-aided search.
   (d) Simulation of the RD setup by assuming chemical and phase equilibrium.
   (e) Sensitivity analysis.

4. Thermodynamic experiments
   (a) Physical properties of pure components.
   (b) Phase equilibrium for binaries of significance.
   (c) Revision of chemical and phase equilibrium.

5. Chemical reaction kinetics.
   (a) Kinetics of uncatalyzed and catalyzed reactions.
   (b) Assessment of selectivity as a function of process variables.
   (c) Catalyst effectiveness.

6. Kinetic design
   (a) Selection of internals.
   (b) Preliminary hydraulic design.
   (c) Kinetic reaction modeling with phase equilibrium.
   (d) Residence-time distribution and backmixing effects.
   (e) Simulation by full rate-based models (optional).

7. Evaluation of alternatives

8. Optimization
   (a) Catalyst distribution.
   (b) Energetic requirements.

9. Detailed design
   (a) Complete configuration of internals.
   (b) Hydraulic design: holdup, pressure drop.
   (c) Design of additional operational units

8.3
Esterification of Lauric Acid with 2-Ethylhexanol

8.3.1
Problem Definition and Data Generation

This section deals with the conceptual design of a catalytic distillation process for the esterification of lauric acid (LA) with 2-ethyl-hexanol (2EtH). Laboratory experiments showed that a superacid sulfated zirconia catalyst exhibits good activity over a large interval, from 130 to 180 °C with no ether formation. On the contrary, the catalyst is sensitive to the presence of free liquid water. Raw materials are lauric acid and 2-ethylhexyl alcohol of high purity. The conversion should be over 99.9%, because the product is aimed at cosmetic applications.

Listing the toxicological properties of species shows no difficult problems in storage and handling. Table 8.2 presents the normal boiling points of components. Vapor-pressure data for 2-ethylhexyl laurate are not available in literature. Estimation by ASPEN Plus by group contributions leads to controversial values of 714 K by Joback and 605 K by Gani. Combining published data and measurements (see later in this chapter) allows the generation of vapor-pressure correlation over a large range of P and T, as displayed in Figure 8.2. It may be observed that the region of volatile species, alcohols and water, is very distinct from the region of heavy species, acid and fatty esters. The volatility with respect to water changes dramatically between methanol and 2-ethylhexanol. Moreover, azeotropes with water appear for the intermediate alcohols, such as C2–C4. Therefore, the separation problem will change qualitatively in a significant manner over the series of alcohols C1–C8.

Table 8.3 lists the properties of the azeotropes that 2-ethylhexanol and lauric acid form with water, both heterogeneous. It is remarkable that the solubility of water in acid and alcohol is large, while the reverse solubility, organic components in water, is extremely low, particularly for the acid. Pressure has little effect. No information about water/fatty ester azeotrope is available, but this should not affect the analysis.

Table 8.2 Normal boiling points for key components in K.

Pressure	Water	2-Ethylhexanol	Lauric acid	2-Ethylhexyl laurate
101.3 kPa	373.15	457.75	571.78	714 [a], 605 [b], 607.6 [c]

Estimated with Aspen Plus by a) Joback and b) Gani methods; c) experimental value.

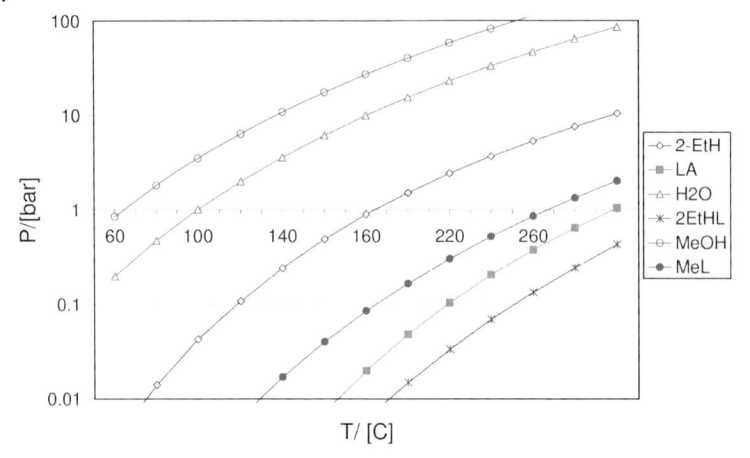

Figure 8.2 Vapor pressures of species by extended Antoine equation.

Table 8.3 Azeotropes of water with 2-ethylhexanol and lauric acid at 1 atm.

	Water/2-ethylhexanol	Water/lauric acid
T_{az}, K	372.25	373.147
y_{az}	0.9679	0.999900
$x_{1,aq}$	0.9996	0.999988
$x_{1,org}$	0.2411	0.2502

8.3.2
Preliminary Chemical and Phase Equilibrium

Inside the reactive zone, chemical and phase equilibrium occur simultaneously. The composition of phases can be found by Gibbs free-energy minimization. The UNIQUAC model is adopted for phase equilibrium, for which interaction parameters are available, except the binary fatty-ester/water handled by UNIFAC-Dortmund.

The composition at equilibrium can be determined from the equilibrium constant based on molar fractions $K_x = K_a/K_\gamma$. This is, in turn, calculated from the equilibrium-constant-based activities K_a and activity coefficients K_γ. Figure 8.3 shows the variation over the range 350–475 K [1, 2]. K_a takes values between 10 and 50, but the correction by K_γ is important bringing K_x between 2 and 10. Above 430 K equilibrium conversion over 80% should be expected. If the water is continuously removed by distillation, then full conversion may be achieved.

Figure 8.4 presents the residue curve map of simultaneous phase and chemical equilibrium at normal pressure. Special coordinates, X_1 (acid + water) and X_2 (acid + ester) enable the representation of all four components in a bidimensional

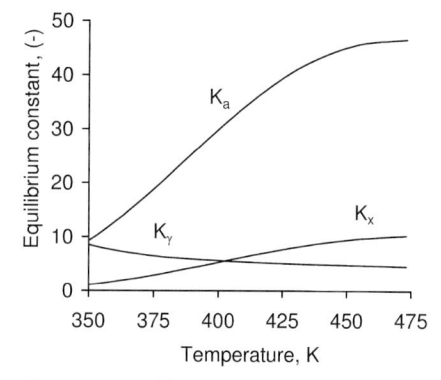

Figure 8.3 Equilibrium constants for esterification of lauric acid with 2-ethylhexanol.

diagram (details in Appendix A and ref. [7]). A particular feature is the segregation in two liquid phases, organic and aqueous. Curve a designates the boundary between homogeneous and heterogeneous regions, as calculated with data originated exclusively from Aspen Plus. Curve b incorporates experimental results, as explained later. On the left side, there is homogeneous organic phase, with the direction of trajectories pointing out from reactants to ester. It may be observed that the water product is located in the heterogeneous region. Note also the presence of a third homogeneous region (only water phase) in the right corner of the diagram, not visualized because of the scale. Since the solubility of organic components in water is so small, quasiquantitative separation of water as a byproduct is possible. However, a large region of liquid-phase segregation appears, for temperatures just below 373 K. The mixing line of acid and alcohol falls in the heterogeneous region, which is not acceptable when working with a heterogeneous catalyst. Consequently, the temperature in the reaction zone must exceed 373 K to avoid the occurrence of a second liquid-water phase.

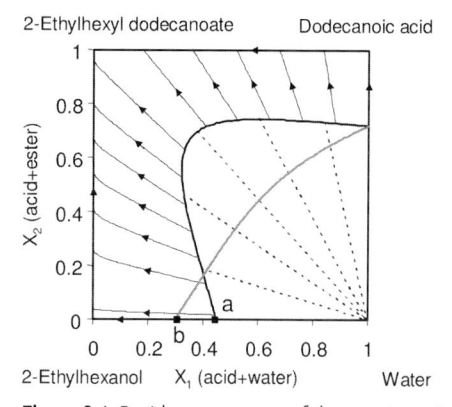

Figure 8.4 Residue curve map of the reactive mixture lauric acid/2-ethylhexanol/water/2-ethylhexyl dodecanoate: a=initial estimation; b=experimental data [2].

Note that there are no reactive azeotropes between the components of the reaction mixture, although the volatile reactants lay between light and heavy products. The explanation may be found in the fact that both acid and alcohol have high boiling points that do not contribute significantly to the total vapor pressure, dominated by alcohol and water.

8.3.3
Equilibrium-based Design

The analysis in the RCM suggests that a RD process may be feasible, as the trajectories converge to the ester vertex, the highest boiler and stable node. The heteroazeotrope of water-alcohol is an unstable node, while the heteroazeotrope water-acid is a saddle. At total conversion, the temperatures at the column's ends at atmospheric pressure would be 373 K in top and about 713 K in bottom. Clearly, the last value is excessive. Here we assume a temperature below 473 K to avoid thermal decomposition, a condition that can be realized working under vacuum at 32 kPa and diluting with about 12 mol% alcohol.

The above considerations lead to the conceptual flowsheet presented in Figure 8.5. Acid and alcohol enter countercurrently at the top and bottom of the reaction zone. The bottom product consisting of ester and some alcohol goes to an evaporator, from which the alcohol is recovered and recycled. The top vapor is condensed, and separated into two phases after decantation. The water leaves the decanter as a byproduct, while the alcohol-rich phase is sent as reflux into the column.

The feasibility of the above setup can be evaluated by simulation with Aspen Plus [19]. The RD column is built-up as a reboiled stripper followed by a condenser and a three-phase flash, with organic phase refluxed to column. The result is that only 3 to 5 reactive equilibrium stages are necessary to achieve over 99% conversion. The stripping zone may be limited at 2–3 stages, while the rectification zone has 1–2 stages.

At this point, it is useful to examine the effect of uncertainty in data by sensitivity analysis. Some physical properties are essential for design, such as for example the vapor pressure of the fatty ester, VLE for binaries involving the lauric acid, alcohol and water, as well as the Gibbs free energy of formation of the fatty ester.

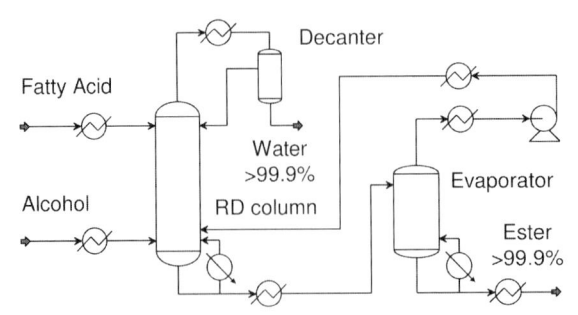

Figure 8.5 Conceptual flowsheet for the synthesis of the 2-ethylhexyl dodecanoate.

By consequence, experimental measurements are necessary to consolidate the design.

8.3.4
Thermodynamic Experiments

VL(L)E measurements for binaries involving water with alcohol and acid have been done, as described elsewhere [2]. Figure 8.6 presents experimental vapor pressure data for 2-ethylhexyl laurate. The normal boiling point (nbp) is 607.6 K, close to the prediction by Gani's method. On the other hand, the prediction of the whole saturation curve by Riedel's method (noted estimation in Figure 8.6) is in large error at lower pressures. This fact can affect the accuracy of chemical equilibrium calculation, but fortunately the errors compensate each other [2].

The solubility of 2-ethylhexyl dodecanoate in water at normal pressure in the range 300 to 360 K can be described by the following polynomial in temperature:

$$S = 1.6908 \times 10^{-5} + 1.4816 \times 10^{-7} T + 3.1158 \times 10^{-10} T^2 \tag{8.1}$$

where S is expressed in wt% and the temperature in K.

The phase diagram for the partially miscible binaries water/2-ethylhexanol and water/lauric acid can be described satisfactorily by UNIQUAC with binary interaction parameters from LLE data plus the azeotropic point. This procedure allows accurate prediction of the liquid–liquid split, while preserving sufficient accuracy for VLE. The interaction parameters are given in Tables 8.5 and 8.6.

A new RCM has been calculated with parameters regressed from experimental data. This plot, b, is to be compared against the initial estimation given by the curve a in Figure 8.4. The general aspect is preserved, but there are differences in the description of the heterogeneous region. The most important is the correct location of the azeotropes 2-ethylhexanol/water and lauric acid/water.

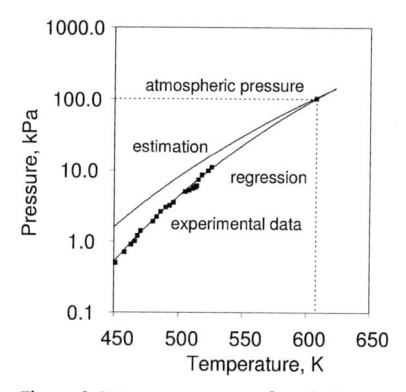

Figure 8.6 Vapor pressure of 2-ethylhexyl laurate; the estimation by Riedel method makes use of the nbp (Gani); experimental data are based on measurements from 450 to 530 K and boiling temperature at normal pressure.

8 Fatty-Ester Synthesis by Catalytic Distillation

Table 8.4 Parameters of vapor-pressure correlation for components.

	2-Ethyl-hexanol	Lauric acid	Water	2-Ethylhexyl dodecanoate	Methanol	Methyl dodecanoate
A	273.62	201.56	73.649	23.578	82.718	155.18
B	18766	20484	7258.2	6247.3	6904.5	14494
C	0	0	0	−89.281	0	0
D	−36.767	−24.334	−7.3037	0	−8.8622	−18.912
E	1.98×10^{-5}	8.06×10^{-18}	4.17×10^{-6}	0	7.47×10^{-6}	7.42×10^{-6}
F	2	6	2	0	2	2

$$\ln P^{sat} = A - \frac{B}{T+C} + D\ln(T) + ET^F, \text{ with } [P] = \text{Pa, and } [T] = \text{K}$$

Table 8.5 UNIQUAC parameters for binaries with lauric acid and 2-ethylhexanol.

Comp. 1	Alcohol	Alcohol	Alcohol	Acid	Acid	Water
Comp. 2	Acid	Water	Ester	Water	Ester	Ester
A12	0	11.517	0	−0.29924	0	−0.04839
A21	0	−10	0	−0.38437	0	−0.44196
B12	49.44	−785.57	90.824	−195.44	97.72	−203.49
B21	−52.77	−2.807	−183.25	−107.62	−140.7	−738.1
C12	0	−1.7187	0	0	0	0
C21	0	1.5911	0	0	0	0

$$\tau_{ij} = \exp\left(A_{ij} + \frac{B_{ij}}{T} + C_{ij}T\right)$$

Table 8.6 Surface (q) and volume (r) parameters for the UNIQUAC model.

Params	Water	Lauric acid	2-Ethyl-hexanol	2-Ethylhexyl dodecanoate	Methanol	Methyl dodecanoate
q	1.400	7.472	5.208	11.752	1.432	7.976
r	0.920	8.931	6.151	14.239	1.431	9.548

8.3.5
Revised Conceptual Design

The preliminary design is revisited taking into account the improved knowledge on thermodynamics. The conclusion is that the conceptual frame does not change fundamentally, and only few equilibrium stages are necessary. Figure 8.7 displays

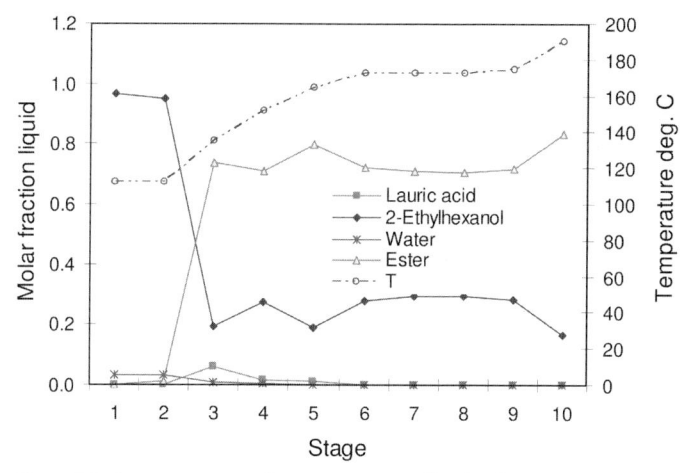

Figure 8.7 Concentration and temperature profiles for the esterification of lauric acid with 2-ethylhexanol with an equilibrium-based model. Top Pressure 0.3 bar. Seven reactive stages, acid feed on 3 and 5 with 0.5 split, alcohol feed on 7.

temperature and composition profiles. Note that the acid can be introduced as two feeds, on stages 3 and 5, in order to obtain a more uniform reaction rate. Otherwise most of the acid conversion would have been achieved only in two stages. It is interesting to note that a minimum reflux (or reboiler duty) can be detected, as the situation corresponding to zero alcohol concentration on the top reactive stage. In this case, the minimum reflux and reboil ratios are about 0.2 and 0.3. At larger reflux the temperature profile shows a monotonic rise over the first top stages, followed by a plateau over the reaction zone, and a sharp climb to the bottom. Excessive bottom temperature can be avoided by using vacuum and diluting the product with alcohol.

In conclusion, the chemical equilibrium model suggests that the process may be feasible. However, the insertion of chemical kinetics is necessary to get a more realistic description.

8.3.6
Chemical Kinetics Analysis

8.3.6.1 Kinetic Experiments

Experiments were executed in an autoclave at temperature between 130 and 180 °C, with alcohol/acid ratios between 1/9 to 27/1, as well as sulfated zirconia catalyst concentration up to 5 wt%. The experimental conditions preserved the chemical equilibrium constraint. Details are given elsewhere [2]. Two contributions in forming the reaction rate can be distinguished: enhancement due to the solid catalyst and an autocatalysis effect by the fatty acid. Consequently, the following expression can be formulated for the overall reaction rate:

$$-r_A = \left(k_A^* \frac{a_{acid}}{V_M} + k_C^* C_{cat}\right)\left(a_{acid} a_{alcohol} - \frac{a_{water} a_{ester}}{K_a}\right)\frac{1}{V_M^2} \quad (8.2)$$

Note the use of activities, as well as of an equilibrium constant based on activities. The kinetic constants for autocatalyzed and catalyzed reactions, k_A^* and k_C^*, were determined from initial reaction rates with liquid activity coefficients calculated by UNIQUAC. Near chemical equilibrium the K_γ is about 6, while K_x is about 5. Table 8.7 gives activation energies and pre-exponential factors obtained by nonlinear regression. The simulation shows that the autocatalysis effect is negligible below 150 °C, but it might increase to 20% at 180 °C.

Table 8.7 Kinetic parameters for the esterification of dodecanoic acid with 2-ethylhexanol catalyzed by sulfated zirconia catalyst.

Parameter	Value	Units
A_{uncat}	1.634×10^5	$m^6 \, kmol^{-2} \, s^{-1}$
A_{cat} b	3.277×10^3	$m^6 \, kmol^{-1} \, kg^{-1} \, s^{-1}$
Ea_{uncat}	65.5	$kJ \, mol^{-1}$
Ea_{cat} b	55.5	$kJ \, mol^{-1}$

However, since the use of concentrations in the simulation software, in the next calculations we consider a simplified concentration-based model, neglecting the autocatalysis term. The reaction rate (kmol/kg catalyst/s) can be reformulated as follows:

$$r = k_1(C_{acid} C_{alcohol} - C_{ester} C_{water}/K_c) \quad (8.3)$$

where the forward reaction constant is given by:

$$k_1 = 54.233 \exp(-6691.843/T) \quad (8.4)$$

and the concentration-based equilibrium constant by:

$$K_x = 0.00008829 \, T^2 - 0.02549 \, T - 0.8424 \quad (8.5)$$

8.3.6.2 Selectivity Issues

Since the acid-to-alcohol ratio inside the RD column varies over three orders of magnitude the superacidic catalyst might promote side reactions, the most probable being the formation of ether by alcohol dehydration. Experimentally, no byproducts were detected at long contact time and higher temperatures. However, a limiting temperature of 200 °C is assigned in this project for the bottom product temperature. This constraint is also important in selecting the operating pressure.

8.3.6.3 Catalyst Effectiveness

The assessment of the internal and external diffusion on the overall reaction rate is necessary before starting kinetic simulation. Criteria for estimating catalyst effectiveness are available for simple irreversible reactions [18]. With some approximation they can be applied in the case of more complex reactions, namely to detect the need of including it in the analysis.

The Weisz–Prater criterion can be used for detecting drop in effectiveness due to internal diffusion. This is a number representing the ratio of actual reaction rate to a diffusion rate, and is given by:

$$C_{\mathrm{WP}} = \eta \phi^2 = \frac{-r_A \rho_c R^2}{D_{A,l} C_{As}} \ll 1 \tag{8.6}$$

In relation 8.6 – r_A is the reaction rate in kmol/(kg catalyst s), ρ_c density of solid catalyst, R particle radius, $D_{A,l}$ diffusion coefficient in liquid and C_{As} reactant concentration at the catalyst surface. If $C_{\mathrm{WP}} \ll 1$ there are no diffusion limitations, but if $C_{\mathrm{WP}} \gg 1$ the catalyst effectiveness is severely affected.

The Mears criterion can be applied for evaluating the influence of the external diffusion the. This is given by the expression:

$$C_M = \frac{-r_A \rho_b R n}{k_m C_{Ab}} < 0.15 \tag{8.7}$$

In addition to the above notations, ρ_b is the bulk density of the catalyst bed, n the reaction order, k_m is the mass-transfer coefficient and C_{Ab} the bulk concentration of the reference reactant. If $C_M < 0.15$ the external mass-transfer resistance can be neglected.

To apply the above relations the most unfavorable situation should be identified. Here, this is near the entering point of the acid where the reaction rate is the highest. By assuming an internal alcohol reflux ratio of 0.5, the ratio acid/alcohol can be taken as 1:0.5. The densities of lauric acid and 2-ethylhexanol at 160 °C calculated by Aspen are 788.5 and 716.6 kg/m³, which give molar concentrations of 2.92 and 1.46 kmol/m³. Consequently, the reaction rate takes the value 5.47E–5 kmol/kg catalyst s. Further, one finds for the fatty acid liquid diffusivity the value 2E–9 m²/s, in accordance with the order of magnitude for such property. The catalyst particle density ρ_c is taken as 1000 kg/m³. The radius R is set by the packing characteristics to 0.425 mm. Replacing the above values in the Weisz–Prater criterion leads to:

$$C_{\mathrm{WP}} = \frac{5.475 \times 10^{-5} \times 1000 \times 0.000425^2}{2 \times 10^{-9} \times 2.92} = 1.693$$

The value 1.693 suggests that the internal diffusion should only slightly affect the catalyst effectiveness. From a Thiele-modulus calculation, the effectiveness should be at least 90%. On the other hand, because of rapid acid consumption on downstream stages the reaction rate will drop considerably, diminishing the effect of the internal diffusion. The above calculation shows that the catalyst effectiveness is not affected seriously by internal diffusion because of very small particles and relatively slow chemical reaction. At the same time, it suggests that loading very active catalyst on the top stages, where the acid concentration is at a maximum, should be not very profitable. Uneven distribution of catalyst activity would permit more uniform reaction rate.

In assessing the role of the external diffusion, the key element is the mass-transfer coefficient. This can be obtained from the Sherwood number as $k_m = Sh D_{A,l}/d_p$. For liquid-phase processes the knowledge of Schmidt number $Sc = \eta_l / \rho_l D_{A,l}$ is necessary. The dynamic viscosity calculated by Aspen is 0.85×10^{-3} N s/m^2 leading to a kinematic viscosity of $\eta_l = 1.1 \times 10^{-6}$ m^2/s, and finally $Sc = 567$. The Reynolds number can be estimated by assuming a liquid load of 10 m^3/m^2/h, optimal for RD processes (see later in this section). Consequently, one gets $Re_p = (10/3600) \times 0.00085/1.1 \times 10^{-6} = 2.08$. Such low Reynolds value indicates pseudostagnant mass-transfer conditions.

The well-known Frössling equation seems applicable, as $Sh = 2 + 0.6 Re_p^{0.5} Sc^{1/3}$. Replacing the values gives $Sh = 10$ and $k_l = 2.37 \times 10^{-5}$ m^2/s. The Mears criterion becomes:

$$C_M = \frac{5.46 \times 1000(1-0.4) \times 0.000425}{2.37 \times 10^{-5} \times 2.92} = 0.2$$

This above value is only slightly above 0.15, suggesting that the external diffusion should not play a role. The above computation may be seen as conservative as other correlations will give higher mass-transfer coefficients [18]. The conclusion is that internal and external diffusion should only slightly affect the process rate.

8.3.7
Kinetic Design

The difficulty in designing a reactive-distillation column is that the device works simultaneously as a reactor and a separator. From the reaction-engineering viewpoint the most influential characteristic is the amount of catalyst. In order to get maximum efficiency the catalyst has to be wetted completely by the liquid reactive mixture. Thus, the amount of catalyst and the liquid holdup are interrelated. Another limitation regards the liquid velocity and the residence-time distribution (RTD), which should be as close as possible to a plug flow. From the separator viewpoint, the operation is bounded by the minimum/maximum flow rates corresponding to acceptable mass-transfer exchange and flooding, respectively. The above requirements are difficult to meet simultaneously. Because some sizing characteristics are needed in simulation, namely the reactive holdup, a preliminary hydraulic design is necessary.

8.3.7.1 Selection of Internals

Unlike conventional distillation, where a large offer of trays and packing exists, the choice of catalytic internals is much more limited. This topic is discussed by Krishna [15]. Among commercial alternatives of catalytic packing we can notice Katapack-S® from Sulzer [11], and Multipack® from Montz [9]. These have the shape of sandwiches manufactured from corrugated wire gauze sheets and catalyst bags, assembled as cylindrical or rectangular boxes. Figure 8.8 shows the picture of a packing element. From the hydraulic viewpoint, the packing structure consists of alternating catalyst bags (CB) and open channel spaces (OC). In order to ensure high efficiency of combined reaction, mass-transfer and diffusion phenomena the catalyst particle should be as small as possible, but large enough to allow complete wetting and acceptable pressure drop. A particle diameter of about 1 mm gives a good compromise.

For this application, we select Multipack® packing, fully characterized by hydraulic correlations [9]. Table 8.8 gives the geometric characteristics. For smaller column diameters Multipack-I seems appropriate. Note that the catalyst

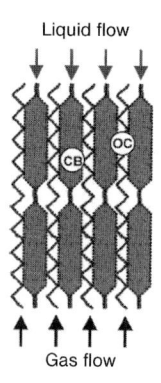

Figure 8.8 Catalytic packing and hydraulic equivalent [9].

Table 8.8 Hydraulic characteristics of the catalytic packing [9].

Parameter	Multipack-I	Multipack-II
Diameter packing d_c, mm	50	100
Specific surface area a m^2/m^3	270	300
Catalyst volume fraction, ψ	0.327	0.398
Void fraction, epsi	0.589	0.545
Inclination angle, θ (deg.)	45	45
Corrugated side, S (mm)	5.6	5.6
Void fraction wire gauze, ε_{WG}	0.72	0.72
Void fraction catalyst bag, ε_{CB}	0.30	0.30
Particle diameter d_p (mm)	0.85	0.85

volume fraction can approach 33%. The bags can host particles of 0.85 to 1 mm for which negligible influence of internal and external diffusion was demonstrated.

8.3.7.2 Preliminary Hydraulic Design

The hydraulics of catalytic packing in the absence of chemical reaction was the object of some experimental studies [9, 11, 14]. However, there are only a few papers dealing with testing the behavior on reactive systems [12, 13]. As mentioned above, the hydraulic design should ensure the operation of the reactive zone as efficient as possible. In contrast with the usual sizing procedure of distillation columns, based on the gas load at the flooding point, the hydraulic computation of a RD column should start with the liquid phase, which is the space of reaction. The catalyst particles in bags have to be completely wetted by liquid. This happens at the "load point". At lower liquid flow rate the wetting is only partial, while at higher load the excess liquid overflows in the open channels and does not take part in the reaction. This picture has been proven by visual experiments [9, 14]. Therefore, a RD column should be operated slightly above the load point. The vapor phase velocity is not important as long as flooding does not occur. It may be concluded that the operation range for a RD column is quite limited with respect to a separation device.

The liquid velocity should be examined also with respect to the residence-time distribution of the liquid phase. Too slow or too fast liquid velocity would increase the variance, as proved experimentally [11]. Values of liquid load of 5 to 20 m^3/m^2 h seem the most practical, with the optimum around 10 m^3/m^2 h.

The liquid velocity at the load point can be calculated by means of the following relations, which is valid for Multipack [9]:

$$U_{LP}^2 = \frac{\varepsilon_{CB}^3 g d_p}{(1-\varepsilon_{CB})\xi_{CB}} \quad \text{with} \quad \xi_{CB} = \frac{160}{Re_{LP}} + \frac{3.1}{Re_{LP}^{0.1}} \quad \text{and} \quad Re_{LP} = \frac{U_{LP} d_p \rho_L}{(1-\varepsilon_{CB})\eta_L} \quad (8.8)$$

With respect to the liquid holdup, only the situation when the catalyst bags are filled with liquid, but do not overflow, has significance from the reaction-engineering viewpoint. Accordingly, the reactive liquid holdup is:

$$h_{l,CB} = \psi \varepsilon_{CB} \quad (8.9)$$

More generally, the liquid holdup is given by three contributions: catalyst bag, open channel and wire gauze. The same holds for pressure-drop calculations. More details are given in [9]. As an illustration, Figure 8.9 presents some experimental results. It can be seen that the liquid holdup is practically independent of the gas load over a large range. Sizing the RD column based only on liquid phase is therefore rational. On the other hand, for good mass-transfer exchange the gas-phase characteristic $F = U_g \sqrt{\rho_G}$ should be in the range 0.5–1.5 Pa$^{0.5}$. The pressure drop varies proportionally with the gas load. A value of 2 mbar/m is recommended for the preliminary design.

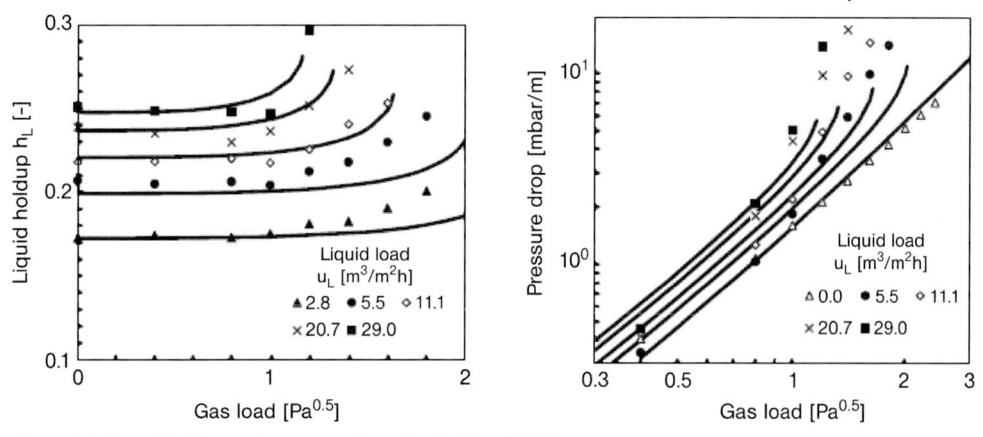

8.3 Esterification of Lauric Acid with 2-Ethylhexanol

Figure 8.9 Liquid holdup and pressure drop for Multipack-II [9].

The last element in analysis is the number of theoretical stages per meter, NTSM. Laboratory experiments indicate a value of 3 for Katapack-SP [11], with 3 to 6 for Multipack-II [9]. On the other hand, NTSM of two seems more appropriate from the industrial viewpoint.

Taking into account the above physical picture, the following approach for the hydraulic designing of a RD column seems sensible.

1. Estimate a volumetric liquid flow rate as representative for operation.

2. Assume an initial value for the virtual liquid velocity at the "load point" (U_{LP}).

3. Assume an initial value for the number of stages per meter (NSTM).

4. Determine the column diameter. On this basis estimate the volume of packing and the catalyst holdup per stage.

5. Introduce the above value in the simulation model in which the reaction rate is expressed per mass of catalyst. Determine the total number of stages needed to achieve the desired conversion. Pay attention to the profiles of temperatures, concentrations, and reaction rate. Extract liquid and gas flows, as well as fluid properties.

6. Recalculate the load-point velocity, the column diameter and the catalyst holdup from the above information by using specific correlations.

7. Check the load point for the gas phase and the pressure drop.

8. Repeat points 4 to 7 until acceptable values are achieved.

The preliminary hydraulic design can be outlined as follows:
The ester production is 10 kmol/h, or 3120 kg/h. An operation time of 8000 h leads to a production of 25 000 tons/year. Assuming an alcohol reflux ratio of 0.5 one may estimate a mean liquid flow of 3120 × 1.5 = 4680 kg/h taking into account

that practically the mass flow of the reaction mixture should not vary too much along the stages. Considering a liquid density of 850 kg/m³ one obtains a volumetric flow of 5.5 m³/h. The most important assumption is the linear liquid velocity, for which 10 m/h is proposed. Thus, one gets the column area 0.55 m² and column diameter of 0.837 m. The catalyst holdup can be determined knowing the catalyst fraction $\psi = 0.327$, NTSM = 3 and catalyst density $\rho_c = 1000 \times (1-0.3) = 700 \, \text{kg/m}^3$. One gets the catalyst volume of $0.55 \times 1 \times 0.327 \times 1000/3 = 60 \, \text{l}$, or a mass holdup $h_m = 60 \times 0.7 = 42 \, \text{kg/stage}$.

8.3.7.3 Simulation

The conceptual flowsheet presented in Figure 8.5 is transposed in a simulation diagram for Aspen Plus, as displayed in Figure 8.10. In order to get more robust convergence the RD column is simulated as a stripper with fixed fresh feeds. The top vapor is condensed and the liquid separated in a two-phase decanter. Water is obtained as aqueous phase, while the organic phase is returned. The RD column has 24 theoretical stages, of which 20 are reactive. The lauric acid is introduced on the top reactive stage as liquid at 160 °C, while the 2-ethyl-hexanol, fresh and recycled from evaporator, enters as vapor at 0.5 bar at the bottom. In order to limit the reboiler temperature to below 200 °C the column is operated under vacuum at 32 kPa. Table 8.9 presents key simulation results. The key result is that very high acid conversion, >99.99%, can be achieved, producing as a consequence an ester purity below 50 ppm.

At this point, it is interesting to compare the liquid load issued from reaction kinetics, with the value calculated from the hydrodynamics of the packing. The following physical properties are known from simulation: liquid density 870 kg/m³, liquid viscosity 0.98 cP. With particles of 0.85 mm and $\varepsilon_{CB} = 0.3$ the Reynolds number of 2.1 is obtained leading to a friction factor of 78. As result, $U_{LP} = 2.03 \, \text{mm/s} = 7.3 \, \text{m/h}$ is obtained. This value is in good agreement with the 7.62 m³/m²/h found by kinetic calculation by Aspen Plus, as well as with the value

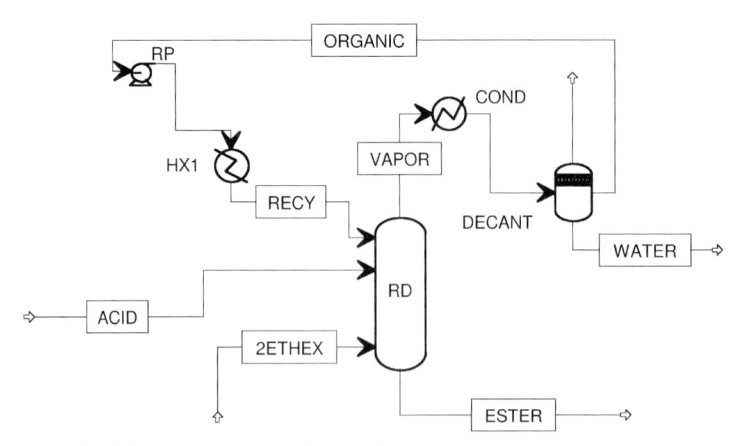

Figure 8.10 Simulation diagram for the RD setup.

Table 8.9 Simulation results.

	Base case	Optimal design
Number of theoretical stages	20	24
Reactive stages	16	22
NSTM	3	3
Diameter, m	0.837	0.837
Cross-sectional area, m^2	0.55	0.55
Catalyst holdup, kg/stage	42	10 (2–5), 20 (0–11), 30 (11–14), 42 (15–23)
Total mass of catalyst, kg	672	606
Lauric acid feed, kmol/h; °C	10; 160	10; 160
Alcohol feed	14; 160	14; 160
Top vapor flow, kmol/h	15	15
Reflux ratio	0.248	0.248
Temperature profile, °C		
• top	113.9	113.5
• entry acid	151.9	151.8
• maximum	178	180
• entry alcohol	175	176.5
• bottom	180	182
Liquid flow[a], kmol/h; kg/h; m^3/h	14.22; 3868.4; 4.2	
Gas flow[a], kmol/h; kg/h; m^3/h	14.08; 1866.2; 1392.4	
Densities liquid/gas, kg/m^3	874.4/1.34	
Liquid load, $m^3/m^2/h$	7.62	
Gas load, $m^3/m^2/s$	0.7	
F factor	0.6	0.6
Ester production, kmol/h	9.89445	9.996
Residual acid, kmol/h	5.52×10^{-3}	4×10^{-3}
Purity (free alcohol ester), ppm	558	290
Productivity, kg ester/kg catalyst	4.64	5.15
Reboiler duty, kW	28.5	35.6

a) Reference to the middle stage of the reactive section.

of 7.72 recommended by Sulzer for a minimum RTD distribution [11]. Note that for a catalyst particle size of 1 mm the hydrodynamic calculation will give a liquid load point at 10 $m^3/m^2/h$, identical with the assumed value of 10 m/h.

Typical profiles are shown in Figure 8.11. The acid concentration in the liquid phase falls rapidly over the first five stages, where most of the reaction takes place. The alcohol concentration is less sensitive, remaining almost constant. The water concentration in the liquid phase is a maximum of 3% mole fraction on the top stage, but only 1% on the first reactive stage, or as a mass fraction less than 1000 ppm. The ester formation takes place mainly on the first half of the reactive zone, the second being necessary to push the conversion over 99.99%. The temperature profile shows a sharp initial increase over the first five stages from 113 °C to a plateau at about 170 °C. The ester production rate shows a maximum on the first two stages.

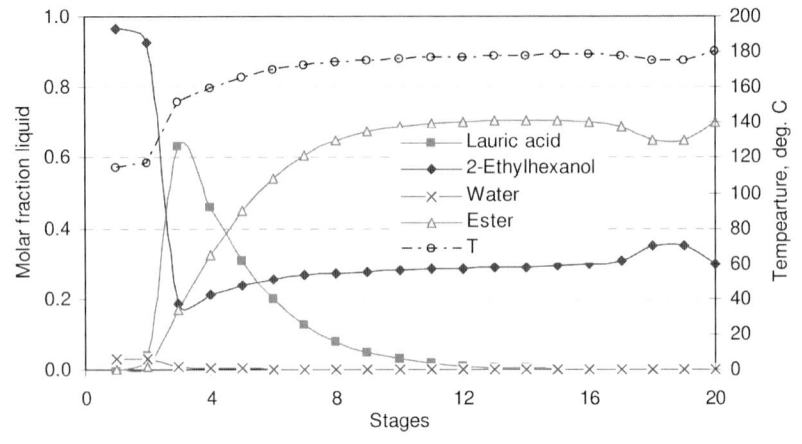

Figure 8.11 Profiles in the RD column by kinetic modeling.

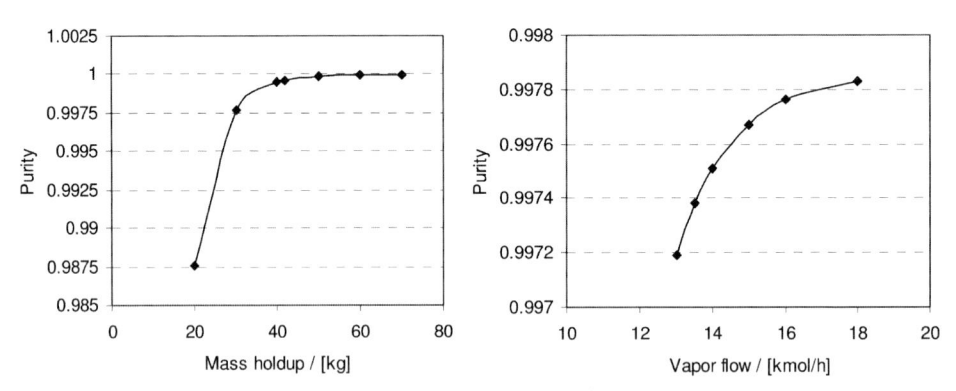

Figure 8.12 Optimization of performance. The effect of catalyst mass holdup and vapor distillate flow on purity.

8.3.8
Optimization

The effect of optimization variables can be investigated by sensitivity analysis applied on the base case defined above. The most influential are catalyst holdup and reflux ratio, here simulated by top vapor distillate. The product purity is particular sensitive to catalyst holdup, as shown in Figure 8.12 (left-hand), up to a purity of 99.8%. From getting even higher purity, the catalyst holdup should increase considerably. Thus, at a holdup of 70 kg/stage (or doubling the stages at constant holdup of 40 kg/stage) a product purity of about 10 ppm can be reached. The purity is also sensitive to the reflux ratio (Figure 8.12, right-hand) in the medium purity range, but without effect at very high purity.

The catalyst distribution on stages may be also seen as an optimization variable, but the effect on productivity is rather small. The real advantage comes from a technological reason. The reaction rate is the highest on the first top stages, and therefore the catalyst could deactivate here more rapidly. Placing less active catalyst on the first reactive stages leads to a more uniform reaction rate. The reduction in productivity can be compensated by only a few reactive stages on the bottom side. The results for the optimal design are presented in Table 8.9. It can be seen that this is not far from the initial design. The main difference is in the catalyst distribution that ensures more uniform reaction rate at maximum 1 kmol/h/stage, which is about half that previously. However, the total amount of catalyst is lower, leading to higher productivity.

Column pressure has a strong effect. Increasing the pressure from 0.2 to 0.5 bar will increase the reaction temperature by about 10 °C, sufficient for doubling the reaction rate. Since the reaction is slightly exothermal, the reboiler duty is only needed to compensate the differences in the enthalpy of components. Consequently, the reboiler duty is small and has no effect on optimization.

In economic terms, an optimization function can be built that maximizes the difference between the benefit as a function of ester purity and the costs involving catalyst and utilities, as a function of catalyst holdup. This aspect is left as an exercise for the reader.

8.3.9
Detailed Design

Preserving the efficiency of the catalyst with variable liquid and gas flows can ensure a high-purity product. A potential drawback would be the maldistribution of the liquid with a loss in conversion and product impurification with acid. Actually, the liquid should flow uniformly and very slowly through the catalyst bed at a linear velocity in the range of a few mm/s. The placement of special redistribution devices is necessary. Besides the industry-proven methods, innovative solutions can be found by taking advantage of the mixing and dispersion properties of the packing itself. An example is "partially flooded beds" designed by Montz and BASF [10] in which standard packing is combined with shorter, specially designed elements to promote a bubbling action similar to that of a tray. In this way, a longer residence time of the liquid can be ensured with minimum backmixing.

8.4
Esterification of Lauric Acid with Methanol

In the following, the strategy presented before will this time be applied for developing a process for the esterification of lauric acid with methanol. All the thermodynamic data for pure components and binary mixtures are available in Aspen Plus. A residue curve map of the reactive mixture at equilibrium can be computed as described in Appendix A. A useful representation can be done in reduced coordinates defined by $X_1 = water + acid$ and $X_2 = acid + ester$. The diagram displayed

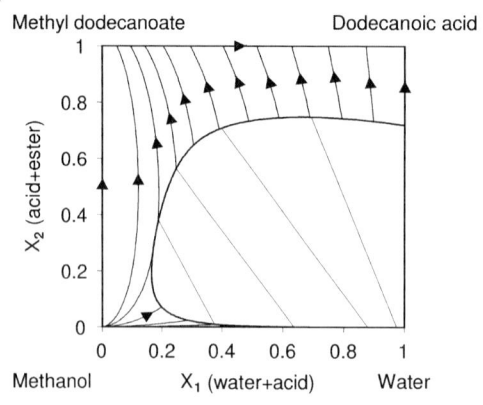

Figure 8.13 Generalized CPE diagram for the esterification reaction of dodecanoic acid with methanol at normal pressure.

in Figure 8.13 (normal pressure) shows two regions, homogeneous organic phase mixtures, and heterogeneous with second water phase. From the separation viewpoint methanol is an unstable node and lauric acid a stable node, while the products, ester and water, are saddles. It can be seen that the reaction represented by a straight line linking the vertices of methanol and lauric acid falls in the heterogeneous region over a large ratio of reactants. Thus, the RCM analysis indicates that a feasible process needs a two-column arrangement: RD for producing the ester, and distillation for methanol recovery. On the contrary, the esterification in a single RD device seems not to be feasible.

However, the unfeasibility at infinite reflux is not necessarily true at finite reflux. Omota et al. [2] solved this issue by means of a tray-by-tray computation based on the solution of MESH equations including chemical reaction with equilibrium assumption. The computation starts from the bottom with composition–temperature specifications corresponding to 99.9% purity and bottom 473 K. The independent parameter is the boilup ratio. Several composition profiles are displayed in Figure 8.14. At boilup ratios higher than 1.25 (Case 1) the composition on the upward stages converges to a binary mixture of methanol and methyl dodecanoate. At only slightly lower boilup ratios (Case 2) the composition profile penetrates into the heterogeneous region. However, at boilup ratios less than 1.2 (Case 3) the composition on the top stages converges to a binary mixture of lauric acid/water. Hence, the process becomes feasible when the boilup ratio is sufficiently low. The column can produce in the bottom an ester of high purity and deliver in the top a vapor mixture of acid and water, from which after condensation and decantation water can be separated quantitatively from acid because of the extremely low solubility.

Thus, the key result from the tray-by-tray calculation is that the column design must ensure complete alcohol consumption in the reactive zone, only lauric acid and water are allowed in the top vapor stream. The column behaves more as a reactive absorber than reactive distillation. A higher number of equilibrium stages

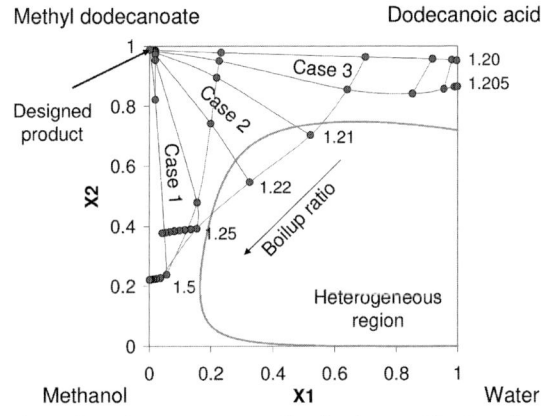

Figure 8.14 Concentration profiles for the esterification of dodecanoic acid with methanol. Case 1: Trajectory leading to a binary mixture methanol/ester; Case 2: Profile finishing in the heterogeneous region; Case 3: Low boilup ratios: feasible design in the homogeneous region.

is necessary, in this case 14. The column pressure may be atmospheric or higher, but the presence of a certain amount of methanol in the bottom product is still necessary to limit the rise in temperature.

This analysis was confirmed by the work of Steinigeweg and Gmehling [17] regarding the esterification of decanoic acid with methanol catalyzed by a strong acid ion-exchange resin (Amberlyst-15) at 393 K and 3 bar. In order to get high conversion a very small reflux 0.01 should be used, which means practically reactive absorption instead reactive distillation. However, for practical reasons, a higher reaction rate and a lower reboiler temperature, they used an excess of alcohol of 2:1 in the initial feed, which in turn implies the secondary distillation of methanol. About 20 stages were necessary to reach high conversion over 95% the reactive part with 5 stages for top and bottom nonreactive zones. The authors compared two packings from Sultzer, Katapak-S (catalytic layer) and Katapack-SP (plus separation layer). The difference is visible at high conversion, where Katapack-S is more efficient due to larger catalyst fraction. Other values of interest are NTSM = 2 for Katapack-SP, vapor and liquid load of 0.8–1 $Pa^{0.5}$ and 3 $m^3/m^2/h$.

Further improvement in the technology of methyl fatty ester synthesis can be achieved by dual esterification [4]. This takes advantage of the fact that the sulfated zirconia catalyst has similar activity for normal alcohols, over the series C1–C8. However, methanol manifests about twice the activity [20]. The removal of water produced by the esterification with methanol is solved simply, by employing a heavy alcohol immiscible with water, such as 2-ethyl-hexanol, which acts simultaneously as a reactant and an entrainer. As a result, the two fatty esters are obtained in the bottom product in the desired ratio by adjusting the feeds. For example, in a preferable operation mode the ratio of fresh feed reactants is acid:methanol:2-ethyl-hexanol 1:0.8:0.2.

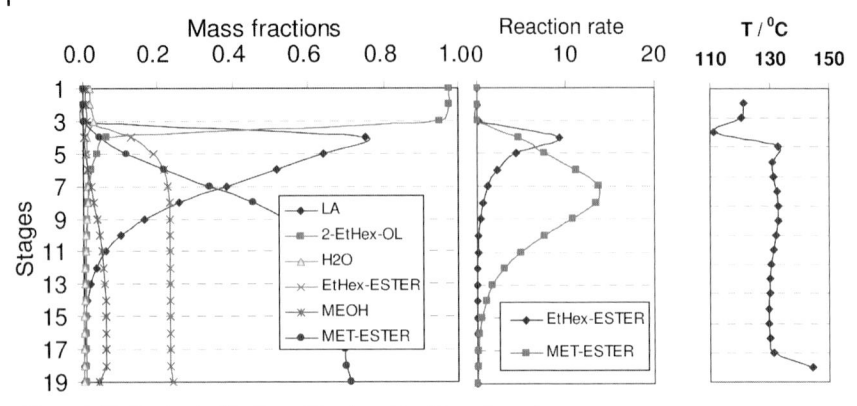

Figure 8.15 Dual esterification of lauric acid with methanol and 2-ethyl-hexanol: liquid concentration, reaction rate and temperature profiles.

The above solution makes sense from the economic viewpoint, because the methyl ester is by far the most demanded, while 2-ethyl-hexanol is a cheap alcohol, in general a waste. The flowsheet in Figure 8.2 is still valid, in which the heavy alcohol is fed on the top stage, as an entrainer. Figure 8.15 shows the liquid concentration, reaction rate and temperature profiles. It can be seen that 2-ethyl-hexanol indeed plays its double role. The concentration of methanol in the top can be maintained at the minimum, such that there is no need for additional columns for methanol recovery and recycle. Note also that the operation takes place at low pressure (1.5 bar), in contrast with 6–12 bar and vacuum with individual light and heavy alcohols.

8.5
Esterification of Lauric Acid with Propanols

When using propanols for esterification, the alcohol and water form azeotropes making necessary secondary recovery and recycling of alcohol. An industrial process that is conducted in this way makes use of homogeneous catalyst (para-toluene sulfonic acid) lost after reaction [5]. A more efficient solution can found by employing an entrainer for breaking the azeotrope water/alcohol and superacid solid catalyst. Besides, the entrainer has a beneficial effect on the reaction rate, by increasing the amount of alcohol recycled to the reaction space [3].

Table 8.10 displays some important thermodynamic properties of the reactive mixture. The components form azeotropes each other. The ester is the highest boiler, followed by the acid, and at large distance by water and 1-propanol. The RD column operates by diluting the product with some alcohol, which can be recycled after product conditioning. Without further treatment, the top distillate would be the azeotrope n-propanol/water.

Table 8.10 Key thermodynamic properties of the reactive mixture.

Normal boiling points	Dodecanoic acid	n-propyl dodecanoate	n-propanol	Water
T, K	571.75	574.95	370.35	373.15
Azeotrope	Acid(1)/ Ester(2) Homogeneous	Acid(1)/ Water(2) Heterogeneous	Ester(1)/ Water(2) Heterogeneous	Alcohol(1)/ Water(2) Homogeneous
T, K Composition	568.9 $y_{az,1} = 0.5784$	373.15 $x_1(w)<0.0001$ $x_1(o) = 0.7195$ $y_{az,1} = 0.0001$	373.13 $x_1(w)<0.0001$ $x_1(o) = 0.9254$ $y_{az,1} = 0.0004$	360.86 $y_{az,1} = 0.4330$

Note: w = water phase, o = organic phase

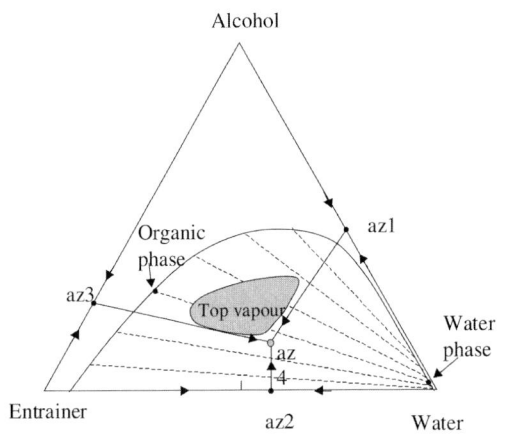

Figure 8.16 Conceptual residue curve map for entrainer selection in fatty acid esterification.

8.5.1
Entrainer Selection

The following three conditions for the entrainer selection can be formulated [3]:

1. Give minimum boiler ternary *heterogeneous* azeotrope with alcohol and water, or most preferably binary heterogeneous azeotrope with water. The entrainer must form also a minimum binary azeotrope with the alcohol.
2. The L–L split of the heterogeneous azeotrope should produce a water-rich phase with low fraction of entrainer and alcohol, while the organic phase should be relatively rich in alcohol but with limited content of water. In addition, the L–L tie lines should point to the vertex of water.

3. The composition of ternary azeotrope should have a reduced amount of entrainer.
4. Give acceptable impurity in the final product.

The screening of entrainers by plotting RCM's highlights the following classes:

1. Hydrocarbons: alkanes and aromatics.
2. Oxygenated components: esters, ethers, ketones.
3. Halogeno-hydrocarbon: 1-chloro-methyl- propane.

Suitable entrainers from the above list are toluene, di-propyl-ether and n-propyl acetate. The first one may be skipped as it is an undesired impurity. Di-propyl-ether is suitable: practically insoluble in water and low reciprocal solubility, of the order of 4%. Moreover, it is a byproduct of the reaction. The choice is n-propyl-acetate is convenient too, such as sharing the same alcohol with the fatty ester.

The process flowsheet is similar to Figure 8.2, except that makeup entrainer is added in the decanter. Figure 8.17 presents results obtained with Aspen Plus™ by considering a stage-equilibrium model with chemical reaction. Note that the reaction rate is similar to the system lauric acid/1-ethyl-hexanol. The design is as follows: 15 reaction stages, 15 rectification stages, acid feed as liquid at 453 K, 1-propanol feed on the stage 30 as vapor, atmospheric pressure.

The examination of profiles (Figure 8.17) shows how the process works. The top vapor composition has to fall in the heterogeneous region alcohol–water–entrainer. The entrainer enhances the water extraction from the reaction zone, and in turn increases the amount of alcohol recycled to the reaction zone. By liquid decantation at 353 K the mixture splits into an aqueous-phase containing 98% water and a recycled organic phase. In the reactor zone the acid concentration decreases sharply after the first 7 stages (Figure 8.17a). The remaining 7 reactive stages are still required to achieve high purity. The temperatures of the reaction and distillation zones are quite different (Figure 8.17b), corresponding to reaction and separation requirements. Therefore, at the junction of the two zones the insertion of a

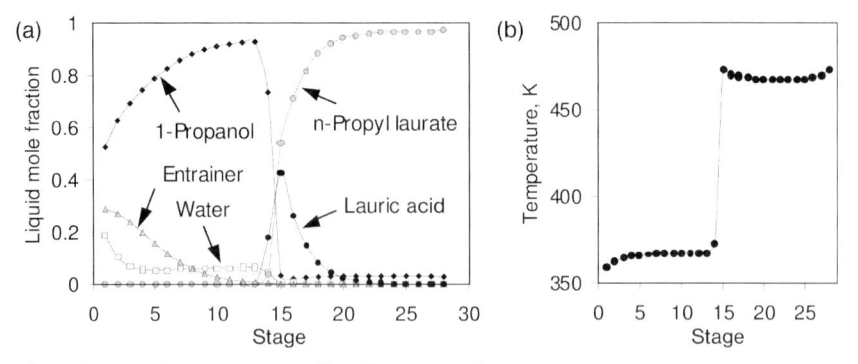

Figure 8.17 (a) Concentration profiles of species in the entrainer-enhanced reactive distillation column; (b) temperature profiles.

8.5.2
Entrainer Ratio

The effect of the entrainer/feed ratio is presented in Figure 8.18, where both liquid and vapor concentration in the rectification zone is displayed in a triangular diagram. A minimum amount of entrainer is necessary, corresponding to the azeotrope composition. Indeed the simulation shows clearly that the composition profile becomes stable above an entrainer ratio of 0.914 (profile A). Increasing further the amount of entrainer pushes more alcohol into the reaction zone, while the water is sent to the top and concentrated in the vapor phase (profiles B and C). At a higher entrainer ratio, in this case 3.8, the profile crosses the distillation boundary set by the azeotrope alcohol/entrainer changing the direction. Now the entrainer concentrates to the bottom of the rectification zone, while the alcohol migrates to the top together with water. Hence, a maximum entrainer ratio is detected.

Hence, the entrainer enhances the water removal, ensuring simultaneously a recycle of alcohol to the reaction zone. As a consequence, the reaction rate can increase substantially. The comparison with a process without entrainer operating as pseudoabsorber shows that the catalyst loading can be reduced up to 50%.

The above approach was confirmed in a recent study on design and control of a process for high-purity isopropyl palmitate by reactive distillation using cyclohexane as entrainer, with substantial energy reduction [21].

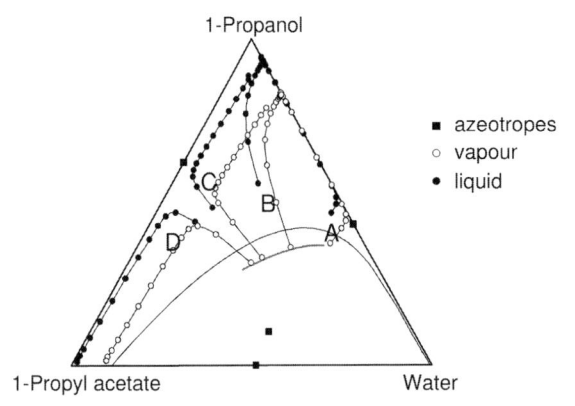

Figure 8.18 Effect of entrainer/feed ratio on the concentration profiles in the RD column displaying in a ternary diagram alcohol–water–entrainer.

8.6
Conclusions

The manufacture of fatty-acid esters of high purity can be achieved by making use of reactive distillation and superacid solid catalysts based on sulfated zirconia. The key constraint is the selective water removal to shift the chemical equilibrium and ensure water-free organic phase. Taking into account that the catalyst manifests similar activities for several alcohols, the design aims to find flowsheet configurations within the same equipment that can be used for producing several products by adjusting only the operation conditions.

The feasibility of design is studied taking as substrates the lauric acid (C12) and several alcohols, from C1 (methanol), C3 (n- and isopropanol) and C6 (2-ethyl-hexanol). The analysis in the residue curve map with an equilibrium reaction brings useful insights. The easiest is the esterification of lauric acid with 2-ethyl-hexanol, since water distillates with alcohol as a heterogeneous azeotrope, and is removed from the system by L–L separation. The dilution of ester with some alcohol and the use of vacuum is necessary to limit the bottom temperature to less than 573 K. Computer simulation by kinetic modeling with phase equilibrium confirms the analysis with chemical equilibrium stages. The hydraulic calculation based on structured catalytic packing indicates that the liquid velocity and the residence time is constrained by the liquid load point.

In the case of esterification of lauric acid with methanol, the RCM analysis shows two alternatives: (1) reflux of alcohol, but secondary methanol distillation, and (2) acid reflux (reactive absorption instead reactive distillation). The last alternative seems to be generic for any type of alcohol, but in practice it is limited by low volatility of fatty acids and high sensitivity to operating conditions. A better alternative is offered by dual reactive distillation, in which 2-ethyl-hexanol is added in the top as a coreactant and a mass-separation agent.

The esterification with propanols raises the problem of breaking the azeotrope that the alcohol forms with water. The solution of this problem passes by the use of an entrainer forming a heterogeneous ternary azeotrope. Suitable entrainers are hydrocarbons and oxygenated species, as esters and ethers. The solution is in principle similar for u-propanol with the notable difference that the reaction rate is much slower for the last when using the same catalyst.

Hence, manufacturing several fatty esters in a reactive distillation setup by using heterogeneous catalysis is possible. The equipment is simple and efficient in operation. Making use of a fast catalyst remains the key aspect in the technology.

References

1 Omota, F., Dimian, A. C., Bliek, A., Fatty-acid esterification by reactive distillation, Part 1: equilibrium-based design, *Chem. Eng. Science*, 58, 3175–3185, 2003

2 Omota, F., Dimian, A. C., Bliek, A., Fatty-acid esterification by reactive distillation, Part 2: kinetic-based design for sulfated zirconia catalyst, *Chem. Eng. Science*, 58, 3175–3185, 2003

3 Dimian, A. C., Omota, F., Bliek, Entrainer-enhanced reactive distillation, *Chem. Eng. Processing*, March, 411–420, 2004

4 Dimian, A.C., Omota, F., Kiss, A.A., Process for fatty-acid methyl esters by dual reactive distillation, Proceedings Escape-17, 2007

5 Bock, H., Wozny, G., Gutsche, B., Design and control of a reaction distillation column including recovery system. *Chem. Eng. Progress*, 36, 101–109, 1997

6 Constantinou, L., Gani, R., New group-contribution method for estimating properties of pure compounds. *AIChE J.*, 40(10), 1697–1710, 1994

7 Doherty, M.F., Malone, M., Conceptual design of distillation systems, McGraw-Hill, New York, USA, 2001

8 Gmehling, J., Menke, J., Krafczyk, J., Fischer, K., Azeotropic Data. Part I. Wiley-VCH, Weinheim, Gemany, 1994

9 Hoffman, A., Noeres, C., Gorak, A., Scale-up of reactive-distillation columns with catalystic packings, *Chem. Eng. Process.*, 43, 383–395, 2004

10 Olujic, Z., Kaibel, B., Jansen, H., Rietfort, T., Zich, E., Frey, G., Internals for process intensification, *Chem. Biochem. Eng.*, Q 17 (4), 301, 2003

11 Gotze, L., Bailer, O., Moritz, C., von Scala, C., Reactive distillation with Katapack, *Catal. Today*, 69, 201–208, 2001

12 Tuchlenski, A., Beckmann, A., Reusch, D., Dussel, R., Weidlich, U, Janowschi, R., Reactive distillation-industrial applications, *Chem. Eng. Sci.*, 56, 387, 2001

13 Peng, J.P., Lextrait, S., Edgar, T.F., Eldrige, R.B., A comparison of steady-state equilibrium and rate based models for packed reactive-distillation columns, *Ind. Eng. Chem. Res.*, 41, 2735, 2002

14 Ellenberger, J., Krishna, R., Counter-current operation of structured catalytically packed distillation columns, *Chem. Eng. Sci.*, 54, 1339–1345, 1999

15 Krishna, R., Hardware selection and design aspects for reactive-distillation columns, Reactive Distillation, eds Sundmacher K. and Kienle, A., Wiley-VCH, Weinheim, Germany, 169–189, 2003

16 Stephenson, R., Stuart, J., Tabak, M., Mutual solubility of water and aliphatic alcohols. *J. Chem. Eng. Data*, 29, 287–290, 1984

17 Steinigeweg, S., Gmehling, J., Esterification of a fatty acid by reactive distillation, *Ind. Eng. Chem. Res.*, 42, 3612–3619, 2003

18 Fogler, H.S., Elements of Chemical Reaction Engineering, Prentice Hall, 1999

19 Aspen Plus, release 12.1, Aspen Technology, 2005

20 Kiss, A. A., Dimian, A.C., Rothenberg, G., Solid acid catalysts for biodiesel production – towards sustainable energy, *Adv. Synth. Catal.*, 348, 75–81, 2006

21 Wang, S-J., Wong, S. H., Design and control of entrainer-aided reactive distillation for fatty ester production, *Ind. Eng. Chem. Res.*, 45, 9041–9049, 2006

9
Isobutane Alkylation

9.1
Introduction

Almost all cars and motorbikes, many boats, and a wide variety of aircrafts and locomotives use internal combustion engines. The internal combustion engines rely on the exothermic reaction of a fuel with air resulting in high temperature and pressure, which are translated into work by the engine. Mixtures of hydrocarbons are the most common fuels. They include diesel, gasoline and liquefied petroleum gas (LPG). The combustion is initiated by means of either electrical (spark-ignition engines) or compression heating (diesel engines) systems. During normal combustion, ignition creates a flame front that moves at high speed through the combustion chamber. The advancing flame front heats and pressurizes the remaining unburned mixture. This may cause an instantaneous ignition in the form of an explosion, called detonation or knocking. As a result, the cylinder pressure rises dramatically, beyond design limits. If allowed to persist, detonation will cause vibration and damage the engine parts.

The octane number is a quantitative, but imprecise measure of the maximum compression ratio at which a particular fuel can be utilized in an engine without "knocking" of the fuel/air mixture. By definition, n-heptane has an octane number of 0, while 2,2,4-trimethyl pentane (isooctane) is 100. Linear combinations of these two components are used to measure the octane number of a particular fuel. Because the performance of a fuel depends upon the way the engine is operated, two octane numbers are used. The research octane number (RON) characterizes fuel performance under low-severity engine operation. The motor octane number (MON) describes operation that might be incurred at high speed or high load. In general, RON values are higher than MON, the difference ranging from 0 to more than 15. The octane number of a gasoline is reported as RON (in Europe) or as the average of RON and MON (in the United States).

The use of a fuel with a high octane rating is essential in preventing detonation. The quality of the fuel is controlled by the ratio in which different components are blended, and by addition of additives such as methyl *tert*-butyl ether (MTBE). Although aromatics and olefins increase the octane number, environmental

Chemical Process Design: Computer-Aided Case Studies. Alexandre C. Dimian and Costin Sorin Bildea
Copyright © 2008 WILEY-VCH Verlag GmbH & Co. KGaA, Weinheim
ISBN: 978-3-527-31403-4

concern is pushing towards their removal from the gasoline. Even MTBE and other ethers used as additives are under pressure to be phased out [1].

Alkylate is a gasoline blending component with exceptional antiknock properties, which seems to avoid the legislative pressure. Alkylate consists exclusively of isoalkanes and is obtained from the C_3–C_4 cut of the FCC units. In many instances, isobutene from the C_3–C_4 fraction is transformed selectively with methanol into methyl *tert*-butyl ether (MTBE). Therefore, a mixture of 1-butene and 2-butene is used for alkylation purposes. The other reactant is isobutane. The major constituents of the alkylate are 2,2,3-, 2,2,4-, 2,3,3- and 2,3,4-trimethyl pentane (TMP). Besides, the alkylate contains other C_8 isoalkanes, such as dimethyl hexane (DMH), 3-ethyl 2-methyl pentane, methyl heptane and ethyl hexane, and even isoalkanes with carbon numbers that are not multiples of 4.

Dimerization of high-purity isobutylene could be a competing chemical route for obtaining trimethyl pentane. However, dimerization is not applicable to the typical mixed butylene compositions found in FCC C_4 fractions, because of the poor octane number resulting from codimers of isobutylene and normal butene. To achieve an effective isobutylene/n-butene ratio in the feed stream, refineries would have to add a costly isobutane dehydrogenation unit and purchase additional isobutane to support isobutylene production. Even with a selective catalyst that dimerizes only isobutylene in a mixed butylene feed, the resulting raffinate rich in n-butene must then be converted to gasoline by alkylation. A standalone alkylation unit appears to be a far more cost effective solution, when compared to the combined dimerization–alkylation solution [2].

This chapter considers the design of an isobutane/butene alkylation plant, following the hierarchical approach methodology presented in Chapter 2. We will pay special attention to the reaction/separation/recycle structure of the flowsheet, showing how plantwide control considerations are introduced during the early stages of conceptual design. More specifically, we will derive a simplified mass balance of the plant using a kinetic model for the reactor, but black-box models for separation units. Degree-of-freedom analysis will be used to suggest plantwide control structures. Among them, one structure will show unfavorable steady-state behavior (the snowball effect) and therefore will be discarded. The steady-state behavior of a promising control structure will be further analyzed. One important result is that, for a given reactor, multiple steady states are possible. Moreover, the operation becomes unfeasible when large production rates are attempted or the combined physical and chemical processes taking place in the reactor are too slow.

An important part of this chapter will be devoted to robustness. The design will ensure that feasibility is preserved when operation and design parameters change or are uncertain. Thus, we will take into account the fact that alkylation plants are often built as onstream units, being required to deal with large changes of the butene feed stream. In addition, we will be aware of large errors of the kinetic data.

The tools used in this chapter include numerical methods for solving systems of nonlinear equations and tracing the dependence upon one parameter by continuation methods. To tackle the robustness of the design, two approaches will be shown, based on the concept of critical manifolds and use of distance to the tangent space and along the normal vector, respectively. After completing the conceptual design, the steady state, dynamic and robustness performances will be assessed using more rigorous tools, such as Aspen Plus and Aspen Dynamics.

9.2
Basis of Design

9.2.1
Industrial Processes for Isobutane Alkylation

In industrial practice, two liquid acids are employed as catalysts for isobutane/butene alkylation, namely sulfuric acid and hydrofluoric acid [3, 19, 20]. Both processes deliver a high-quality gasoline component. The catalyst consumption in the H_2SO_4 process is high, typically 70–100 kg/t. The spent sulfuric acid contains tarry hydrocarbons and water and has to be processed externally. On the other hand, corrosiveness and toxicity of HF are reasons of concern that require use of additives that lower the HF vapor pressure and minimize the amount of HF released in the case of an accident. However, in many industrialized countries, new HF alkylation processes are no longer approved by authorities.

During the last 40 years, an enormous effort was put into searching for a solid catalyst [4, 5]. The main obstacle still to be overcome is the formation of acid-soluble oils (ASO, also known as conjunct polymers or red oil) which accompanies the alkylation process. This material contains highly unsaturated cyclic hydrocarbons, which rapidly passivate the catalyst. When liquid catalysts are used, they can be easily withdrawn from the process and replaced, without interrupting the alkylation operation. UOP has developed the Alkylene technology, which uses the proprietary HAL™ 100 catalyst in a process that is claimed to be commercially competitive [6].

Taking into account the environmental concerns raised by using HF as a catalyst and the lack of data concerning the solid-catalyzed processes, in this chapter we will design an alkylation process based on the H_2SO_4 technology. The process will produce 30 kmol/h alkylate (about 25 000 ton/year), with a selectivity of 85%.

9.2.2
Specifications and Safety

The alkylate is a mixture of hydrocarbons. The main components are 2,2,3-, 2,2,4-, and 2,3,4-trimethyl pentane, which account for about 70% of the alkylate composition. A typical butene alkylate has the octane ratings RON = 94–98, and MON = 92–95. The alkylation yield is 1.70–1.78 m^3 alkylate/m^3 butene, while the isobutene and acid consumptions are 1.10–1.16 m^3 isobutane/m^3 butene and 0.12 kg H_2SO_4/

10^{-3} m³ alkylate [7]. The undesired byproducts include light paraffins, acid-soluble oils and isoparaffines with low octane number known as pseudoalkylates [8].

From the hazard point of view [9], the process presents two problems: the acid catalyst that is corrosive and toxic and the hydrocarbons that are flammable and potentially explosive.

H_2SO_4 is contained in carbon steel, but may become very corrosive if diluted with water. Therefore, frequent analysis of acid strength is required. Before sending the spent acid to regeneration, care must be taken to remove the dissolved hydrocarbons. Moreover, explosive conditions in the vapor space of the tanks must be avoided by blanketing with inert gases. The alkylate leaving the reaction section is normally neutralized. However, the piping area where acidic effluent and aqueous alkali are contacted is susceptible to corrosion. In the plant, pressure-relief systems that are exposed to acidic material should be inspected regularly.

In alkylation plants, the large volumes of LPG that are stored should be taken into account by the training procedures. A key element of managing the hazard is preventing the release. Additionally, the facilities should be protected against the impact of fire.

9.2.3
Chemistry

The reaction occurring during alkylation can be explained by carbenium ion chain mechanism initiated by the protonation of the olefin. The main steps of the mechanism are [7]:

Alkylation of isobutane giving 2,2,3 TMP, 2,2,4 TMP, 2,3,4 TMP and DMH

$$C_4H_8 + H^+ \leftrightarrow C_4H_9^+$$

$$C_4H_9^+ + C_4H_8 \leftrightarrow C_8H_{17}^+ \quad (2,2,3-TMP^+, 2,2,4-TMP^+, \ldots 2,4-DMH^+ \ldots)$$

$$C_8H_{17}^+ + iC_4H_{10} \leftrightarrow C_8H_{18} + iC_4H_9^+$$

Multiple olefin additions giving heavy products

$$C_8H_{17}^+ + C_4H_8 \leftrightarrow C_{12}H_{25}^+$$

$$C_{12}H_{25}^+ + C_4H_8 \leftrightarrow C_{16}H_{33}^+$$

Breakdown of heavy ions to give C_5, C_6, C_7, ... paraffins and olefins

$$C_{12}H_{25}^+ \to \begin{cases} iC_5^+ + iC_7^= \\ iC_6^+ + iC_6^= \end{cases}$$

$$C_{16}H_{33}^+ \to \begin{cases} iC_7^+ + iC_9^= \\ iC_8^+ + iC_8^= \end{cases}$$

Proton transfer to paraffinic carbocations and self-alkylation of light olefins

$$C_x^+ + iC_4H_{10} \leftrightarrow C_x + iC_4H_9^+$$
$$C_x^= + 2iC_4H_{10} \leftrightarrow C_x + 2, 2, 4 - TMP, x = 5, 6, 7, 8, 9, ..$$

9.2.4
Physical Properties

The basic physical properties of the main species involved in the process are presented in Tables 9.1 and 9.2. It may be observed that there are large differences between the boiling points of the inert (propane), reactants (isobuane and butene), product (isooctane) and byproduct (dodecane). Also taking into account the ideal behavior of the hydrocarbons mixture at low and moderate pressures, no separation difficulties are expected. The data of Table 9.1 is useful for choosing the operating pressure. At the reaction temperature, the mixture should be in the liquid phase. Concerning the distillation columns, the boiling point of the light components should be high enough to allow cooling using water. Therefore, we choose 8 bar as the operating pressure. However, the last split (isooctane/heavies) will be performed at atmospheric pressure.

Table 9.1 Boiling points of the components found in reactor effluent.

Component (reactor effluent)	Boiling point [°C]				Destination
	1 Bar	5 Bar	8 Bar	12 Bar	
Propane (I)	−42.5	1.8	18.3	34.3	Inert, removed
Isobutane (B)	−12.0	38.0	56.6	74.7	Recycle
1-butene (A)	−6.7	43.4	62.02	80.0	Recycle
Isooctane (P)	98.7	167.4	192.9	217.4	Product
Dodecane (R)	214.4	295.6	325.4	353.8	Byproduct
H_2SO_4 (immiscible with the organic phase)					Catalyst, removed

9.2.5
Reaction Kinetics

Although the chemistry of alkylation is very complex, we will use the following reactions to capture the essential features of the process, where 1-butene (A), isooctane (B) 2,2,4 tri-methyl pentane (P) and dodecane (R) are used to lump the chemical species involved in the process:

$$C_4H_8 + iC_4H_{10} \rightarrow C_8H_{18}$$
$$\text{(A)} \quad \text{(B)} \quad \text{(P)}$$

$$C_4H_8 + C_8H_{18} \rightarrow C_{12}H_{26}$$
$$\text{(A)} \quad \text{(P)} \quad \text{(R)}$$

The nominal production rate and selectivity of the process are:

$$F_P = 30 \, \text{kmol/h} \tag{9.1}$$

$$\sigma_{P/A} = \frac{F_P}{F_{A0}} = 0.85 \tag{9.2}$$

The process to be designed will be based on H_2SO_4 as the catalyst. In the alkylation reactor, hydrocarbon droplets are dispersed in a H_2SO_4 continuous phase. Isobutane and butane are transferred to the interface between the two immiscible liquids. The rate of this step is influenced by the agitator design and speed, which determines the size of the droplets, the interfacial area and the flow pattern within the droplet. The chemical reactions take place around the interface [10]. A reasonable assumption is that reaction rate is proportional to the reactant concentration and follows an Arrhenius-type temperature dependence. In this study, we will use the following global pseudohomogeneous process rate:

$$r_1 = \varphi k_{10} e^{-\frac{E_{a,1}}{RT}} c_A c_B \tag{9.3a}$$

$$r_2 = \varphi k_{20} e^{-\frac{E_{a,2}}{RT}} c_A c_P \tag{9.3b}$$

Table 9.2 Molecular volumes, molecular weights and densities.

	$V_{\mu,i}$ (−5 °C), m^3/kmol	μ_i, kg/kmol	ρ, kg/m^3
1-butene (A)	0.089	56	629
Isobutane (B)	0.098	58	591
Isooctane (P)	0.158	114	721
Dodecane (R)	0.222	170	765
Propane (I)	0.082	44	536

Table 9.3 Kinetic parameters and heat of reaction for isobutane/1-butene alkylation.

Parameter	$C_4H_8 + iC_4H_{10} \rightarrow C_8H_{18}$	$C_4H_8 + C_8H_{18} \rightarrow C_{12}H_{26}$
Pre-exponential factor/[m^3/kmol s]	1.62×10^9	4.16×10^{12}
Activation energy/[kJ/kmol]	6.5×10^4	8.1×10^4
Heat of reaction/[kJ/kmol]	−90 722	−65 133

In the above relation c_K, $K = A$, B and P are concentrations in the dispersed organic phase. The pre-exponential factors k_{10}, k_{20} and activation energies $E_{a,1}$, $E_{a,2}$ are taken from [11] and are presented in Table 9.3. The parameter φ is introduced to describe the effect of the physical phenomena on the overall process rate. We consider $\varphi = 1$ during nominal operation, but we will account for the rather large uncertainty of this parameter.

9.3
Input–Output Structure

The input-output structure of the flowsheet is presented in Figure 9.1. Butene (feed rate $F_{A,0}$) and isobutane (feed rate $F_{B,0}$) are the raw materials. The butene feed is impure with quite large amounts of propane ($F_{I,0}$). The main product is the alkylate C_8H_{18}, at the rate F_P. The selectivity of the process is not 100%, therefore heavy products are formed at the rate F_R. The inert fed into the process must also leave the plant, the flow F_I including light byproducts that are formed in secondary reactions. Often, significant quantities of n-butane are mixed with the isobutane fresh feed. For this case, development of the flowsheet and the design of the main units is left as an exercise for the reader.

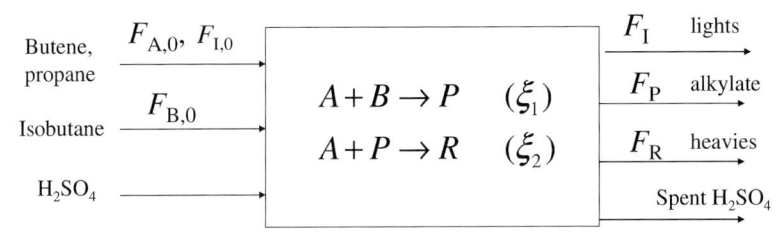

Figure 9.1 Input–output structure of the plant.

The butene feed rate can be calculated from the known production rate and selectivity:

$$F_{A,0} = \frac{F_P}{\sigma_{P/A}} = 35.3 \, \text{kmol/h} \tag{9.4}$$

Neglecting the light byproducts, and assimilating all the heavy byproducts with dodecane, the input–output mass balance can be written using the reaction extents ξ (kmol/h):

$$F_{A,0} = \xi_1 + \xi_2 \tag{9.5a}$$

$$F_{B,0} = \xi_1 \tag{9.5b}$$

$$F_P = \xi_1 - \xi_2 \tag{9.5c}$$

$$F_R = \xi_2 \tag{9.5d}$$

$$F_I = F_{I,0} \tag{9.5e}$$

From Eqs. (9.5a) and (9.5c), and the definition of selectivity (9.2), the isobutene feed and heavies production rate can be calculated:

$$F_{B,0} = \xi_1 = \frac{1}{2}(F_{A,0} + F_P) = \frac{1}{2}(1 + \sigma_{P/A})F_{A,0} = 32.65 \, \text{kmol/h} \tag{9.6}$$

$$F_R = \xi_2 = \frac{1}{2}(F_{A,0} - F_P) = \frac{1}{2}(1 - \sigma_{P/A})F_{A,0} = 2.65 \, \text{kmol/h} \tag{9.7}$$

9.4 Reactor/Separation/Recycle

The reactor/separator/recycle structure is decided by considering the physical properties of the species found in the reactor effluent (Table 9.1). The catalyst and the organic phase are immiscible. Therefore, they can be separated by liquid–liquid splitting. The separation of the organic components by distillation seems easy. In a direct sequence, the inert and any light byproduct will be removed in the first column. The second column will separate the reactants, which have adjacent volatilities. Therefore, there will be only one recycle for both reactants. The third column will separate the product from the heavies. The reactor/separation/recycle structure of the flowsheet is presented in Figure 9.2.

9.4.1 Mass-Balance Equations

At this design stage, the mole balance model of the plant is given by the following equations, where the reactor and separation performances are described by the reaction extents ξ_1 and ξ_2 and the recoveries β_P and β_I, respectively.

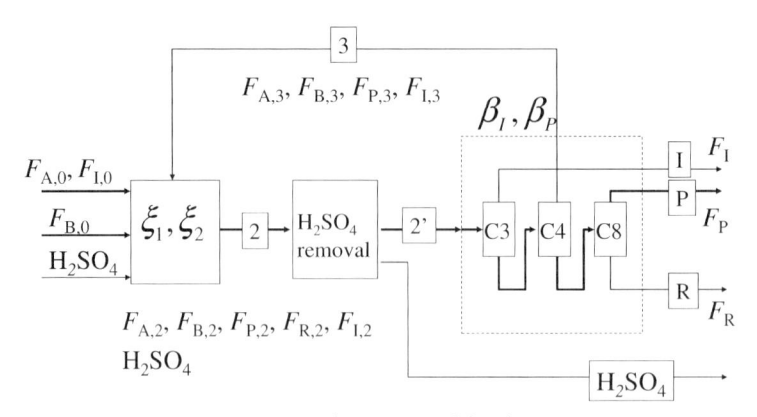

Figure 9.2 Reactor/separation/recycle structure of the plant.

9.4 Reactor/Separation/Recycle

Reactor

$$F_{A,0} + F_{A,3} - F_{A,2} - \xi_1 - \xi_2 = 0 \tag{9.8a}$$

$$F_{B,0} + F_{B,3} - F_{B,2} - \xi_1 = 0 \tag{9.8b}$$

$$F_{P,3} - F_{P,2} + \xi_1 - \xi_2 = 0 \tag{9.8c}$$

$$\xi_2 - F_{R,2} = 0 \tag{9.8d}$$

$$F_{I,0} + F_{I,3} - F_{I,2} = 0 \tag{9.8e}$$

Separation

$$F_{A,3} = F_{A,2} \tag{9.9a}$$

$$F_{B,3} = F_{B,2} \tag{9.9b}$$

$$F_{P,3} = \beta_P F_{P,2} \tag{9.9c}$$

$$F_{I,3} = \beta_I F_{I,2} \tag{9.9d}$$

$$F_P = (1 - \beta_P) F_{P,2} \tag{9.9e}$$

$$F_R = F_{R,2} \tag{9.9f}$$

$$F_I = (1 - \beta_I) F_{I,2} \tag{9.9g}$$

The system (9.8) and (9.9) contains 12 equations and 19 variables. We can assign values to only 5 variables: $F_{A,0}$, $F_{I,0}$, $F_{B,0}$, β_I, β_P. Therefore, the model is underdetermined and we have to look for additional equations. First, we replace the partial flow rates $F_{A,0}$ and $F_{I,0}$ by the total flow rate F_0 and the mole fraction $z_{A,0}$:

$$F_{A,0} = z_{A,0} F_0 \tag{9.10a}$$

$$F_{I,0} = (1 - z_{A,0}) F_0 \tag{9.10b}$$

After simple algebraic manipulations, we are left with the following equations, where the variables ξ_1, ξ_2, $F_{P,2}$, $F_{I,2}$, $F_{R,2}$ are unknown. Note that Eqs. (9.11) do not allow calculation of the reactor-outlet reactant flow rates $F_{A,2}$ and $F_{B,2}$.

$$F_0 z_{A,0} - \xi_1 - \xi_2 = 0 \tag{9.11a}$$

$$F_{B,0} - \xi_1 = 0 \tag{9.11b}$$

$$-(1 - \beta_P) F_{P,2} + \xi_1 - \xi_2 = 0 \tag{9.11c}$$

$$-F_{R,2} + \xi_2 = 0 \tag{9.11d}$$

$$F_0 (1 - z_{A,0}) - (1 - \beta_I) F_{I,2} = 0 \tag{9.11e}$$

At this point, an important design decision is taken: we choose the type of reactor. Taking into account the mixing required for achieving the hydrocarbons–H_2SO_4

emulsion, a continuous stirred tank (CSTR) model is adequate. Now, we are able to write the additional relations needed, as the reactor's characteristic equations:

$$\xi_1 = k_1 c_{A,2} c_{B,2} V \quad (9.12a)$$

$$\xi_2 = k_2 c_{A,2} c_{P,2} V \quad (9.12b)$$

New variables have been introduced (concentration $c_{A,2}$, $c_{B,2}$ and $c_{P,2}$) but they can be easily expressed as functions of partial flow rates:

$$c_{A,2} = \frac{F_{A,2}}{F_2^{vol}} \quad (9.13a)$$

$$c_{B,2} = \frac{F_{B,2}}{F_2^{vol}} \quad (9.13b)$$

$$c_{P,2} = \frac{F_{P,2}}{F_2^{vol}} \quad (9.13c)$$

$$F_2^{vol} = F_{A,2} V_{\mu,A} + F_{B,2} V_{\mu,B} + F_{P,2} V_{\mu,P} + F_{R,2} V_{\mu,R} + F_{I,2} V_{\mu,I} \quad (9.14)$$

According to the assumptions implied by the kinetic equation (9.3), the volume V and the concentrations $c_{K,2}$ refer to organic phase. Moreover, Eq. (9.14) assumes that there is no change of volume due to mixing. This is a reasonable assumption in view of the data presented in Table 9.2. The system of Eqs. (9.11) to (9.14) is square and can be solved numerically, for example using the Newton–Raphson method. Table 9.4 presents typical results, for a reactor of $10\,m^3$ operated at various temperatures. A large excess of i-butane (B) is necessary to achieve the required transformation. Butene is almost completely converted, while isobutane conversion is much lower. For this reason, the recycle contains mainly the excess isobutane. Moreover, the main reaction is favored by low temperatures.

At this point, we could select the operating conditions, for example corresponding to a reactor of volume $V = 10\,m^3$ that will be operated at $-10\,°C$. As we are concerned with the robustness of our design, we will study the dependence of the operating point versus changes in the operating or design parameters. We note that the set of specifications employed is equivalent to a plantwide control structure that fixes the plant feed rates F_0, $z_{A,0}$, and $F_{B,0}$, the reactor volume V and the

Table 9.4 Molar flow rates at reactor inlet and reactor outlet.

	Reactor inlet (kmol/h)					Reactor outlet (kmol/h)				
	A	B	P	R	I	A	B	P	R	I
−10°C	72.99	673.41	0.61	0	9.01	37.69	640.75	30.61	2.65	9.01
0°C	50.18	870.48	0.61	0	9.01	14.88	837.8	30.61	2.65	9.01
10°C	41.9	1107.6	0.61	0	9.01	6.60	1074.9	30.61	2.65	9.01

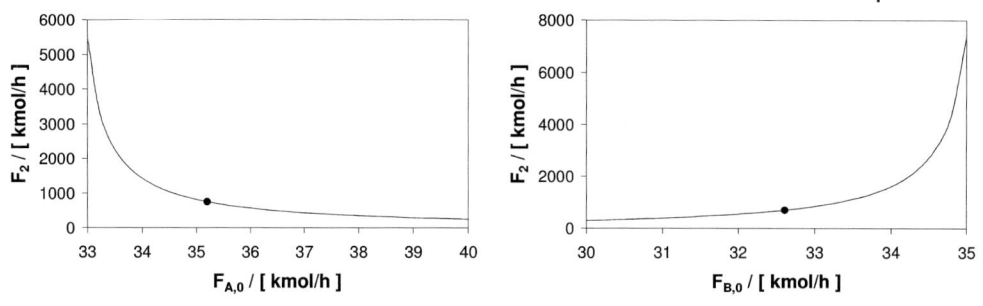

Figure 9.3 Snowball effect. Small changes of feed rate yield huge changes of internal flow rates.

separation performance β_I, β_P. Among these variables, 10% changes of the flow rates are in the range of minimal flexibility requirement and practical measurement and control actuation errors. Figure 9.3 presents the dependence of reactor-outlet flow rate F_2 versus the feed rates $F_{A,0}$ and $F_{B,0}$, all other parameters being constant. It can be seen that less than 10% change of any of the feed flow rates leads to a huge change of the reactor-outlet flow rate (more then 500%). Such a disturbance cannot be tolerated by the separation section. This extreme sensitivity of the internal flow rates, which has been observed in many other recycle systems, is known as the snowball effect [12].

The conclusion of this exercise is that the plant cannot be operated by keeping constant feed rates F_0, $z_{A,0}$, and $F_{B,0}$, reactor volume V and separation performance β_I, β_P. Bildea and Dimian [13] recommend to keep the reactor inlets on flow control and to supply the fresh reactant in any inventory device from the recycle loop. When the reactants are recycled together (as in our plant), the recommended strategy is to design the plant for high conversion of one reactant (butene), set its

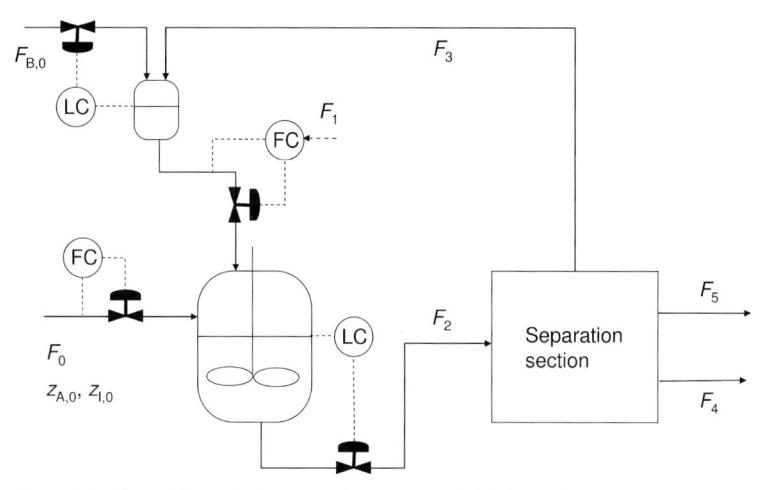

Figure 9.4 Plantwide control strategy fixing reactor-inlet flow rates.

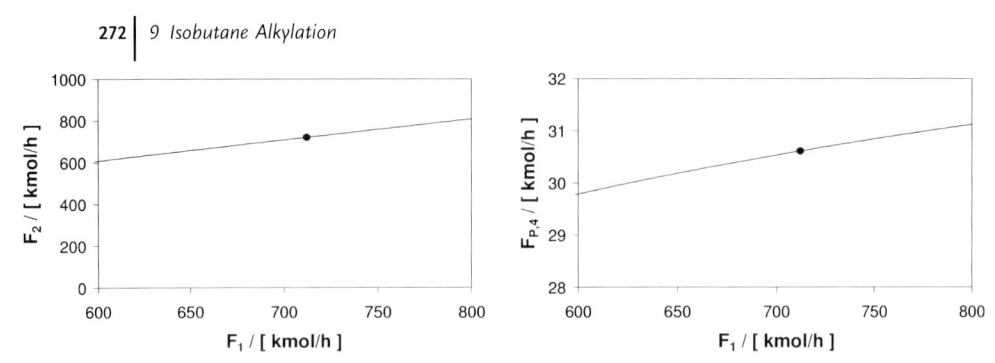

Figure 9.5 Dependence of the reactor-outlet flow and production rate versus the recycle rate.

feed on flow control, fix the total recycle flow, and add the makeup of the other reactant somewhere in the recycle. This strategy is shown in Figure 9.4.

In the mass-balance model of the plant, the flow rate $F_{B,0}$ becomes an unknown, and the following fixed-flow equation is added:

$$F_1 - (F_{B,0} + F_{A,2} + F_{B,2} + \beta_P F_{P,2} + \beta_I F_{I,2}) = 0 \tag{9.15}$$

The first diagram of Figure 9.5 shows that the flow rate of the reactor-outlet stream is almost insensitive to variations of the recycle rate F_1. A similar picture is obtained when variations of the feed rate F_0 are considered (not shown). The second diagram shows that, as expected, increasing the excess of isobutane by larger recycle has a beneficial effect on selectivity, and more alkylate is obtained.

9.4.2
Selection of a Robust Operating Point

Figure 9.6 presents the dependence of process selectivity $\sigma_{P/A}$ and per-pass conversion x_A versus the reaction volume, for different values of the reaction temperature. The plots show a region of unfeasibility (no steady state) at low reactor volume. Here, the capacity of the reactor to transform reactants into products is small and exceeded by the amount of butene fed in the process. In the feasibility region, multiple steady states exist, only the high-conversion, high-selectivity branch being stable. Around the nominal operating point, the selectivity is almost constant for quite a large range of reactor volumes. Moreover, the dependence versus the kinetic parameter φ follows the same pattern, because the V and φ appear only as the product $V\varphi$ in the model (Eq. 9.12). Note that for fixed butene feed rate, $z_{A,0}F_0$, the selectivity is a direct measure of production rate (see Eq. 9.4).

From Figure 9.6, we can see that a smaller reactor could be used. At 263 K, even a 2-m³ reactor will work, although the operating point is dangerously close to the feasibility limit. A good alternative seems to be $T = 268$ K and $V = 3$ m³ (white dot). Note that because the feed and recycle rates F_0 and F_1 are fixed, the production rate and product distribution remain unchanged. Therefore, the load of the separation

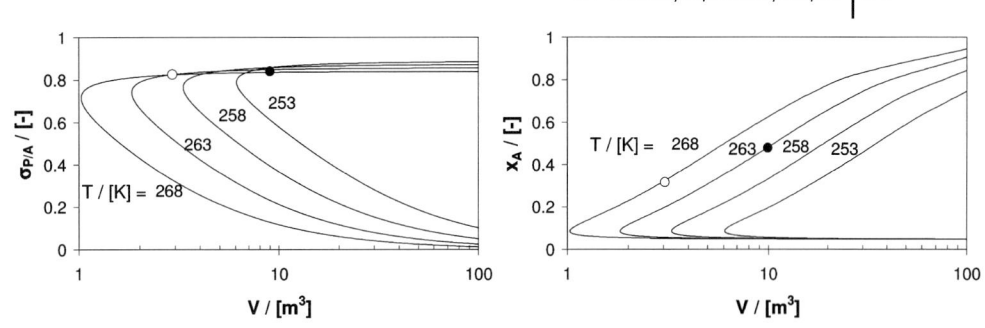

Figure 9.6 Selectivity and conversion versus reaction volume, at different reaction temperatures.

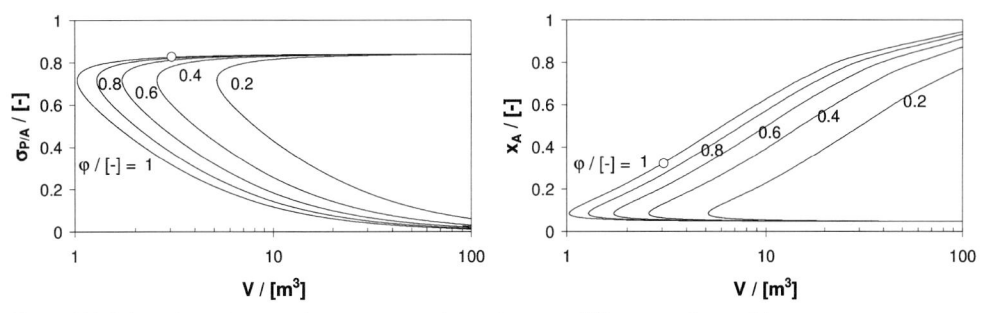

Figure 9.7 Selectivity and conversion versus reaction volume, at different catalyst activity.

section is not affected if a smaller reactor is used. Lower conversion does not have an adverse effect on recycle flow, although the recycle composition changes.

Figures 9.7 and 9.8 show the effect of the kinetic factor φ and feed purity $z_{A,0}$. The 3-m^3 reactor operated at 268 K will work even if φ drops to 40% of the initial value. The feed purity is not a "bad" disturbance, having a small influence. Lower purity means less reactant to be converted and therefore enlarges the region of feasibility.

From the previous analysis, we conclude that a robust plantwide control structure will fix the combined isobutane + recycle (F_1) and fresh butene flow (F_0), as illustrated in Figure 9.4. The desired production rate and selectivity could be achieved in a 3-m^3 reactor, operated at 268 K. The operating point shows low sensitivity to errors in the manipulated variable F_1 (Figure 9.5). This design seems to ensure feasible operation even if the temperature decreases to 260 K (Figure 9.6) or the catalyst activity becomes 40% of the initial value (Figure 9.7), irrespective of the purity of the butene feed stream (Figure 9.8).

The approach we used in this section to choose an operating point that is "far" from the feasibility boundary is difficult to apply when more parameters are involved. Firstly, plotting diagrams is restricted to the two-dimensional space. Secondly, the number of diagrams to be traced increases exponentially with the number of parameters. The normal-space approach, presented in the next section,

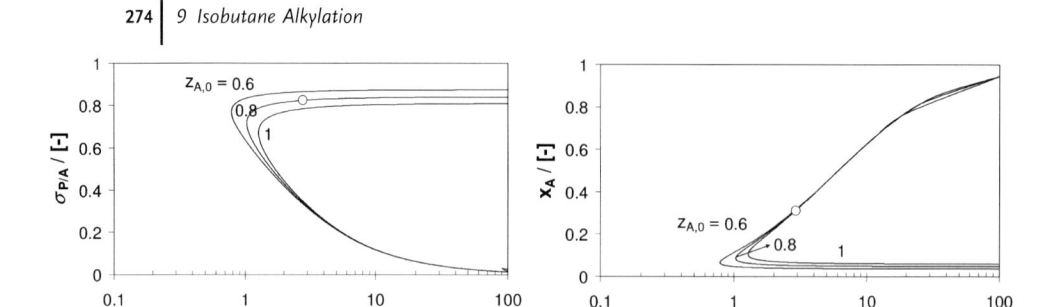

Figure 9.8 Selectivity and conversion versus reaction volume, at different purity of the butene feed.

provides a solution by calculating the shortest distance from the nominal operating point to the manifold of critical points, and imposing a certain lower bound on this distance.

9.4.3
Normal-Space Approach

In this section, we will present the normal-space approach to selection of a robust operating point. For illustration purposes, we will restrict our analysis to the case of two uncertain parameters, namely the temperature T and the kinetic parameter φ. The method, however, can be applied to any number of parameters, with a linear increase in complexity. The approach can be integrated in an optimization problem. This will be illustrated here by choosing a very simple objective function. It should be noted that the approach is not limited to problems where the critical manifold is represented by a turning point in the state vs. parameter diagram (known also a fold bifurcation). It can be applied to other problems involving, for example, Hopf bifurcations, feasibility constraints, constraints on eigenvalues, or higher codimension singularities. The reader is referred to the work of Mönnigmann [14] for a more detailed presentation.

The problem we will approach in this section is stated as: "For the operating conditions presented in Table 9.5, find the reactor of minimum volume such that a feasible operating point exists. Additionally, feasibility should be preserved for any temperature in the range $T_0 \pm \Delta T$, and any catalyst activity in the range $\varphi_0 \pm \Delta\varphi$".

9.4.3.1 Critical Manifolds

Loosely speaking, a manifold of dimension $(n-k)$ is a set of points in an n-dimensional space defined by $k < n$ equations. Suppose that, after some algebraic manipulations, the plant model given by Eqs. (9.11) to (9.15) and the sensitivity definition (9.4) are reduced to one equation with one unknown $\sigma_{P/A}$. Then, the dependence of the state variable $\sigma_{P/A}$ versus one parameter (the volume V) as defined by the model equation can be graphically depicted, for example as in Figure 9.6. The plot is a one-dimensional manifold in the two-dimensional space

Table 9.5 Nominal operating point.

Variable	Symbol	Value
Butene feed rate/[kmol/h]	F_0	44.12
Butene purity/[–]	$z_{A,0}$	0.8
Recycle + fresh isobutane flow rate/[kmol/h]	F_1	712
Product recovery in the recycle stream/[–]	β_P	0.02
Inert recovery in the recycle stream/[–]	β_I	0.02
Reaction temperature/[K]	T_0	268
Catalyst activity/[–]	φ_0	1.0
Maximum change of reaction temperature/[K]	ΔT	5
Maximum change of catalyst activity/[–]	$\Delta\varphi$	0.5

$V - \sigma_{P/A}$. The turning point of each diagram is special: there are no steady states to the left of this point; two steady states exist on the right. We say that the turning point is a critical point. If we consider the family of $V - \sigma_{P/A}$ diagrams traced for different values of the additional parameter T, a surface is obtained. This is a two-dimensional manifold in the three-dimensional $V - \sigma_{P/A} - T$ space. As each diagram has its own turning point defined by one additional equation, the locus of critical points is a line, a one-dimensional manifold in the $V - \sigma_{P/A} - T$ space. It is easy to see that the dimension of the critical manifold is $n_\alpha - 1$, where n_α is the number of parameters included in the analysis.

From the previous analysis we conclude that, when T and φ are considered as uncertain (varying) parameters, the locus of turning points is a one-dimensional manifold in the $\sigma_{P/A} - T - \varphi$ space. This can be found by solving the extended system:

$$f(x, \alpha, p) = 0 \qquad (9.16a)$$

$$f_x(x, \alpha, p) \cdot v = 0 \qquad (9.16b)$$

$$v^T \cdot v - 1 = 0 \qquad (9.16c)$$

$f(x, \alpha, p)$ is a short notion for the model Eqs. (9.11) to (9.15)
$f_x(x, \alpha, p)$ is the Jacobian of f with respect to unknowns x
$x = [F_{A,2}, F_{B,2}, F_{P,2}, F_{1,2}, F_{R,2}, \xi_1, \xi_2, c_{A,2}, c_{B,2}, c_{P,2}, F_{B,0}, F_2^{vol}]$ is the vector of unknowns
$\alpha = [T, \varphi]$ is the vector of uncertain parameters
$p = [V, F_0, F_1, z_{A,0}, F_{B,0}, \beta_P, \beta_I]$ is the vector of fixed parameters.
v is a vector of auxiliary unknowns.

9.4.3.2 Distance to the Critical Manifold

The manifold of critical points can be projected onto the space of the uncertain parameters $T - \varphi$, as shown in Figure 9.9. Each $V =$ constant line divides the $T - \varphi$ plane into two regions. In the right-upper region, operation of a reactor of

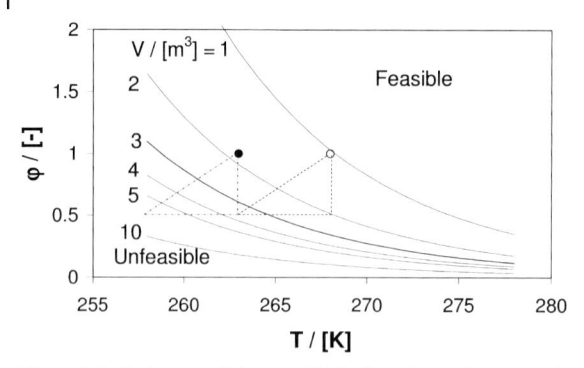

Figure 9.9 Projection of the manifold of turning points onto the T–φ space.

volume V is feasible. In the left-lower region, the reactor is too small and no steady state exists. Two operating points are also shown in the drawing. If the reactor is designed for operation at 268 K (white dot), it must be at least 4 m³ in size to withstand a combined 5 K and 50% decrease of temperature and the kinetic factor, respectively. When the nominal operating point is chosen at 263 K, the reactor must be about 7.5 m³ in size to accept these disturbances.

The approach presented in Figure 9.9 can be generalized to problems involving any number of parameters. The idea is to measure, in the space of uncertain parameters, the distance between a candidate point of operation and the critical manifold. If this distance exceeds a certain predefined quantity, the operating point is robust with respect to parameter uncertainty. To define a distance in the space of parameters having different units, we introduce the scaled variables α_i^s:

$$\alpha_i^s = \frac{\alpha_i}{\Delta \alpha_i} \tag{9.17}$$

where $\Delta\alpha_i$ is the maximum uncertainty or expected change of the parameter.

The distance between an operating point α_0^s and the turning points locus can be found by solving the following system of equations:

$$f(x, \alpha^s, p) = 0 \tag{9.18a}$$

$$f_x(x, \alpha^s, p) \cdot v = 0 \tag{9.18b}$$

$$v^T \cdot v - 1 = 0 \tag{9.18c}$$

$$r - f_{\alpha^s}^T \cdot v = 0 \tag{9.18d}$$

$$\alpha_0^s - \left(\alpha^s + \rho \frac{r}{\|r\|} \right) = 0 \tag{9.18e}$$

Equations (9.18a) to (9.18c) define the manifold of turning points. Equations (9.18d) and (9.18e), allow calculation of the normal vector r and of the distance ρ from the operating point α_0 to the critical manifold (Figure 9.10).

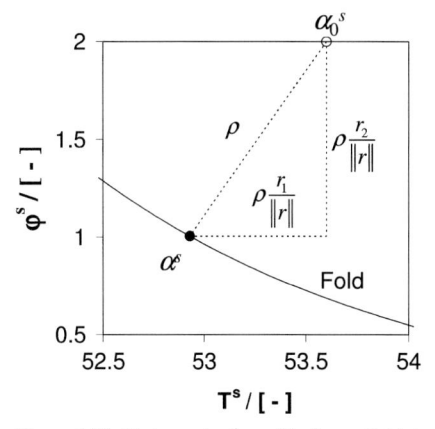

Figure 9.10 Distance to the critical manifold, in the space of scaled uncertain parameters.

9.4.3.3 Optimization

Calculation of normal vector and the distance to the critical manifold can be included in an optimization methodology. Here, we are looking for the reactor of minimum volume that ensures feasible operation of the reactor/separation/recycle system, for temperatures in the range $268 \pm 5\,\text{K}$, and kinetic uncertainties in the range 1 ± 0.5. Using the scaled variables, the optimization problem can be written as:

$$\min V$$
$$\text{subject to Eqs. (9.20)} \tag{9.19a}$$
$$\rho \geq \sqrt{n_a}$$

Solving the optimization problem (9.19) gives a reactor volume $V = 4\,\text{m}^3$ (in agreement with Figure 9.9). The composition of reactor inlet and outlet streams is presented in Table 9.6. This operating point will be used in the further design of the plant.

Table 9.6 Composition of reactor inlet and outlet streams. $V = 4\,\text{m}^3$; $T = 268\,\text{K}$.

	Reactor inlet	Reactor outlet
$c_A/[\text{kmol}/\text{m}^3]$	1.229	0.745
$c_B/[\text{kmol}/\text{m}^3]$	8.971	8.501
$c_P/[\text{kmol}/\text{m}^3]$	0.0032	0.407
$c_R/[\text{kmol}/\text{m}^3]$	0	0.041
$c_I/[\text{kmol}/\text{m}^3]$	0.123	0.123
$F^{\text{mass}}/[\text{kg}/\text{s}]$	12.09	12.09
$\rho/[\text{kg}/\text{m}^3]$	595	593

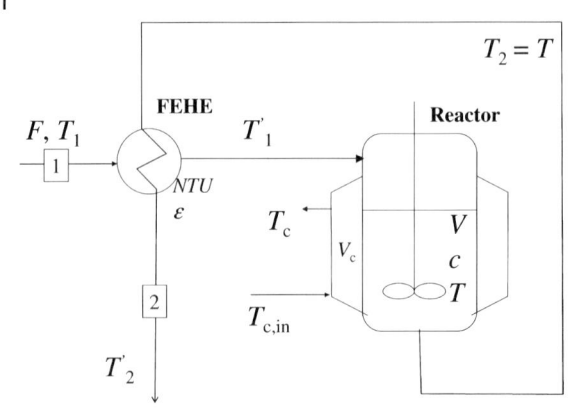

Figure 9.11 Cooled alkylation reactor with feed-effluent heat exchanger.

9.4.4
Thermal Design of the Chemical Reactor

In the previous sections, it was decided that the reaction will be operated at −5 °C. At this temperature, a good selectivity can be achieved with a reasonable excess of isobutane. Because the reaction is exothermic, cooling is needed. One alternative is to use a mixture of isobutene and alkylate as coolant, and cooling coils as the heat-transfer hardware. Another option is to achieve cooling by partial vaporization of light hydrocarbons [15]. This alternative, however, increases the range of hydrocarbon compositions [10]. In the following, we will evaluate the first option. Note that the amount of cooling can be reduced if a feed-effluent heat exchanger (FEHE) is used, as in Figure 9.11.

Sizing the reactor implies calculating, besides reactor volume, the heat-transfer area of the cooling coils and of the FEHE. We assume constant volumetric flows and coolant temperature, and neglect the temperature dependence of the reaction heat and physical properties. With these assumptions, the mathematical model of the reactor which includes the energy balance is given by:

$$\frac{F^{vol}}{V}(c_{A,1} - c_{A,2}) - (r_1 + r_2) = 0 \tag{9.20a}$$

$$\frac{F^{vol}}{V}(c_{B,1} - c_{B,2}) - r_1 = 0 \tag{9.20b}$$

$$\frac{F^{vol}}{V}(c_{P,1} - c_{P,2}) + (r_1 - r_2) = 0 \tag{9.20c}$$

$$\frac{F^{vol}}{V}(c_{R,1} - c_{R,2}) + r_2 = 0 \tag{9.20d}$$

$$\frac{F^{vol}}{V}(c_{1,1} - c_{1,2}) = 0 \tag{9.20e}$$

$$\frac{F^{vol}}{V}(T_1' - T) + \frac{(-\Delta^r H_1)r_1 + (-\Delta^r H_2)r_2}{\rho c_p} - \frac{UA}{V\rho c_p}(T - T_c) = 0 \tag{9.20f}$$

$$\frac{F_c^{vol}}{V_c}(T_{c,in} - T_c) + \frac{UA}{V_c \rho_c c_{p,c}}(T - T_c) = 0 \tag{9.20g}$$

$$T_1' = (1-\varepsilon)T_1 + \varepsilon T \tag{9.20h}$$

The coolant temperature is chosen as $T_c = 258\,\text{K}$ ($-15\,°\text{C}$). To estimate the FEHE heat transfer efficiency ε, we assume the heat-transfer area $A_{\text{FEHE}} = 500\,\text{m}^2$, and the heat-transfer coefficient $U_{\text{FEHE}} = 500\,\text{W/m}^2/\text{K}$. This gives:

$$\text{NTU} = \frac{U_{\text{FEHE}} A_{\text{FEHE}}}{F^{vol}\rho c_p} = \frac{500\,\dfrac{\text{W}}{\text{m}^2\text{K}} 500\,\text{m}^2}{12\,\dfrac{\text{kg}}{\text{s}} 2200\,\dfrac{\text{J}}{\text{kg\,K}}} \tag{9.21}$$

$$\varepsilon = \frac{\text{NTU}}{1+\text{NTU}} = \frac{9.4}{1+9.4} \approx 0.9 \tag{9.22}$$

The model is solved numerically. Figure 9.12 presents the reaction temperature and conversion versus heat-transfer capacity of the reactor, UA. From Figure 9.12, we find that we need to provide $UA = 95\,\text{kW/K}$ heat-transfer capacity, if we want to reach the operating point $T = 268\,\text{K}$, $x_A = 0.4$. Figure 9.12 also shows that the combined FEHE–reactor system does not show multiple steady states, which should always be a concern when designing heat-integrated reactors [16].

For a heat-transfer coefficient of $500\,\text{W/m}^2/\text{K}$, the necessary area is $A \approx 190\,\text{m}^2$. This requires use of cooling coils, or performing the cooling operation in an external heat exchanger. The second alternative is presented in Figure 9.13.

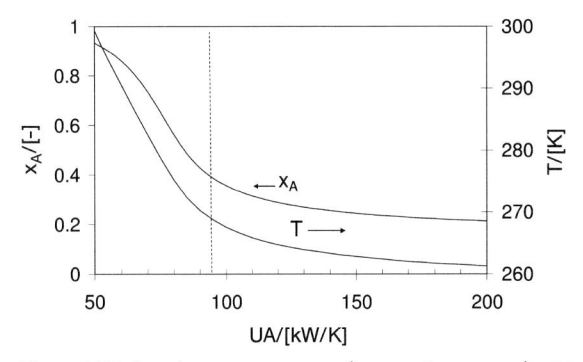

Figure 9.12 Reaction temperature and conversion versus heat-transfer capacity.

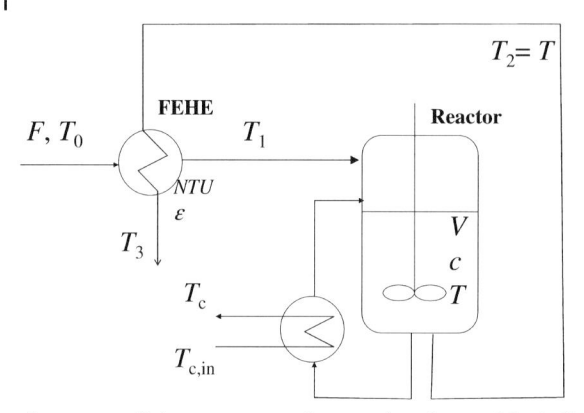

Figure 9.13 Alkylation reactor with external cooling and feed-effluent heat exchanger.

9.5
Separation Section

The reactor-outlet stream contains a dispersion of hydrocarbons in sulfuric acid. The first separation step is therefore a liquid–liquid split. The sulfuric-acid phase contains some amounts of *sec*-butyl acid sulfate, which decomposes at higher temperature (15 °C) to produce conjuct polymers dissolved in the acid and a mixture of C_4–C_{16} isoparaffins with low octane number (pseudoalkylate) that separates as a second liquid phase. The hydrocarbon phase contains a small amount of di-isoalkyl sulfates. These need to be removed before entering the distillation units; otherwise they will decompose and release sulfuric acid. The sulfates are removed by washing with either dilute caustic or sulfuric acid. In the first case, sulfates are converted to salts that are discarded. With sulfuric acid, sulfates are converted to isoalkyl acid sulfates that can be recycled to the alkylation reactor [15, 10].

The sequence of distillation units separating the hydrocarbon mixture can be easily designed. The first column (C3, Figures 9.3 and 9.14) will separate the propane, together with all low-boiling byproducts. In the second column (C4), isobutane and butene are taken as distillate and recycled. The bottom stream is sent to the last column (C8) where isooctane is separated from the heavy byproduct. The first two distillation columns are operated at high pressure (8 bar), which allows use of water for cooling purposes. The last column is operated at ambient pressure. Note that the sensitivity studies performed in the previous sections showed that high recovery of propane or isooctane is not needed, as recycling these components does not have an unfavorable effect on plant operation.

For each column, the recoveries of the light and heavy key components have been specified. Then, the minimum number of trays N_{min} and the minimum reflux ratio R_{min} have been calculated using the shortcut distillation model DSTWU with Winn–Underwood–Gilliland method in Aspen Plus. The reflux ratio was set to

Table 9.7 Distillation columns.

	C3-column	C4-column	C8-column
No of trays	28	29	24
Feed tray	18	10	11
Reflux ratio	110.7	0.0545	0.205
Distillate:feed	0.017	0.954	0.909
Diameter/[m]	2.6	1.6	0.6
Reflux drum/[m^3]	10.6	6.75	0.5
Column sump/[m^3]	20.1	7.7	0.25
Temperature control stage	3	27	4 and 23

$1.2 R_{min}$ and the corresponding number of trays calculated (~$2 N_{min}$). The shortcut models were replaced by rigorous RADFRAC units, where the reflux and distillate: feed ratio were adjusted by means of design specifications, in order to meet the desired separation. The trays were sized using Aspen's facilities. Finally, the dimensions of the reflux drum and column sump were found based on a residence time of 5 min and aspect ratio H:D = 2:1. Table 9.7 presents the results of distillation column sizing.

9.6
Plantwide Control and Dynamic Simulation

The basic plantwide control loops (Figure 9.14) follow easily from the conceptual design. In the previous section, we have concluded that we need to keep on flow control the butene feed and the combined isobutane feed + recycle, in order to achieve robustness with respect to production-rate changes and errors in measuring or setting the flow rates. The butene feed is used as a throughput manipulator, or it is a disturbance initiated by the upstream unit. Isobutane is fed as a makeup stream controlling the level in the buffer vessel. The buffer vessel can be a distinct unit as in Figure 9.14 or the reflux drum of the C4 column. We also know that the plant can tolerate 50% uncertainty in reaction kinetics combined with a 5-K decrease of the reaction temperature. For this reason, reaction temperature will be controlled, using the coolant flow rate for this purpose. The sensitivity analysis indicated that recycling small amounts of propane (inert) and isooctane (product) does not have a detrimental effect on plant operation. Therefore, control of the columns C3 and C4 consists of one-point temperature control, in addition to the inventory control loops (pressure, levels in the reflux drum and column sump). For the product column C8 we will control the temperatures both in the top and in the bottom of the column.

A steady-state model was developed in Aspen Plus. Knowledge gained from the reactor/separation/recycle study proved very helpful in choosing the correct design

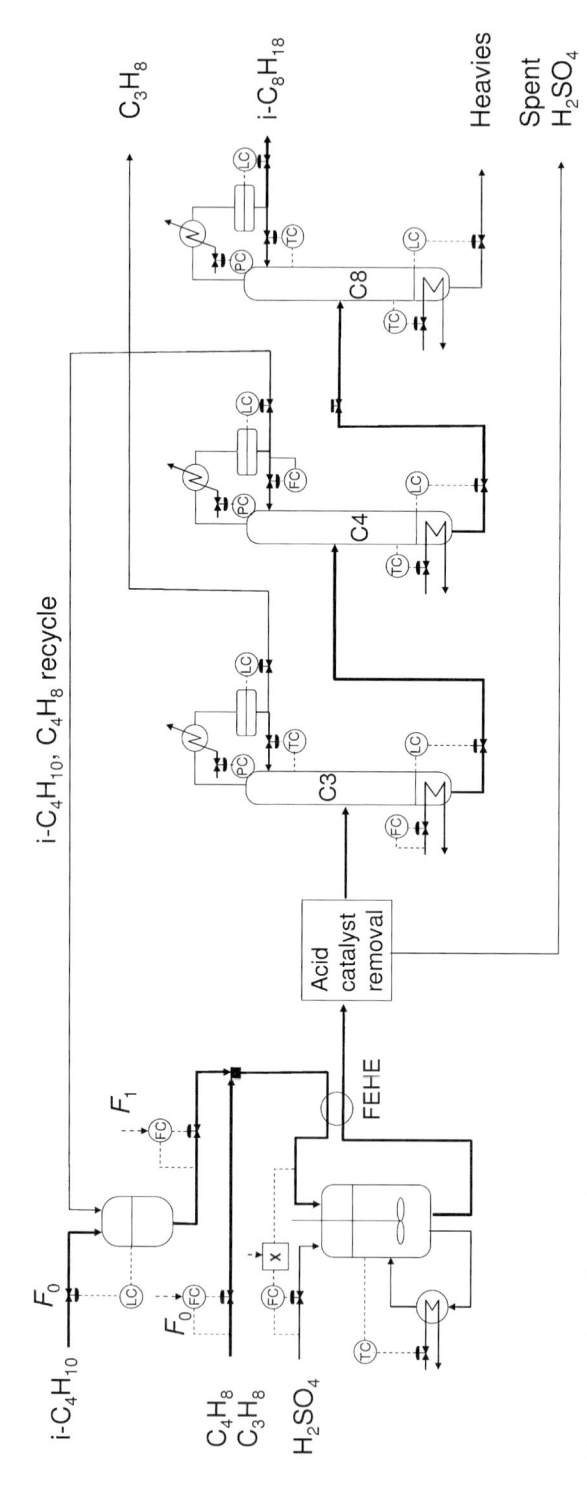

Figure 9.14 Plantwide control strategy.

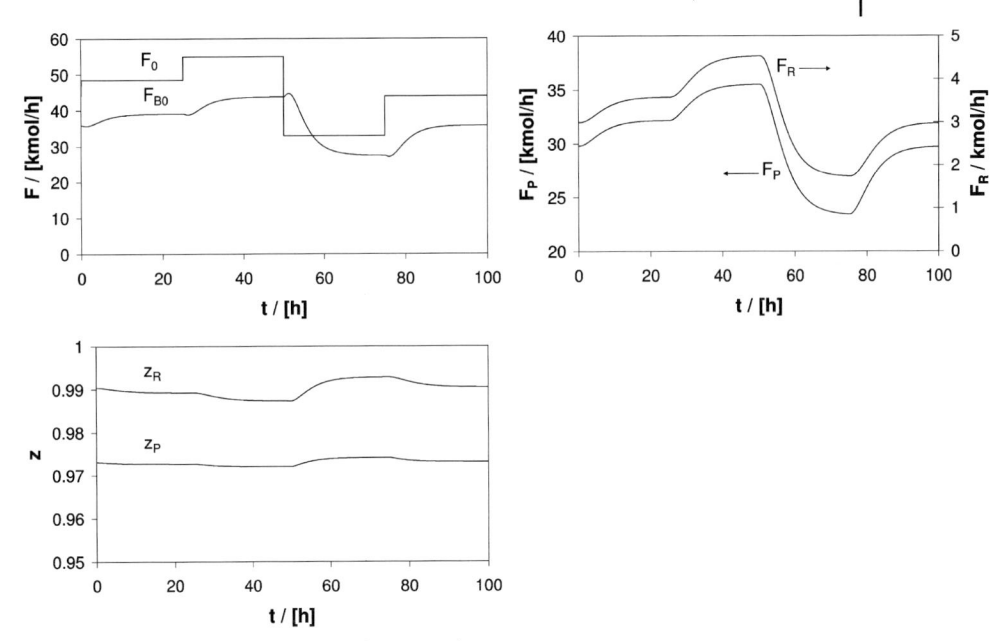

Figure 9.15 Closed-loop dynamic simulation results.

specifications for a fast convergence. A dynamic simulation of the plant was exported to Aspen Dynamics. The reactant makeup strategy described above and temperature controllers were implemented. In all control loops, a gain of 1%/% was used. For temperature controllers, an integral time of 20 min was used. The range of the controlled variable was set to twice the nominal value for levels, and 10 °C for temperatures. The range of the manipulated variable (flow rate or duty) was twice the nominal value.

Figure 9.15 presents results of dynamic simulation. Starting from the nominal steady state, the butene feed F_0 is increased by 10% at time $t = 0$. Fresh isobutane flow follows the butene feed. The small inverse response can be explained as follows. When more butene is added, the reactor flow rate increases, leading to a shorter residence time. The per-pass conversion decreases, which results in larger recycle. Consequently, the isobutane feed is reduced. All these events increase the concentration of butene in the system, leading to a higher reaction rate, up to the point where the effect of larger residence time is overcome. The dynamics of the plant is very slow, as it takes more than 20 h until a new steady state is established. Subsequently, the butene feed is further increased to 125% of the nominal value (at time 25 h), then drastically reduced to 75% (at time 50 h). Finally, at $t = 75$ h, the butene feed is brought back to the initial value. The production rate follows slowly the change of the feed. During all these changes, the process selectivity (given by the ratio isooctane:heavies) remains high. Also, the purity of the product stays around the design value of 97%.

9.7
Discussion

In the alkylation process, the main reaction involves the olefin and isobutane. In contrast, the secondary reactions consist of olefin polymerization or the reaction between the olefin and C_8 paraffins. For this reason, a high concentration of isobutane in the reactor is necessary. In our design the isobutane:olefin ratio is 7.3:1, while typical values are in the range 5:1 to 10:1 [7]. Note that the reactants are fed to the process in a nearly stoichiometric proportion, the high excess of isobutane being accomplished by recycling.

The main reaction has a lower activation energy, compared to the secondary reactions (Table 9.3). Therefore, low temperature favors the desired reaction. Moreover, equipment corrosion increases at high temperatures. On the other hand, low temperature slows the settling of the acid from the alkylate. In industrial processes, the temperature is around 10 °C. The process designed in this chapter works at −5 °C. This lower temperature could be explained by the inaccuracies of the kinetic data.

In this study, the acid concentration was not explicitly taken into account. At concentrations below 85%, olefins react preferentially with olefins, to produce polymers. Due to their unsaturated nature, these polymers are soluble in sulfuric acid, which leads to a further concentration decrease. Additionally, polymers are oxidized, while sulfuric acid is reduced to water and sulfur dioxide. These positive-feedback effects are known as "acid runaway" [17]. Acid runaway is a major concern in alkylation plants, as the acid strength can no longer be controlled and the olefin feed must be cut out.

In general, high ratios olefin/sulfuric acid increase acid consumption and decrease octane number. This ratio is often presented as "olefin space velocity", defined as volume of olefin charged divided by the volume of H_2SO_4 in the reactor. Typical values are in the range 0.25–0.5 h^{-1}. For our process, the olefin space velocity is 0.5 h^{-1}, calculated for an emulsion containing 50% acid (45–65% is the recommended range).

Mixing is an important parameter that influences the emulsion of hydrocarbons into sulfuric acid. Because the reaction occurs at the acid/hydrocarbon interface, better emulsion means smaller droplets and therefore faster reaction.

Sulfuric acid alkylation units have excellent safety reports [18]. The main concerns are the H_2SO_4 and the large inventories of light hydrocarbons. The incidents that have been reported include emulsion-separation difficulties leading to acid carryover, acid runaway resulting in low-quality alkylate and requiring shut down of the olefin feed, or tight emulsion in the net effluent water wash with the consequence of carryover of the alkaline water into the separation section.

9.8
Conclusions

In this chapter, the design of an isobutane/butene alkylation plant was performed following the methodology presented in Chapter 2. Important points concerning the reactor/separation/recycle level are illustrated. At this level, a simplified but nevertheless realistic and very useful mass balance of the plant is derived. Plantwide control structures are easily generated based on degrees of freedom analysis applied to this model. Sensitivity studies allow recognition of undesired behavior, such as the snowball effect, and rejection of unsuitable control strategies (Section 9.4.1). The nonlinear phenomena described in Chapter 4, such as unfeasibility and state multiplicity, are illustrated. A robust design ensures that the plant remains operable even when it is faced with large disturbances or design parameter uncertainties (Section 9.4.2). The concept of distance to critical manifold and its use for robust design is exemplified in Section 9.4.3. Finally, it is shown that heat-integration issues can be addressed, such as heat transfer between the feed and the effluent of the chemical reactor (Section 9.4.4). We emphasize that traditional procedures tackle plantwide control and heat integration toward the end of the design. The newly introduced reactor/separation/recycle level (Chapters 2 and 4) allows an early solution to these problems, with the result of avoiding unnecessary loops in the design process. Rigorous design and closed-loop dynamic simulation prove the effectiveness of the approach (Section 9.6).

References

1 Mata, T., Smith, R., Young, D. and Costa, C., 2003, Life cycle assessment of gasoline blending options, *Environ. Sci. Technol.*, 37, 3724–3732

2 Stratco Inc, 2000, Alkylation vs. dimerization a comparative overview, http://www.stratco.dupont.com/alk/pdf/alkyvsdimecomparitive.pdf (last visited February 2006)

3 Hommeltoft, S.I., 2001, Isobutane alkylation. Recent developments and future perspectives, *Appl. Catal. A: General*, 221, 421–428

4 Weitkamp, J. and Traa, Y., 1999, Isobutane/butene alkylation on solid catalysts. Where do we stand? *Catal. Today*, 49, 193–199

5 Marcilly, C., 2003, Present status and future trends in catalysis for refining and petrochemicals, *J. Catal.*, 216, 47–62

6 Roeseler, C., 2003, UOP AlkyleneTM process for motor fuel alkylation, in Meyers, R. (ed), Handbook of Petroleum Refining Processes, 3rd edn, McGraw-Hill, New York, USA

7 Kranz, K., 2003, Alkylation chemistry. Mechanisms, operating variables, and olefin interactions, http://www.stratco.dupont.com/alk/pdf/alkylation_chemistry_2003.pdf (last visited February 2006)

8 Albright, L., 2002, Alkylation of Isobutane with C3–C5 Olefins: Feedstock Consumption, Acid Usage, and Alkylate Quality for Different Processes, *Ind. Eng. Chem. Res.*, 41, 5627–5631

9 Scott, B., 1991, Alkylation process hazards management. Does it matter which acid you use? http://www.stratco.dupont.com/alk/pdf/AlkylationProcessHazardsMan.pdf (last visited February 2006)

10 Albright, L., 2003, Alkylation of Isobutane with C3–C5 Olefins to produce high-quality gasolines: Physicochemical sequence of events, *Ind. Eng. Chem. Res.*, 42, 4283–4289

11 Mahajanam, R.V., Zheng, A. and Douglas J.M., 2001, A shortcut method for controlled variable selection and its application to the butane alkylation process. *Ind. Eng. Chem. Res.*, 40, 3208–3216

12 Luyben, W.L., 1994, Snowball Effects in Reactor/Separator Processes with Recycle, *Ind. Eng. Chem. Res.*, 33, 299–305

13 Bildea, C.S. and Dimian, A.C., 2003, Fixing flow rates in recycle systems: Luyben's rule revisited. *Ind. Eng. Chem. Res.*, 42, 4578–4585

14 Mönnigmann, M., 2004, Constructive Nonlinear Dynamics for the Design of Chemical Engineering Processes, Fortschritt-Berichte VDI, Reihe 3, Nr. 801; VDI Verlag, Düsseldorf, Germany, 2004

15 Branzaru, J., 2001, Introduction to sulfuric acid alkylation unit process design, http://www.stratco.dupont.com/alk/pdf/AlkyUnitDesign2001.pdf (last visited February 2006)

16 Bildea, C.S. and Dimian, A.C., 1998, Stability and Multiplicity Approach to the Design of Heat-Integrated PFR, *AIChE J.*, 44, 2703–2712

17 Liolios, G., 2001, Acid runaways in a sulfuric acid alkylation unit, http://www.stratco.dupont.com/alk/pdf/AcidRunaway2001.pdf (last visited February 2006)

18 Parker, T., 2000, Atypical alkylation anecdotes. An Anthology. http://www.stratco.dupont.com/alk/pdf/AlkyAnecdotes2001.pdf (last visited February 2006)

19 Lerner, H., 1997, Exxon sulfuric-acid alkylation technology, in Meyers, R. (ed.), Handbook of Petroleum Refining Processes (2nd edn), McGraw-Hill, New York, USA.

20 Shah, B.R., 1986, UOP HF Alkylation Process, in Meyers, R. (ed.), Handbook of Petroleum Refining Processes, McGraw-Hill, New York, USA

10
Vinyl Acetate Monomer Process

10.1
Basis of Design

The vinyl acetate monomer (VAM) is large-scale commodity chemical mostly used in manufacturing polyvinyl acetate, the basic ingredient in water-soluble acrylic paints. Other applications are coatings for textile and paper industries, laminated safety glass, packaging, automotive fuel tanks and acrylic fibers. The worldwide production of vinyl acetate was about 5 million tonnes per year in 2005, with rapid growth in the emerging markets. Higher efficiency can be achieved by upstream integration with the production of low-cost acetic acid, as well as by downstream integration with the manufacturing of polyvinyl acetate and polyvinyl alcohol.

10.1.1
Manufacturing Routes

Three main routes for vinyl acetate manufacturing are employed today, as follows [1, 2]:

1. Acetic acid and acetylene
   The process is based on the reaction:

   $HC \equiv CH + CH_3COOH \rightarrow H_2C=CH\text{-}O\text{-}(CO)CH_3$ with $\Delta H = -118 \, KJ/mol$

   The operating conditions are gas phase at 170–250 °C and $Zn(OAc)_2$ catalyst impregnated on charcoal. Per-pass acetylene conversion is 60–70% with a selectivity of 93% acetylene and 99% acetic acid. High acetylene cost and safety problems make this process less competitive today.

2. Acetaldehyde and acetic anhydride
   The process takes place in two stages. Firstly acetaldehyde and acetic anhydride form ethylidene diacetate in liquid phase at 120–140 °C with $FeCl_3$ as a catalyst:

   $CH_3CHO + (CH_3CO)_2O \rightarrow CH_3CH(OCOCH_3)_2$

Chemical Process Design: Computer-Aided Case Studies. Alexandre C. Dimian and Costin Sorin Bildea
Copyright © 2008 WILEY-VCH Verlag GmbH & Co. KGaA, Weinheim
ISBN: 978-3-527-31403-4

In the second step the intermediate decomposes at 120 °C with acid catalyst:

$$CH_3CH(OCOCH_3)_2 \rightarrow H_2C=CH\text{-}O\text{-}(CO)CH_3 + CH_3COOH$$

Note that this process may rely completely on renewable raw materials.

3. Acetic acid, ethylene and oxygen

   This route dominates today and it will be adopted in this project. In older technologies the reaction was conducted in liquid phase at 110–130 °C and 30–40 bar in the presence of a redox catalyst $PdCl_2/CuCl_2$, but corrosion raised problems. Modern processes operate in gas phase with Pd-based catalysts. A highly undesired secondary reaction is the combustion of ethylene to CO_2. With modern Pd/Au catalysts the selectivity may reach 94%, based on ethylene and 98–99% based on acetic acid. The removal of CO_2 – usually by a wash with hot KOH solution – negatively affects the overall economics. Hoechst/Celanese and Bayer/DuPont are the most widespread processes, the main difference being in the formulation of the catalyst. With respect to reaction engineering a multitubular fixed-bed reactor is employed, where the operational difficulty is mastering the occurrence of excessive temperature rise (hot spot). Recently, fluid-bed reactor technology was developed with better productivity and 30% lower investment [13].

Searching for low-cost acetic acid sources is important since this takes about 70 wt% in the end product. From this viewpoint two processes developed by the Halcon Company can be mentioned [2]:

1. Integration of vinyl acetate and ethylene glycol manufacturing through the intermediate 1,2-diacetoxyethane.
2. Hydrogenative carbonylation of methyl acetate to 1,1-diacetoxyethane followed by cleavage to vinyl acetate and acetic acid. Only syngas is involved as raw materials.

10.1.2
Problem Statement

The project deals with a VAM plant capacity of 100 kton per year for an effective operation time of 8400 h. The process will be based on the acetoxylation of ethylene conducted in gas phase in the presence of a palladium-based solid catalyst. The case study will tackle the problem of process synthesis and energy integration, as well as the dynamics and control for ensuring flexibility in production rate of ±10%, while preserving safety and environment protection.

Table 10.1 presents typical specifications for a polymerization-grade product, as well as some physical properties. Prohibited impurities refer to inhibitors (crotonaldehyde, vinyl acetylene), chain-transfer agents (acetic acid, acetaldehyde, acetone) and polymerizable species (vinyl crotonate), while methyl and ethyl acetate impurities are tolerated.

Table 10.1 Properties for industrial vinyl acetate [1].

Property	Value
Molecular weight	86.09
Vinyl ester content	≥99.9%
Distillation range (101.3 kPa)	72–73 °C
Freezing point	−93 °C
Water	Max. 400 ppm
Acid content (acetic acid)	Max 50 ppm
Acetaldehyde	Max 100 ppm
Inhibitor content	3–5 ppm
n_D^{20}	1.369
d_{20}^{20}	0.93
Liquid viscosity at 20 °C	0.41 cP
Solubility VAM-water/water-VAM at 25 °C	0.9%/2.3%
Solubility in organic solvents	complete
Upper/lower explosion limit vapor in air at 20 °C	2.6/13.4% vol.

10.1.3
Health and Safety

Vinyl acetate is slightly or moderately toxic to humans and animals. The vapor irritates the eyes starting with 20 ppm, while the detection threshold is reported to be about 0.5 ppm. Released into the environment the vinyl acetate evaporates easily, being degraded rapidly by photochemical reactions, as well as biodegraded by either anaerobic or aerobic mechanisms. Therefore, the bioaccumulation of vinyl acetate in the ecosphere is unlikely.

Vinyl acetate is dangerous when exposed to heat, flame or oxidizers, and as a consequence, it requires adhering to the safety measures when stored or manipulated by operators. The same precaution is valid for the raw materials. Ethylene is highly explosive in mixture with oxygen, the explosion limit being at 10% vol. The acetic acid is a highly toxic and corrosive substance. Stainless steel of Cr/Ni/Co type is employed for operations involving acetic-acid solutions in boiling conditions, but normal stainless steel may be used for vapor-phase operations.

10.2
Reactions and Thermodynamics

10.2.1
Reaction Kinetics

The manufacturing of vinyl acetate by the oxyacetylation of ethylene is described by the following stoichiometric reaction:

$$C_2H_4 + CH_3COOH + 0.5O_2 \rightarrow C_2H_3OOCCH_3 + H_2O \tag{10.1}$$

Gas-phase reaction is preferred because of better yield and less corrosion problems.

The combustion of ethylene to CO_2 is a highly undesired secondary reaction since it lowers the yield and complicates the removal of the reaction heat:

$$C_2H_4 + 3O_2 \rightarrow 2CO_2 + 2H_2O \tag{10.2}$$

Note that the standard heat of reaction is of −176.2 and −1322.8 kcal/kJ per mol for vinyl acetate and ethylene combustion, respectively. If both reactions are considered the mean exothermal effect may be estimated at about −250 kJ/mol [1].

The catalyst plays a crucial role in technology. Previously, catalysts were based on palladium of 1 to 5 wt% impregnated on silica with alkali metal acetates as activators. Modern catalysts employ as enhancers noble metals, mostly gold. A typical Bayer-type catalyst consists of 0.15–1.5 wt% Pd, 0.2–1.5 wt% Au, 4–10 wt% KOAc on spherical silica particles of 5 mm diameter [14]. The reaction is very fast and takes place mainly inside a thin layer on the particle surface (egg-shell catalyst).

Typical catalyst lifetime is 1–2 years. Preferred operation conditions are temperatures around 150 to 160 °C and pressures 8 to 10 bar. Hot spots above 200 °C lead to permanent catalyst deactivation. The reactant ratio should ensure an excess of ethylene to acetic acid of about 2:1 to 3:1. Due to an explosion danger the oxygen concentration in the reaction mixture should be kept below 8%, based on an acetic-acid-free mixture [1]. The above figures formulate design constraints in the present project. In addition, a small amount of water in the initial mixture could be necessary for catalyst activation.

Because of the highly exothermic effect, measures to moderate the temperature increase are necessary, such as the dilution of the reaction mixture with some inert gas. Because of selectivity and heat-removal constraints the reactor is designed at low per-pass conversion, generally 15–35% for the acetic acid and 8–10% for ethylene [2].

Analyzing the mechanism of the catalytic chemical reaction allows the identification of the major factors that could affect the reactor design. Early in 1970 Samanos et al. [11] demonstrated that the reaction mechanism in gas phase shows great similarity to a liquid-phase reaction. This viewpoint has been adopted by a more general concept of supported liquid-phase catalysis (SLPC), in which the same reaction mechanism may be employed to explain both homogeneous and heterogeneous processes. A salient example is the class of selective oxidation reactions, including the present ethylene acetoxidation [12]. Under typical plant conditions, the adsorption of acetic acid and water on the catalyst can be substantial, the acetic acid forming about three monolayers. The promoter, in general an alkali metal acetate, plays an important role too. For example, KOAc gives a salt with water with the melting point at 148 °C. This salt contributes to the formation of the acetic acid layer, as described by the reaction KOAc + AcOH → KH(OAc)$_2$, which further prevents the combustion of ethylene. Moreover, the KOAc enhances the formation of actives centers through the solvatation of palladium complexes as Pd(OAc)$_2$ + KOAc → KPd(OAc)$_3$. Therefore, a probable reaction mechanism could be formulated as follows:

$$Pd + 0.5O_2 + 2AcOH \leftrightarrow Pd(OAc)_2 + H_2O \tag{1}$$

$$Pd(OAc)_2 + AcO^- \leftrightarrow Pd(OAc)_3^- \tag{2}$$

$$Pd(OAc)_3^- + C_2H_4 \leftrightarrow VAM + AcOH + AcO^- + Pd \tag{3}$$

As a result, from a reactor-design viewpoint the reaction kinetics is not sensitive to the concentration of the acetic acid, but the presence of some water is necessary to activate the catalyst. On the contrary, ethylene and oxygen are involved in kinetics through a complex adsorption/surface-reaction mechanism. This behavior was confirmed by both academic and industrial research [8, 9].

Modern catalysts for vinyl-acetate synthesis contain Au in the chemical formulation, which manifests in much higher activity and selectivity. This is reflected by fundamental changes in the kinetics, such as for example switching the reaction order of ethylene from negative to positive [8]. As a consequence, in more recent studies the formation of vinyl acetate can be described conveniently by a power-law kinetics involving only ethylene and oxygen:

$$r_{VA} = k_1 p_{C_2H_4}^{\alpha_1} p_{O_2}^{\beta_1} \tag{10.3}$$

The exponent α_1 is 0.35–0.38 and β_1 0.18–0.21 indicating strong adsorption limitations. The reaction constant is given by $k_1 = A_1 \exp(-E_1/RT)$ in which the energy of activation depends on the Pd content, for example 39 kJ/mol for Pd 1% and 17.3 kJ/mol for Pd 5%.

Similar kinetics has been found to describe the secondary combustion reaction, but with reaction orders very different from the main reaction:

$$r_{CO_2} = k_2 p_{C_2H_4}^{\alpha_2} p_{O_2}^{\beta_2} \tag{10.4}$$

Table 10.2 presents the kinetic information for the main reactions, in which the frequency factors have been calculated from turnover-frequency (TOF) data [8, 9]. This term, borrowed from enzymatic catalysis, quantifies the specific activity of a catalytic center. By definition, TOF gives the number of molecular reactions or catalytic cycles occurring at a center per unit of time. For a heterogeneous catalyst the number of active centers can be found by means of sorption methods. Let us consider that the active sites are due to a metal atom. By definition [15] we have:

$$TOF = \frac{\text{volumetric rate of reaction}}{\text{number of center/volume}} = \frac{\text{moles\_A}}{\text{l\_cat}} \frac{\text{l\_cat}}{\text{l\_cat} \times \text{s moles\_Me}} = s^{-1} \tag{10.5}$$

The catalyst is characterized by the metal weight fraction w_{Me} with MW_{Me} the atomic weight, the metal dispersion coefficients D giving the fraction of active centers, and ρ_{cat} the grain catalyst density. By replacing into the relation (10.5) the following formula can be obtained for the calculation of TOF from kinetic experiments:

$$\text{TOF} = \frac{\text{moles\_A/s}}{\text{l\_cat}} \cdot \frac{1}{\frac{g\_Me}{g\_cat} \times \frac{D}{MW_{Me}} \cdot \frac{g\_cat}{l\_cat}} = R_A \bigg/ \left(w_{Me} \times \frac{D}{MW_{Me}} \right) \quad (10.6)$$

Conversely, transforming TOF into reaction rate data can be done with the relation:

$$-r_A = \text{TOF} \times w_{Me} \times \frac{D}{MW_{Me}} \times \rho_{cat} \quad (10.7)$$

As an example, let us consider the reported value TOF = $6.5 \times 10^{-3}\,\text{s}^{-1}$. Catalyst data are: ρ_{cat} = 1000 g/l, MW_{Pd} = 106.4, w_{Pd} = 0.01, D = 0.4. One gets $-r_A$ = $6.5 \times 10^{-3} \times 0.01 \times (0.4/106.4) \times 1000$ = 2.44×10^{-4} moles VA/l_cat/s = 2.10×10^{-2} g/l/s = 7.52×10^{-2} kg/l/h. Note that this TOF value is obtained at a low partial pressure of ethylene. In industrial conditions the pressure is an order of magnitude higher. Thus, the reaction rate would be about 0.7 kg/l/h. This value is in agreement with STY reported in some recent patents. For example, a modern Pd-gold catalyst on supported 4–6 mm silica spheres (patent US 649229931) gives a space-time yield (STP) of 700 g VAM/l/h and a selectivity of 92–94% at 150 °C and 7.8 bar with a gas mixture of ethylene 53.1/acetic acid 10.4/oxygen 7.7 and inert 28.6 (vol.%).

Note that expressing the catalyst activity in terms of TOF allows tailoring the catalyst activity to the requirements of process design. Because the activation energy remains constant, the only affected parameter is the pre-exponential factor A, which in turn is proportional to the weight fraction of the active center, in this case the metal. Table 10.2 shows two situations. In the first case the pre-exponential factor is taken from the original TOF data, which corresponds to a fast

Table 10.2 Kinetic parameters for VAM synthesis over a Pd/Au/SiO$_2$ catalyst [8, 9].

Reactions	Power-law kinetics		Kinetic constants		
	C_2H_4	O_2	E (J/mol)	A_1	A_2
$C_2H_4 + CH_3COOH + 0.5O_2$ $\rightarrow C_2H_3OOCCH_3 + H_2O$	0.36	0.20	15000	2.65×10^{-4}	7.95×10^{-5}
$C_2H_4 + 3O_2 \rightarrow 2CO_2 + 2H_2O$	−0.31	0.82	21000	7.50×10^{-4}	2.25×10^{-4}

Reaction rate in mol/(liter catalyst·s).

Table 10.3 Typical operation conditions for the VAM reactor [14].

Parameter	Typical values		
Pressure	5–12 atm		
Temperature	140–180 °C		
GHSV	2000–4000 h^{-1}		
Reaction mixture	**Composition mol%**	**Conversion %**	**Selectivity**
Ethylene	50	8–10	91–94
Acetic acid	10–20	15–30	>99
Oxygen	6–8	60–70	60–70
CO_2	10–30	–	–
Inert (balance)	N_2 or Ar	–	–

catalyst. Preliminary simulation showed that in this case the occurrence of hot spot is very likely. For this reason in the second case the catalyst activity is lowered by a factor of three, by less metal impregnation. This catalyst is adopted for the reactor design and flowsheet simulation.

Table 10.3 presents typical operation conditions for a multitubular chemical reactor. Higher pressure has a positive effect on productivity, but affects the selectivity negatively since increased adsorption of ethylene on the catalytic sites favors the combustion reaction. The reaction temperature should be above 150 °C in order to bring the salt activator into the molten state. However, temperatures higher than 150 °C have only a moderate effect on the reaction rate because of low activation energy, but it may decrease considerably the selectivity. Preventing the hot spot and operating close to isothermal conditions is desirable. This may be achieved by diluting the catalyst in the entry zone and/or by manipulating the pressure of the raising steam. But most of all, by means of compositional and structural effects there is the possibility to modify the activity of a particular catalyst so as to fulfil an optimal reactor design.

10.2.2
Physical Properties

Table 10.4 presents basic physical properties of the key components. By boiling point the acetic acid is the heaviest. Vinyl acetate is a light species with a normal boiling point at 72.6 °C. Of major interest is the low-boiler heterogeneous azeotrope vinyl acetate/water with 25 mol% water and nbp at 65.5 °C. The very low solubility of vinyl acetate in water, less than 1 wt%, is to be noted. Low reciprocal solubility can be exploited for separating the mixtures vinyl acetate/water by azeotropic distillation. In addition the densities of water and vinyl acetate are sufficiently distinct to ensure good liquid–liquid decanting.

Table 10.4 Basic physical properties of the main components.

Name	Ethylene	Acetic acid	VAM	Water
Molecular formula	C_2H_4	$C_2H_4O_2$	$C_4H_6O_2$	H_2O
Molecular weight	28.05	60.05	86.09	18.015
Normal boiling point, K	162.42	391.04	345.95	373.15
Melting point, K	104	289.93	180.35	273.15
Critical temperature, K	282.34	594.45	519.15	674.14
Critical pressure, bar	50.41	57.90	40.3	220.64
Critical volume, cm^3/mol	131.1	171.0	270	55.95
Liquid density, kg/m^3 (°C)	577 (−110)	1049.2 (20)	934 (20)	1000 (15)
Vaporization heat at nbp, kJ/mol	13.553	23.7	31.49	40.66

10.2.3
VLE of Key Mixtures

It can be anticipated that the liquid-separation system should handle the mixture vinyl acetate/acetic acid/water. Figure 10.1 displays the residue curve map at 1 atm calculated by Aspen Split™ [16] with the thermodynamic NRTL/HOC and VLLE option. The phase equilibrium is dominated by the binary heterogeneous azeotrope vinyl acetate/water with the composition 0.745/0.255 mol% or 0.933/0.0667 wt% and boiling points at 65.6 °C. It may be observed that the relative position of the azeotrope, as well as reciprocal solubility and the tie-line direction depend on the compositional coordinate, in molar or mass fractions, because of the large difference in the molar weight of the two components. The residue curve map suggests as a separation strategy the distillation in top of the VAM/water azeotrope followed by the separation of VAM by L–L decanting, while the acetic acid can be obtained as bottom product.

10.3
Input–Output Analysis

The selection of raw materials takes into account the price variation versus purity, with constraints on undesired species. Table 10.5 shows the choice. Acetic acid is of high purity with small amounts in acetaldehyde and formic acid. Ethylene is of high purity too, with severe specifications on CO and sulfur in order to protect the catalyst, but small amounts of ethane are allowed.

10.3.1
Preliminary Material Balance

Taking the acetic acid as reference, its molar feed is 100 000/8400/86 = 138.43 kmol/h, which is rounded further to 140 kmol/h. A key decision regards the selectivity.

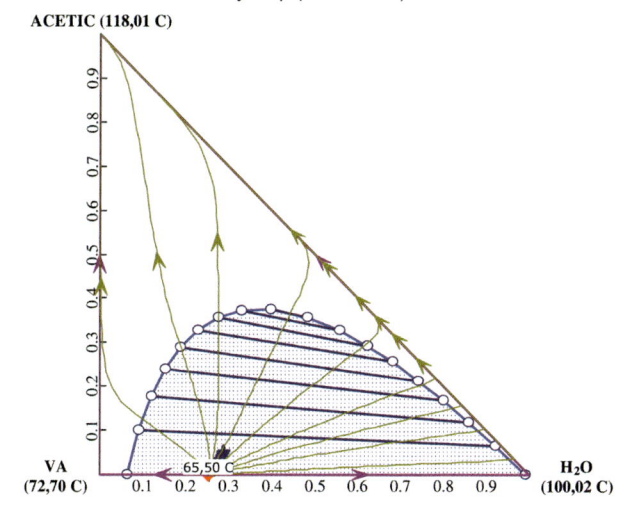

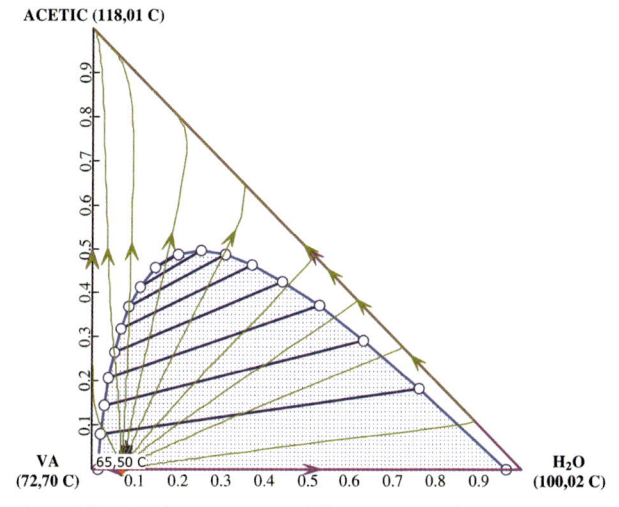

Figure 10.1 Residue curve map of the mixture vinyl acetate/acetic acid/water.

Table 10.5 Purity specifications for the raw materials.

	Ethylene	Acetic acid
Purity	99.9%	99.5%
Impurities	Ethane < 1%	Formic acid 50 ppm
	CO 2–10 ppm	Acetaldehyde 50 ppm
	Sulfur 1–10 ppm	Water < 0.5%

Table 10.6 Ideal input/output material balance.

	Input		Output	
	kmol/h	kg/h	kmol/h	kg/h
C_2H_4	148.936	4170.213		
Acetic acid	140.000	8400.000		
O_2	96.809	3097.872		
VAM			140.000	12 040.000
Water			157.872	2841.702
CO_2			17.872	786.383
Total	385.745	15 668.085	315.745	15 668.085

In a preliminary stage we select a value of 94%, at the upper limit but realistic for modern catalysts. We deliberately set zero targets for losses. With this assumption the preliminary input/output material balance looks as in Table 10.6. Some observations of significance for design may be immediately drawn. Thus, the large amount of wastewater will involve non-negligible costs for neutralization and biological treatment. Similarly, the process develops inherently CO_2. Therefore, some ecological penalties have to be taken into account when estimating the economic potential.

10.4
Reactor/Separation/Recycles

The kinetic analysis has shown that because of incomplete conversion two recycles have to be considered, for both ethylene and acetic acid. Preferably, the recycle policy should ensure an ethylene/acetic acid ratio of about three. A strong safety constraint is the concentration of oxygen in the reaction mixture. In general, the oxygen concentration at the reactor inlet should be below 8 vol% in order to avoid ignition, based on an acetic-acid-free mixture [1]. Note that in the stoichiometric mixture this is of 20%. The presence of a gaseous inert is recommended for better safety and control of the reactor temperature. Some reports indicate the use of ethane, since this is present as impurity in the fresh feed [6], but this solution is not adopted here. Since CO_2 is produced by reaction in large amount its use as an inert would be the most economical. Indeed, several technology reports indicate this possibility in an amount of 10–30% vol [2, 14]. However, since CO is a catalyst poison its presence in the recycled gas has to be prevented.

The chemical reactor will be designed in the context of a recycle system, as explained in Chapters 2 and 4. The strategy is fixing the flow rate and composition of the reactor-inlet mixture at values that are compatible with the operation requirements of the catalyst, as given in Table 10.3, for example 50% mol C_2H_4, 20% mol acetic acid, 6% mol oxygen and 24% mol CO_2. Figure 10.2 presents the simulation

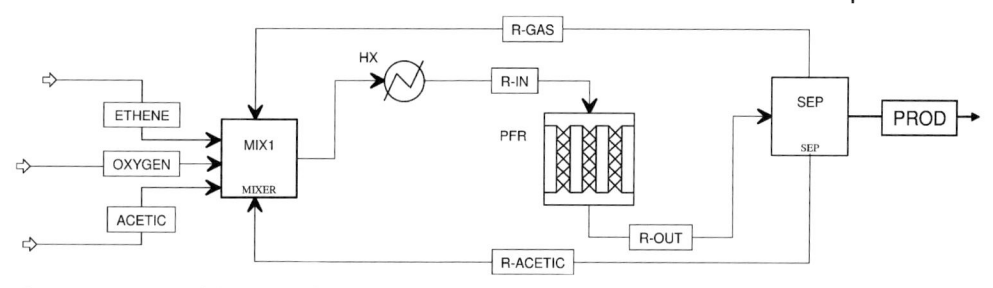

Figure 10.2 Design of the reactor for vinyl-acetate manufacturing in a recycle system.

flowsheet with a PFR reactor and black-box separation unit, as well as with recycles for the gas (ethylene and CO_2) and liquid (acetic acid). All other components, namely VAM, water and byproduct CO_2, are lumped into a product stream sent to further separation.

The reactor is of the heat-exchanger type with catalyst and tubes and rising steam outside. In this project we consider spherical catalyst particles of 5 mm diameter and a bed void fraction of 45%, which offers a good trade-off between efficiency and lower pressure drop. The gas inlet pressure is 10 bar. We aim at a pressure drop less than 15% of the operating pressure, namely a maximum of 1.5 bar.

For predesign calculations we consider a gas velocity of 0.5 m/s [4]. Mean physical properties for the above reaction mixture are: density 11.5 kg/m³, viscosity 1.5 $\times 10^{-5}$ N s/m², thermal conductivity 2.9×10^{-2} W/m K. The calculation of the heat-transfer coefficient follows the relations given in Chapter 5. Applying the relation (5.9) leads to $Re_p = 2090$ and $Nu = 412$, from which the partial heat-transfer coefficient on the gas side is $\alpha_w = 350$ W/m K. Taking into account other thermal resistances we adopt for the overall heat-transfer coefficient the value 250 W/m² K. For the cooling agent we consider a constant temperature of 145 °C, which is 5 °C lower than the inlet reactor temperature. This value is a trade-off between the temperature profile that avoids the hot spot and the productivity.

By using the kinetic data given in Table 10.2 (slow catalyst with pre-exponential factor A_2) the simulation leads to the following reactor design: 4900 tubes of internal diameter of 37 mm and 7.5 m length. The pressure drop calculated by the Ergun relation gives 0.93 bar, under the constraint of 1.5 bar.

Table 10.7 presents the stream table around the chemical reactor. As mentioned above, the inlet mixture is fixed because of control requirements. In this way, the outlet reflects the transformation of composition due to the chemical reactor. The molar conversions for ethylene and acetic acid are in agreement with industrial data. The amount of CO_2 formed by reaction is quite limited, denoting good catalyst selectivity; only about 11% from ethylene is consumed by combustion, the rest, 89%, going into vinyl acetate. Finally, the computation gives a catalyst productivity of 326.6 kg VAM/m³-catalyst · h.

Note that other designs are possible using other values for the activity of that catalyst and the rate of heat removal. The challenge is to find a temperature profile

10 Vinyl Acetate Monomer Process

Table 10.7 Stream table around the chemical reactor.

Mole flow kmol/h	Reactor inlet	Reactor outlet	Conversion
C_2H_4	1450	1291.04	0.109
O_2	180	76.97	0.572
CO_2	750	768.84	
Vinyl acetate	0	149.54	
H_2O	0	168.38	
Acetic acid	600	450.46	0.249

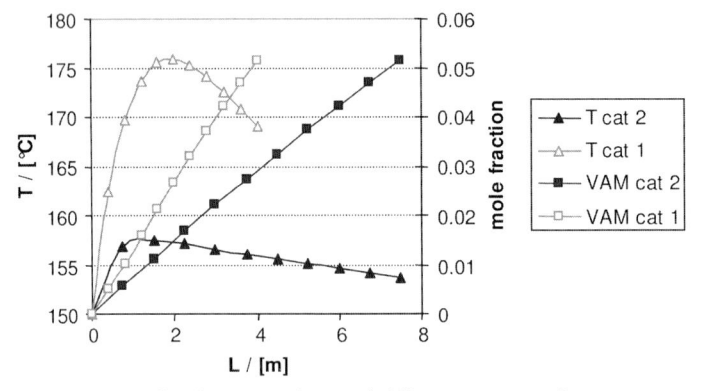

Figure 10.3 Profiles for two catalysts and different reactor configuration.

that preserves the catalyst integrity but ensures high productivity and selectivity. Figure 10.3 shows a comparison of the temperature profiles with the actual catalyst (2) against a fast catalyst (1), which is three times faster, as described by the values in Table 10.2. The number of tubes is reduced from 4900 to 2590, and the length from 8 to 4 m. The cooling-agent temperature is lowered to 140 °C, while the heat-transfer coefficient is raised to 300 W/m² K. Since the gas velocity is doubled, the pressure drop would increase about four times, but the effect is limited by reducing the tube length. Another solution is employing a monolith-type catalyst, which can be justified by the fact that the reaction zone is limited to a thin layer. Because the catalyst volume diminishes from 39.5 to 10.3 m³ the increase in productivity is remarkable, from 326 to 1260 kg/m³ catalyst h, which corresponds to the best modern technologies [1, 14].

10.5
Separation System

The reactor-exit mixture has roughly the following composition 1290 C_2H_4, 770 CO_2, 150 VAM, 170 H_2O and 450 acetic acid, all in kmol/h. To this we should add some lights, such as ethyl acetate, and heavies, such as di-acetyl ethylidene.

However, these are disregarded since these should not affect the structure of the separations. The strategy consists of decomposing the separation problem in two subproblems, for gas and liquid separations, respectively, by designing a suitable first separation step.

10.5.1
First Separation Step

To get an idea about the relative volatilities of components we proceed with a simple flash of the outlet reactor mixture at 33 °C and 9 bar. The selection of the thermodynamic method is important since the mixture contains both supercritical and condensable components, some highly polar. From the gas-separation viewpoint an equation of state with capabilities for polar species should be the first choice, as SR-Polar in Aspen Plus™ [16]. From the liquid-separation viewpoint liquid-activity models are recommended, such as Wilson, NRTL or Uniquac, with the Hayden O'Connell option for handling the vapor-phase dimerization of the acetic acid [3]. Note that SR-Polar makes use of interaction parameters for C_2H_4, C_2H_6 and CO_2, but neglects the others, while the liquid-activity models account only for the interactions among vinyl acetate, acetic acid and water. To overcome this problem a mixed manner is selected, in which the condensable components are treated by a liquid-activity model and the gaseous species by the Henry law.

Table 10.8 presents a comparison of SR-Polar EOS and Wilson-HOC with Henry components. The predictions by the two methods are in good agreement, although surprisingly for the ability of SR-Polar to account for liquid-phase nonideality.

Because by single flash only two-thirds of the vinyl acetate passes in liquid phase, a multistage equilibrium separation is necessary for its advanced recovery, namely an absorption unit. A suitable solvent is the acetic acid itself. Consequently, after cooling the reaction gas is treated in countercurrent in the column (C-1) with acetic acid fed on the top stage (Figure 10.4). To achieve higher recovery a liquid pump-around with intermediate cooling is employed. The simulation in Aspen Plus™ [16] indicates that 20 stages are sufficient for >99.9% recovery with a pump-around between stages 4 and 2.

Table 10.8 Flash of the outlet reaction mixture at 9 bar and 33 °C.

Species	Molar fraction	SR-Polar		Wilson-HOC + Henry components	
	Initial	Vapor	Liquid	Vapor	Liquid
C_2H_4	0.5400	0.5321	7.88E–03	0.5398	1.23E–04
O_2	0.0200	0.0199	3.95E–05	0.0199	1.88E–05
CO_2	0.2200	0.2136	6.33E–03	0.2159	4.08E–03
VAM	0.0428	5.67E–03	0.0371	4.92E–03	0.0379
H_2O	0.0486	1.54E–03	0.0470	1.32E–03	0.0472
Acetic acid	0.1285	1.76E–03	0.1268	2.15E–03	0.1264

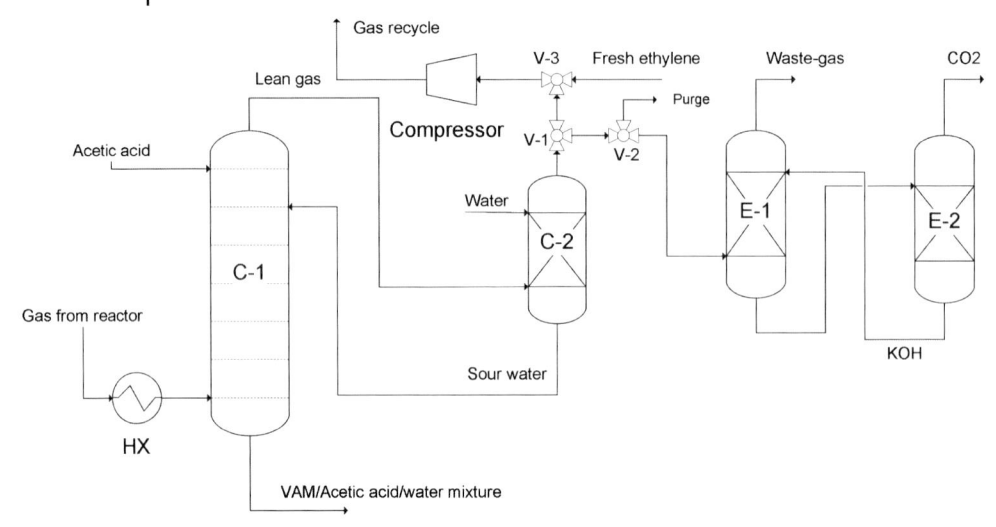

Figure 10.4 First separation step and gas separation section at vinyl acetate synthesis.

10.5.2
Gas-Separation System

Because some acetic acid is entrained in the top vapor of the absorber, this is captured with water in a small absorption column, the resulting liquid being sent back to the column C-1. The gas containing unreacted C_2H_4 and O_2 as well as CO_2 as an inert is recycled to the reaction section. However, removing the excess of CO_2 is necessary to prevent its accumulation. This operation is done in industry by standard techniques, such as washing with amines or potassium carbonate solutions. With these elements the flowsheet for the gas-separation section can be sketched, as illustrated in Figure 10.4. Note that by the valve V-1 the gas split is adapted to the operation conditions, namely to constant gas composition of the gas recycle. A second valve V-2 can withdraw a gas purge on the line to the washing section. Finally, the valve V-3 ensures the makeup of ethylene into the gaseous recycle to the reaction section.

10.5.3
Liquid-Separation System

The first separation step produces essentially the liquid ternary mixture vinyl acetate, water and acetic acid with some dissolved gases. Other light and heavy components are neglected. The RCM analysis indicated as feasible the separation of the heterogeneous azeotrope VAM/water in top followed by quantitative separation of components by decantation. The flowsheet configuration is shown in Figure 10.5. The feed of the column (C-3) collects the ternary mixture from the absorber combined with the water solution from the wash column. The column

10.5 Separation System

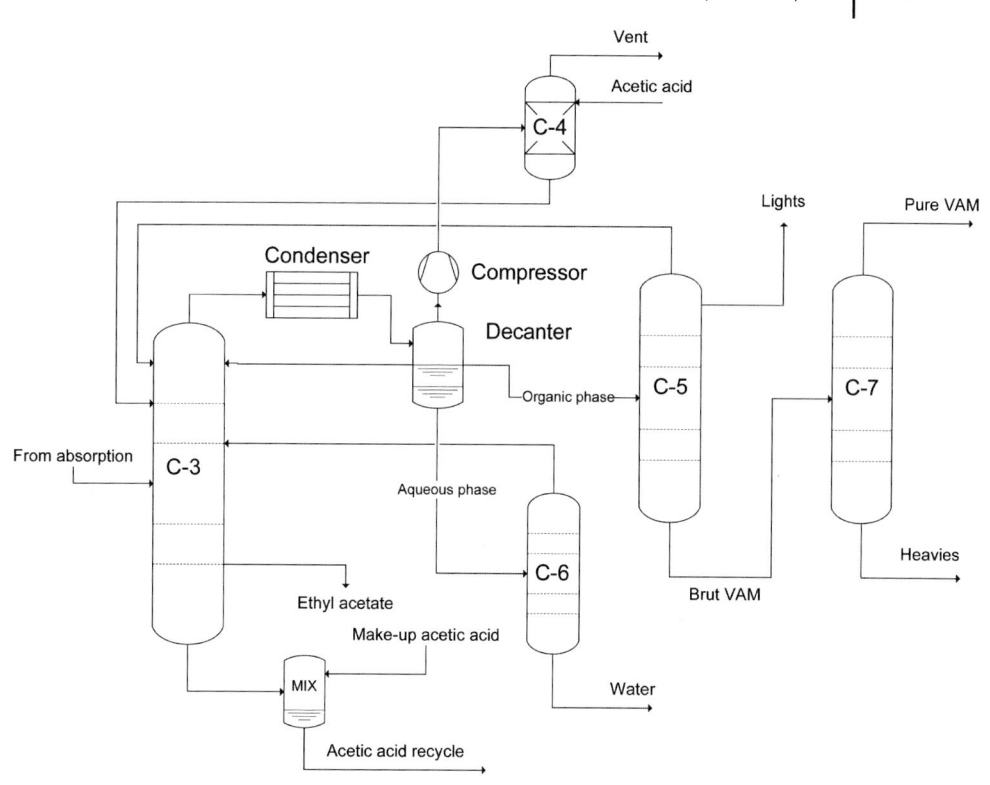

Figure 10.5 Liquid-separation system.

can be designed to deliver in bottom water-free acetic acid. Small amounts of ethyl acetate can be removed as a side stream to prevent its accumulation, if necessary. The top vapor will contain mainly VAM and water, which after condensation separates into two liquid phases. The organic phase from the decanter containing raw VAM product is submitted to downstream purification. The column C-5 separates in top the azeotrope VAM/water recycled to C-3, with simultaneous removal of lights. The column C-7 delivers pure VAM in top with removal of heavies in bottoms. Note that the decanter also supplies the reflux for the column C-3. The aqueous phase is submitted to VAM recovery by stripping in the column C-6. The VAM/water azeotrope is recycled to separation in C-3, while the waste is water obtained in bottoms, which will be further sent to a biological treatment. The off-gas stream from the decanter is compressed to 8–10 bar and sent to absorption in acetic acid in the column C-4, from which the VAM is recovered in bottoms and recycled to C-3. In this way, a very high yield of vinyl acetate can be ensured with practical no losses in the gas vent or wastewater. As shown in Figure 10.5 the makeup of the acetic acid needed for reaction can be done directly in the recycle. Consequently, the structure of separations developed so far can fulfil the requirements of the plantwide control of reactants, as will be explained in Section 10.8.

10.6
Material-Balance Simulation

By merging the separation schemes from Figures 10.4 and 10.5 one gets the flowsheet displayed in Figure 10.6, submitted to simulation in Aspen Plus™. The key columns are C-1 for gas absorption, C-3 for VAM/acetic acid separation and C-5 for VAM purification. Additional separators ensure high material efficiency, such as C-2 for acetic-acid recovery with water, C-4 for VAM recovery with acetic acid and C-6 for VAM recovery from water. Accordingly, several recycle streams appear, which can raise convergence problems. Building the flowsheet starts by dealing first with the key columns so as to ensure good performance. Then, the additional columns are inserted and recycles closed. As thermodynamics, one can use SR-Polar and liquid-activity models with similar results, exempting the three-phase flash, where NRTL or Uniquac should be applied. Using the Henry components option with a liquid-activity model helps greatly the convergence. The specification mode of units also deserves some comments. For C-1 and C-2 there are zero degrees of freedom; only the number of stages and the solvent rate can be manipulated by design. For C-3 the reflux is dictated by the azeotrope VAM/water. Because this has 25% mol water, a substantial amount of VAM has to be recycled. This can be manipulated by the split fraction, for example in this case set at 0.85. Finally, for C-5 a good specification is the ratio bottom product/feed, which ensures fast and robust convergence. The same is valid for C-6.

Table 10.9 presents the main results. The absorption unit C-1 working under pressure ensures high recovery of 99.9% vinyl acetate and minimum 98% acetic acid with 20 theoretical stages. Furthermore, the components are sharply sepa-

Table 10.9 The characteristiques of key separators by the vinyl acetate plant.

Columns	C-1	C-3	C-5
Function	Absorption VAM and acetic acid	VAM/acetic acid separation	VAM purification
Split			
• top	0.984 C_2H_4, 0.970 CO_2	0.999 VAM as azeotrope	
• bottom	0.999 VAM, 0.980 acid	0.979 acid	
Stages (feed)	20	23 (8)	20
Pressure (bar)	8	1.3	1.1
T top (°C)	31.8	75.1	72.3
T bottom (°C)	35.2	129.5	76.0
Reflux	None	0.85	0.3
Q_{cond} (MW)	None	29.2	3.0
Q_{reb} (MW)	None	30.3	3.1
Internals	Sieves	Koch flexitrays	Valves
Diameter	2.2 m	5.35 m	1.7

Figure 10.6 Flowsheet of the material balance.

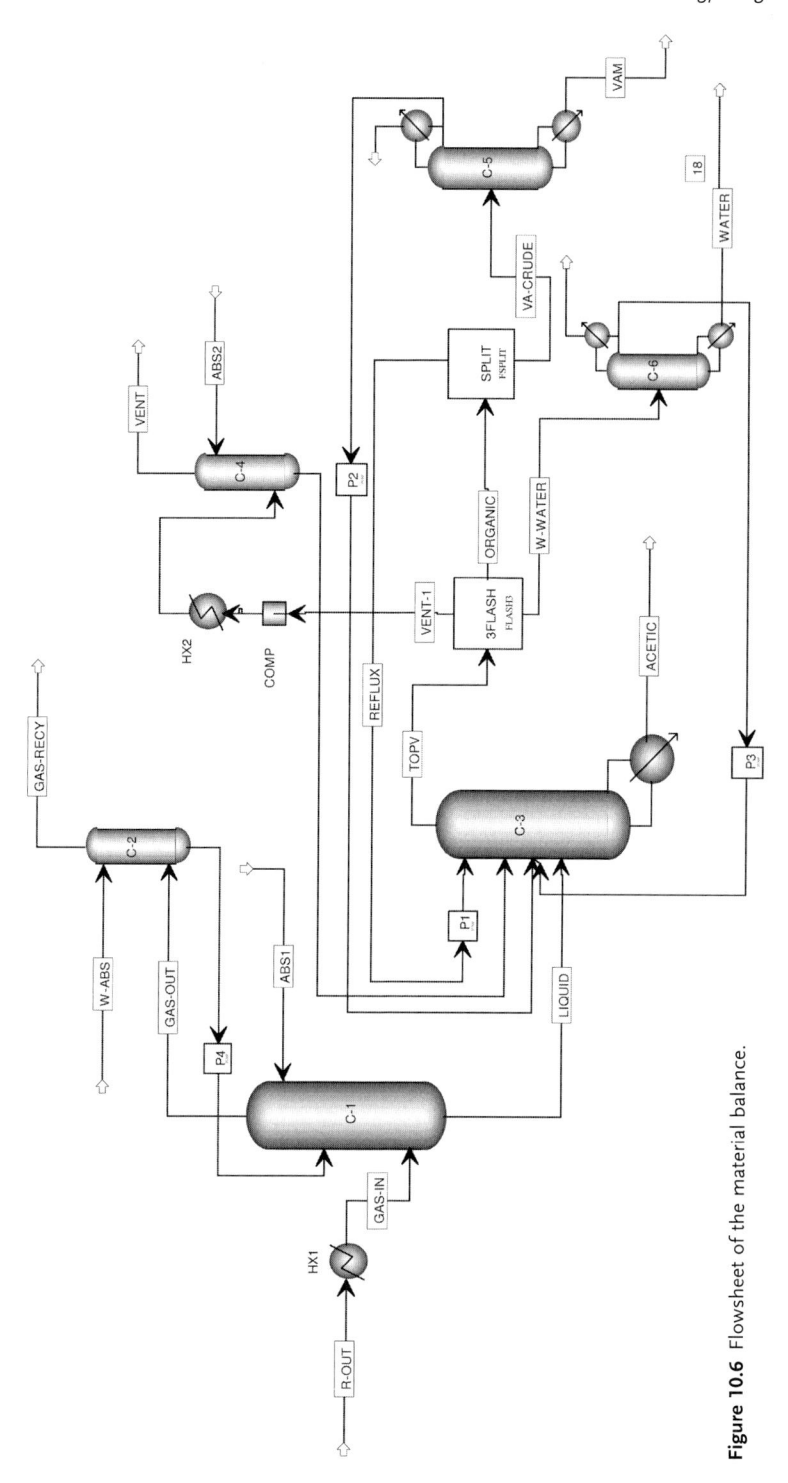

rated in C-3, which has 23 theoretical stages and operates at nearly atmospheric pressure. This column has a large diameter that can be reduced if high-efficiency internals are used, such as Koch flexitrays. Note also the large reboiler duty.

10.7
Energy Integration

Because of the highly exothermic chemical reaction, the reactor makes free about 10 MW energy for energy integration. Since the reaction temperature is in the range 150–160 °C, the heat of reaction can be recovered as low-pressure steam at 140–145 °C and 3–3.5 bar. This amount can cover a substantial part of the energy needed in the process, for distillation or evaporation. In addition, the enthalpy of the exit stream may be used for feed preheating via a classical FEHE heat exchanger, as shown in Figure 10.7.

Most of the energy is consumed in the distillation section, namely for VAM recovery and purification. The reboiler duty for the azeotropic distillation of VAM is particularly high, of about 30 MW. It can be observed that this is due to the large recycle of VAM necessary to carry out the water formed by reaction (3 mole VAM per mol water). Thus, any measure is welcome that can reduce the water content in the crude VAM/acetic acid mixture. Figure 10.8 shows an ingenious method known as "gas dehydration" [1]. The reactor outlet, cooled up to the dew point,

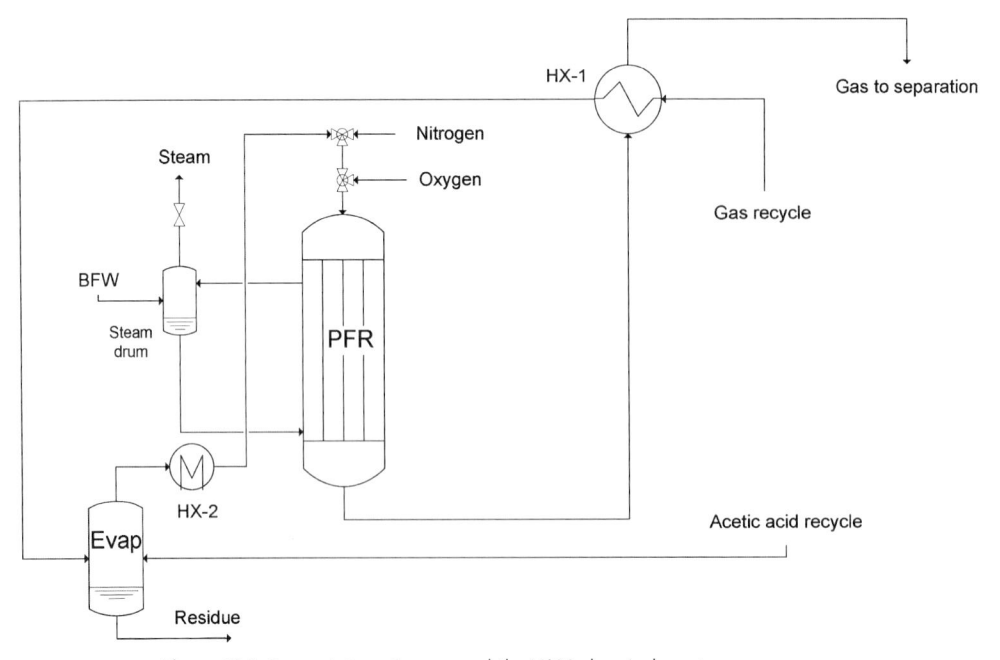

Figure 10.7 Energy integration around the VAM chemical reactor.

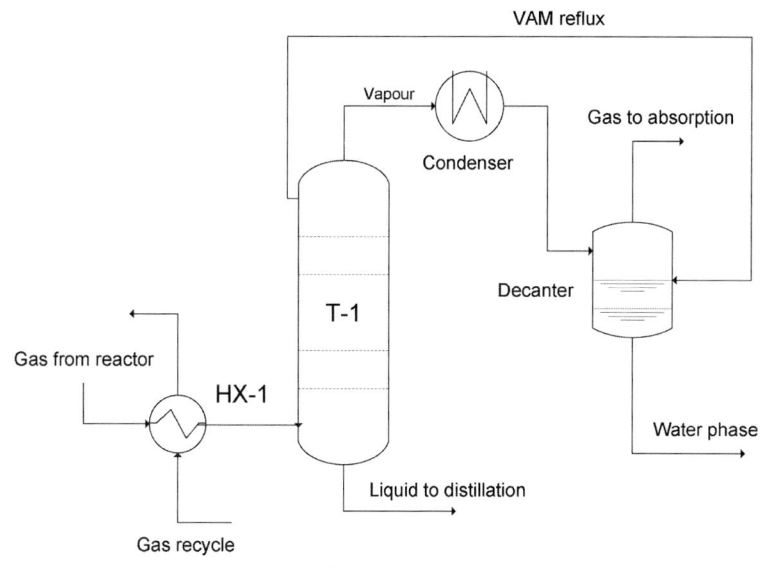

Figure 10.8 Gas-dehydration setup for energy saving.

enters the absorption column T-1 operated with liquid VAM in countercurrent. After cooling, the top mixture is submitted to separation by a three-phase flash. The lean-gas phase consists of ethylene, CO_2 and some VAM is practically free of acetic acid. The organic liquid, in majority VAM, is returned as reflux to T-1. The aqueous phase from the decanter contains in large majority water with only small amounts of VAM. The bottom stream from the absorber ensures the recovery of 99.9% acetic acid and 75% VAM, but taking only 50% water with respect to the initial gas.

The separation section will change accordingly (Figure 10.11). The lean gas with only 25% VAM from the initial gas and free of acetic acid is sent to the absorption column (C-1). The column (C-2) is no longer necessary. The bottom from (T-1) is led directly to (C-3). Less water means lower reflux from the three-phase decanter. In consequence, the duty of the column (C-3) can be reduced from 30 MW to about 8.5 MW, which represents a considerable energy saving of about 70%. In this way the reboiler duty of (C-3) can be covered entirely by the steam produced in the reactor.

10.8
Plantwide Control

The main goals of the plantwide control system are to ensure safe operation, the desired production rate and product quality, and to optimize the efficiency of using the material and energetic resources.

Figure 10.10 presents the control structure around the chemical reactor, where the main safety issues arise. Firstly, the reactor-inlet mixture must not contain

306 | *10 Vinyl Acetate Monomer Process*

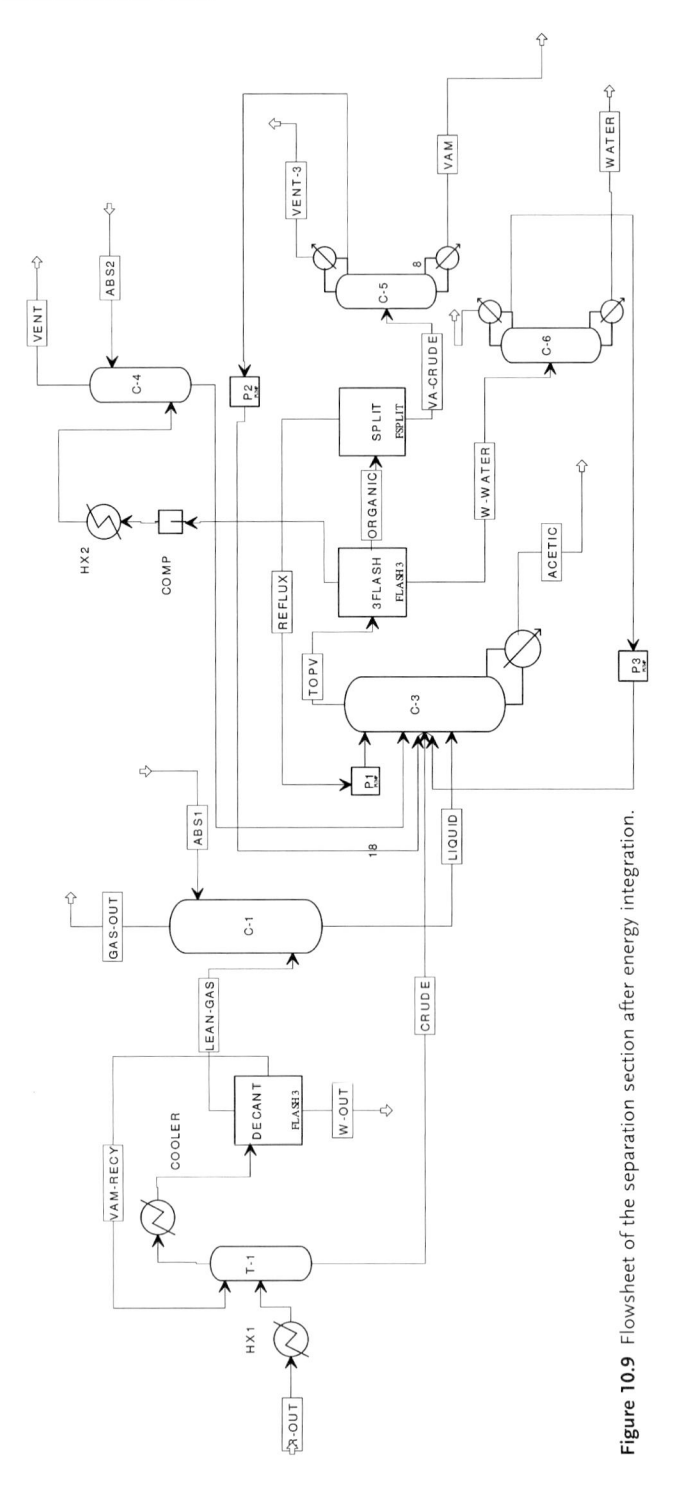

Figure 10.9 Flowsheet of the separation section after energy integration.

10.8 Plantwide Control | 307

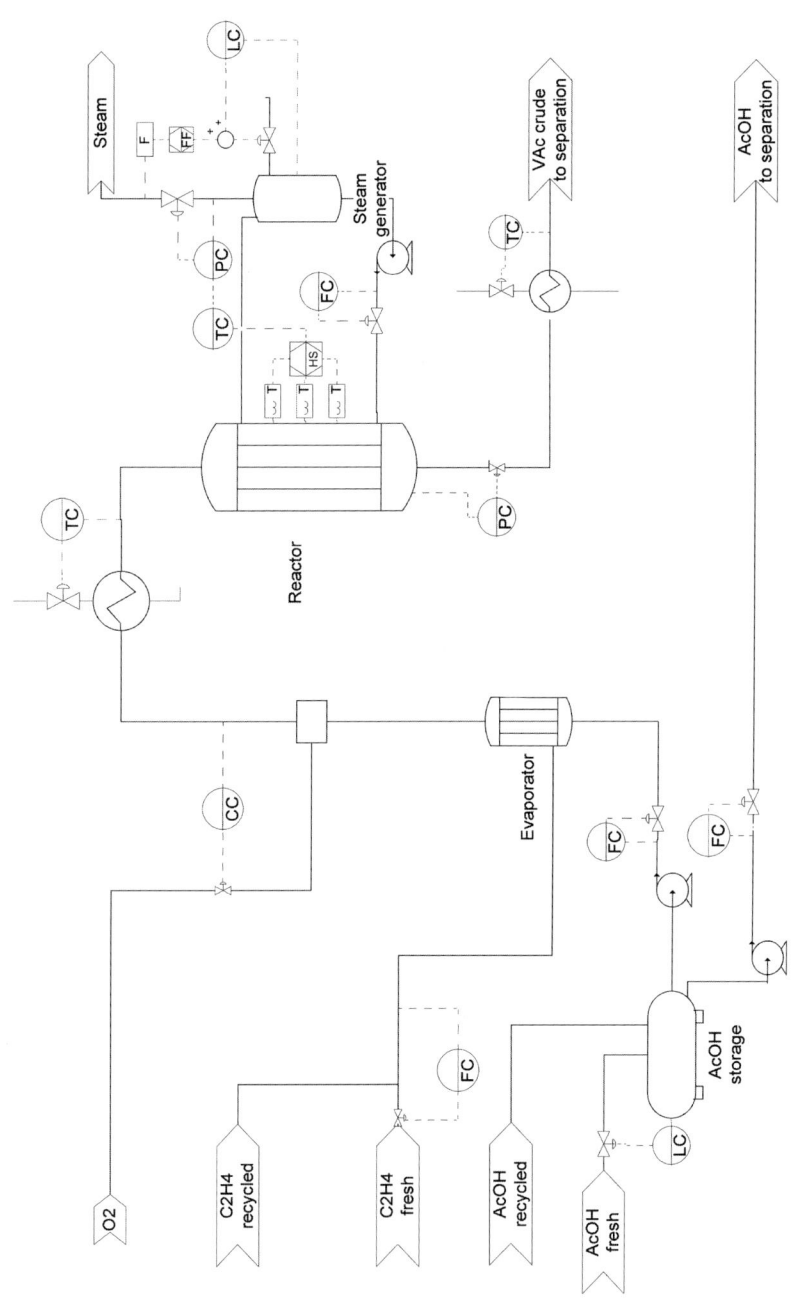

Figure 10.10 Plantwide control: control loops around the reactor and the feed policy.

more than 8% oxygen base in the acid-free mixture in order to avoid explosion risks. Therefore, the oxygen is added under concentration control, in a mixing chamber placed behind concrete walls. Secondly, the cooling should avoid reaction runaway. In a runaway situation, the excessive temperature leads to the danger of explosion, catalyst deactivation, and a drastic decrease of the selectivity. The coolant is circulated at a constant rate. Several temperature measurements are placed along the reactor bed and the highest value is selected as the process variable. The manipulated variable of the control loop is the steam-generator pressure, which directly influences the coolant temperature. The water level in the steam generator is controlled by the water makeup. Note that using a simple feedback loop may not work. When the steam rate increases, the correct action is to add more water makeup. However, the pressure simultaneously decreases. The lower pressure means that, initially, the steam bubbles will occupy a larger volume, and the liquid level will increase. A feedback level controller will wrongly decrease the water makeup rate. Therefore, the steam rate is measured and the required water makeup is calculated. This feedforward action is combined with the feedback provided by the level controller.

As recommended in Chapter 4, the inventory of reactants in the plant is maintained by fixing the reactor-inlet flows. Acetic acid is taken with constant rate from a storage tank, and the fresh feed is added on level control. The gas rate going to the evaporator is a good estimation of the ethylene inventory. Therefore, this flow is kept constant by adjusting the fresh ethylene feed. The fresh oxygen rate is manipulated by a concentration control loop, as previously explained.

The control of the separation section is presented in Figure 10.11. Although the flowsheet seems complex, the control is rather simple. The separation must deliver recycle and product streams with the required purity: acetic acid (from C-3), vinyl acetate (from C-5) and water (from C-6). Because the distillate streams are recycled within the separation section, their composition is less important. Therefore, columns C-3, C-5 and C-6 are operated at constant reflux, while boilup rates are used to control some temperatures in the lower sections of the column. For the absorption columns C-1 and C-4, the flow rates of the absorbent (acetic acid) are kept constant. The concentration of CO_2 in the recycle stream is controlled by changing the amount of gas sent to the CO_2 removal unit. The additional level, temperature and pressure control loops are standard.

In Chapter 4, two ways for achieving production-rate changes were presented. The first strategy, manipulating the reactor inlet flows, does not work here. The acetic acid does not influence the reaction rate, the per-pass conversion of ethylene is very low (10%), while the reactor-inlet oxygen concentration is restricted by the safety concerns. Therefore, the second strategy of manipulating the reaction conditions should be applied.

The reactor model available in Aspen Dynamics™ [16] only provides the possibility of changing the coolant temperature. Figure 10.12 shows results dynamic simulation results, for the following scenario: the plant is operated at the nominal steady state for 1 h. Then, the coolant temperature is increased from 413 to 425 K and simulation is continued for 2 h. The maximum temperature inside the reactor

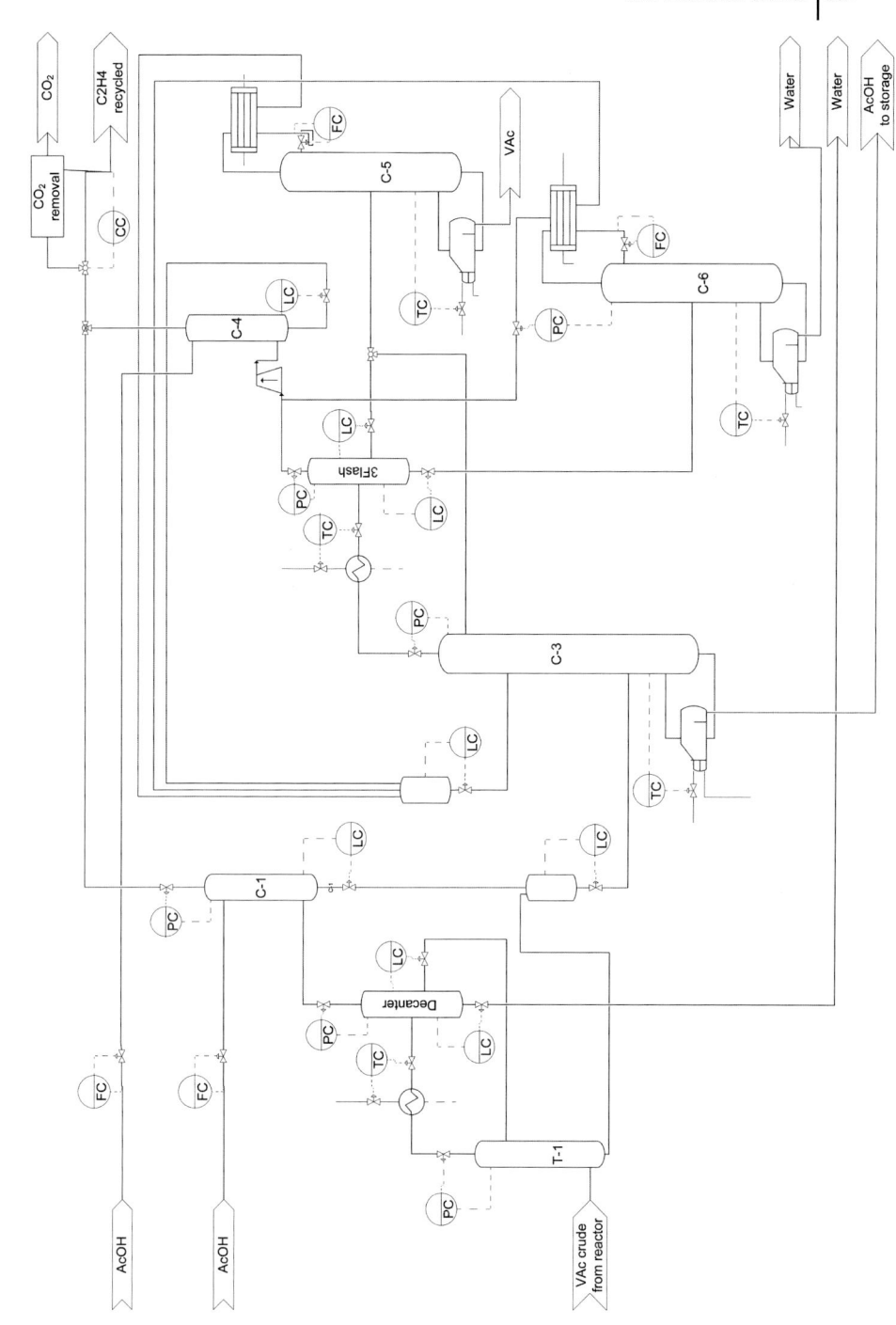

Figure 10.11 Plantwide control: control of the separation section.

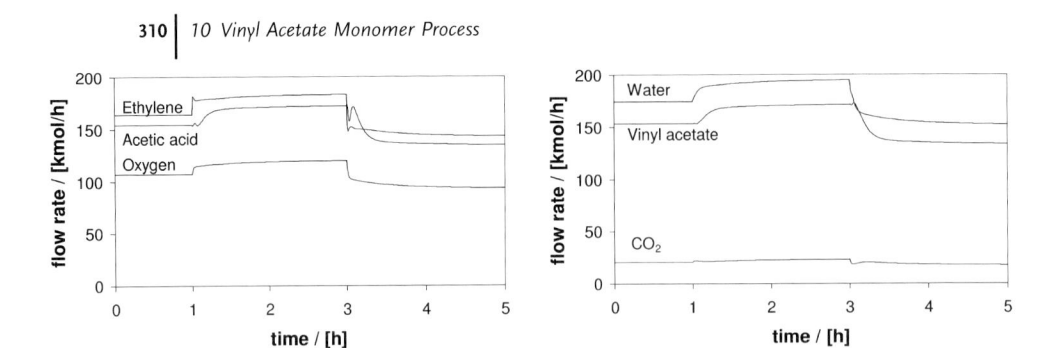

Figure 10.12 Dynamic simulation results.

increases from 455 K (at 0.8 m from reactor inlet) to 469 K (at 1.2 m from inlet). The higher temperature results in higher reaction rates, less reactants being recycled. The gas recycle is the fastest, and the ethylene feed is the first to be adjusted. Then, more oxygen is added by the concentration controller. The dynamics of the liquid recycle is slower and it takes about 0.5 hours until the acetic acid feed reaches the new stationary value. The vinyl acetate production-rate increases from 154 to 171 kmol/h. At time $t = 3$ h, the coolant temperature is reduced to 400 K, and the simulation is run for another 2 h. The maximum reactor temperature drops to 452 K (near the reactor inlet) and the production rate is decreased to 134 kmol/h. During the entire simulation, the oxygen concentration stays very close to the setpoint of 6%. Moreover, the concentration of the vinyl acetate product is above the 99.98% specification.

A vinyl acetate plant with a different structure of the separation section has been suggested as test case for plantwide control design procedures [5, 6]. Similarly to our results, Luyben and co-workers [7, 17] proposed to fix the reactor-inlet flow rate of acetic acid and to use the fresh feed to control the inventory in the bottom of the acetic-acid distillation column. The two control strategies are equivalent from a steady state point of view. However, Olsen et al. [18] showed that Luyben's structure has an unfavourable dynamics due to the large lag between the manipulated and controlled variables. The other important control loops in [7] paired the oxygen feed with oxygen concentration, and ethylene feed with pressure in the system. The production rate was also manipulated by the setpoint of reactor temperature controller. Chen and McAvoy [19] applied a methodology where several control structures, generated on heuristic grounds, were evaluated using a linear dynamic model and optimal control. Their results also indicate that fixing the reactor-inlet flows is the recommended strategy.

10.9
Conclusions

The case study of the synthesis of vinyl acetate emphasizes the benefits of a systematic design based on the analysis of the reactor/separation/recycles structure. The core of the process is the chemical reactor, whose behavior in recycle depends on the kinetics and selectivity performance of the catalyst, as well as the safety and technological constraints. Moreover, the recycle policy depends on the reaction mechanism of the catalytic reaction.

Thus, for palladium/Au catalysts the adsorption of the acetic acid on the active sites is fast and not rate limiting, while ethylene and oxygen are involved through a LHHW mechanism. In consequence, the designer should respect the composition of the reaction mixture at the reactor inlet that is compatible with the experimental conditions in which the kinetics of the catalytic process has been studied. From the reactor-design viewpoint this implies keeping the recycles of reactants, ethylene and acetic constant on a ratio of 3–4:1, while oxygen is made up to the maximum limit allowed by safety reasons. Because of selectivity reasons, low per-pass conversion has to be kept for both ethylene and acetic acid. The secondary combustion reaction can be limited by keeping both the pressure and temperature on the lower side. Two designs are compared, for slow and fast catalysts. Productivity higher than 1000 kg VAM/m^3 catalyst h can be achieved working at higher temperature and shorter residence time, as well as with good temperature control.

The separation section takes advantage of the heterogeneous azeotrope formed by vinyl acetate and water, with very low reciprocal solubility. However, because of lower water content the recovery of VAM requires a high reflux rate and a large amount of energy. Significant energy saving, up to 70%, can be obtained by making use of a dehydration gas pretreatment. In this way, the exothermic reaction can cover up to 90% from the energy requirements of the distillations.

The approach in steady-state reactor design finds a dynamic equivalent in the plantwide control strategy. Because low per-pass conversion of both ethylene and acetic acid, manipulating the reactant feed to the reactor has little power in adjusting the production rate. The reaction temperature profile becomes the main variable for manipulating the reaction rate and hence ensuring the flexibility in production. The inventory of reactants is adapted accordingly by fresh reactant makeup directly in recycles. This approach can be seen as generic for low per-pass reactions.

References

1 Roscher, G., Vinyl esters in Ullmans's Encyclopedia of Industrial Chemistry, Wiley-VCH, Weinheim, Germany, 2002

2 Weissermel, K., Arpe, H. J., Industrial Organic Chemistry, 4th edn, Wiley-VCH, Weinheim, Germany, 2003

3 Dimian, A. C., Integrated Simulation and Design of Chemical Processes, CACE series, No. 13, Elsevier, Amsterdam, The Netherlands, 2003

4 Rase, H. F., Fixed-based Reactor Design and Diagnosis, Butterworth, Boston, USA, 1990

5 Chen, R., Dave, K., McAvoy, T. J., A nonlinear dynamic model of a vinyl acetate process, *Ind. Eng. Chem. Res.*, 42, 4478–4487, 2003

6 Luyben, W., Tyreus, B., An industrial design/control study for vinyl acetate monomer plant, *Comput. Chem. Eng.*, 22, 867, 1998

7 Luyben, W., Tyreus, B., Luyben, M., Plantwide Process Control, McGraw-Hill, New York, USA, 1999

8 Han, Y. F., Wang J. H., Kumar, D., Yan, Z., Goodman D. W., A kinetic study of

vinyl acetate synthesis over Pd-based catalysts, *J. Catal., 232*, 467, 2005

9 Han, Y. F., Kumar, D., Sivadinarayana C., Goodman, D. W., Kinetics of ethylene combustion in the synthesis of vinyl acetate, *J. Catal., 224*, 60, 2004

10 Process for the production of vinyl acetate (BP International), US Patent 6.492.299 B1, 2002

11 Samanos, B., Boutry, P., The mechanism of vinyl acetate formation by ethylene acetoxidation, *J. Catal.*, 1971

12 Reilly, C. R., Lerou, J. J., Supported liquid phase catalysis in selective oxidation, *Catal. Today*, 433, 1998

13 Fiorentino, M., et al. (BP chemicals), Oxidation process in fluidized bed reactor, USP 0030729 A1, 2006

14 Renneke, R. et al., Development of a high performance catalyst for the production of vinyl acetate monomer, *Top. Catal., 38(4)*, 279–287, 2006

15 Hagen, J., Industrial Catalysis – A Practical Approach, 2nd edn, Wiley, New York, USA, 2005

16 Aspen Plus®, Aspen Dynamics™, Aspen Split™, Aspen Technology Inc., Cambridge/MA, USA, release 12

17 Luyben, M.L., Tyreus, B.D., Luyben, W.L., Plantwide control design procedure, *AIChE Journal, 43 (12)*, 3161–3174, 1997

18 Olsen, D., Svrcek, W., Young, B., Plantwide control study of a vinyl acetate monomere process design, *Chem. Eng. Comm, 192 (10)*, 1243–1257, 2005

19 Chen, R. and McAvoy, T., Plantwide control system design: methodology and application to a vinyl acetate process, *Ind. Eng. Chem. Res., 42*, 4753–4771, 2003

11
Acrylonitrile by Propene Ammoxidation

11.1
Problem Description

Acrylonitrile (AN) is one of the leading chemicals with a worldwide production of about 6 million tonnes in 2003. The most important applications are acrylic fibers, thermoplastics (SAN, ABS), technical rubbers, adiponitrile, as well as speciality polymers. Details about technology can be found in the standard textbooks [1, 2].

About 90% of the worldwide acrylonitrile (AN) is manufactured today by the ammoxidation of propene, as described by the reaction:

$$CH_2=CH\text{-}CH_3 + NH_3 + 3/2 O_2 \rightarrow CH_2=CH\text{-}CN + 3H_2O \qquad (11.1)$$

The reaction is highly exothermal ($\Delta H = -123$ kcal/mol) and takes place in gaseous phase over a suitable catalyst at temperatures of 300–500 °C and pressures of 1.5–3 bar in fluid-bed or fixed-bed reactors with efficient cooling. The catalyst employed makes the difference in technologies. The first commercial plant built by Sohio (now BP International) used a catalyst based on $Bi_2O_3.MoO_3$ [3]. Since then, numerous chemical formulations have been patented. The catalyst should be multifunctional and possess redox properties. The most commonly employed contain molybdenum or antimonium oxides mixed with transition metals, such as Fe, Ni, Co and V, activated by alkali- and rare-earth elements [4–9].

The reaction rate is high enough to achieve almost total per-pass conversion at ratios of reactants close to stoichiometry. High selectivity in acrylonitrile remains a challenge. Today the best catalysts can give a yield in acrylonitrile of 80–82%, mainly because of losses in propene by combustion. Significant amounts of highly toxic species form, such as HCN, acetonitrile (ACN) and heavy nitriles. Their removal from aqueous mixtures is difficult, as reflected in elevated water-treatment and energy costs.

Using much cheaper propane instead of propene was proposed in recent years. The difference in price should allow drop in manufacturing costs of about 15–20%. On the other hand, the problem of selectivity becomes even more challenging, because the catalyst should perform *in-situ* dehydrogenation simultaneously with

Chemical Process Design: Computer-Aided Case Studies. Alexandre C. Dimian and Costin Sorin Bildea
Copyright © 2008 WILEY-VCH Verlag GmbH & Co. KGaA, Weinheim
ISBN: 978-3-527-31403-4

Table 11.1 Specification of technical acrylonitrile [1].

Property	Value
Molecular weight	53.09
Acrylonitrile content	≥99.4%
Distillation range (101.3 kPa)	74.5–78.5 °C
Refractive index (25 °C)	1.3882–1.3891
Freezing point	−93 °C
Water	0.25–0.45%
HCN	5 ppm max
Acroleine	10 ppm max
Acid content (acetic acid)	20 ppm max.
Acetaldehyde	20 ppm max.
Acetonitrile	300 ppm max.
Inhibitor content	35–45 ppm
Liquid density at 20 °C	0.8060
Liquid viscosity at 25 °C	0.34 cP
Solubility in/of water at 20 °C	7.3%/3.08%
Solubility in organic solvents	complete

ammoxidation. Numerous catalysts have been patented, but their performance is still insufficient to justify the massive replacement of current technologies. A comparison of two processes will be discussed at the end of this chapter.

In this project the target plant capacity is 120 000 ton/yr, corresponding to a reactor production of 272.34 kmol/h or 14 461.25 kg/h acrylonitrile polymer-grade purity. Accordingly, the propene feed is 340 kmol/h. Table 11.1 lists the quality specifications. The most stringent regards the levels of HCN, acroleine and acetaldehyde. Although the acetonitrile specification seems much looser, this is hard to meet because of difficult separation. The gaseous emissions must be free of highly toxic nitriles and bad-odor acroleine. For water and soil protection, highly toxic heavy impurities must be destroyed or disposed of by deep landfill. The treatment of large amounts of polluted water is necessary before recycling or dumping. In summary, the safety, health and environmental protection put severe constraints on the design of an acrylonitrile process.

11.2
Reactions and Thermodynamics

11.2.1
Chemistry Issues

The ammoxidation of propene to acrylonitrile described by the global equation (11.1) actually involves a very complex reaction mechanism. More generally, the reaction of ammoxidation refers to the interaction of ammonia with a hydrocarbon

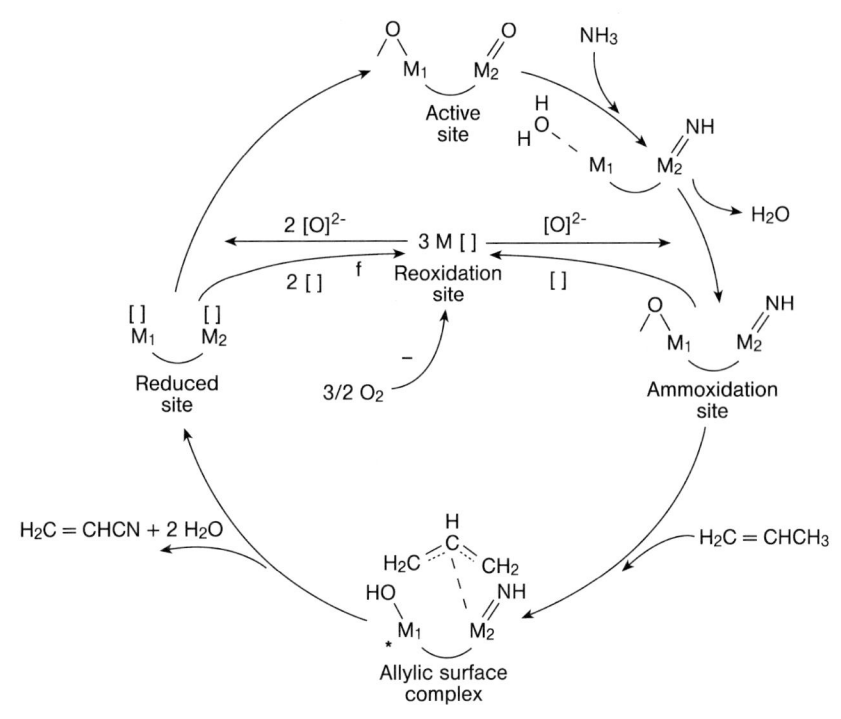

Figure 11.1 Mechanistic cycle for alkene ammoxidation [5, 6].

partner (alkene, alkane or aromatic) in the presence of oxygen and suitable catalyst. An ammoxidation catalyst must fulfil two conditions: possess redox properties and be multifunctional. Figure 11.1 shows the major steps in a catalytic cycle [5, 7]. Firstly, ammonia interacts with the bifunctional active centers, here M_1 and M_2, generating an extended "ammoxidation site". The first active species forms from ammonia as =NH, then on this site the alkene inserts as an "allylic complex" by α-hydrogen abstraction. After the rearrangement of atoms the surface complex is transformed in the product $H_2C=CH\text{-}CN$, which further desorbs from the surface. The result of this process is a reduced surface site whose regeneration takes place by the oxygen (O^{2-}) coming exclusively from the catalyst lattice. Subsequently, the lattice has to be filled-in with oxygen coming from the gas phase. Thus, the overall reaction takes place via a common solid-state lattice capable of exchanging electrons, anion vacancies and oxygen transmission. The above mechanism is consistent with the concept of "site isolation" proposed by Grasselli and Callahan [3, 5, 6] the inventors of the SOHIO catalyst, which states that an (amm)oxidation catalyst becomes selective when the reacting oxygen species at the active centers are spatially isolated from each other.

The knowledge of the reaction mechanism is important for process design. Firstly, only olefins with activated methyl groups may undergo ammoxidation reactions to nitriles. Otherwise, oxidative dehydrogenation takes place preferentially.

For example, from the isomers = C_4 only isobutene can give methacrylic nitrile. Toluene and xylenes can be converted to the corresponding nitriles too. Secondly, the role of ammonia as chemisorbed species = NH is primordial in reaction, because they start the catalytic cycle before propene. Therefore, sufficient ammonia has to be present in the reaction mixture, slightly above the stoichiometric amount. Otherwise, the sites are occupied by oxygen and the combustion prevails. The oxygen should be fed so as to replace only the amount consumed in the lattice, in slight excess above the stoichiometry. As a result, the reaction mechanism suggests that propane and ammonia should be mixed and fed together, while the oxygen should enter the reaction space independently in order to fill the lattice. This principle is applied in the reactor technology.

Among secondary reactions the most important losses are by oxidations, namely by propene combustion:

$$CH_2=CH\text{-}CH_3 + 3/2O_2 \rightarrow 3CO_2 + 3H_2O \tag{11.2}$$

$$CH_2=CH\text{-}CH_3 + 3O_2 \rightarrow 3CO + 3H_2O \tag{11.3}$$

As a consequence, the overall exothermic effect rises to about 160 kcal/mol propene.

In the absence of ammonia the active sites are oxidic leading to acroleine:

$$CH_2=CH\text{-}CH_3 + 1/2O_2 \rightarrow CH_2=CH\text{-}CHO + H_2O \tag{11.4}$$

Partially, the oxidation may progress to alylic alcohol.

Other byproducts of significance are HCN and acetonitrile, whose formation may be expressed by the overall reactions:

$$CH_2=CH\text{-}CH_3 + 3NH_3 + 3O_2 \rightarrow 3HCN + 6H_2O \tag{11.5}$$

$$2CH_2=CH\text{-}CH_3 + 3NH_3 + 3/2O_2 \rightarrow 3CH_3\text{-}CN + 3H_2O \tag{11.6}$$

The stoichiometry indicates a complex reaction mechanism. The amount of HCN is generally larger than that of acetonitrile, the ratio depending on the catalyst formulation and reaction conditions. Both reactions are favored by higher temperature and pressure, as well as by longer residence time.

It is interesting to note that supplementary reactions leading to impurities may take place outside the reaction space, mostly in the aqueous phase during the first separation steps of quench and absorption in water. Typical examples are the formation of propion-cyanhydrine and dinitrile-succinate favored by a basic pH:

$$CH_2=CH\text{-}CHO + HCN \rightarrow NC\text{-}CH_2\text{-}CH_2\text{-}CHO \tag{11.7}$$

$$CH_2=CH\text{-}CN + HCN \rightarrow NC\text{-}CH_2\text{-}CH_2\text{-}CN \tag{11.8}$$

The reaction (11.7) may be exploited to convert acroleine, which is difficult to remove, into heavier species. Reaction (11.8) may take place during the distillation of acrylonitrile. More generally, the separation/purification of acrylonitrile is complicated by secondary chemical reactions in which the pH of liquid phase plays an important role. These aspects will be examined later.

In addition, undesired species may originate from reactions with impurities present in the fresh feed, such as ethylene giving acetaldehyde and acetic acid, or butenes leading to heavies. For this reason the concentration of non-C_3 alkene in the fresh propylene feed has to be limited to a maximum of 0.5%.

11.2.2
Physical Properties

Table 11.2 presents fundamental physical properties for the key components implied in separations. The difference in the boiling points favors the separation, except of acrylonitrile and acetonitrile. The differences in the freezing point are also sensitive, but they did not justify the investment in a separation by crystallization. It remains that distillation-based separation methods should be tried in the first place. However, the formation of azeotropes of components with water will present difficulties.

An important property for process design is the limited reciprocal solubility of acrylonitrile in water. Table 11.3 shows the dependence against temperature. The solubility of AN in water is around 7% w/w, while water in AN about 3% at 20 °C. Therefore, liquid–liquid separation by decantation can be combined advantageously with azeotropic distillation for acrylonitrile purification.

Table 11.2 Caracteristic physical properties of the key components.

Component	Unit	AN	HCN	ACN	Acroleine	Succinic nitrile
Molecular weight	g/mol	53.1	27.0	41.1	56.1	80.1
Nbp	K	350.5	298.9	354.8	325.8	540.2
Freezing point	K	189.6	259.9	229.3	185.5	331.3
Tc	K	535.0	456.7	545.5	506.0	770.0
Pc	atm	44.2	53.2	47.7	49.3	34.9
Vc	cm^3/mol	212.0	139.0	173.0	197.0	300.0
Zc	–	0.214	0.197	0.184	0.234	0.166
Liquid volume	cm^3/mol	66.2	53.6	52.9	67.2	81.0
Enth. vaporization	cal/mol	7418.7	6431.2	7212.3	6833.9	13 030.7

Table 11.3 Solubility of acrylonitrile in water g/100 g solution.

Temperature, °C	0	20	40	60	80
AN/water	7.15	7.30	7.90	9.10	11.1
Water/AN	2.10	3.08	4.85	7.65	10.95

11.2.3
VLE of Key Mixtures

The VLE of acrylonitrile/acetonitrile indicates a quasi-ideal system, but very low relative volatility, below 1.15, which makes necessary a large number of stages and high reflux. Actually, both components are involved in the process as mixtures with water. Figure 11.2 presents T–x–y phase diagrams for the binaries acrylonitrile/water and acetonitrile/water calculated by Aspen Plus [23] with Uniquac/Redlich–Kwong as the thermodynamic option. Note that acrylonitrile/water forms a heterogeneous azeotrope, while the azeotrope acetonitrile/water is homogeneous. Both azeotropes show close compositions (0.68 vs. 0.65 mole fraction nitrile) and narrow boiling points (345 vs. 350 K). The prediction of VLE is accurate. Although the solubility of water in acrylonitrile at higher temperature is somewhat underestimated, this will not alter the conceptual design.

The conclusion is that the separation by simple distillation of the ternary mixture acrylonitrile/acetonitrile/water is not possible because of azeotropes. Changing the VLE behavior by a mass-separation agent is desirable. The literature search

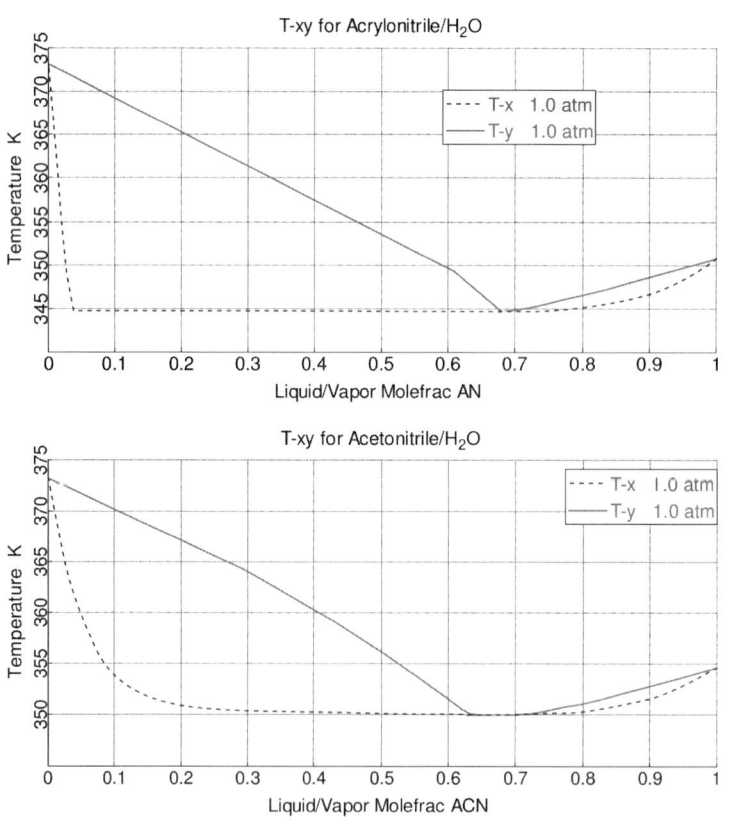

Figure 11.2 VLE of binaries acrylonitrile/water and acetonitrile/water.

indicates that water itself may play this role, which is quite surprising! Some flash calculations help to understand this phenomenon. Consider a mixture AN/ACN in ratio 10:1 and variable water amount. At lower water/organic ratio there are two-liquid phases, while the composition of the vapor phase is fixed by the binary azeotrope. At ratios larger than 10 the heterogeneous azeotrope acrylonitrile/water disappears since there is only one liquid phase. The K values for AN and ACN are about 20 and 4, respectively, giving a relative volatility of 5, while the K value for water is below 1. The increase in the relative volatility AN/ACN is sensible up to water/mixture ratios of about 20. As result, the separation of the acrylonitrile/acetonitrile mixture by extractive distillation with water may be feasible in a single column but with a large amount of entrainer.

11.3
Chemical-Reactor Analysis

Figure 11.3 presents the sketch of a fluid-bed reactor for ammonia oxidation of propylene. The reactor is a large-diameter cylindrical vessel provided with a gas-distribution grid for supporting the fluid bed, as well as with injection devices for feeding the gaseous reactants. The optimal catalyst particles size is in the range 40 to 100 μm, in which the presence of a certain amount of fines is necessary for ensuring homogeneous fluidization. The gas velocity is slightly above the minimum, in general between 0.4 to 0.5 m/s. Trays or screens, usually between 5 and 15, can be placed transversally in order to reduce the negative effect of backmixing. This modification gives much better performance in term of acrylonitrile yield. Because of the highly exothermal reaction cooling coils are immersed in the fluid bed. Since the temperature of reaction is around 420–450 °C high-pressure steam of 30 to 40 bar can be raised.

The feeding strategy of reactants should take into account the reaction mechanism. Usually, the oxygen (air) is introduced below the bottom grid, with the mixed propylene and ammonia through "spiders" positioned above the grid. The catalyst plays an important role in preserving the safety as scavenger for oxygen radicals. No explosion was ever encountered over decades of operation [7].

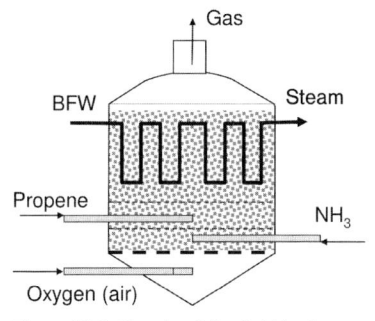

Figure 11.3 Sketch of the fluid-bed reactor for acrylonitrile synthesis.

The operating pressure should be as low as possible to prevent the formation of byproducts. On the other side higher pressure would be preferable for quenching and scrubbing of gases. Overpressures of 0.5 to 2 bar are preferable. As indication, Table 11.4 shows the test of catalyst activity as a function of pressure [9]. Almost complete conversion of propylene may be seen and selectivity around 80% in acrylonitrile can be obtained. The data are representative for modern catalysts.

The residence time in the reactor is between 2 and 20 s, with an optimal range from 5 to 10 s. Longer residence time gives more byproducts. A more sophisticated design of the fluid-bed reactor requires advanced modeling and simulation capabilities and is beyond the scope of this case study [10, 11].

For the assessment of the separation system a simple but realistic stoichiometric analysis is sufficient. Representative reactions are listed in Table 11.5 with stoichiometric coefficients from Table 11.4 at 2 bar. The resulting gas composition is given in Table 11.6 for a mixture of propene/ammonia/air of 1/1.2/9.5.

Table 11.4 The efect of pressure on the catalyts activity for ammoxidation of propene.

P (MPa)	Conversion propene	AN	ACT	HCN	Acrolein + acrylic acid	CO_2 + CO
0.18	97.8	79.6	2.1	2.3	4.1	9.6
0.2	98.3	80.1	2.1	2.7	2.7	10.7
0.25	97.5	78.2	3.2	2.5	2.1	11.2

Table 11.5 Chemical reactions at ammoxidation of propene in a fluid-bed reactor.

	Reactions	Conversion
1	$CH_2=CH\text{-}CH_3 + NH_3 + 3/2O_2 \rightarrow CH_2=CH\text{-}CN$ (AN) $+ 3H_2O$	0.801
2	$2CH_2=CH\text{-}CH_3 + 3NH_3 + 3/2O_2 \rightarrow 3CH_3\text{-}CN$ (ACT) $+ 3H_2O$	0.021
3	$CH_2=CH\text{-}CH_3 + 3NH_3 + 3O_2 \rightarrow 3HCN + 6H_2O$	0.027
4	$CH_2=CH\text{-}CH_3 + 3/2O_2 \rightarrow 3CO_2 + 3H_2O$	0.107
5	$CH_2=CH\text{-}CH_3 + 1/2 O_2 \rightarrow CH_2=CH\text{-}CHO$ (ACR) $+ H_2O$	0.027
6	$CH_2=CH\text{-}CN + HCN \rightarrow NC\text{-}CH_2\text{-}CH_2\text{-}CN$ (SCN)	0.005

Table 11.6 Material balance around the chemical reactor for 1 kmol/h propylene.

	= C3	NH_3	O_2	N_2	AN	HCN	ACN	ACR	CO_2	H_2O
I	1	1.2	1.9	7.6	–	–	–	–	–	–
O	0.017	0.286	0.077	7.6	0.796	0.076	0.031	0.027	0.32	3.07

11.4
The First Separation Step

Before separation, the reactor off-gas must be quenched quickly to prevent thermal degradation and secondary reactions. At the same time, this operation removes the excess of ammonia. Two methods can be employed:

1. one step quench and acidic treatment (acidic quench),
2. two-step separate quench and ammonia removal (basic quench).

In the "acidic quench", the gas is brought into contact with a solution of sulfuric acid of 30–40%. Recycled water is added to compensate the losses by evaporation. Precooling down to the dew point could be employed for maximum energy recovery. The entrained catalyst fines are recovered by filtration. Ammonium sulfate is separated, purified by crystallization and finally obtained as a saleable byproduct. Note that the acidic pH helps keep the formation of heavy impurities and polymeric residues low and maximizes the yield in acrylonitrile.

By a "basic quench" the removal of catalyst fines and ammonia take place separately. Due to the neutral pH in the first stage the loss in acrylonitrile is higher than previously. On the contrary, the production of ammonium sulfate is increased.

After quench, the cleaned gas is submitted to acrylonitrile absorption in cold water. The vent gas containing mainly carbon oxides, nitrogen and unreacted propylene, as well as light organics, is sent to catalytic oxidation. The liquid mixture is sent to acrylonitrile recovery in a stripping column. The top product is raw acrylonitrile, while the water separated in bottoms is recycled to quench and absorption, eventually with some pretreatment. Figure 11.4 displays the flowsheet, including heat exchangers for performing the energy saving-analysis.

Next, we present results obtained by using Aspen Plus [23]. Using the Uniquac-RK model with Henry components for supercritical gases ensures correct description of the absorption-desorption process. Table 11.7 shows the composition of streams around the reactor and the first separation step.

The gas is cooled in HX1 from 420 °C to the 220 °C. This operation delivers a heat duty Q_{HX1} = 7.8 MW that can be used for raising steam of 12 bar. Furthermore, the gas is treated in a quench tower with H_2SO_4 solution, where both cooling and neutralization of the excess ammonia take place. This operation is simulated by the black-box unit QUENCH, in which the water is regulated to a flow rate of 3600 kg/h by a design specification so as to get the gas outlet at the dew point. Then, the cleaned gas is cooled to 30 °C before absorption. To ensure high acrylonitrile recovery of 99.5%, cold water at 5 °C is used. The column has 10 theoretical stages. A three-phase option is used to account for the possible occurrence of two-liquid phases with a convergence algorithm for "strong nonideal liquid". A design specification is employed to adjust the water flow rate to achieve 99.5% AN recovery. The result of simulation is a water flow rate of 180.5 t/h or 12.6 kg water/kg AN. This value, corresponding to a solubility of 7.37 g AN/100 g water at 25 °C, is in excellent agreement with the equilibrium

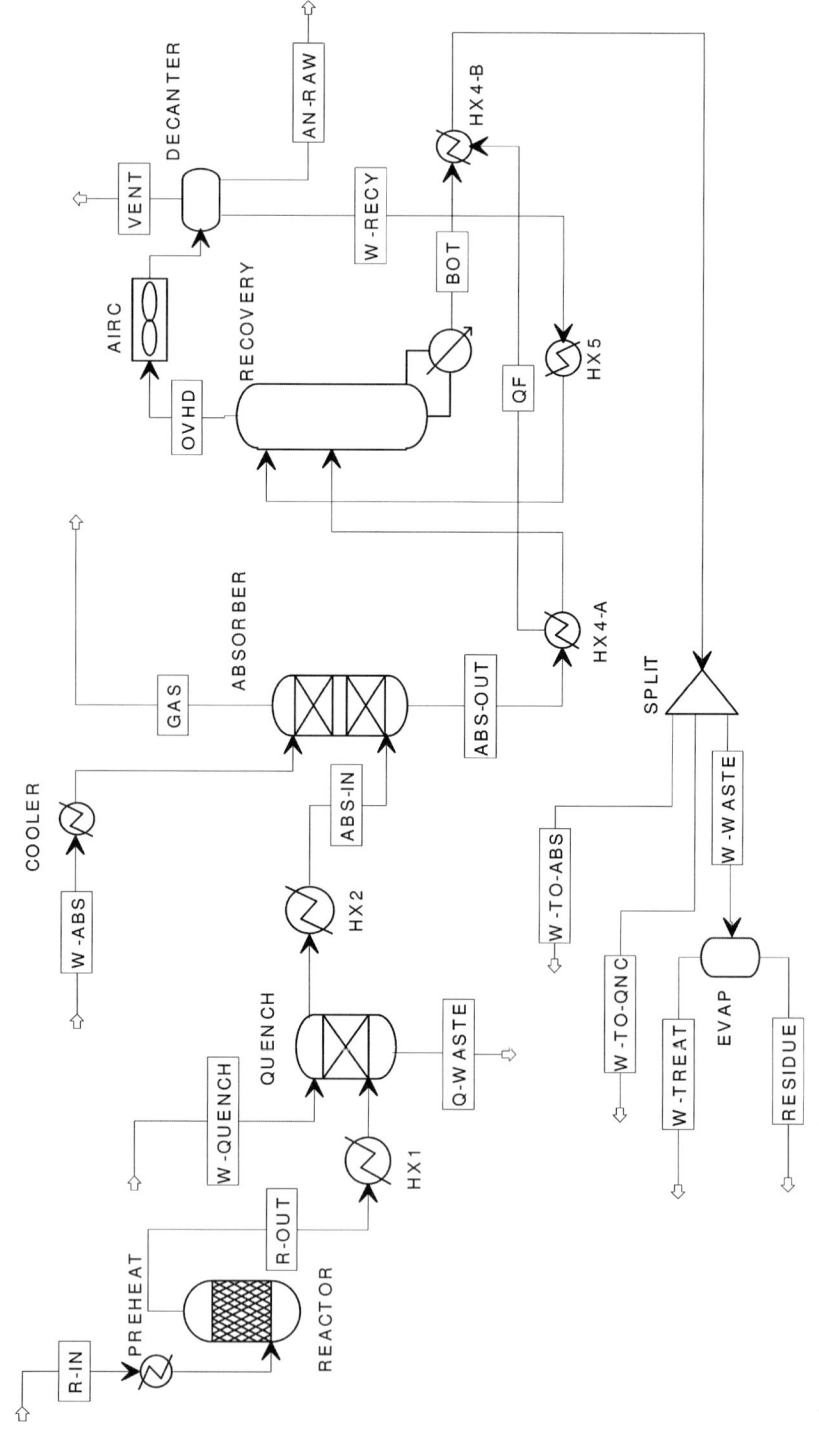

Figure 11.4 Reactor section and first separation step by acrylonitrile manufacturing.

Table 11.7 Stream table around chemical reactor and first separation step.

	Reactor inlet	Reactor outlet	Absorber inlet	Absorber outlet	AN brut
Mole flow kmol/h					
C_3H_6	340	5.78	5.78	0.0046	1.1E–03
O_2	646	26.35	26.35	0.0049	1.8E–04
N_2	2584	2584	2584	0.2523	5.5E–03
AMMONIA	408	97.41	0	0	0
CO_2	0	109.14	109.14	0.54	0.277
HCN	0	27.54	27.54	27.54	26.93
AN	0	272.34	272.34	271.0	271.0
ACROLEIN	0	9.18	9.18	9.04	9.03
ACN	0	10.71	10.71	10.71	10.44904
H_2O	30	1041.84	993.47	10998.8	49.4
Total flow kmol/h	4008	4184.29	4038.51	14870.56	367.07
Total flow kg/h	114854.4	114854.4	112324.1	278250.6	1694.5
Temperature K	623.15	693.15	303.15	301.15	303.15
Pressure atm	2.2	2	1.7	1.6	1.5

value of 7.3 g at 20 °C from Table 11.2. In this way, the thermodynamic model is validated.

The residual gas leaving, containing unconverted propene, CO_2 and other VOC, is usually sent directly to flare after catalytic combustion

Next, the recovery of useful components from the aqueous solution is targeted. The working assumption is high recovery, over 99.9% for acrylonitrile, and over 95% for both HCN and acetonitrile as valuable byproducts. An appropriate separation technique is reboiled stripping. After top vapor condensation, phase separation takes place by a L–L split, the water phase being refluxed in the stripping column, while the organic phase recovers the acrylonitrile. Simulation shows that 10 theoretical stages are sufficient. The separation performance can be followed in terms of component split at variable distillate rate or reboiler duty. One observes that high acrylonitrile recovery can be obtained rather easily in the top product, while HCN and acetonitrile have the tendency to pass in the bottom. If the last two species are waste, the stripping should be conducted at low boilup and lower feed temperature. If these are to be recovered in top then the reboiler duty and feed temperature should increase. The option considered at this stage is full recovery of AN, HCN and ACN in a single stream for downstream processing. Other alternatives will be considered later.

The performance of the first separation step is illustrated by the Table 11.7. The raw acrylonitrile stream contains approximately 85% acrylonitrile and 5% water, the rest being organic impurities, namely HCN, acroleine and acetonitrile. The bottom stream consists of water with nitrile impurities and heavies. This stream is further split to 91% for recycling to absorption, 2% for recycling to quench and the rest as wastewater. The highly toxic material can be removed by using multieffect evaporation. The vapor stream containing light nitriles is condensed and sent to a biological treatment station. The liquid concentrates heavies and oligomers that cannot be economically separated. These are sent to post-treatment, which usually takes place by "water burning" and deep landfill. Note that bleed streams (nonrepresented) are taken off periodically from recycles to prevent the accumulation of impurities.

11.5
Liquid-Separation System

11.5.1
Development of the Separation Sequence

In this section we will handle the treatment of the raw acrylonitrile stream recovered by absorption-stripping as described previously. Figure 11.5 presents the flowsheet. Following the heuristics in Chapter 3, the first split C-1 removes the hazardous and toxic species, such as HCN and acroleine. The second step is the removal of acetonitrile in the unit C-2, purified further in C-3. This task is particularly difficult, but feasible, by extractive distillation with water, as discussed before. Finally, the acrylonitrile is purified, firstly by dewatering in the column C-4 and then by removal of heavier impurities in C-5. This scheme is in agreement with the industrial practice [12].

Problems in operating the distillation columns may arise because of side reactions occurring between various unsaturated species, in which the pH plays a significant role [13, 14]. Some patents indicate that maintaining neutral pH in the recovery column allows carbonyl species to react with HCN, forming heavy components soluble in water, for example converting the acroleine in cyano-acroleine. The presence of nonsaturated components raises the risk of polymer formation. For this reason, an inhibitor is added during the operations involving the distillation of acrylonitrile, starting with the recovery column. Similarly, oligomers appear from HCN in aqueous solutions. Prevention measures are lower distillation temperature (−10 to −20 °C) under vacuum, using large reflux and supplementary stages in the stripping zone. Another solution is keeping an alkaline pH by adding suitable compounds.

After recovering the acetonitrile the problem is breaking its azeotrope with water. VLE investigation shows that this is sensitive to the pressure change. For example, at 0.4 bar the azeotropic point is $x(ACN) = 0.80$ and $T = 326$ K, while at 6 bar this shifts to $x(ACN) = 0.57$ and $T = 412$ K. Consequently, pressure-swing distillation may be applied as indicated in a recent patent [15].

11.5.2
Simulation

The simulation of the above separation scheme is performed with Aspen Plus software [23]. The starting mixture is the acrylonitrile stream given in the last

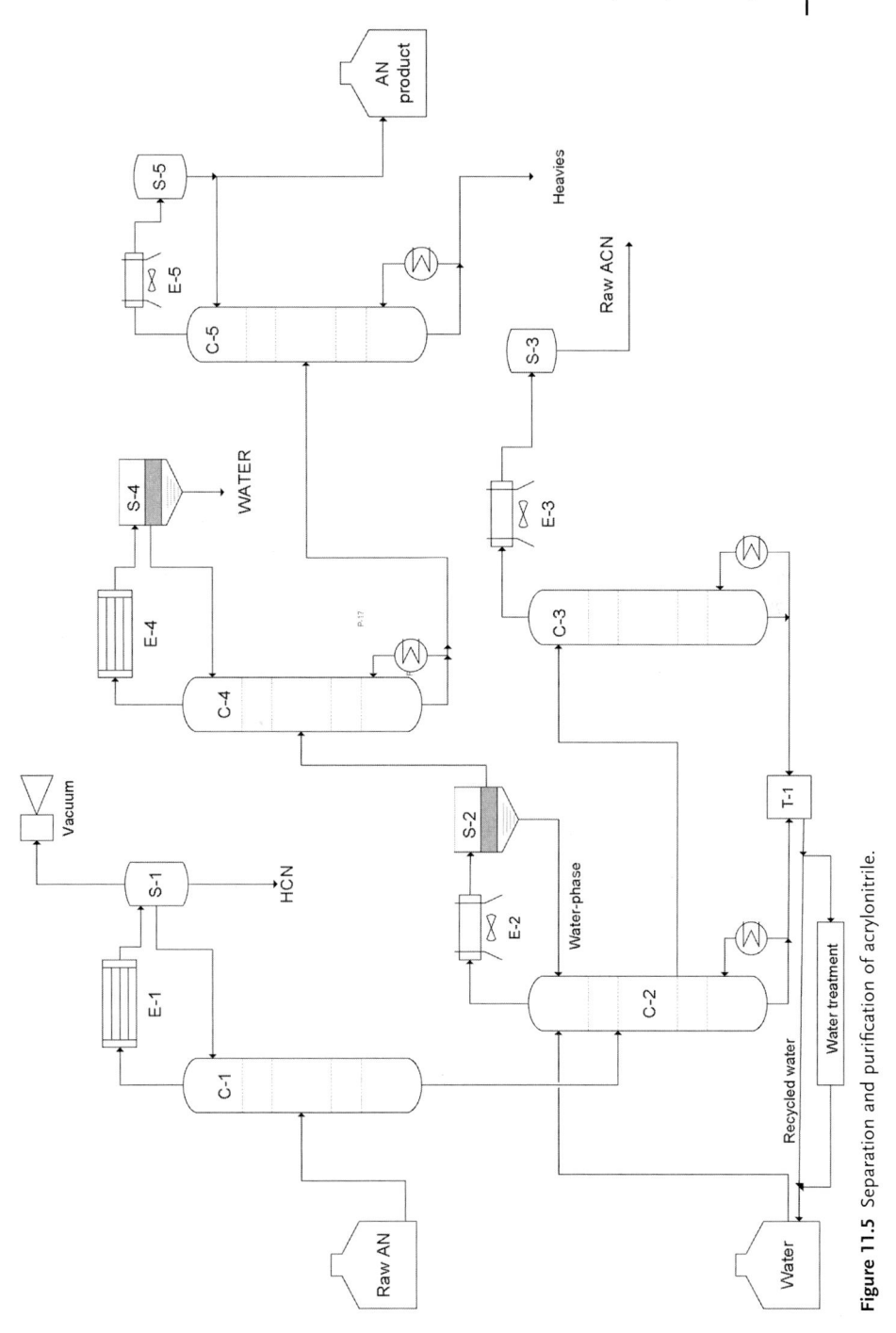

Figure 11.5 Separation and purification of acrylonitrile.

column of Table 11.7, disregarding the dissolved gases. The thermodynamic model is UNIQUAC-RK. Note that all interaction parameters are retrieved for VLE data, except the binary acrylonitrile-water from Aspen LLE. Figure 11.6 presents the process-simulation diagram, which reflects the separation scheme developed before. Table 11.8 presents some sizing elements of the columns. A reasonable number of stages were assumed and the energy consumptions were adapted to meet the target performance in terms of component recovery and product purity. In addition, the diameter of trays and packing columns are compared.

Column C-1 has the target of removing the "heads" including light impurities as HCN and acroleine. HCN separates easily in top, although over 99% recovery demands tall columns and a high reflux ratio. The removal of acroleine is much more difficult. A solution of this problem is chemical conversion to a heavier species by reaction with HCN, as given by reaction (11.7). The reaction may take place in a pretreatment vessel, or directly on the stages of the distillation column. Since a sharp HCN split cannot be done without some acrylonitrile loss, the separation is done in two columns, HCN removal C-1A and acrylonitrile recovery C-1B, respectively.

The next column C-2 handles the separation of acrylonitrile/acetonitrile binary by extractive distillation. A large amount of water is necessary to modify the volatility of components. The simulation indicates a ratio solvent/mixture of 10:1, which corresponds roughly to the complete dissolution of acrylonitrile in water. The column has 40 theoretical stages, being simulated as reboiled stripping. Water is introduced on the top stage, the organic feed in the middle. Purified acrylonitrile leaves in top, while acetonitrile is drawn off as a liquid side stream.

Table 11.8 Unit design and performance of the acrylonitrile separation section.

Item/Unit	C-1A	C-1B	C-2	C-3	C-4
Task	Distillation lights	Purification HCN	Separation AN/ACN	Recovery ACN	Distillation AN
Pressure, bar	1.1	1.013	1.22	0.5	1.1
Temp. top	38.9	25.6	87.1	56.5	70.6
Temp. bottom	82.7	75.2	111.3	128.3	81.9
Stages (feeds)	40 (10)	30 (10)	40(1)	30 (5)	30 (15)
D (trays)	1.26 (Sieve)	0.62 (Sieve)	2.5 (Sieve)	0.95 (Sieve)	1.5 (Valve)
D (packing)	1.5 (Sieve)	0.65 (IMTP)	2.5 (IMTP) 3.2 (Pall)	1.22 (Pall)	1.7 (IMTP) 2.2 (Pall)
Q cond MW	−2.60	−0.64	(−5.3)	−1.7	(−2.74)
Q reb MW	+2.54	+0.64	+13.0	+2.1	+3.02
Recovery	99.3% HCN	99.9% HCN	99.9% AN 97.6% ACN	100% ACN	99.6% AN
Remarks	Conversion of acroleine	100% purity HCN product	ACN in side stream 25	Azeotrope ACN/w in top	99.9% purity AN product

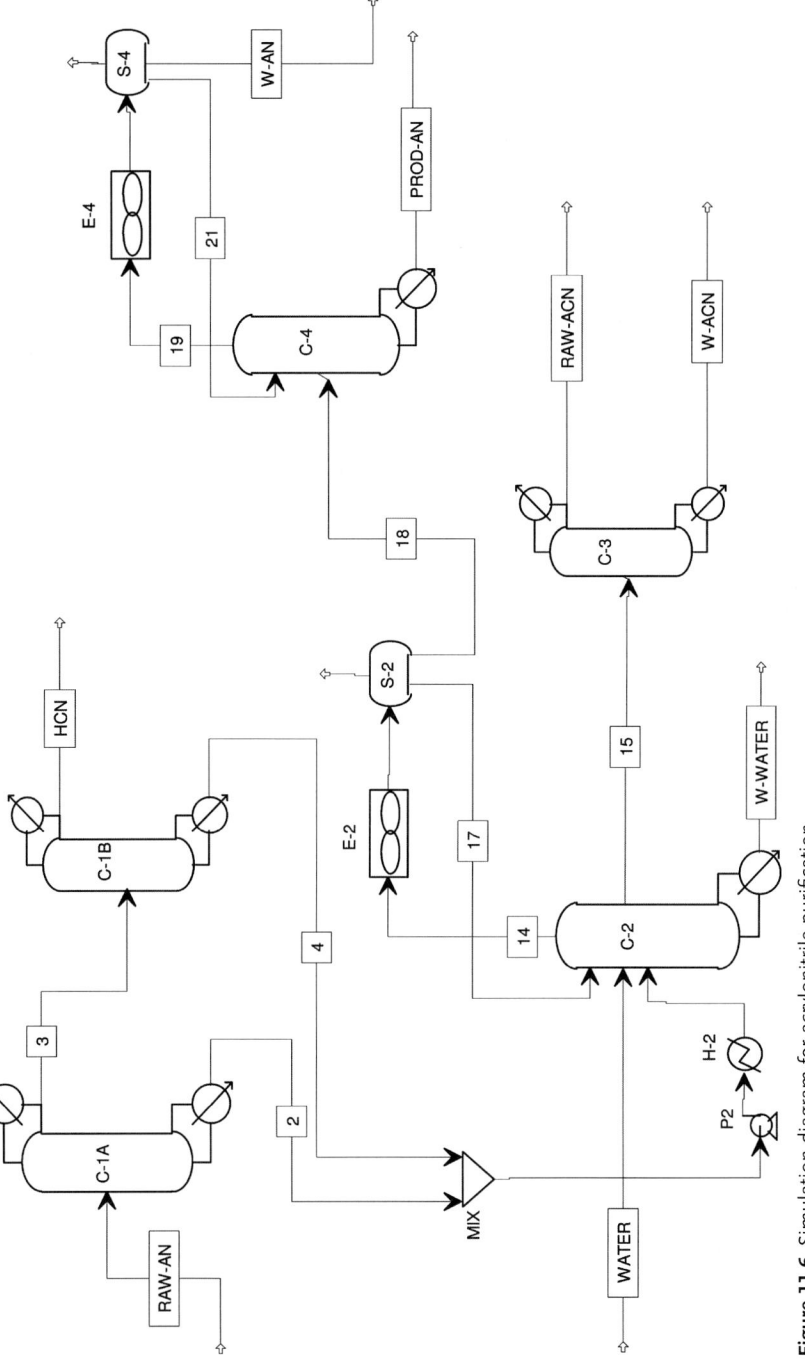

Figure 11.6 Simulation diagram for acrylonitrile purification.

After condensation, the top vapor is separated in a decanter, the water phase being refluxed to extractive distillation, while the organic phase goes to the next purification. The simulation indicates a high sensitivity of acetonitrile recovery with the temperature of the water feed, which should be about 10 °C below the top temperature.

Because a large amount of water is entrained in the side stream, this is removed in the column C-3. Raw acetonitrile, namely a binary azeotrope with 20% water, separates in top. The bottom stream contains water with heavy impurities. Vacuum distillation at 0.5 bar is adequate to limit the bottom temperature. In the next step pure acetonitrile can be obtained by using pressure-swing distillation.

The water-free acrylonitrile is obtained as bottoms in the column C-4. Water leaves in top as a binary azeotrope, followed by decantation and reflux of organic phase. The water phase also removes some light impurities. The final product meets closely the specifications indicated in Table 11.1. Since heavy impurities inevitably appear, a final vacuum distillation of acrylonitrile is performed in practice before shipping.

An interesting aspect is the relation between the design of units and the quality specifications. The path of each impurity can be traced by paying attention to generation, exit points and accumulation in recycles. In this respect the "component split" matrix available in Aspen Plus [23] gives very useful information and is highly recommended.

Let us consider the HCN. High recovery in the separator C-1A is desirable, since any amount left in bottoms should be found in the final acrylonitrile. However, by examining the behavior of C-2 it is clear that most of the HCN leaves in the side stream with acetonitrile. This behavior is counterintuitive.

The simulation shows that acroleine is the most difficult to isolate. If not removed in C-1A it will be found in the top of C-2 and further in the end product. The column C-1A is designed with a ratio stripping/rectification 3:1 to ensure over 99.9% HCN recovery. Despite a K value of 1.7 the acroleine concentrates in the middle of the stripping zone, from which a quantitative removal by a large side stream or secondary recovery column is not efficient. The best solution is chemical conversion in heavies.

The separation of acrylonitrile involves a large amount of water. This is obtained as bottom stream of the distillation columns, with heavy impurities and traces of HCN, ACN and AN. Before reuse, the wastewater is cleaned by multieffect evaporation. The concentrated residual in organics is burned. The water amount produced by reaction is sent to advanced purification in a biological unit.

11.6
Heat Integration

The very high exothermal reaction of propylene ammoxidation develops a large amount of heat, in this case 61.5 MW or 15.35 GJ/t. Up to 30% can be exported on site, depending on the technology [21]. For assessing the opportunities for

Table 11.9 Sources and sinks for heat integration.

	Source			Sink			
Unit	Duty MW	T1 °C	T2 °C	Unit	Duty MW	T1, °C	T2, °C
Reactor	61.5 (HP steam)	420	419				
Reactor effluent	11.8 (7.7 MP steam)	420	120	Feed preheating	7.1	25	210
Recovery condenser	22.1	104.8	70.0	Recovery reboiler	26.3	111.7	117.8
C1-A condenser	2.69	38.9	38.8	C-1A reboiler	2.54	82.7	82.8
C-2 condenser	5.3	87.1	87.0	C-2 reboiler	13	111.3	111.4
C-3 condenser	1.7	56.5	56.4	C-3 reboiler	2.1	128.3	128.4
C-4 condenser	2.74	70.6	70.5	C-4 reboiler	3.02	81.9	82.0

energy saving Table 11.9 presents the main sources and sinks. The chemical reactor (inside reaction plus effluent) can produce 70 MW, equivalent to 141 t/h steam at 30 bar. The reactor off-gas has an important energy potential for producing supplementary medium-pressure steam, while the lower-temperature segment may serve for feed preheating. As consumers the most important are acrylonitrile recovery and extractive distillation cumulating 26.3 + 13 = 39.3 MW. The purification columns have moderate duties and relatively low bottom temperatures. The condensers of the distillation columns may be counted as sources too, especially when their temperature is slightly below or above 100 °C.

Inspecting the sources and sinks suggests possibilities for thermal coupling. The temperature of the condenser E-2 is above the reboilers of columns C-1A and C-4. Increasing only slightly the pressure in C-2 at 2 bar can ensure at least ΔT_{min} = 10 °C covering the duties of the columns C-1A and C-4 while saving 5.3 MW LP steam. Another combination could be the condenser of the product column C-4 and the reboiler of the HCN column C-1.

However, more drastic measures for energy saving should regard the main consumers, the AN recovery and extractive-distillation columns. An interesting attempt would be to integrate them tightly, eventually as a single column. Figure 11.7 presents a heat-integrated flowsheet. Column RECOVERY receives the crude acrylonitrile from absorption, delivering acrylonitrile-rich vapor (OVHD) in top and acetonitrile/water mixture in bottoms. The top vapor is submitted to condensation and separation in two liquid phases. The water phase is returned as reflux, while the organic phase stream (RAW-AN) recovers quantitatively the acrylonitrile

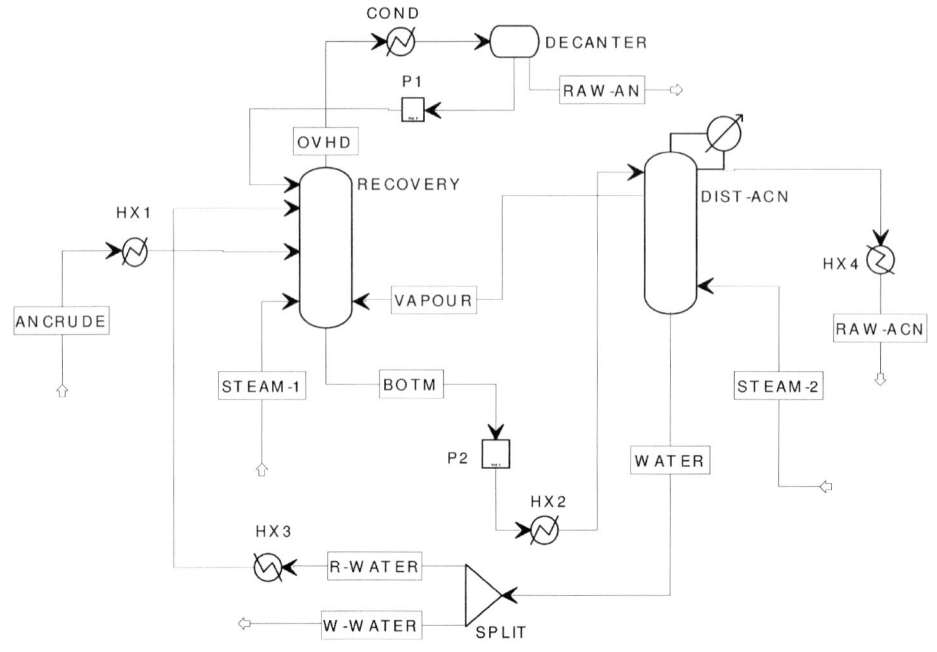

Figure 11.7 Integrating acrylonitrile recovery with extractive distillation.

with some water and light impurities. Hence, this column performs both recovery and extractive distillation tasks. In a second stage the acetonitrile solution is sent to a stripping column named DIST-ACN, where the acetonitrile concentrates in top with some water and light impurities. The majority of water leaves as bottoms with heavy impurities, the largest portion being recycled to the recovery column, while the excess is sent to wastewater treatment. Note that the column RECOVERY has neither a reboiler nor a condenser, while DIST-ACN has only a condenser. Both columns are driven by direct injection of LP steam. A remarkable feature is that the two columns are thermally linked by a vapor stream. Robust simulation is obtained when the vapor is drawn off near the top of the stripping column and injected close to the bottom of the recovery column. The following design is convenient:

- RECOVERY: 1.4 bar, 30 theoretical stages, crude feed on stage 18 at 73 °C, water feed on top at 70 °C, steam 3 bar 900 kmol/h.
- DIST-CAN: 1.8 bar, 20 theoretical stages, vapor draw on stage 2 at 1800 kmol/h, steam 3 bar 2500 kmol/h.

In order to achieve a good split of acrylonitrile/acetonitrile a large amount of water should be recycled, in this case about 70 000 kg/h. In this way over 99.9% acrylonitrile recovery may be achieved with acetonitrile under 250 ppm (purity target). The interesting fact is that more than 90% HCN goes to the bottom with acetonitrile. On the contrary, the acroleine seems to be entrained completely with acrylonitrile (thermodynamic data subject to uncertainty). Hence, the separation scheme of impurities developed previously remains valid, with the difference that this time the HCN should be recovered from both acrylonitrile and acetonitrile streams.

The cumulative consumption of steam is equivalent to 37 MW, which, compared with the previous 39.3 MW, represents a rather modest saving. Note that the energy for feed preheating (heat exchangers HX1) with a duty over 12 MW may be covered by the side-stream cooling to stripping (heat exchanger HX2). However, in this case the real saving is in capital, since one column replaces two columns, recovery and extractive distillation. Moreover, this is thermally integrated with the acetonitrile stripping column (C-3). Eventually, both may be brought in a single shell.

Regarding the operation, the variables that may be manipulated for control are the temperature of feeds, the steam flow rate in both columns, as well as the side-stream vapor flow rate. The simulation shows that the stripping column controls the acetonitrile recovery and needs more steam than the recovery column, although injecting steam in the latter is necessary for achieving high acrylonitrile recovery. Lower temperature of feeds is beneficial, but this should be optimized in agreement with steam consumption. Hence, the above device can be adapted conveniently to a particular operational environment. Note that the above heat-integration scheme is largely applied in industry, as described in a patent from Monsanto [16].

Power consumers in the acrylonitrile process are the air compressors, as well as the compressors of the refrigeration units for HCN condensation. These can be largely covered by HP steam produced in the reactor in combined heat and power cycles.

Cryogenic facilities are needed to support low-temperature operations. A major consumer is cold water of 5 °C for acrylonitrile absorption, with a duty of 7.7 MW. The power W needed to extract the heat duty Q_E can be roughly estimated by assuming a reversible Carnot cycle and global efficiency of 0.6, by using the relation:

$$W = \frac{Q_E}{0.6} \frac{T_C - T_E}{T_E} \qquad (11.9)$$

T_C and T_E are the temperatures of condenser and evaporator on the side of the thermal fluid. If a minimum temperature difference of 10 °C is assumed, on the evaporator side (cold water) the refrigerant temperature should be at least $T_E = 278 - 10 = 268$ K, while on the condenser side (heat rejected to air at 25 °C) the temperature should be at $T_C = 298 + 10 = 308$ K. The coefficient of performance of a reversible ideal Carnot cycle is:

$$COP_r^{id} = \frac{268}{308 - 268} = 6.7 \qquad (11.10)$$

In consequence, the power consumption for driving the cryogenic plant results is:

$$W = 7.7/(6.7 \times 0.6) = 1.915 \text{ MW} \tag{11.11}$$

11.7
Water Minimization

The acrylonitrile manufacturing process by the ammoxidation of propylene is characterized by a large use of processing water, namely in the primary separation stages. Obviously, every measure for water saving will have a significant impact on both energetic and ecological performances of the process as a whole. The large water consumption starts with the first separation step, acrylonitrile recovery from the reactor off-gas. However, computer simulation shows that by performing a simple flash at lower temperature a substantial amount of acrylonitrile could be recovered. This idea leads to the flowsheet presented in Figure 11.8. By cooling and condensation at 10 °C followed by L–L decanting roughly 50% acrylonitrile can be recovered in the organic phase, which is further sent directly to separation. The remaining gas is submitted to absorption in cold water. If the gas pressure remains low the simulation indicates that the water consumption is still large. A significant advantage is obtained only by raising the gas pressure to compensate lower acrylonitrile molar fraction. For example, if the gas pressure is raised from 1.6 to 4.5 bar the water for absorption drops from 180 000 to 71 000 kg/h, which represents a saving of 60%. The compression energy of 3.3 MW can be covered by a steam engine in a combined heat–power cycle. Supplementary saving is obtained in the recovery stage, which now asks only for 10 MW, compared with 26.7 MW in the previous scheme. In this way both water and energy savings are substantial. Note that this principle of partial condensation has been patented [17]. In addition to this, the heat-integrated scheme (Figure 11.7) may be adopted for simultaneous acrylonitrile recovery and acetonitrile separation.

The above development raises conceptually the problem of steam distillation versus reboiled stripping. The first has the advantage of simpler equipment, but the disadvantage of contamination. With the second the situation is reversed. The choice depends largely on local conditions.

Another problem of significance is the optimum policy of water recycling. This subject is in itself substantial and cannot be handled here. An economical approach involves optimal allocation of streams, both as flow rates and contaminant concentration. The analysis may be performed systematically with tools based on the concept of "water pinch" and "mass-exchange networks". This subject is treated thoroughly in specialized works, as in the books of El-Halwagi [19] and Smith [20]. A source-sink mapping technique developed around the acrylonitrile plant may be found in the book of Allen and Shoppard [21].

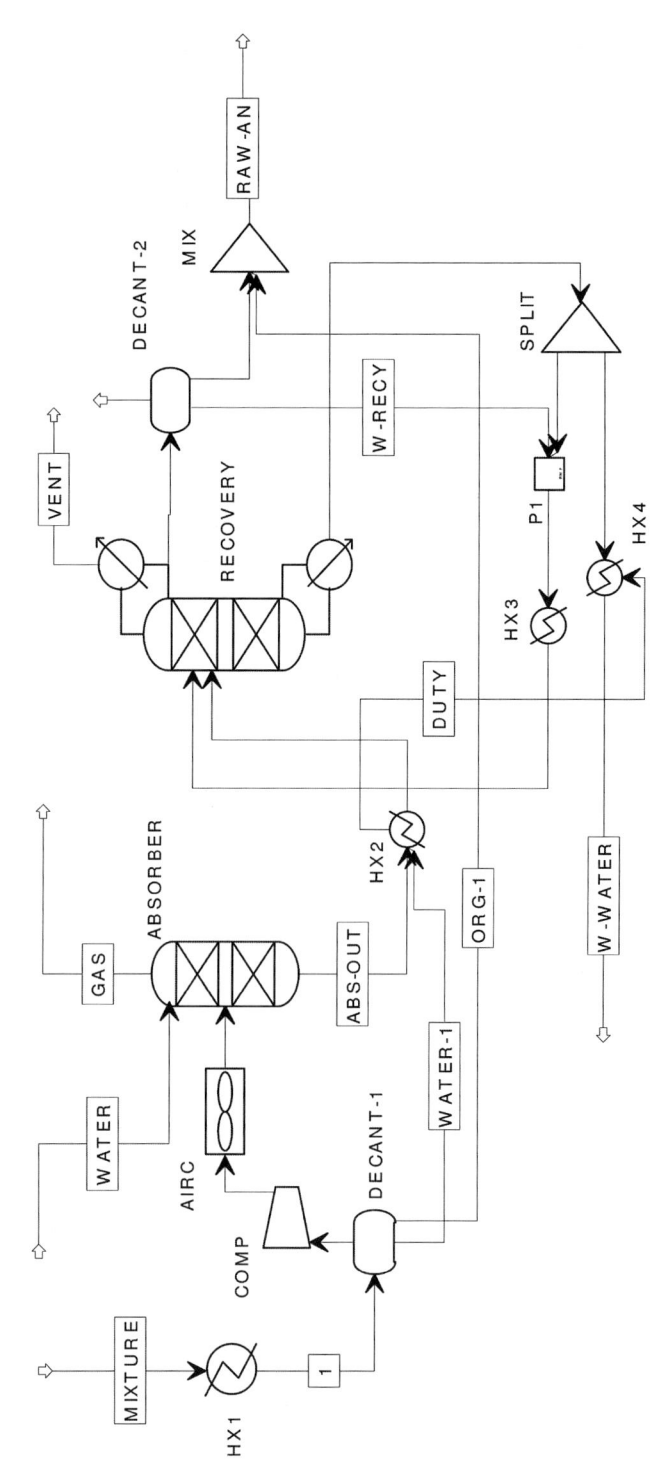

Figure 11.8 Inproved first separation step of acrylonitrile by partial condensation.

11.8
Emissions and Waste

The nature, amount and concentration of emissions are regulated by public norms and rules specific for each country. The European Commission released recommendations and example of implementation in different countries in the frame of reference documents [22] known as the best available technique (BAT). One of the most useful document deals with large-volume organic chemicals (LVOC). The following short discussion regards specifically the acrylonitrile, but the approach may be applied to other situations.

11.8.1
Air Emissions

The emissions in the air originate from several sources, such as the combustion of off-gases, the burning of residues, fugitive emissions from storage tanks and vent streams. The admitted limits are regulated by norms regarding the air quality, such as TA Luft in Germany. For example, for organics the most restrictive regards chloromethane at $30\,mg/N\,m^3$, while the value for carcinogenics (benzene, acrylonitrile) is of $5\,mg/N\,m^3$.

In this case, a major source of pollution may be the off-gas leaving the acrylonitrile absorber. Ensuring high recovery of nitriles is desirable, which can be realized by using low-temperature water, as well as an efficient absorber design. Thermal or catalytic oxidation can ensure 99.9% destruction of escaped toxics. The same techniques can be used to complete the incineration of other toxic liquid and solid residues.

The storage and handling of acrylonitrile and intermediates requires specific safety and pollution preventions measures. Fugitive emissions due to leaks of vessels are limited because of low operating pressure, but the vents streams are treated by water scrubbing. Small amounts of inhibitors, such as hydroquinone derivatives, are added to avoid losses and fouling by polymerization. Since hydrogen cyanide and acrylonitrile are highly toxic (permitted limits of 2 ppm and 10 ppm), the rapid detection of such species in and around the plant area is compulsory.

11.8.2
Water Emissions

Emission limits for discharging wastewater into running water are specific for each country. As an example, Table 11.10 show some key parameters sampled for German standards.

Summing up, the application of BAT measures can lead to a low emission and waste level that could be estimated at 0.4 kg total organic carbon per ton acrylonitrile.

Table 11.10 Limits of emissions for discharge to running water [22].

Parameter	Limit
Temperature	30 C
Filterable matter	30 mg/l
pH	6.5–8.5
Chemical oxygen demand (COD)	75 mg/l
5-day biochemical oxygen demand (BOD5)	20
Total nitrogen	40
Total hydrocarbons	5
Adsorbable organic halogen (ac Cl)	0.5 mg/l
Cyanide	0.1 mg/l
BTEX	0.1 mg/l
Organics (EB, EDC, VCM, VAM, *etc.*)	1 mg/l
Mercury	0.01 mg/l

11.8.3
Catalyst Waste

Since the cyclone system is not 100% efficient, a certain amount of the catalyst used in the fluid-bed reactor is entrained in the off-gas and captured in the quench system. The loss is in the range 0.3–0.7 kg/t acrylonitrile.

11.9
Final Flowsheet

The considerations developed so far allows setting up the final conceptual flowsheet, as displayed in Figure 11.9. After reaction and quench the off-gas is submitted to a first separation of acrylonitrile by low-temperature cooling, at 10 °C. In the decanter the liquid splits into two phases. If the acetonitrile concentration is negligible, the organic phase containing acrylonitrile can be sent directly to the first purification column (Heads). The aqueous phase is sent to the acrylonitrile recovery. The off-gas from flash is compressed at 4.5 bar and submitted to absorption in cold water of 5 °C. In this way higher acrylonitrile recovery may be achieved (over 99.8%) with reduced water consumption.

Next, the separation of acrylonitrile from the aqueous solution takes place simultaneously with the separation from acetonitrile by extractive distillation in a single column (AN recovery). Because a large amount of water is necessary as well as a high energy consumption, this operation is done in combination with the stripping by live steam of acetonitrile (ACN distillation). The two columns are thermally integrated by a vapor side stream and can be hosted in the same shell. The raw acrylonitrile is submitted to purification from lights, dewatering and final distillation in a series of distillation columns. Note that both HCN and acetonitrile

336 | *11 Acrylonitrile by Propene Ammoxidation*

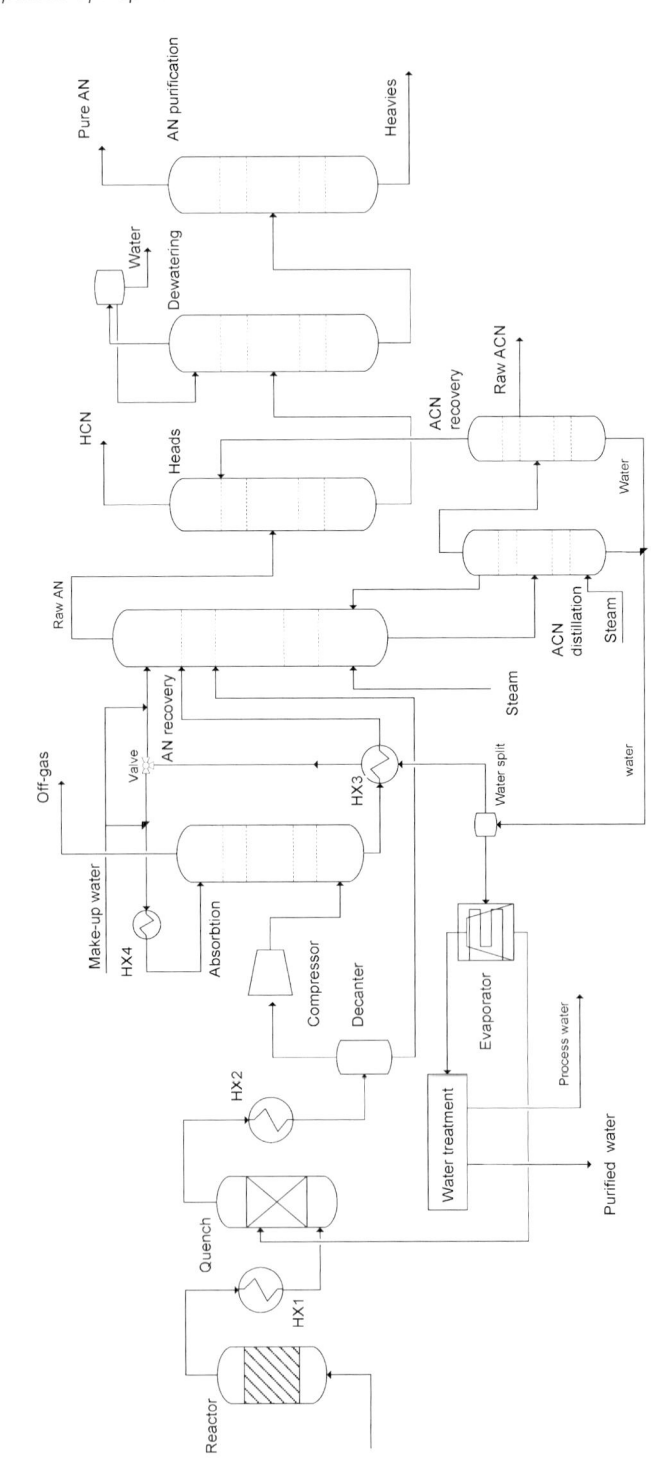

Figure 11.9 Final flowsheet for acrylonitrile process by the ammoxidation of propene.

11.10
Further Developments

An economic analysis can show that the most effective factor in reducing the manufacturing costs could be replacing the propene by propane, which is 30% cheaper. An industrial process is attributed to Asahi [18]. The key competitive element is the availability of a suitable catalyst. Despite intensive research the performance of today's catalysts remains rather modest. The conversion should be kept low, below 50%, while the selectivity cannot be pushed beyond 60%. Because the recycle of unconverted propane and larger spectrum of byproducts the advantage of lower price seems to be not sufficient for a technological breakthrough.

An interesting nonpetrochemical alternative developed by Monsanto [2] is based on synthesis gas and NH_3. In the first step, acetonitrile is obtained with selectivity of 85% at 300–600 °C and pressures up to 35 bar by using Mo/Fe oxide catalysts:

$$NH_3 + 2CO + 2H_2 \rightarrow CH_3CN + 2H_2O$$

In the second step acetonitrile is converted to acrylonitrile by catalytic oxidative methylation, as described by the global reaction:

$$CH_3CN + CH_4 + O_2 \rightarrow H_2C=CHCN + 2H_2O$$

At this stage the conversion is of 45% and selectivity up to 70%. An additional advantage is the valorization of acetonitrile byproduct directly in aqueous solution.

11.11
Conclusions

The manufacturing of acrylonitrile by ammoxidation of propene remains highly competitive because of the high performance achieved with the modern catalysts based on molybdenum/antimonium oxides. The conversion of propene is practically complete, while the ammonia and oxygen are used in amounts close to stoichiometry. Fluid-bed-reactor technology allows short reaction times and very high heat-transfer coefficients to be achieved, by preserving safety despite the potential explosive reaction mixture and very high exothermic effect.

The conceptual interest of this case study is the development of a complex separation scheme. The absorption of acrylonitrile in water, apparently trivial, needs a large amount of cold water, and accordingly a large amount of low-temperature energy. Important saving can be achieved by a two-stage scheme. The first step is a simple flash at lower temperature by which a substantial amount of acrylonitrile is separated. In the second step, the remaining gas is compressed to 4.5 bar before water absorption. In this way the water consumption can drop up to 60% with the supplementary advantage of much lower energy in the distillations implying water solutions.

The separation of acetonitrile from acetonitrile by extractive distillation with water can be done in a more efficient two-column heat integrated setup. The separation of acrylonitrile from water, which is hindered by the existence of an azeotrope, can actually take advantage of the large immiscibility gap. Valuable byproducts, such as HCN and acetonitrile can be efficiently separated. Chemical conversion can solve the separation of difficult impurities, such as acroleine.

References

1 Langvardt, P., Acrylonitrile, Ullmann's Encyclopedia of Industrial Chemistry, Wiley-VCH, Weinheim, Germany, 2002
2 Weissermel, K., Arpe, H. J., Industrial Organic Chemistry, Wiley-VCH, Weinheim, Germany, 2003
3 Callahan J. L., Milberg, E. C., Process for preparing olefinically, unsaturated nitriles, USP 3230246, 1966
4 Grasselli, R. K., Selectivity in (amm)oxidation catalysis, Catal. Today, 25, 2005
5 Grasselli, R. K., Fundamental principles of selective heterogeneous oxidation in catalysis, Top. Catal., 79, 2002
6 Grasselli, R. K., Advances and future trends in selective oxidation and ammoxidation catalysis, Catal. Today, 49, 141, 1999
7 Grasselli, R. K., in Handbook of Heterogeneous Catalysis (Ertl, G., Knötzinger, H., Weitkamp, J. eds), Chap 4.6.6., 2302, 1997
8 Chen, Q., Chen, X., Mao, L., Cheng, W., Recent advances of commercial catalysts, Catal. Today, 141, 1999
9 Guan et al., Catalyst for producing acrylonitrile, USP, 6596987, 2003
10 Chen, B. H., Dai, Q. L., Wu, D. W., Modelling a loop fluidized bed reactor for propylene ammoxidation, Chem. Eng. Sci., 51, 11, 298–88, 1996
11 Stergiou, L., Laguerie, CF., Gilot, B., Some reactor models for ammoxidation of propylene, Chem. Eng. Sci., 39(4), 713, 1984
12 Godbole, S. P., Acrylonitrile recovery process, USP 6054603, 2000
13 Godbole, S. P., Operation of heads column, USP 6793776 B2, 2004
14 Godbole, S. P., Process for recovering acrylonitrile, USP, 0181086A1, 2004
15 Godbole, S. P., Process for the purification of acetonitrile, EP 1301471B1, 2005
16 Lovett, G. H., Monsanto, USP 3399120, 1968
17 Wu, H. C., Recovery of acrylonitrile by condensation, USP 4232519, 1980
18 Midorikawa, H., Sugiyama, N., Hinago, H., Asahi-Japan, Process for producing acrylonitrile from propane by ammoxidation, USP 5973186, 1999
19 El-Halwagi, M., Pollution Prevention through Process Integration: Systematic Design Tools, Academic Press, San Diego, CA, 1997
20 Smith, R., Chemical Process Design and Integration, John Wiley, Chichester, 2005
21 Allen, D. T., Shonnard, D. R., Green Engineering, Prentice Hall, Upper Saddle River, NJ, USA, 2002
22 European Commission, Integrated Pollution Prevention and Control (IPPC). Best available techniques in the Large Volume Organic Chemical Industry, Feb. 2003
23 Aspen Plus®, Aspen Technology Inc., Cambridge/MA, USA, release 12

12
Biochemcial Process for NO_x Removal

12.1
Introduction

Reducing the nitrogen oxides (NO_x) content of gases is an important industrial issue since over a long time scale they contribute to climate changes. NO_x trigger the occurrence of acid rains and act indirectly as greenhouse gases by producing ozone during their breakdown in the atmosphere. In addition, nitrogen oxides are highly toxic and play a role in the development of several illnesses, such as Parkinson's disease and asthma. About 20% of the anthropogenic NO_x emissions come from flue gases. Therefore, the removal of NO_x from flue gases is of paramount importance for preserving a clean and healthy environment. As a result, governments are implementing stricter environmental policies for industrial waste emissions to the environment.

Most of the existing processes for nitrogen oxide removal are chemically based requiring high temperature or expensive catalysts. The main techniques involve either selective noncatalytic reduction (SNCR) or selective catalytic reduction (SCR). SNCR uses ammonia for conversion of NO_x to N_2 and H_2O at elevated temperatures (550–850 K). SCR can use catalysts such as TiO_2 with active coatings of V_2O_5 and WO_{-3}.

Biotechnology can provide an alternative for chemical-based cleaning processes, this time the function of the catalyst being played by micro-organisms [1]. Milder operating conditions give a definite economical advantage. However, the low solubility of NO_x in water hinders the transport over the membrane into the cell. To overcome this barrier, one can take advantage of the formation of a complex between NO_x and various metal complexes to increase the solubility of NO_x in water.

In this chapter, we will focus on the NO complexation using $Fe(II)EDTA^{2-}$ as chelating agent:

$$Fe(II)EDTA^{2-} + NO \rightarrow Fe(II)EDTA\text{-}NO^{2-} \qquad (12.1)$$

This reaction can be performed in an absorber. In a second step, the absorbed NO is converted into N_2 in a biochemical reactor [1, 3, 4]. The reaction is:

Chemical Process Design: Computer-Aided Case Studies. Alexandre C. Dimian and Costin Sorin Bildea
Copyright © 2008 WILEY-VCH Verlag GmbH & Co. KGaA, Weinheim
ISBN: 978-3-527-31403-4

$$6Fe(II)EDTA\text{-}NO^{2-} + C_2H_5OH \rightarrow 6Fe(II)EDTA^{2-} + 3N_2 + 2H_2O + 2CO_2 \quad (12.2)$$

However, the extent of reaction (12.1) is diminished by the oxidation of Fe(II)EDTA^{2-} complex with the oxygen dissolved in the liquid phase:

$$4Fe(II)EDTA^{2-} + O_2 + 4H^+ \rightarrow 4Fe(III)EDTA^- + 2H_2O \quad (12.3)$$

Therefore, Fe(II) must be regenerated, for example by means of a Fe(III) reducing bacteria using ethanol as an electron donor [4]:

$$12Fe(III)EDTA^- + C_2H_5OH + 5H_2O \rightarrow 12Fe(II)EDTA^{2-}$$
$$+ 2H_2O + 2CO_2 + 12H^+ \quad (12.4)$$

The biochemical processes of denitrification and Fe(III) reduction have been experimentally investigated in a batch reactor [5]. The feasibility of the integrated absorption–bioreaction process was demonstrated on a laboratory-scale setup [6].

Figure 12.1 presents the principle of a biochemical process for NO$_x$ removal (BioDeNOx).

In this case study we will model, simulate and design an industrial-scale BioDeNOx process. Rigorous rate-based models of the absorption and reaction units will be presented, taking into account the kinetics of chemical and biochemical reactions, as well as the rate of gas–liquid mass transfer. After transformation in dimensionless form, the mathematical model will be solved numerically. Because of the steep profiles around the gas/liquid interface and of the relatively large number of chemical species involved, the numerical solution is computationally expensive. For this reason we will derive a simplified model, which will be used to size the units. Critical design and operating parameters will be identified

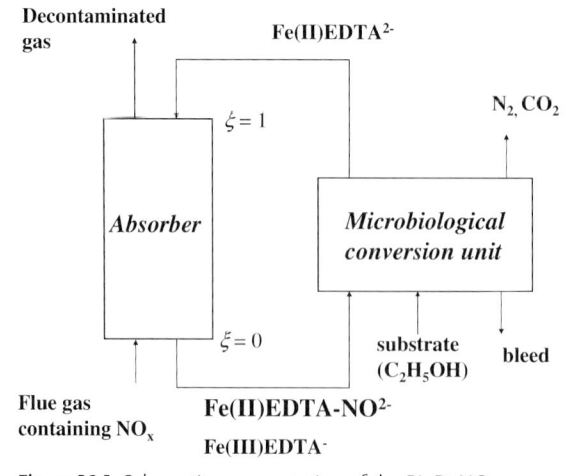

Figure 12.1 Schematic representation of the BioDeNOx proces

by sensitivity studies. This chapter will demonstrate that bringing together chemistry, microbiology and engineering results in the development of an efficient process for removal of NO_x from flue gases.

12.2
Basis of Design

The problem addressed in this chapter concerns the design of a process for NO_x removal from flue gases to levels below 10 ppm vol. The process will be based on the formation of the Fe(II)EDTA-NO^{2-} complex, followed by microbial denitrification. The process should be flexible enough to handle large variations in the flue-gas load originating from a 10 MW gas-turbine, with the following characteristics (nominal values in **bold**):

- Flow rate: 10 000–**20 000** N m³/h
- Temperature: **400 °C**, ±50 °C
- Pressure: 1.1 bar
- Composition:
    - NO_x 100–**300** ppm (vol)
    - CO_2 13 vol.%
    - O_2 2.0–5 vol.%
    - Particles   5 mg/m³

12.3
Process Selection

Separation of NO_x from the flue gas requires the existence of a G/L or G/S interface. The conversion of nitrogen oxide bounded to the Fe(II)EDTA^{2-} complex into N_2 is catalyzed by denitrifying bacteria, according to the reaction (12.2). The microbial communities are living organisms needing an aqueous environment for prevention of dehydration, supply of nutrients and removal of the wastes. Because the Fe(II)EDTA-NO^{2-} complex must come into contact with the cell membrane, we choose a process where NO is removed through a G/L interface by using an aqueous solution of Fe(II)EDTA^{2-}, followed by denitrification of the nitrosyl-complex also taking place in liquid phase.

The second decision concerns integration of both reactions (12.1) and (12.2) into a single unit, or dividing the process into two subsystems. The integrated system has advantages such as smaller size and fewer or no streams between units. However, the micro-organisms catalyzing reactions (12.2) and (12.4) need very specific conditions, such as temperature, pH, REDOX potential. These might be incompatible with the optimal conditions of the absorption step. Moreover, if we take into account the required flexibility and the extensive experience regarding design and operation of absorption units and bioreactors, the two-step process is the preferred choice.

The absorption can be performed in different types of unit, for example spray tower, packed bed, plate tower, bubble column, static mixer, and bubble tank. Gas-liquid interfacial area and liquid fraction are criteria for equipment selection.

According to the film theory, in reactive-absorption processes the resistance to mass transfer is concentrated in a small region near the gas/liquid interface. The ratio between the rate of chemical reaction and liquid-phase mass transfer is given by the Hatta number. For a second-order reaction (12.1), the Hatta number is defined as:

$$\text{Ha}_{NO} = \frac{\sqrt{k_1 C_{ref}^L D_{NO}}}{k_{L,NO}} \tag{12.5}$$

Typically, the following values can be considered $D_{NO} = 8.6 \times 10^{-9}\,\text{m}^2/\text{s}$, $k_{L,NO} = 3.9 \times 10^{-4}\,\text{m/s}$, $k_1 = 1.5 \times 10^5\,\text{m}^3/\text{mol/s}$, $C_{ref}^L = 50\,\text{mol/m}^3$ [6, 8, 10, 13], which leads to a value $\text{Ha}_{NO} = 651$. This indicates a very fast reaction taking place in a thin region of the film, adjacent to the gas/liquid interface. Therefore, reaction (12.1) is favored by large gas/liquid interfacial area but not influenced by the bulk liquid.

Similarly, for reaction (12.3), the Hatta number is defined as:

$$\text{Ha}_{O_2} = \frac{C_{ref}^L \sqrt{k_2 D_{O_2}}}{k_{L,O_2}} \tag{12.6}$$

The reaction rate constant is $k_2 = 0.016\,\text{m}^6/\text{mol}^2/\text{s}$ [9, 10, 11, 14]. Considering similar mass transfer and diffusion coefficients of NO and oxygen, $\text{Ha}_{O2} = 1.5$ is obtained. This small value shows that the reaction is slow and takes place mainly in the bulk of the liquid. We recall that reaction (12.3) is undesired, because it consumes the iron-EDTA complex.

To conclude, the desired reaction (12.1) is favored by a large gas/liquid interfacial area and a small liquid fraction. This can be obtained in a spray column. Another consideration favoring spray-tower over packed-bed absorbers is the large gas flow to be processed.

The choice of the bioreactor type for denitrification of NO to N_2 and the regeneration of the chelating agent Fe(II)EDTA^{2-} is another important decision. A properly designed and controlled bioreactor should ensure high biomass concentration, constant temperature and pH, optimal mixing and correct shear conditions. Other important aspects are listed below:

- The phases present in the bioreactor are: (1) solid phase of particulates and biomass, (2) liquid phase of water and ethanol, with the reacting agents and nutrition for the biomass and (3) gas phase with the products of the reaction (N_2, and CO_2).
- The absorber generates a continuous stream of water with the nitrosyl-complex that has to be treated in the bioreactor. The bioreactor generates a continuous stream of Fe(II)EDTA^{2-} solution required by the absorption unit.
- Because the oxygen modifies the Fe(II)EDTA^{2-} complex, its level inside the bioreactor must be as low as possible.

- The solids must be removed from the system, otherwise they accumulate.
- A turndown ratio of 50% is desirable.

These considerations point toward a continuous, stirred-tank reactor.

The BioDeNox process is based on two chemical reactions within the absorber and two microbial reactions within the reactor. There are different kinds of denitrifying bacteria that are capable to reduce NO to N_2. Van der Maas et al. [12] investigated the reducing capacity of four different bacterial sludges (named after the plant location in the Netherlands): "Veendam", from a full-scale denitrification reactor handling surface water; "Emmen", denitrifying sludge originated from a biological wastewater treatment; "Nedalco", anaerobic granular sludge from distillery wastewater treatment; "Eerbeek", sludge used for treating paper-mill wastewater.

Due to the oxygen in the flue gas, the Fe(II)EDTA^{-2} complex will be oxidized to the inactive Fe(III)EDTA$^-$. The regeneration of Fe(II)EDTA^{-2} is possible with methanogenic bacteria and ethanol as the electron donor, according to reaction (12.4). Experimental studies [6, 12] showed that the original "Veendam" inoculum has a rather low potential for the reduction of Fe(III)EDTA$^-$. However, a combination of the "Veendam" and "Eerbeek" sludges resulted in a strong increase of the Fe(III)$^-$ reduction capacity

Ethanol, acetate and methanol are suitable electron donors for the reduction of Fe (III)EDTA$^-$ to Fe(II)EDTA^{2-}. From experimental measurements [5], it appeared that the lag phase with methanol is substantially longer than with ethanol or acetate. Therefore, ethanol will be used as an electron donor.

12.4
The Mathematical Model

In this section, a mathematical model is developed, solved and used to find the optimal sizes of the absorption and bioreactor units. First, the equations describing the mass transfer and chemical reaction around the gas/liquid interface are derived. By solving these equations, we prove that the main reaction (12.1) is fast and takes place near the G/L interface, while reaction (12.3) is slow and takes place in the liquid bulk. This allows the development of a simplified model, which has analytical solution. The film model can then be integrated into the model of the absorption tower. Finally, the bioreactor model is added. By solving the integrated model, we find the optimal size of the absorption unit that guarantees the best performance and derive the necessary conditions concerning the size of the bioreactor. Finally, the flexibility of the design is checked.

12.4.1
Diffusion-Reaction in the Film Region

In the absorption unit, the mass transfer between gas and liquid phases is described by the film theory, which assumes that the resistance to mass transfer

12.2 Figure

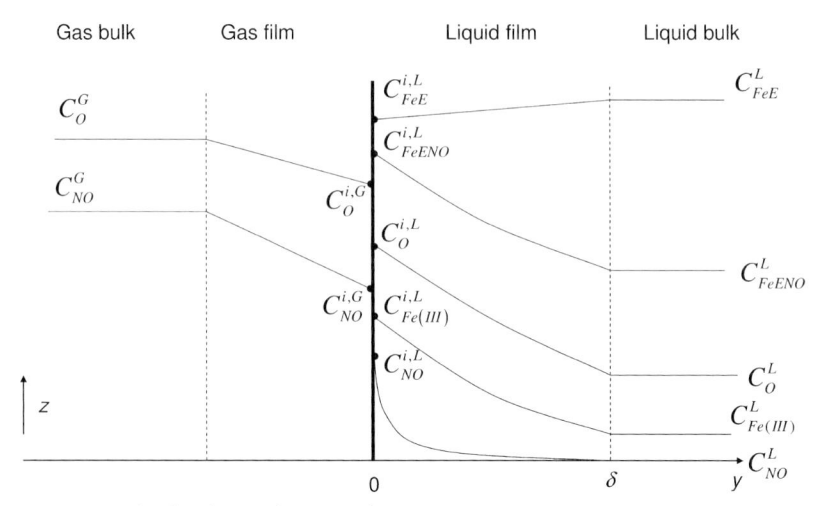

Figure 12.2 The film theory of reactive absorption processes.

is concentrated in two thin films adjacent to the gas/liquid interface (Figure 12.2). Within the two films, the mass transfer occurs only by steady-state molecular diffusion. The chemical reactions take place both in the film and bulk regions.

In the liquid film, the NO diffusion and reaction is described by the following differential equation:

$$D_{NO}\frac{d^2 C_{NO}}{dy^2} - k_1\left(C_{NO}C_{FeE} - \frac{1}{K_{eq}}C_{FeENO}\right) = 0 \tag{12.7}$$

The boundary conditions express the flux continuity at the gas/liquid interface and the concentration continuity at the film/bulk boundary:

$$k_{G,NO}(C_{NO}^G - C_{NO}^{i,G}) = -D_{NO}\left.\frac{dC_{NO}}{dy}\right|_{y=0} \tag{12.8}$$

$$C_{NO}|_{y=\delta} = C_{NO}^L \tag{12.9}$$

In addition, the gas and liquid phases are in thermodynamic equilibrium at the interface:

$$C_{NO}^{i,G} = H_{NO}C_{NO}|_{y=0}$$

In Eqs. (12.7) to (12.9), C_K is the concentration of species K, k_G is the mass-transfer coefficient in the gas film, D is the diffusion coefficient, k_1 and K_{eq} are reaction rate and equilibrium constants, respectively, of the reversible reaction (12.1).

12.4 The Mathematical Model

To derive a dimensionless model, we introduce the reference concentrations C_{ref}^L and C_{ref}^G and define the dimensionless concentrations c_K by the following relationships:

$$C_{NO} = c_{NO}C_{ref}^L; \quad C_{FeE} = c_{FeE}C_{ref}^L; \quad C_{FeENO} = c_{FeENO}C_{ref}^L$$
$$C_{NO}^G = c_{NO}^G C_{ref}^G; \quad C_{NO}^{i,G} = c_{NO}^G C_{ref}^G \qquad (12.10)$$

At this point, we anticipate the derivation of the dimensionless model for the gas bulk. This model should be capable of predicting the change of the NO molar flow Φ_{NO} along the absorption column. This can be expressed as:

$$\Phi_{NO} = F_V^G C_{ref}^G c_{NO}^G = \Phi_{ref}\varphi_{NO} \qquad (12.11)$$

The identification of the terms leads to the following equivalences:

$$F_V^G C_{ref}^G \equiv \Phi_{ref} \qquad (12.12)$$

$$c_{NO}^G \equiv \varphi_{NO} \qquad (12.13)$$

The definition of the mass-transfer coefficient according to the film model is:

$$k_{L,NO} = \frac{D_{NO}}{\delta} \qquad (12.14)$$

After inserting dimensionless variables into Eq. (12.7), the following dimensionless model is obtained:

$$\frac{d^2 c_{NO}}{dx^2} = Ha_{NO}^2 c_{NO} c_{FeE} - \frac{Ha_{NO}^2}{k_{eq}} c_{FeENO} \qquad (12.15)$$

with the boundary conditions:

$$\left.\frac{dc_{NO}}{dx}\right|_{x=0} = -\kappa_{NO}(\varphi_{NO} - h_{NO}c_{NO}|_{x=0}) \qquad (12.16)$$

$$c_{NO}|_{x=1} = c_{NO}^L \qquad (12.17)$$

The dimensionless parameters are defined as:

$$Ha_{NO}^2 = k_1 \frac{C_{ref}^L D_{NO}}{k_{L,NO}^2}; \quad k_{eq} = K_{eq}C_{ref}^L; \quad \kappa_{NO} = \frac{k_{G,NO}C_{ref}^G}{k_{L,NO}C_{ref}^L}; \quad h_{NO} = H_{NO}\frac{C_{ref}^L}{C_{ref}^G} \qquad (12.18)$$

Similarly, the diffusion-reaction processes of the other species are described by the following partial-differential equations with the corresponding boundary conditions:

Oxygen

$$\frac{d^2 c_{O_2}}{dx^2} = Ha_{O_2}^2 c_{O_2} c_{FeE}^2 \tag{12.19}$$

$$\left.\frac{dc_{O_2}}{dx}\right|_{x=0} = -\kappa_{O_2}(\varphi_{O_2} - h_{O_2} c_{O_2}|_{x=0}) \tag{12.20}$$

$$c_{O_2}|_{x=1} = c_{O_2}^L; \tag{12.21}$$

Fe(II)EDTA^{2-}

$$\frac{d^2 c_{FeE}}{dx^2} = Ha_{NO}^2 \beta_{FeE}\left(c_{NO} c_{FeE} - \frac{1}{k_{eq}} c_{FeENO}\right) + 4Ha_{O_2}^2 \beta_{O_2} c_{O_2} c_{FeE}^2 \tag{12.22}$$

$$\left.\frac{dc_{FeE}}{dx}\right|_{x=0} = 0 \tag{12.23}$$

$$c_{FeE}|_{x=1} = c_{FeE}^L \tag{12.24}$$

Fe(II)EDTA-NO^{2-}

$$\frac{d^2 c_{FeENO}}{dx^2} = -Ha_{NO}^2 \beta_{FeENO}\left(c_{NO} c_{FeE} - \frac{1}{k_{eq}} c_{FeENO}\right) \tag{12.25}$$

$$\left.\frac{dc_{FeENO}}{dx}\right|_{x=0} = 0 \tag{12.26}$$

$$c_{FeENO}|_{x=1} = c_{FeENO}^L \tag{12.27}$$

Fe(III)EDTA$^-$

$$\frac{d^2 c_{Fe(III)}}{dx^2} = -4Ha_{O_2}^2 \beta_{O_2} c_{O_2} c_{FeE}^2 \tag{12.28}$$

$$\left.\frac{dc_{Fe(III)}}{dx}\right|_{x=0} = 0 \tag{12.29}$$

$$c_{Fe(III)}|_{x=1} = c_{Fe(III)}^L \tag{12.30}$$

In Eqs. (12.22), (12.25) and (12.28), parameters β are ratios of diffusion coefficients:

$$\beta_{FeE} = \frac{D_{NO}}{D_{FeE}}; \quad \beta_{O_2} = \frac{D_{O_2}}{D_{FeE}}; \quad \beta_{FeENO} = \frac{D_{NO}}{D_{FeENO}} \tag{12.31}$$

12.4.1.1 Model Parameters

Sada et al. [13] found the value $k_1 = 1.23 \times 10^5 \, m^3/mol/s$. Further, Yih and Lii [17] determined values ranging from $1.24 \times 10^5 \, m^3/mol/s$ (at 298 K) to $1.47 \times 10^5 \, m^3/mol/s$ (353 K). In the following, we will consider $k_1 = 1.4 \times 10^5 \, m^3/mol/s$.

The equilibrium constant K_{eq} has values from $218\,\text{m}^3/\text{mol/s}$ (at 288 K) to $252\,\text{m}^3/\text{mol/s}$ (at 348 K) [8]. The dependence with temperature can be expressed as [9, 14]:

$$K_{eq} = \exp\left(\frac{4702}{T} - 8.53\right) [\text{m}^3/\text{mol/s}] \tag{12.32}$$

The kinetic constant of the oxidation reaction is [14]:

$$k_{ox} = 5.3 \times 10^3 \exp\left(-\frac{4098}{T}\right) [\text{m}^6/\text{mol}^2/\text{s}] \tag{12.33}$$

The mass-transfer coefficients are taken from [14], assuming that they are the same order of magnitude in a packed bed and spray tower:

$$k_{L,NO} = 3.88 \times 10^{-4}\,\text{m/s};\; k_{L,O} = 1.83 \times 10^{-4}\,\text{m/s};\; k_{G,NO} = k_{L,NO} = 0.0314\,\text{m/s} \tag{12.34}$$

The Henry coefficients are:

$$H_{NO} = 26.25;\; H_O = 40.4 \tag{12.35}$$

The following values were used for the diffusion coefficients [14, 15]:

$$D_{FeE} = 3.3 \times 10^{-10}\,\text{m}^2/\text{s};\; D_O = 4 \times 10^{-10}\,\text{m}^2/\text{s};\; D_{NO} = 5 \times 10^{-9}\,\text{m}^2/\text{s} \tag{12.36}$$

The dimensionless parameters of the model are as follows:

$$Ha_{NO} = 483;\; Ha_{O_2} = 1.34 \tag{12.37}$$

$$\kappa_{NO} = 0.011;\; \kappa_{O_2} = 0.0204 \tag{12.38}$$

$$h_{NO} = 194\,600;\; h_{O_2} = 295\,600 \tag{12.39}$$

$$\beta_{FeE} = 6.848;\; \beta_{O_2} = 3.36;\; \beta_{FeENO} = 7.316$$
$$k_{eq} = 12450 \tag{12.40}$$

The following boundary conditions are assumed:

$$\varphi_{NO} = 1;\; \varphi_{O_2} = 250 \tag{12.41}$$

$$c^L_{NO} = 0;\; c^L_{O_2} = 0;\; c^L_{FeE} = 1.0;\; c^L_{FeENO} = 0.0;\; c^L_{Fe(III)} = 0.0 \tag{12.42}$$

The model (12.15) to (12.17), (12.19) to (12.30) is solved in Matlab using the bvp4c function, which implements a collocation technique for solving partial-differential equations [16].

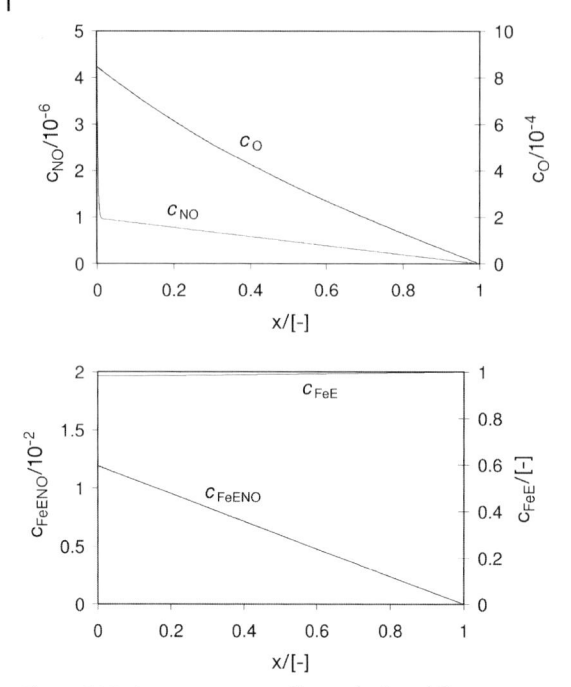

Figure 12.3 Concentration profiles in the liquid film.

Figure 12.3 presents concentration profiles in the liquid film. The following conclusions can be drawn:

- The reaction between NO and Fe(II)EDTA^{2-} is very fast, and takes place in a narrow region near the G/L interface. This in agreement with the large value of the Hatta number Ha$_{NO}$.
- The complex Fe(II)EDTA-NO^{2-} formed near the G/L interface is transported to the liquid bulk by diffusion.
- The reaction between O$_2$ and Fe(II)EDTA^{2-} is slow. The oxygen concentration profile is almost linear, characteristic of pure diffusion. Therefore, we can safely assume that the oxidation reaction takes place mainly in the liquid bulk.
- The reactant Fe(II)EDTA^{2-} is in large excess. For this reason, its concentration in the film can be considered constant.

12.4.2
Simplified Film Model

In view of the only reaction between NO and Fe(II)EDTA^{2-}, and assuming a constant concentration of Fe(II)EDTA^{2-}, the model describing the concentration profiles in the liquid film contains only the balance equations for NO and Fe(II)EDTA-NO^{2-}.

$$\frac{d^2 c_{NO}}{dx^2} = (Ha^*_{NO})^2 \left(c_{NO} - \frac{1}{k^*_{eq}} c_{FeENO} \right) \tag{12.43}$$

$$\frac{d^2 c_{FeENO}}{dx^2} = -\beta_{FeENO}(Ha^*_{NO})^2 \left(c_{NO} - \frac{1}{k^*_{eq}} c_{FeENO} \right) \tag{12.44}$$

The Hatta number and the dimensionless equilibrium constant that appear in Eqs. (12.43) and (12.44) include the concentration of Fe(II)EDTA^{2-}, assumed to be constant, $c_{FeE} = c^L_{FeE}$:

$$Ha^*_{NO} = Ha_{NO} \cdot \sqrt{c^L_{FeE}} \tag{12.45}$$

$$k^*_{eq} = k_{eq} c^L_{FeE} \tag{12.46}$$

The model (12.43) and (12.44) consists of two second-order linear differential equations whose solution has the following form:

$$\begin{bmatrix} c_{NO} \\ c_{FeENO} \end{bmatrix} = \begin{bmatrix} C_1 \\ C_2 \end{bmatrix} \cdot e^{\lambda x} \tag{12.47}$$

where the eigenvalues λ and the constants C_1 and C_2 have to be determined.

Substitution of Eq. (12.47) into Eqs. (12.43) and (12.44) leads to the following system of homogeneous linear equations in the unknowns C_1 and C_2.

$$(\lambda^2 - (Ha^*_{NO})^2) \cdot C_1 + \frac{(Ha^*_{NO})^2}{k^*_{eq}} C_2 = 0 \tag{12.48}$$

$$\beta_{FeENO}(Ha^*_{NO})^2 \cdot C_1 + \left(\lambda^2 - \beta_{FeENO} \frac{(Ha^*_{NO})^2}{k^*_{eq}} \right) C_2 = 0 \tag{12.49}$$

Equations (12.48) and (12.49) have a nontrivial solution $C_1 \neq 0$, $C_2 \neq 0$ if, and only if, the determinant of the system is nonzero. This leads to a fourth-order polynomial equation in the unknown λ. Solving for λ and inserting the results into Eqs. (12.48) and (12.49) results in the following conditions:

$$\lambda_1 = 0 \qquad C_2 = k^*_{eq} C_1 \tag{12.50}$$

$$\lambda_2 = 0 \qquad C_2 = k^*_{eq} C_1 \tag{12.51}$$

$$\lambda_3 = Ha^*_{NO} \sqrt{1 + \frac{\beta_{FeENO}}{k^*_{eq}}} \qquad C_2 = -\beta_{FeENO} C_1 \tag{12.52}$$

$$\lambda_4 = -Ha^*_{NO} \sqrt{1 + \frac{\beta}{k^*_{eq}}} \qquad C_2 = -\beta_{FeENO} C_1 \tag{12.53}$$

The general solution of the differential equations (12.43) and (12.44) is a linear combination of the particular solutions defined by Eqs. (12.50) to (12.53):

$$\begin{bmatrix} c_{NO} \\ c_{FeENO} \end{bmatrix} = \begin{bmatrix} 1 \\ k^*_{eq} \end{bmatrix}(\alpha_1 + \alpha_2 x) + \begin{bmatrix} 1 \\ -\beta_{FeENO} \end{bmatrix}\alpha_3 e^{\lambda_3 x} + \begin{bmatrix} 1 \\ -\beta_{FeENO} \end{bmatrix}\alpha_4 e^{\lambda_4 x} \tag{12.54}$$

The constants $\alpha_1, \alpha_2, \alpha_3, \alpha_4$ can be determined from the boundary conditions (12.16), (12.17), (12.26) and (12.27). The derivative of NO concentration at the G/L interface, which gives the rate of the reactive absorption process, can be calculated, leading to the following result:

$$-\frac{dc_{NO}}{dx}\bigg|_{x=0} = \kappa_{NO} Ha^*_{NO} \frac{\varphi_{NO}(k^*_{eq} + \beta_{FeENO}) - h_{NO}(\beta c^L_{NO} + c^L_{FeENO})}{\kappa_{NO} h_{NO}(\beta_{FeENO} Ha^*_{NO} + k^*_{eq}) + Ha^*_{NO}(\beta_{FeENO} + k^*_{eq})} \tag{12.55}$$

Numerical solution of the full model gives the value $-\dfrac{dc_{NO}}{dx}\bigg|_{x=0} = 1.6308 \cdot 10^{-3}$, while Eq. (12.55) predicts $-\dfrac{dc_{NO}}{dx}\bigg|_{x=0} = 1.6454 \times 10^{-3}$. The agreement is remarkably good and enables one to use Eq. (12.55) for calculating the absorption rate along the column, needed for modeling the bulk gas (next section).

The expression for the flux of NO at the film/bulk interface, needed in the model of the bulk liquid, can be also obtained as the functional dependence (too complex to be reproduced here):

$$\frac{dc_{NO}}{dx}\bigg|_{x=1} = f(\varphi_{NO}, c^L_{NO}, c^L_{FeENO}, p) \tag{12.56}$$

where the set of model parameters p includes the modified Hatta number

$$\tilde{Ha} = Ha^*_{NO}\sqrt{1 + \frac{\beta_{FeENO}}{k^*_{eq}}} \tag{12.57}$$

The other fluxes through the liquid film–liquid bulk boundary can be calculated based on stoichiometry considerations:

$$\frac{dc_{FeE}}{dx}\bigg|_{x=1} = -\beta_{FeE}\left(\frac{dc_{NO}}{dx}\bigg|_{x=0} - \frac{dc_{NO}}{dx}\bigg|_{x=1}\right) \tag{12.58}$$

$$\frac{dc_{FeENO}}{dx}\bigg|_{x=1} = \beta_{FeENO}\left(\frac{dc_{NO}}{dx}\bigg|_{x=0} - \frac{dc_{NO}}{dx}\bigg|_{x=1}\right) \tag{12.59}$$

$$\frac{dc_{O_2}}{dx}\bigg|_{x=0} = \frac{dc_{O_2}}{dx}\bigg|_{x=1} = c^*_{O_2} - c_{O_2}\big|_{x=0} \tag{12.60}$$

$$\frac{dc_{Fe(III)}}{dx}\bigg|_{x=1} = 0 \tag{12.61}$$

The relationships (12.56) to (12.61) will be used in the model of the bulk liquid, presented in the next section.

12.4.3
Convection-Mass-Transfer Reaction in the Bulk

12.4.3.1 Bulk Gas

The mass-balance equation for the gas bulk describes the evolution of NO flow rate, Φ_{NO}, along the absorption tower coordinate z.

$$\frac{d\Phi_{NO}}{dz} = -N_{NO}Sa_v \tag{12.62}$$

In Eq. (12.62), N_{NO} is the flux of NO through the gas/liquid interface. S and a_v are the column cross-sectional area and the specific gas/liquid interfacial area, respectively.

The flux of NO through the gas/liquid interface can be expressed as:

$$N_{NO} = -D_{NO}\frac{dc_{NO}}{dy}\bigg|_{y=0} = -\frac{D_{NO}C^L_{ref}}{\delta}\frac{dc_{NO}}{dx}\bigg|_{x=0} = -k_{NO,L}C^L_{ref}\frac{dc_{NO}}{dx}\bigg|_{x=0} \tag{12.63}$$

To derive dimensionless equations, we introduce the reference flow rates Φ_{ref} (Eq. (12.12)) and define the following dimensionless parameters:

$$Da = \frac{k_1 V}{\Phi^L_V}\varepsilon_L C^L_{ref} \tag{12.64}$$

$$\gamma_{NO} = \frac{k_{g,NO}a_v C^G_{ref}\Phi^L_V}{\Phi_{ref}\varepsilon_L k_1 C^L_{ref}} \tag{12.65}$$

Then, Eq. (12.62) becomes:

$$\frac{d\varphi_{NO}}{d\xi} = -Da\frac{\gamma_{NO}}{\kappa_{NO}}\frac{dc_{NO}}{dx}\bigg|_{x=0} \tag{12.66}$$

The introduction of parameters Da, γ_{NO} and k_{NO} in the gas-phase balance equations might seem unnecessarily complicated. However, it allows inclusion of the column volume V in only one dimensionless parameter (Da), and leads to simple form for the bulk-liquid model.

The mass balance for oxygen in the bulk film is derived in a similar way, taking into account the assumption that there is no reaction in the liquid film:

$$\frac{d\Phi_{O_2}}{dz} = -Sa_v\frac{(C^G_{O_2} - H_{O_2}C^i_{O_2})}{\frac{1}{k_{g,O_2}} + \frac{H_{O_2}}{k_{L,O_2}}} \tag{12.67}$$

Equation (12.67) has the following dimensionless form:

$$\frac{d\varphi_{O_2}}{dz} = -Da\gamma_{O_2} \frac{(\varphi_{O_2} - h_{O_2}c^L_{O_2})}{1 + h_{O_2}\kappa_{O_2}} \tag{12.68}$$

12.4.3.2 Bulk Liquid

The NO mass balance in the liquid bulk can be written as:

$$\Phi^L_V \frac{dC^L_{NO}}{dz} = Sa_v D_{NO} \left.\frac{dC_{NO}}{dy}\right|_{y=\delta} + S\varepsilon_L k_1 \left(C^L_{NO} C^L_{FeE} - \frac{1}{K_{eq}} C^L_{FeENO} \right) \tag{12.69}$$

In Eq. (12.69) Φ^L_V and ε_L represent the volumetric flow rate and the volumetric fraction of the liquid bulk.

After introducing the dimensionless parameters:

$$\lambda_{NO} = \frac{k_{L,NO}a_v}{k_1 \varepsilon_L C^L_{ref}}; \quad Da = \frac{k_1 V}{\Phi^L_V} \varepsilon_L C^L_{ref} \tag{12.70}$$

the following dimensionless form is obtained:

$$\frac{dc^L_{NO}}{d\xi} = Da \left(\lambda_{NO} \left.\frac{dc_{NO}}{dx}\right|_{x=1} + \left(c^L_{NO} c^L_{FeE} - \frac{1}{k_{eq}} c^L_{FeENO} \right) \right) \tag{12.71}$$

Similar equations can be written for the other species involved:

Oxygen

$$\frac{dc^L_{O_2}}{d\xi} = Da \left(\lambda_{O_2} \kappa_{O_2} \frac{\varphi_{O_2} - h_{O_2}c^L_{O_2}}{1 + h_{O_2}\kappa_{O_2}} + \Gamma c^L_{O_2} (c^L_{FeE})^2 \right) \tag{12.72}$$

with

$$\lambda_{O_2} = \frac{k_{L,O_2}a_v}{k_1 \varepsilon_L C^L_{ref}}; \quad \Gamma = \frac{k_{ox} C^L_{ref}}{k_1} \tag{12.73}$$

FeEDTA(2−)

$$\frac{dc^L_{FeE}}{d\xi} = Da \left(\lambda_{FeE} \left.\frac{dc_{FeE}}{dx}\right|_{x=1} + \left(c^L_{NO} c^L_{FeE} - \frac{1}{k_{eq}} c^L_{FeENO} + 4Da\Gamma c^L_{O_2} c^2_{FeE} \right) \right) \tag{12.74}$$

FeENO

$$\frac{dc^L_{FeENO}}{d\xi} = Da \left(\lambda_{FeENO} \left.\frac{dc_{FeENO}}{dx}\right|_{x=1} + \left(c^L_{NO} c^L_{FeE} - \frac{1}{k_{eq}} c^L_{FeENO} \right) \right) \tag{12.75}$$

Fe(III)

$$\frac{dc^L_{Fe(III)}}{d\xi} = Da\left(\lambda_{Fe(III)} \left.\frac{dc_{Fe(III)}}{dx}\right|_{x=1} - \Gamma c^L_{O_2}(c^L_{FeE})^2\right) \tag{12.76}$$

Note that numerical values for the dimensionless parameters λ_{FeE} and λ_{FeENO} are not needed, because of the following equivalences derived from Eqs. (12.31), (12.58) and (12.59):

$$\lambda_{FeE} \left.\frac{dc_{FeE}}{dx}\right|_{x=1} = -\lambda_{Fe}\beta_{FeE}\left(\left.\frac{dc_{NO}}{dx}\right|_{x=0} - \left.\frac{dc_{NO}}{dx}\right|_{x=1}\right) = \lambda_{NO}\left(\left.\frac{dc_{NO}}{dx}\right|_{x=0} - \left.\frac{dc_{NO}}{dx}\right|_{x=1}\right) \tag{12.77}$$

$$\lambda_{FeENO} \left.\frac{dc_{FeENO}}{dx}\right|_{x=1} = \lambda_{FeENO}\beta_{FeENO}\left(\left.\frac{dc_{NO}}{dx}\right|_{x=0} - \left.\frac{dc_{NO}}{dx}\right|_{x=1}\right)$$
$$= \lambda_{NO}\left(\left.\frac{dc_{NO}}{dx}\right|_{x=0} - \left.\frac{dc_{NO}}{dx}\right|_{x=1}\right) \tag{12.78}$$

In addition, the value of $\lambda_{Fe(III)}$ is not important because of the assumption (12.61).

The model can be solved by discretizing the axial coordinate ξ using finite differences. Considering $k = 1 \ldots N$ points along the column, where $k = 1$ corresponds to the bottom (gas inlet) and $k = N$ corresponds to the top location (liquid inlet), the differential equations are transformed into a set of $7 \times N$ algebraic equations. For example, after applying a backward differentiation scheme, Eq. (12.66) becomes:

$$\frac{\varphi_{NO,k+1} - \varphi_{NO,k}}{\Delta\xi} = -Da\frac{\gamma_{NO}}{\kappa_{NO}}\left.\frac{dc_{NO,k+1}}{dx}\right|_{x=0}, \quad k = 2 \ldots N \tag{12.79}$$

For the liquid phase, a forward scheme must be employed. For example, Eq. (12.71) becomes:

$$\frac{c^L_{NO,k+1} - c^L_{NO,k}}{\Delta\xi} = Da\left(\lambda_{NO}\left.\frac{dc_{NO,k}}{dx}\right|_{x=1} + \left(c^L_{NO,k}c^L_{FeE,k} - \frac{1}{k_{eq}}c^L_{FeENO,k}\right)\right), \quad k = 1 \ldots N-1 \tag{12.80}$$

The remaining 7 equations needed to solve the discretized model are provided by the boundary conditions: 2 concentrations of the gas-inlet stream (known from the condition of the flue gas), and 5 concentrations of the liquid-inlet streams (calculated from the model of the bioreactor).

Typical values for an absorption tower are volume $V = 30\,m^3$, liquid fraction $\varepsilon_L = 0.05$, interfacial area $\alpha_v = 400\,m^2/m^3$, liquid volumetric flow rate $\Phi^L_V = 0.03\,m^3/s$. We also consider the NO flow in the gas feed $\Phi_{ref} = 2.4 \cdot 10^{-2}\,mol/s$; the reference concentration in the liquid phase $C_{ref}^L = 30\,\text{mol/m}^3$; reference concentration in the gas phase $C_{ref}^G = 4.15 \times 10^{-3}\,\text{mol/m}^3$. These give the following values for the parameters of the column model:

$$\text{Da} = 6.85 \times 10^8;\ \Gamma = 3.5 \times 10^{-6};\ \gamma_{NO} = \gamma_{O_2} = 6.68 \times 10^{-7};\ h_{O_2} = 40.4;$$
$$\lambda_{NO} = 7.39 \times 10^7;\ \lambda_{O_2} = 3.485 \times 10^{-7} \tag{12.81}$$

12.4.4
The Bioreactor

The task of the bioreactor is to regenerate the absorbent Fe(II)EDTA^{2-}, by carrying on the reactions (12.2) and (12.4) that, for convenience, are reproduced here:

$$6\text{Fe(II)EDTA-NO}^{2-} + C_2H_5OH \rightarrow 6\text{Fe(II)EDTA}^{2-} + 3N_2 + 2H_2O + 2CO_2$$

$$12\text{Fe(III)EDTA}^- + C_2H_5OH + 5H_2O \rightarrow 12\text{Fe(II)EDTA}^{2-}$$
$$+ 2H_2O + 2CO_2 + 12H^+$$

The first reaction is catalyzed by a mixed microbial population in which *Bacillus azotoformans* is the main denitrifying bacteria, and ethanol is the electron donor. Fe(II) is also regenerated from Fe(III) by means of reducing bacteria *Deferribacteres*, using ethanol as the electron donor. These bacteria are found in the "Veendam" and "Eerbeek" sludges, as described in a previous section.

Knowing the concentrations in the absorber-outlet liquid stream $C_K^L|_{\xi=0}$, the model of the bioreactor allows calculation of the concentrations in the reactor-outlet stream $C_K^L|_{\xi=1}$. The model consists of mass-balance equations for FeEDTA complexes:

$$C_{FeE}^L|_{\xi=1} - (C_{FeE}^L|_{\xi=0} + r_1 + r_2 + C_{FeE,add}^L) = 0 \tag{12.82}$$

$$C_{FeENO}^L|_{\xi=1} - (C_{FeENO}^L|_{\xi=0} - r_1) = 0 \tag{12.83}$$

$$C_{FeE(III)}^L|_{\xi=1} - (C_{FeE(III)}^L|_{\xi=0} - r_2) = 0 \tag{12.84}$$

In Eqs. (12.82) to (12.84), r_1 and r_2 represent the rate of FeEDTA^{2-} regeneration from FeEDTA-NO^{2-} and Fe(III)EDTA$^-$, respectively.

The following equation assumes that a concentration controller keeps the concentration of Fe complexes in the reactor-outlet stream at the constant value C_{total}:

$$C_{FeE}^L|_{\xi=1} + C_{FeENO}^L|_{\xi=1} + C_{FeE(III)}^L|_{\xi=1} - C_{total} = 0 \tag{12.85}$$

We note that our model does not include removal of (wet) solids from the reactor and other secondary reactions to nonregenerable byproducts. Therefore, the model must predict no loss of FeE complexes and the condition $C_{FeE,add}^L = 0$ can be used to check the correctness of the combined absorber–bioreactor–recycle model.

Little is known about the kinetics of the bioprocesses. A reasonable assumption is that the reaction rates are proportional to the amount of micro-organisms catalyzing the reactions. The influence of the other reactants is more complex, for example a nutrient in high concentration often has an inhibiting effect. Moreover, factors such as pH, salt concentrations, temperature, can have effects that are difficult to quantify. For this reason, we assume first-order kinetics and include all the other factors influencing the process rate in two Damköhler numbers. The following dimensionless reactor model is obtained:

$$c_{FeE}^L|_{\xi=1} - (c_{FeE}^L|_{\xi=0} + r_1 + r_2 + c_{FeE,add}^L) = 0 \qquad (12.86)$$

$$c_{FeENO}^L|_{\xi=1} - (c_{FeENO}^L|_{\xi=0} - r_1) = 0 \qquad (12.87)$$

$$c_{FeE(III)}^L|_{\xi=1} - (c_{FeE(III)}^L|_{\xi=0} - r_2) = 0 \qquad (12.88)$$

$$c_{FeE}^L|_{\xi=1} + c_{FeENO}^L|_{\xi=1} + c_{FeE(III)}^L|_{\xi=1} - C_{total} = 0 \qquad (12.89)$$

$$r_1 = Da_1 \cdot c_{FeENO}^L|_{\xi=1} \qquad (12.90)$$

$$r_2 = Da_2 \cdot c_{FeE(III)}^L|_{\xi=1} \qquad (12.91)$$

12.5
Sizing of the Absorber and Bioreactor

The main parameters having an influence on the performance of the process are the absorber volume, the interfacial area, the flow rate and the total concentration of Fe complexes in the absorber-inlet liquid stream, and the performance of the bioreactor reflected in the Damköhler numbers Da_1 and Da_2. We start by building a base-case simulation, where $\Phi_V^L = 0.03\,m^3/s$, $C_{total} = 40\,mol/m^3$, and the large values $Da_1 = Da_2 = 100$ ensure that the regeneration step does constrain the performance.

Figure 12.4 presents the NO concentration in the gas-outlet stream, as a function of the absorber size, for different values of the interfacial area. We observe that there is an optimum absorber size for getting maximum performance. This can be explained by the fact that the secondary reaction (12.3) leads to $FeEDTA^{2-}$ deactivation if the absorber size is too large. The second observation concerns the necessary interfacial area. In order to reduce the NO concentration below 10 ppm (10% of the initial value), 300 m^2/m^3 are needed when the $FeEDTA^{2-}$ complex is completely regenerated in the bioreactor. Expecting that the regeneration step will be incomplete, we choose an interfacial area of 400 m^2/m^3. The maximum performance is obtained for an absorber of 25 m^3, and corresponds to 5 ppm NO in the gas-outlet stream. For a liquid fraction $\varepsilon_L = 0.05$, the required interfacial area corresponds to droplets of 0.75 mm diameter. We also note the concentrations of Fe complexes in the absorber-outlet liquid flow: $c_{FeE} = 0.41$, $c_{FeENO} = 0.025$, $c_{Fe(III)} = 0.89$, showing that the oxidation reaction still has an important effect. At the same time, the concentrations in the absorber-inlet flow are $c_{FeE} = 1.32$, $c_{FeENO} = 2.5 \times 10^{-4}$, $c_{Fe(III)} = 8.8 \times 10^{-3}$, showing that the regeneration is complete.

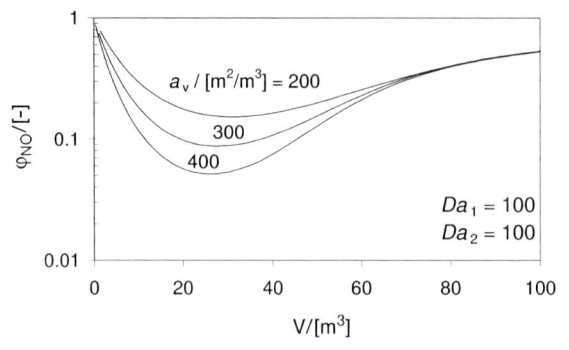

Figure 12.4 Performance of the BioDeNOx process as a function of the absorber size, for different values of the interfacial area.

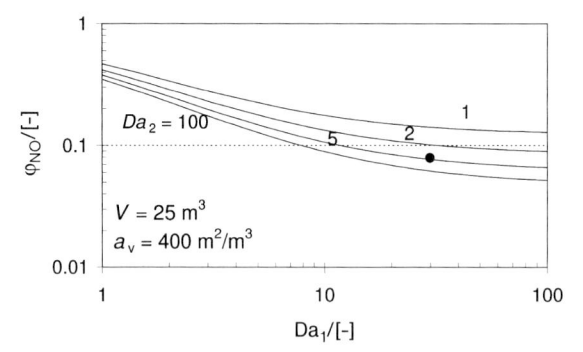

Figure 12.5 Performance of the BioDeNOx process as a function of the bioreactor performance. The nominal operating point is represented by the black dot.

Figure 12.5 presents the performance of the process when the bioreactor Damköhler numbers, Da_1 and Da_2, are varied. As expected, when the activity of either microbial population is reduced below a certain limit, the requirement of NO concentration in the gas-outlet stream can no longer be fulfilled. In order to ensure robustness with respect to uncertain rates of the bioreactions, we choose $Da_1 = 30$, $Da_2 = 5$. From Figure 12.5 it follows that the requirements will be met even when the activity of both microbial populations is reduced to half of the design value.

Figure 12.6 presents concentration profiles along the absorption column. In the gas phase (top plot), NO is absorbed and its concentration drops below the 0.1 limit at the top of the column ($\xi = 1$). Although only a small fraction of O_2 in the gas-inlet stream is absorbed, the amount is significant because of the large concentration. Profiles in the liquid phase (bottom plot) show that the performance of the bioreactor is very good, as the liquid-inlet stream ($\xi = 1$) contains mainly Fe(II)EDTA^{2-}. The secondary reaction (12.3) is very important, and therefore the concentration of Fe(III)EDTA$^-$ in the liquid-outlet stream ($\xi = 0$) is quite high.

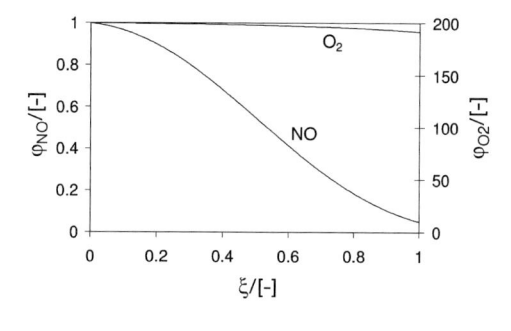

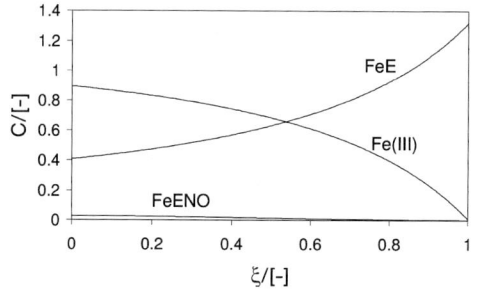

Figure 12.6 Concentration profiles along the absorption column. $V = 25 \, m^3$; $a_v = 400 \, m^2/m^3$; $Da_1 = 30$; $Da_2 = 5$.

Figure 12.7 presents the gas- and liquid-outlet concentrations when the gas feed rate is reduced. It should be remarked that the gas turbines produce higher NO and O_2 concentrations when the gas flow rate is lower. Linear relationships between concentrations and flow rate are assumed. It turns out that the most demanding conditions are around $14\,000 \, N \, m^3/h$. However, the NO concentration in the gas-outlet stream remains below the 10 ppm limit (0.1 dimensionless concentration).

12.6
Flowsheet and Process Control

Figure 12.8 presents a simplified flowsheet of the BioDeNOx process, together with the main control loops. Several units are needed, in addition to the absorber and the bioreactor. Firstly, the temperature of the flue gas is very high and energy can be recovered in the form of high-pressure steam. Moreover, the absorption process is faster at lower temperature. For these reasons, two heat exchangers are inserted on the gas-inlet stream. Secondly, obtaining a large interfacial area requires a small diameter of the liquid-spraying nozzles. Because the liquid coming from the bioreactor contains biomass (solids), it has to be carefully filtered. Finally, there might be other secondary reactions not taken into account in the modelling part. These may lead to degradation of Fe-complexes into products that cannot be regenerated. To avoid their accumulation, a liquid bleed stream is provided.

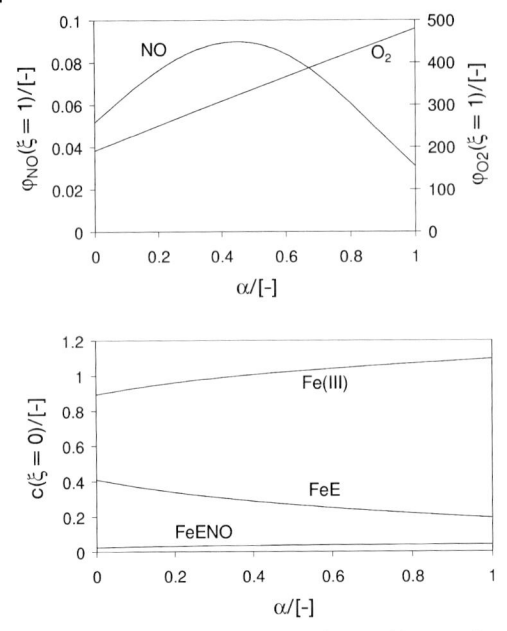

Figure 12.7 Outlet concentrations for variable gas-inlet conditions. $V = 25\,m^3$; $a_v = 400\,m^2/m^3$; $Da_1 = 30$; $Da_2 = 5$; $\Phi_v^G/[N\,m^3/h] = (1-\alpha)\cdot 20000 + \alpha\cdot 10000$; $c_{NO,\,\xi=0}/[ppm] = (1-\alpha)\cdot 100 + \alpha\cdot 300$; $c_{O_2,\,\xi=0}/[\%] = (1-\alpha)\cdot 2 + \alpha\cdot 5$.

The process-control scheme contains loops for maintaining concentration of the absorbent (by manipulating the fresh FeEDTA^{2-} feed), the pH and the Redox potential in the bioreactor (by addition of acid/base and ethanol, respectively). Other control loops maintain the liquid level in the sump of the absorber, and the level in the bioreactor.

12.7
Conclusions

This case study presents the design of a biochemical process for NO$_x$ removal from flue gases, where an absorber and a bioreactor are the main units. Based on a rough estimation of Hatta numbers, it was concluded that a spray tower offering a large G/L interfacial and a small liquid fraction is the best type of equipment, favoring the main chemical reaction. The bioreactor was chosen as a CSTR.

The reactive absorption was modeled based on the film theory. After transforming in dimensionless form, the mathematical model was solved numerically. This required a considerable computing effort. The results obtained in this way allowed the formulation of reasonable assumptions, which were the starting point of a

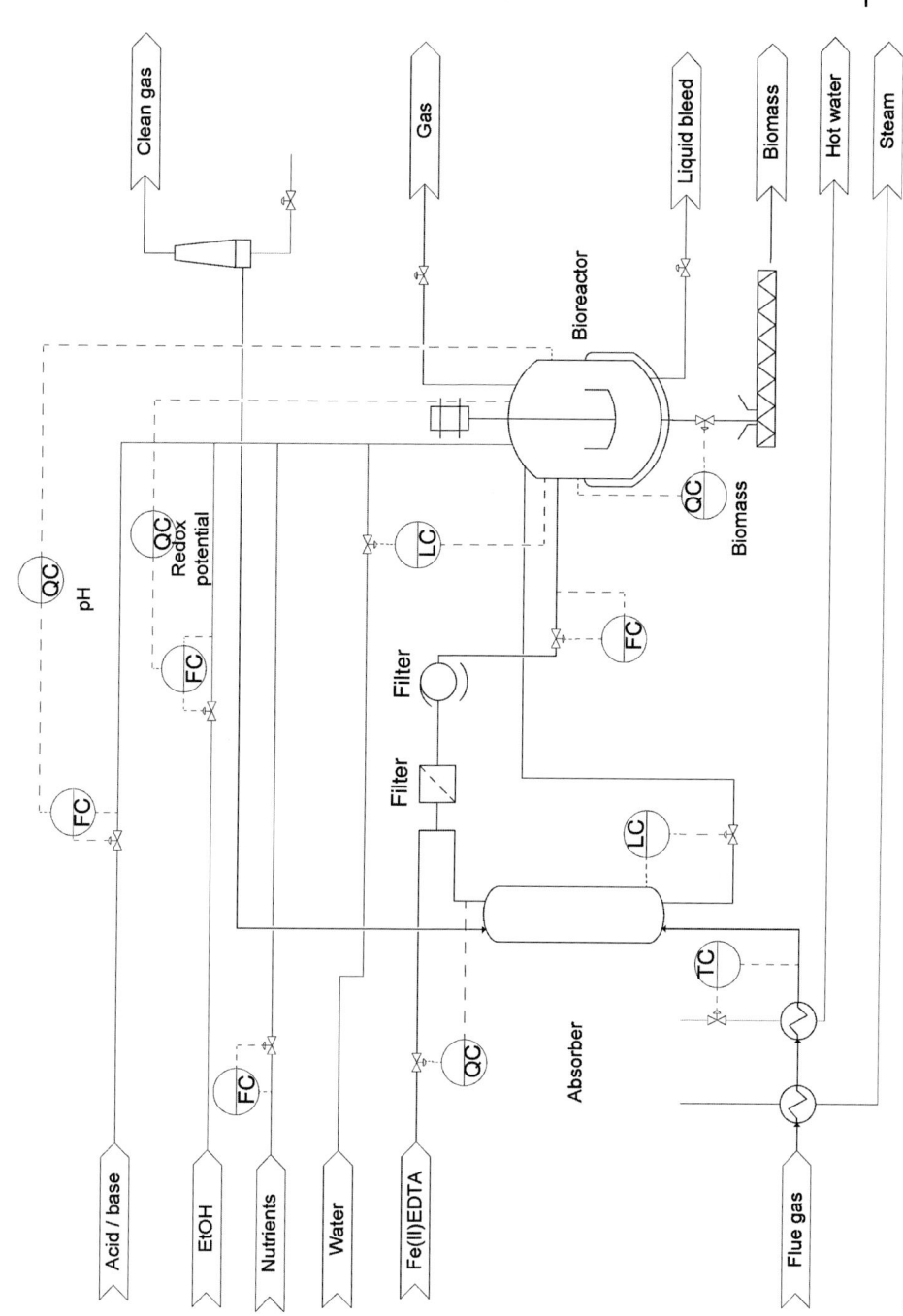

Figure 12.8 Flowsheet and process control of the BioDeNOx process.

simplified model. An analytical formula for the absorption rate was derived, which was incorporated in the model of the absorber.

A simple model of the bioreactor was used, assuming first-order kinetics with respect to FeEDTA species and lumping in two Damköhler numbers the effects of various factors such as micro-organisms concentration, pH, salt concentrations and temperature.

Sizing of the absorption column started from a base case that assumed complete recovery of FeEDTA^{2-} in the bioreactor. Then, sensitivity studies provided the values of the G/L interfacial area and of the absorber volume giving maximum performance. The values of the two Damköhler numbers characterizing reactor performance were found after relaxing the assumption of complete FeEDTA^{2-} recovery. Finally, the specification of NO$_x$ concentration in the purified gases was checked, for different feed conditions.

The process-control scheme contains loops for maintaining concentration of the absorbent, the pH and the Redox potential in the bioreactor. Other control loops maintain the liquid level in the sump of the absorber, and the level in the bioreactor.

References

1 Jin, Y., Veiga, M.C., and Kennes, C., 2005, Bioprocesses for removal of nitrogen oxides from polluted air, *J. Chem. Tech. Biotech.*, 80, 483–494

2 van der Maas, P., van de Sandt, T., Klapwijk, B., and Lens, P., 2003, Biological reduction of nitric oxide in aqueous Fe(II)EDTA solutions, *Biotechnol. Prog.*, 19(4), 1323–1328

3 Kumaraswamy, R., Muyzer, G., Kuenen, J.G., and van Loosdrecht, M.C.M., 2004, Biological removal of NOx from flue gas, *Water Sci. Tech.* 50(6), 9–15

4 Kumaraswamy, R., van Dongen, U., Kuenen, J.G., Abma, W., van Loosdrecht, M.C.M., and Muyzer, G., 2005, Characterization of microbial communities removing nitrogen oxides from flue gas: the BioDeNOx process, 2005, *Appl. Env. Microbiol.*, 71(10), 6345–6352

5 van der Maas, P., Harmsen, L., Weelink, S., Klapwijk, B., and Lens, P., 2004, Denitrification in aqueous FeEDTA solutions, *J. Chem. Tech. Biotech.*, 79, 835–841

6 van der Maas, P., van den Bosch, P., Klapwijk, B., and Lems, P., 2005, NOx removal from flue gas by an integrated physicochemical absorption and biological denitrification process, *Biotech. Bioeng.*, 90(4), 433–441

7 Demmink, J.F., van Gils, I.C.F, and Beenackers, A.C.M., 1997, Absorption of nitric oxide into aqueous solutions of ferrous chelates accompanied by instantaneous reaction, *Ind. Eng. Chem. Res.*, 36, 4914–4927

8 Huasheng, L. and Wenchi, F., 1988, Kinetics of absorption of nitric oxide in aqueous Fc(II)-EDTA solution. *Ind. Eng. Chem. Res.* 27, 770

9 Gambardella, F., Alberts, M., Winkelman, J., and Heeres, E. J., 2005, Experimental and modeling studies on the adsorption of NO in aqueous EDTA solutions, *Ind. Eng. Chem. Res.*, 44, 4234–4242

10 Zang, V. and van Eldik, R., 1990, Kinetics and mechanism of the autoxidation of iron(II) induced through chelation by ethylenediaminetetraacetate and related ligands, *Inorg. Chem.*, 29, 1705–1711

11 Wubs, J.H. and Beenackers, A.C.M., 1993, Kinetics of the oxidation of ferrous chelates of EDTA and HEDTA in aqueous

solution, *Ind. Eng. Chem. Res., 32*, 2580–2594
12 van der Maas, P., van de Sandt, T., Klapwijk, B., and Lens, P., 2003, Biological reduction of nitric oxide in aqueous Fe(II)EDTA solutions, *Biotechnol. Prog., 19 (4)*, 1323–1328
13 Sada, E., Kumazawa, H., and Takada, Y., 1984, Chemical reactions accompanying absorption of NO into aqueous mixed solutions of Fe(II)-EDTA and Na2SO3. *Ind. Eng. Chem. Res. 26*, 1468
14 Gambardella, F., NO and O2 absorption in aqueous FE(II)EDTA solutions, PhD dissertation, State University Groningen, 2005
15 Wise, D.L. and Houghton, G., The diffusion coefficient of ten slightly soluble gases in water at 10–60°C, *Chem. Eng. Sci.*, 1996, *21*, 999–1010
16 Kierzenka, J. and Shampine, L.F., 2001, A BVP Solver based on Residual Control and the MATLAB PSE, *ACM TOMS, 27*, No. 3, 299–316
17 Yih, S.-M. and Lii, C.-W., Absorption of NO and SO_2 in Fe-(II)-EDTA solutions I: absorption in a double stirred vessel, *Chem. Eng. Commun*, 1998, *73*, 43

13
PVC Manufacturing by Suspension Polymerization

13.1
Introduction

13.1.1
Scope

The selected case study illustrates a typical problem of product design. The key issue is not the flowsheet configuration and the sizing of equipment, as in continuous processes, but the optimal operation procedure of an expensive hardware that has to ensure the quality of a variety of products at the most cost-effective level. At the same time this application pinpoints the most important features of a batch process, such as may be encountered in other industries as fine chemicals and biotechnologies. Here, the variety of products asks for great flexibility in the operation of the batch reactor, each grade being associated with a recipe consisting of a certain amount of ingredients and of a particular temperature profile. In addition, the supervision by computer control plays a determinant role.

In particular, this application illustrates the key features of a product design in the field of polymer technology, such as building the polymerization recipe and the reactor operating procedures necessary to achieve different polymer grades. The link between these aspects is realized by means of detailed kinetic modeling of the polymerization reaction, including the molecular-weight distribution. Reactor-design issues are examined from the viewpoint of heat-transfer intensification. These elements are imbedded in a dynamic reactor model with control-system implementation, simulated in Matlab. The results are expressed as profiles in time of temperature, concentrations of monomer, initiator and MWDs. From the simulation, important decisions can be taken regarding technical measures for improving the technology, as well as its optimization.

13.1.2
Economic Issues

Polyvinylchloride (PVC) is among the largest commodity plastics with a world production of about 30 million tonnes in 2006. The yearly growth in the decade

1990–2000 was about 3.8%, of which 4.5% was in Asia, 3% was in North America and 1.7% was in Western Europe. New large-scale facilities are today in construction in China and India.

PVC has a broad range of applications. The largest fraction is used in civil engineering, as tubing, profiles, cables, wallpapers and floor coverings. In the automotive industry PVC is used extensively for dashboards, interior modules, seating, wiring, and body protection. In everyday life, PVC is involved in countless applications, from synthetic leather, food packaging, bottles and toys, to medical objects, such as blood and medication containers, tubing, medical gloves, *etc.*

The presence of chlorine in PVC raised some environmental concerns, particularly the formation of HCl and dioxine during fire outbreak or waste incineration. However, scientific studies demonstrated that the PVC is no more hazardous than other plastic materials, the danger resulting mostly from the inherent high temperature of combustion and CO formation and not from HCl. On the contrary, PVC develops proportionally less CO_2 by combustion than other plastics. Moreover, reusing large PVC parts in the car industry is possible by employing modern recycling techniques.

From the manufacturing viewpoint the progresses achieved in technology makes the modern VCM and PVC processes among the most safe and environmental friendly. Because of the very large production scale, even incremental improvements are imperative for competitiveness. A first series of measures aims at the lowest cost of the monomer, by employing highly integrated processes with minimum waste and maximum energy recovery, or by making use of cheaper raw materials, such as ethane instead ethylene. Secondly, a high efficiency of the polymerization process can be achieved by adopting the technology of large reactors, multi-initiator recipes and advanced process control.

From the process-design perspective the versatility of applications of PVC demands a precise adjustment of material properties to quality requirements, which are for the most part determined during the polymerization stage, namely by molecular-weight distribution (MWD) and the morphology of particles.

Today 80% of PVC is manufactured by the technique of batchwise suspension polymerization (S-PVC), the remaining part being shared between emulsion and bulk polymerization. A distinctive characteristic of the S-PVC processes is the large size of the reactors, 50 to 200 m^3, and operation at 10 to 12 bar pressure. The reaction time has been reduced from 18 h in the 1960s to only 3.5–5 h today by making use of fast initiation systems and recipe optimization.

Table 13.1 presents the evolution of economic indices in the period 1960–2000. The technology of large reactors was developed in the period 1960–1980, while very small batches and energy saving occurred after 1980. With respect to the cost structure one may estimate that the VCM contribution is about 80%, while the plant fixed fees are under 10%. The conclusion is that the suspension PVC process may be considered as a mature and advanced technology.

13.2 Large-Scale Reactor Technology

Table 13.1 Improvement of technology for a large size S-PVC plant [1].

Item	Unit	1960s	1980s	2000s
Capacity	t/h	34 000	114 000	120 000
Reactor volume	m^3	17 × 13.3	2 × 170	2 × 150
Cycle time	h	18	7-11	5.5
VCM	t/t	1.045	1.020	1.003
Chemicals	kg/t	1.2	1.2	1.2
Electric power	kW h/t	340	350	230
Steam	t/t	1.7	2.2	0.9
Pure water	t/t	5.0	2.2	3.1
Productivity	t/m^3/month	13.6	30.4	36.4

13.1.3
Technology

A standard flowsheet for PVC manufacturing is presented in Figure 13.1. The key unit is the polymerization autoclave provided with a heating/cooling jacket and a condenser. Initially, water, VCM, dispersant agent and initiator are charged under mixing, the ratio water/monomer being about 1.2 (weight). The role of water is to ensure efficient removal of the heat developed by polymerization, since the reaction is highly exothermic, ~100 kJ/mol. Uniform temperature is fundamental for the process safety and product quality. The reaction mass is heated to the reaction temperature, usually 50 to 60 °C. A total pressure of about 8 to 10 bar is necessary to bring the mixture under boiling conditions. The pressure remains constant as long as there is free monomer in drops. At a certain point the fall tendency of the pressure is noticed, when the so-called *critical conversion* is reached. From this point on the reaction takes place only in the solid grains that are swollen with monomer. The batch is stopped when the pressure drops below 5–6 bar, or in conversion terms at about 85–90%. Then, the mixture is evacuated through a blow-down drum for VCM degassing and polymer separation by centrifugation, followed by drying, screening and storage. The unconverted VCM is recycled. A typical polymerization recipe is shown in Table 13.2.

13.2
Large-Scale Reactor Technology

The progress of technology for PVC reactors is one of the most interesting in chemical engineering. Figure 13.2 presents the evolution of shape from 4 m^3 in 1950 to 60–200 m^3 today. The medium-size autoclaves of 20–30 m^3 in the 1960s were provided with a glass-lined wall, a cooling jacket and a top mixer. Larger reactors of 30–60 m^3 were equipped with a bottom mixing device to avoid shaft

13 PVC Manufacturing by Suspension Polymerization

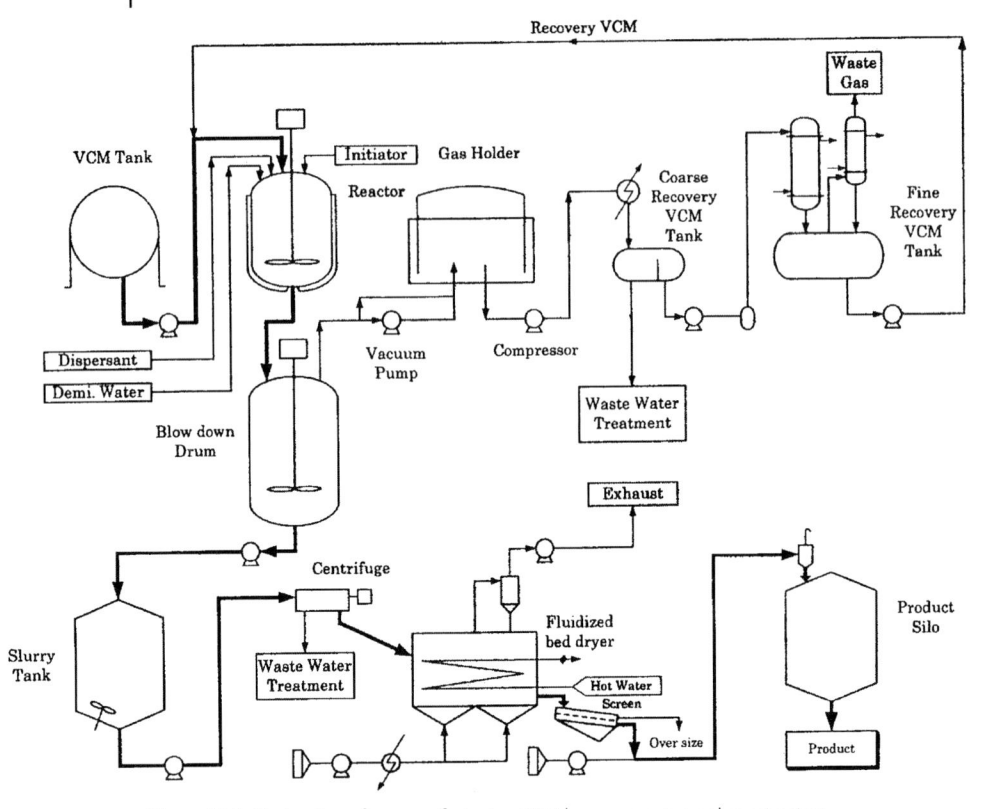

Figure 13.1 Technology for manufacturing PVC by suspension polymerization.

Table 13.2 Typical polymerization recipe for PVC manufacturing.

VCM	100 parts
Water	120 parts
Suspending agent (PVA, etc.)	0.05–0.10 parts
Initiator (peroxy compounds, etc.)	0.03–0.2 parts
Polymerization temperature	55–60
Polymerization time	6–8 h
Conversion	85–90%
Initial pressure	8–10 bar
Final pressure	5 bar

vibration. In the 1970s very large reactors were designed by different companies, such as Shinetsu in Japan and Hüls in Germany, of 130 and 200 m^3, respectively. Today, the technology of large reactors is mature and largely adopted for the new units because of significant lower costs of hardware, construction, instrumentation and manpower.

(d) 60~200 m³   **(c) 30~60 m³**   **(b) 10~30 m³**   **(a) 4 m³**

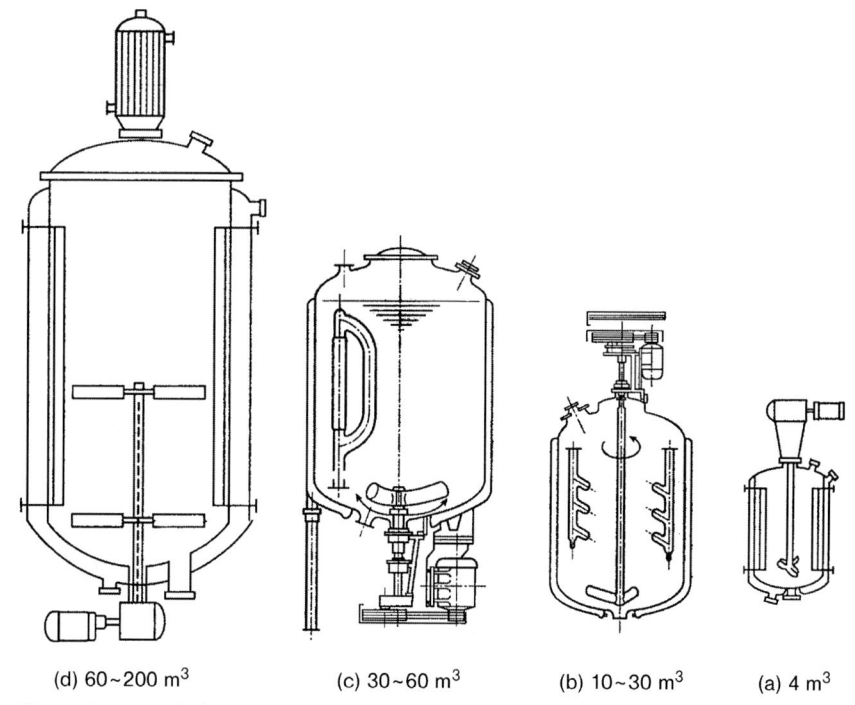

Figure 13.2 Typical shape of the polymerization reactor.

13.2.1
Efficient Heat Transfer

A key aspect in the polymerization reactor design is achieving good heat transfer. The specific heat-transfer area diminishes drastically with the increasing volume:

$$S/V \approx 1/D \approx 1/V^{1/3} \tag{13.1}$$

For example, scaling up from 30 to 100 m³ reduces the S/V from 1.5 to 1. The effective heat-transfer area depends on the shape factor H/D, which in turn is determined by the technology of the mixing device. As rule of thumb, a reactor of 100 m³ has an external heat-transfer surface of about 100 m² [1].

A fundamental aspect in the reactor design is the contribution of different thermal resistances in achieving a highly efficient heat transfer. The overall heat coefficient U is given by the relation:

$$1/U = 1/h_i + 1/h_{if} + 1/h_t + 1/h_{of} + 1/h_o \tag{13.2}$$

The partial heat-transfer coefficients correspond to inside film transfer (h_i), inside fouling (h_{if}), wall conduction (h_t), outside fouling (h_{of}) and outside jacket (h_o). The coefficient h_o in the jacket may rise to over 6500 W/m² K ensuring a high water velocity by a coil-type construction. The reactor heat-transfer coefficient h_i should have large values too, since high turbulence is ensured by mixing. As an order of magnitude this may take values of about 3500 W/m² K. A possible decline with time might take place because of the increased viscosity of the suspension with conversion. The thermal resistance of the wall is of significance too because stainless steel has a low thermal conductivity. For example, the wall thickness is about 20 mm for a volume of 10 m³, but 40 mm for 100 m³, which results in a decrease of the overall heat-transfer coefficient of about 25%. An interesting solution was found by developing an inner jacket so as to form a spiral channel where the cooling water can circulate at higher speed under pressure. The thicker wall calculated to support high pressure is outside the jacket, not against the reaction space, as shown in Figure 13.3. In this way, the wall thickness may be reduced to about 10 mm. By considering an average conductivity of 30 W/m K metal (8 mm carbon steel and 2 mm stainless steel) the heat transfer through the wall is equivalent to 3000 W/m² K. Inserting the above values in Eq. (13.2) leads to $U = 1(1/6500 + 1/3500 + 1/3000) \approx 1300$ W/m² K, an exceptionally high value for such large reactors.

At this stage, we should mention another important factor that can affect drastically the heat-transfer capability of a polymerization reactor: fouling by polymer

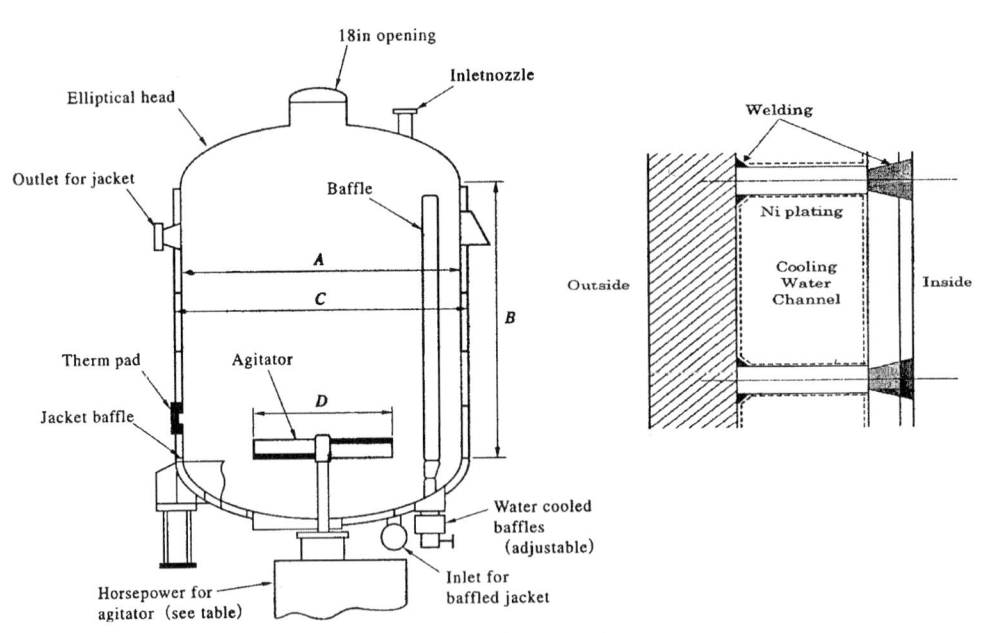

Figure 13.3 Geometric characteristics of a large-scale reactor with a bottom mixing device, segmented heating/cooling jacket, internal cooled baffles and external condenser. The right picture illustrates the cooling-water channel designed for higher pressure in order to reduce the wall thickness.

deposit (scale). To get an order of magnitude let us consider a thin uniform polymer layer of 200 μm. Taking for PVC conductivity the value 0.20 W/m K leads to a thermal resistance corresponding to 1000 W/m² K. The result is that the thermal resistance by polymer fouling might control the overall heat-transfer coefficient. In fact, the scale prevention was the most important obstacle in developing the large reactor technology. In addition, the contamination of scale into PVC causes fish-eye of product. Today, this problem is largely solved by using special antifouling formulation, electrical polishing of stainless steel surface and water jet cleaning. These measures can ensure the operation over several batches without the need for cleaning [1].

Even with these improvements the heat-removal capacity of the reactor vessel is not sufficient when using fast initiators. By the addition of an external condenser the cooling capacity may increase by about 30% to 50% [1].

13.2.2
The Mixing Systems

Besides intensive heat transfer, the mixing system has the key task of ensuring an adequate quality of the polymer product. The energy dissipated by agitation controls the particle formation and the distribution of sizes. During the polymerization the structure of the particles evolves from a simple liquid drop to a solid grain with complicated morphology. This results in physical properties such as bulk density, porosity and grain size, which in turn are determinant for the end-use polymer properties, such as rheological behavior, plasticizer adsorption and fish-eye formation. The grains of 50–250 μm (average 130) are formed by the agglomeration of several microparticles, as illustrates by Figure 13.4. Microparticles of polymer precipitate very early from the monomer phase, already at about 2% conversion. From this point on the polymerization takes place by a two-phase mechanism, as explained later.

Initially, the droplets have a size of approximately 40 μm. Experimental studies in reactors of different sizes showed that the droplet mean size can be correlated

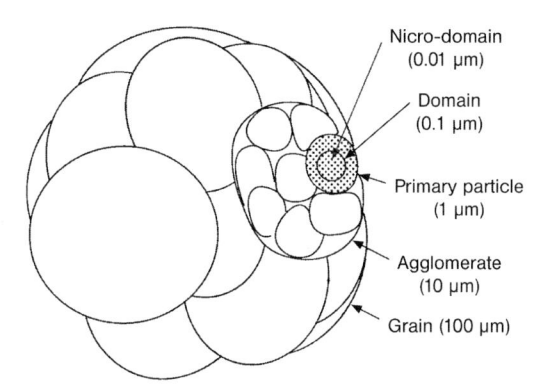

Figure 13.4 Two-phase model for vinyl-chloride polymerization in suspension.

13 PVC Manufacturing by Suspension Polymerization

by the Weber number. This is defined by We = $\rho N^2 D^3/\sigma$, in which ρ is the density of the mass, N the rotation speed, D the mixer diameter and σ the interfacial tension. The average droplet is obtained for We in the range 8×10^5 to 10×10^5. Accordingly, the following rules can be applied for the scale-up of the mixing system [1]:

1. The power consumption per volume must be kept constant.
2. The pumping capacity per volume also must be kept constant.
3. A relatively larger ratio of blade to tank diameter is needed for larger reactors.

13.2.3
Fast Initiation Systems

A large variety of initiators is employed in PVC manufacturing. The most common are dialkyl peroxydicarbonates, such as di(2-ethylhexyl) peroxydicarbonate, or peroxyesters. The temperature profile may also follow an optimal pattern according to the desired quality grade. This aspect may become a subtle problem of operation and control. Figure 13.5 presents a typical reaction profile for a fast polymerization batch, where the goal is to achieve high productivity. The use of a single fast initiator (dotted line) gives a sharp peak of the reaction rate, which in turn causes excessive heat generation and difficulty in controlling the temperature constant. Employing a cocktail of initiators (continuous line) can ensure a more uniform reaction rate and avoid the development of a hot spot. The cooling capacity of the reactor can be used more efficiently over most of the batch time. The conclusion

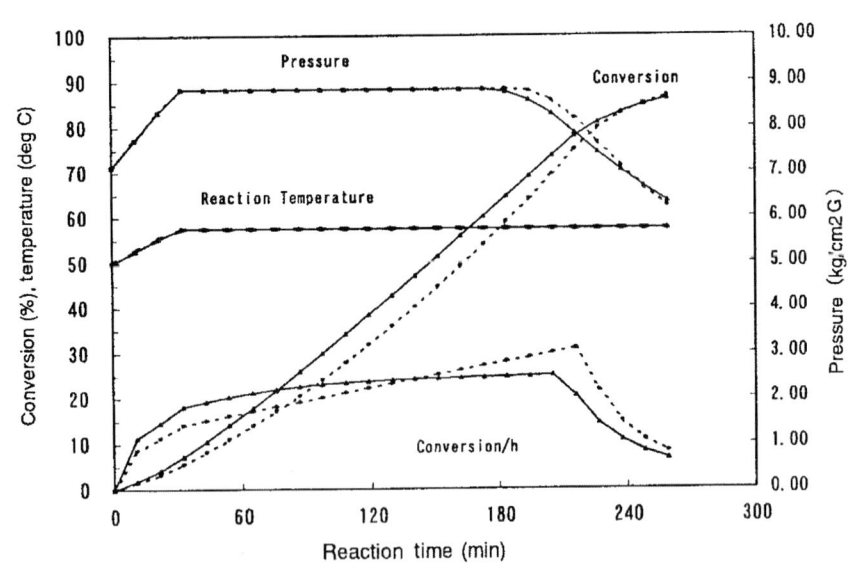

Figure 13.5 Reaction profile of a fast polymerization batch; dotted line: di-2-ethylhexyl peroxy dicarbonate (OPD); solid line: OPD + cumyl peroxy neodecanate (after Saeki et al. [1]).

is that the thermal design of the reactor and the development of an optimal polymerization recipe should be co-ordinated. This goal can be achieved by employing computer simulation, as illustrated later in this chapter.

13.3
Kinetics of Polymerization

From the physical viewpoint the polymerization of vinyl chloride in suspension implies firstly the formation of the droplets of monomer in water, which finally are converted in solid particles. As order of magnitude the PVC grains are ~130 µm in size. The polymerization starts in the monomer droplets similarly as the bulk polymerization. A particular feature of this process is that PVC is not soluble in its own monomer. By precipitation and agglomeration of macromolecules a polymer-rich phase appears, even shortly after the onset of reaction, for example at conversion under 0.1% at 50°C, in which a certain amount of monomer is trapped. Consequently, a two-phase mixture may be imagined as coexisting in each droplet, respectively, the monomer-rich phase and the polymer-rich phase. The polymerization reaction takes place in the both phases, the first diminishing and the second growing. This situation holds up to a certain critical conversion X_f, where the monomer-rich phase disappears. The value of X_f depends on temperature, for example it is 73% at 50°C, but it can be physically recorded by the incipient fall in pressure. Beyond this point the polymerization proceeds only by consuming the monomer captured in the polymer-rich phase up to the final conversion, of about 90%.

The above physical assumption known as *the two-phase model* was proposed forty years ago by Talamini [2] and reviewed more recently by the same author [3] in the light of experience accumulated. With different modifications this model has been used in numerous studies issued from academia, but has proved to be a good support also for describing the behavior of real industrial systems. The key assumptions of this model are:

1. The two phases are in thermodynamic equilibrium, thus the composition of each phase remains constant throughout the reaction.
2. The ratio between the steady-state concentrations of free radicals in the two phases is constant.
3. The partition ratio of the initiator between the two phases does not vary throughout the reaction.

These assumptions will be adopted in the present study. The main conclusion is that monomer consumption and the development of polymer species may be considered as taking place independently for each phase, but with different rates. If R_m is the specific polymerization rate in the monomer-rich phase and R_c in the polymer-rich phase one can write:

$$R_c = QR_m \tag{13.3}$$

Q is larger than one expressing the fact that the polymerization rate in the polymer-rich phase is sensibly higher than in the monomer-rich phase. The physical explanation comes from the so-called "gel effect", which in this case consists of a slower termination reaction rate because of higher viscosity or of hindered diffusion of macroradicals. Consequently, a higher concentration of live radicals and a higher polymerization rate is obtained. The molecular-weight distribution is affected accordingly.

A typical polymerization mechanism by free radicals includes the following steps: initiation, propagation, chain transfer, formation of chain defect structure, termination and inhibition, as listed below:

1. $I \xrightarrow{k_d} 2A$ initiator decomposition
2. $A + M \xrightarrow{k_i} R_1$ formation of primary radicals
3. $R_x + M \xrightarrow{k_p} R_{x+1}$ overall propagation (head-to-tail, etc.)
4. $R_x + M \xrightarrow{k_{tm}} P_x + R_1$ chain transfer to monomer
5. $R_x + P_y \xrightarrow{k_{tp}} P_x + R_y$ chain transfer to polymer
6. $R_x \xrightarrow{k_b} R_x$ intramolecular transfer (backbiting)
7. $R_x + R_y \xrightarrow{k_{td}} P_x + P_y$ termination by disproportionation
8. $R_x + R_y \xrightarrow{k_{tc}} P_{x+y}$ termination by combination
9. $A + H \xrightarrow{k_{inh}} D$ inhibition

Other reactions may be taken into consideration, with an effect on polymer structure, namely the formation of short- and long-chain branches. A complete list of reactions in S-PVC polymerization may be found in Kiparissides et al. [5]. On the above basis kinetic equations may be written. To keep it simple the chain transfer, back-biting and inhibition reactions are disregarded, while termination is considered to occur only by disproportionation. The elementary reaction rates for initiator decomposition and free radicals generation are as follows:

$$\frac{d[I]}{dt} = -\frac{1}{2}\frac{d[A]}{dt} = -k_d[I] \tag{13.4}$$

$$\frac{d[R_1]}{dt} = r_{in} + k_{tm}[M]\sum_{y=1}^{\infty}[R_y] - k_p[R_1][M] - k_{tm}[R_1][M] - k_t[R_1]\sum_{y=1}^{\infty}[R_y] \tag{13.5}$$

$$\frac{d[R_x]}{dt} = k_p[R_{x-1}][M] - k_p[R_x][M] - k_{tm}[R_x][M] - k_t[R_x]\sum_{y=1}^{\infty}[R_y] \tag{13.6}$$

In the above scheme $\sum_{y=1}^{\infty}[R_y]$ denotes the total concentration of x-mer radicals. In order to compute the distribution of radicals the following function may be introduced:

$$H(z,t) = \sum_{x=1}^{\infty} z^x [R_x] \tag{13.7}$$

By successive multiplication of Eqs. (13.4) to (13.6) with z^x and summation, one gets:

$$\frac{dH(z,t)}{dt} = zr_{in} + k_p[M](z-1)H(z,t) - k_t H(1,t)H(z,t) \tag{13.8}$$

It may be observed that $H(1,t) = \sum_{x=1}^{\infty}[R_x]$ represents the total concentration of radicals. For $z = 1$, Eq. (13.8) becomes:

$$\frac{dH(1,t)}{dt} = r_{in} - k_t H^2(1,t) \tag{13.9}$$

Equation (13.9) describes the evolution of radical concentration. Numerical calculation shows that the stationary state is reached very fast in most cases, in fractions of a second. Therefore, the hypothesis of constant concentration of radicals is justified. At steady state one obtains:

$$\sum_{x=1}^{\infty}[R_x] = (r_{in}/k_t)^{1/2} \tag{13.10}$$

Equation (13.10) indicates that the total concentration of macroradicals depends on the ratio of initiation to termination rates, but not on the propagation rate. In numerous studies only the initiator decomposition is considered as rate limiting for the formation of first radicals, and therefore we may write:

$$r_{in} = 2fk_d[I] \quad \text{with} \quad [I] = [I_0]\exp(-k_d t)/(1 - \varepsilon X) \tag{13.11}$$

where ε is a volume-contraction factor due to the difference in density between monomer and polymer defined as:

$$\varepsilon = (\rho_p - \rho_m)/\rho_p \tag{13.12}$$

The formation of dead polymers is described by the following relation, which considers only the disproportionation:

$$\frac{d[P_x]}{dt} = k_t[R_x]\sum_{y=1}^{\infty}[R_y] \tag{13.13}$$

The monomer consumption rate is given by:

$$\frac{d[M]}{dt} = k_p[M]\sum_{y=1}^{\infty}[R_y] = k_p[M](r_{in}/k_t)^{1/2} \tag{13.14}$$

The above mechanism captures the main features of a generic radical polymerization, but has to be adapted in the case of PVC. Following Talamini [2–4], the acceleration of polymerization in the polymer-rich phase is due to a much slower termination reaction because of the more difficult diffusion of polymer radicals. Accordingly, the concentration of radicals increases, as expressed by Eq. (13.10).

This phenomenon is called the "gel effect". The above physical picture was contradicted by several researchers, in the first place by Xie and Hamielec [5, 7, 8], who proposed a mechanism based on the precipitation and migration of living macroradicals to micropolymeric domains. Although physically attractive, the model based on radical migration contains a number of parameters that are difficult to measure. Dimian et al. [9] showed that a similar model is appropriate for simulating the dynamics of large industrial reactors, but with kinetic parameters tuned on experimental data. More recently, Talamini [3, 4] reviewed the criticism of his two-phase model, arriving at the conclusion that the objections raised by the other works have limited practical consequences.

Summing up, the two-phase model is physically consistent and may be applied for designing industrial systems, as demonstrated in recent studies [10, 11]. Modeling the diffusion-controlled reactions in the polymer-rich phase becomes the most critical issue. The use of free-volume theory proposed by Xie et al. [6] has found a large consensus. We recall that the free volume designates the fraction of the free space between the molecules available for diffusion. Expressions of the rate constants for the initiation efficiency, dissociation and propagation are presented in Table 13.3, together with the equations of the free-volume model.

Equations (13.4) to (13.6), (13.13) and (13.14) can be upgraded to consider a two-phase model, as will be presented in Section 13.6.2.

13.3.1
Simplified Analysis

A simplified analysis can be useful for checking the kinetic constants. Based on Eq. (13.3) the following equation for conversion rate may be written:

$$\frac{dX}{dt} = R_m(1 - X - AX) + QR_m AX = R_m(1 + qX) \tag{13.15}$$

A is the weight ratio of monomer to polymer in the polymer-rich phase and Q an acceleration factor due to the gel effect. The following relations hold:

$$A = \frac{1 - X_f}{X_f} \quad \text{and} \quad q = QA - A - 1 \tag{13.16}$$

Values of q and Q have been regressed by Talamini [3] from experimental data. These are, for example, 4.0–6.0 at 50°C for suspension polymerization initiated by lauroil peroxide (LPO). R_m is the specific polymerization rate in the monomer phase given by

$$R_m = \frac{k_p}{\sqrt{2k_{tm}}} r_{in}^{0.5} \quad \text{with} \quad r_{in} = 2 f k_d [I_0] \exp(-k_d t) \tag{13.17}$$

Table 13.3 Kinetic constants for the suspension polymerization of vinyl choride.

Before the critical conversion

Dissociation	$k_{d,1} = 2.31 \times 10^{15} \times 60 \times \exp\left(-\dfrac{14653}{T}\right)$	min^{-1}
	$k_{d,2} = k_{d,1}$	
Propagation	$k_{p,1} = 3 \times 10^9 \times 60 \times \exp\left(-\dfrac{3320}{T}\right)$	m^3/kmol/min^{-1}
	$k_{p,2} = k_{p,1}$	
Termination	$k_{t,1} = 1.3 \times 10^{12} \times 60 \times \exp\left(-\dfrac{2114}{T}\right)$	m^3/kmol/min^{-1}
	$k_{t,2} = \dfrac{k_{t,1}}{\alpha^2}$	
	$\alpha = 24 * \exp\left[1007\left(\dfrac{1}{T} - \dfrac{1}{333.15}\right)\right]$	

After the critical conversion

Initiation efficiency	$f_{2,2}^{1/2} = \left(\dfrac{k_{p,1}(X_f)}{k_{p,2,2}(X)}\right)^2 f_2^{1/2} \exp\left(-\dfrac{D}{V_{corr}}\right)$
	$D = 16.04 \times 10^4 \times \exp\left(-\dfrac{3464}{T}\right)$
Dissociation	$k_{d,2,2} = k_{d,2} \exp\left(-\dfrac{C}{V_{corr}}\right)$
	$C = 4.77 \times 10^2 \times \exp\left(-\dfrac{2295}{T}\right)$
Propagation	$k_{p,2,2} = k_{p,2} \exp\left(-\dfrac{B}{V_{corr}}\right)$
	$B = 1.85 \times 10^3 \times \exp\left(-\dfrac{2595}{T}\right)$
Termination	$k_{t,2,2} = k_{t,2}$
	$T_{gp} = 87.1 - 0.132 \cdot T$ (with T in °C)
	$T_{gm} = 70$
	$V_{fm} = 0.025 + 9.98 \cdot 10^{-4}(T - T_{gm})$
	$V_{fp} = 0.025 + 5.47 \cdot 10^{-4}(T - T_{gp})$
	$V_f = (1 - \phi_2) V_{fm} + \phi_2 V_{fp}$
	$\dfrac{1}{V_{corr}} = \dfrac{1}{V_f} - \dfrac{1}{V_{fc}}$

Equation (13.15) becomes:

$$\frac{dX}{dt} = \kappa (r_{in,0})^{0.5} \exp(-k_d t/2)(1+qX) \tag{13.18}$$

where $\kappa = k_p/\sqrt{2k_{tm}}$ is a global polymerization constant and $r_{in,0} = 2fk_d I_0$ the initial initiation rate. By integrating between 0 and X the following relation is obtained:

$$X = \frac{1}{q}\left\{\exp\left[\frac{2}{k_d}\kappa r_{in,0}^{0.5}(1-\exp(-k_d t/2))\right] - 1\right\} \tag{13.19}$$

13.4
Molecular-Weight Distribution

The utilization properties of PVC are intimately linked to the molecular-weight distribution (MWD) of macromolecules. The MWD may be measured by appropriate techniques, such as gel-permeation chromatography, but also predicted by computation. Comparing experimental and calculated MWD allows the validation of a kinetic model as well as the tuning of parameters. On this basis, the operation procedure necessary to get a target MWD may be simulated.

Let us consider a polymer species of x monomer units of molecular weight $M_x = xM_0$. If the distribution of x-mers is characterized by n_x molar fractions and g_x weight fractions then the MWD is defined by the following averaged values:

M_n, the number-averaged molecular weight:

$$M_n = \frac{\sum M_x n_x}{\sum n_x} = \frac{M_0 \sum x n_x}{\sum n_x} = M_0 \tau_n \tag{13.20}$$

M_w, the weight-averaged molecular weight:

$$M_w = \frac{\sum g_x M_x}{\sum g_x} = \frac{\sum M_0 x n_x \cdot x M_0}{\sum M_0 x n_x} = M_0 \frac{\sum x^2 n_x}{\sum x n_x} = M_0 \tau_w \tag{13.21}$$

Accordingly, polymerization degrees may be defined as number-averaged τ_n and weight-averaged τ_w. The polydispersity index is defined as the ratio of weight-average to number-average values, $D = M_w/M_n$. The above measures may be defined as instantaneous values, as well as time-cumulative values.

The MWD may be related mathematically to the so-called *moments* of a continuous or discrete distribution. If u is a random variable and $F(u)$ is its distribution function then the ith-order moment may be defined by the relation:

$$\mu_i = \int_0^\infty u^i F(u) du \cong \sum_1^\infty u^i f(u) \tag{13.22}$$

13.4 Molecular-Weight Distribution

In the hypothesis of long-chain polymerization the distribution function is proportional to the amounts of x-mers, R_x or P_x. Accordingly, the moments of discrete distribution for live and dead polymer species may be defined by the relations:

$$\lambda_i = \sum x^i R_x \quad \text{and} \quad \mu_i = \sum x^i P_x \quad i = 0, 1, 2, 3 \tag{13.23}$$

For example, for dead-polymer macromolecules the moments of distribution are:

$$\mu_0 = \sum P_x, \mu_1 = \sum x P_x, \mu_2 = \sum x^2 P_x \quad \mu_3 = \sum x^3 P_x \tag{13.24}$$

The MWD may be expressed in terms of moments as:

$$M_n = M_0 \frac{\mu_1}{\mu_0}, \quad M_w = M_0 \frac{\mu_2}{\mu_1}, \quad M_z = M_0 \frac{\mu_3}{\mu_2} \tag{13.25}$$

in which M_n is the number-average molecular weight, M_w the mass-average molecular weight, and M_z the z-average molecular weight, obviously with $M_n \ll M_w \ll M_z$. The average values are not sufficient for characterizing the shape of distribution. In addition, two indices should be used:

Polidispersity index $D = M_w/M_n = \mu_2/\mu_1$ \hfill (13.26)

$$\text{Asymmetry index } \alpha_3 = \frac{(M_z M_w M_n - 3M_n^2 M_w + 2M_n^3)}{(M_w M_n - M_n^2)^{2/3}} \tag{13.27}$$

The differential equations used to calculate the evolution in time of the moments of the molecular-weight distribution will be presented in Section 13.6.2.

13.4.1
Simplified Analysis

The analysis of Talamini et al. [3, 4] is presented below. In the case of free-radical homogeneous polymerization the kinetic chain length is given by the relation:

$$\nu = \frac{k_p \ [M]}{2\sqrt{k_t \, f k_d} \ [I]^{0.5}} \tag{13.28}$$

Applying the above relation for the monomer and polymer rich phases one gets:

$$\frac{\nu_{m0}}{\bar{\nu}} = \frac{2(\beta + q)}{\varepsilon + 2q} \frac{\ln(1 + qX)}{qX} + \frac{\varepsilon - 2\beta}{\varepsilon + 2q} \frac{\ln(1 - \varepsilon X/2)}{-\varepsilon X/2} \tag{13.29}$$

where $\bar{\nu}$ is the mean kinetic length and ν_{m0} the initial value given by Eq. (13.28). The parameters ε (relative volume contraction) and q (two-phase kinetics) were defined before. The parameter β is given by

$$\beta = \frac{1}{X_f}\left(1 - \frac{f_p}{f_m} \frac{\rho_m}{\rho_p} K_I\right) \tag{13.30}$$

in which f_p and f_m are the initiation efficiency in polymer and monomer phase, and K_I the initiator partition coefficient. By knowing the kinetic chain length the polymerization degree P_n may be calculated by the well-known Mayo equation:

$$\frac{1}{P_n} = \left(1 - \frac{C}{2}\right)\frac{1}{\bar{v}} + C_M \tag{13.31}$$

where C is the fraction of terminations by combination and $C_M = k_{fm}/k_p$ is the ratio of the chain transfer to monomer to propagation. The computations showed that the results are insensitive to f_p/f_m as well to K_I, which may all be set to one. For estimation purposes one may assume $\varepsilon = \beta = 0$, $C = 0.5$ and $C_M = 5.78\exp(-2768.1/T)$ [3, 4].

13.5
Kinetic Constants

Table 13.3 presents the expressions for the rate constants applied in this work. The parameters are taken mostly from the work of Xie et al. [6]. A distinctive feature of the numerical simulation of the influence of gel effect on the termination in the polymer-phase is described by a relation proposed by Kipparisides et al. [5]. This combination of parameters gives realistic results on modeling both the reaction dynamics and the development of the molecular-weight distribution, reproducing closely experimental data (see Figure 13.6). The subscript 1 refers to the monomer phase, 2 to the polymer phase, and 22 to the polymer-phase after the critical conversion X_f. In addition, Table 13.4 presents first-order constants for usual initiators.

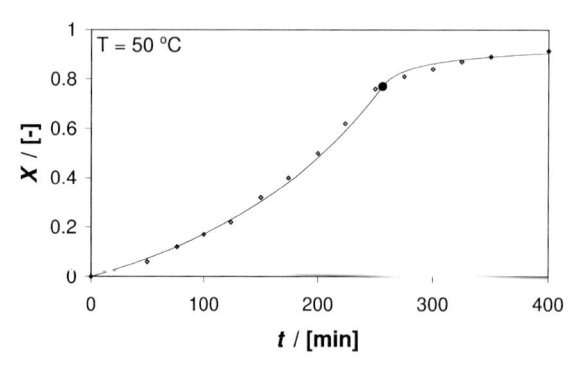

Figure 13.6 Conversion vs. time in an isothermal reactor operated at 50°C: comparison of experimental data of Xie et al. [7] with simulation in this work.

13.6
Reactor Design

This section presents a detailed reactor model that includes mass balance, molecular-weight distribution, the equation for the control system, and detailed calculations of the heat-transfer rate. In Section 13.7 the model will be used for designing the reactor and the operating procedure.

Table 13.4 Kinetic constants for initiator decomposition.

Initiator	A (s^{-1})	E/R (K)
Di(sec-butyl) peroxidicarbonate	3.69×10^{14}	13 588
AIBN	1.05×10^{15}	15 460
Perkadox 16-W40	2.31×10^{15}	14 653
Tert-butyl peroxineodecanoate (TBPD)	1.52×10^{14}	13 895

13.6.1
Mass Balance

Assuming instantaneous thermodynamic equilibrium, the VCM monomer distributes between the suspension, water and gas phases. In the subsequent analysis we consider as representative for the reactor productivity only the monomer involved in droplets, as representing more than 99.9% of the overall balance.

The mass balance is written using as state variables the amount of initiator I (kmol), monomer M (kg) and polymer P (kg). The reaction rates are in kmol/s, and kmol/m^3, respectively. Subscripts 1 and 2 refer to the monomer-rich and polymer-rich phases, respectively. For clarity, we will use symbols as $c_{K,j}$ to denote the concentration (kmol/m^3) of species K in phase j.

$$\frac{dI}{dt} = -r_{in,1} - r_{in,2} \tag{13.32}$$

$$\frac{dM}{dt} = -r_1 - r_2 \tag{13.33}$$

$$\frac{dP}{dt} = r_1 + r_2 \tag{13.34}$$

Initial conditions:

$$\text{At } t = 0: I = I_0; M = M_0; P = 0; \tag{13.35}$$

Note that in addition to the monomer, the initial reaction mixture also contains water in a mass ratio $W_0/M_0 = 1.2$.

Calculation of the reaction rates requires, for each phase, the amounts of initiator and monomer, and the concentration of radicals.

Before the critical conversion:

$$\begin{aligned} r_{in,1} &= k_{d,1} I_1; \quad r_{in,2} = k_{d,2} I_2 \\ r_1 &= k_{p,1} M_1 c_{R,1}; \quad r_2 = k_{p,2} M_2 c_{R,2} \end{aligned} \tag{13.36}$$

After the critical conversion:

$$r_{in,1} = 0; \quad r_{in,2} = k_{d,22} I$$
$$r_1 = 0; \quad r_2 = k_{p,22} \cdot M \cdot c_{R,2}$$

(13.37)

where the critical conversion is given by the following empirical equation:

$$X_f = 0.8075 + 5 \times 10^{-4} T - 2.5 \times 10^{-5} T^2 \quad \text{(with T in °C)} \tag{13.38}$$

The two-phase model assumes that the amount of monomer dissolved in the polymer-rich phase is proportional to the amount of polymer present in the reactor:

Before the critical conversion:

$$M_2 = A \cdot P; \quad M_1 = M - M_2 \tag{13.39}$$

After the critical conversion:

$$M_2 = M; \quad M_1 = 0 \tag{13.40}$$

The value of the proportionality constant A follows from the requirement that M_1 vanishes at the critical conversion X_f

$$A = \frac{1 - X_f}{X_f} \tag{13.41}$$

The volumes of the two phases and the volume occupied by the monomer in the polymer-rich phase are calculated as:

Before the critical conversion:

$$V_1 = \frac{M_1}{\rho_m}; \quad V_2 = \frac{M_2}{\rho_m} + \frac{P_2}{\rho_p}; \quad V_{2,m} = \frac{M_2}{\rho_m} \tag{13.42}$$

After the critical conversion:

$$V_{2m} = \frac{M}{\rho_m}; \quad V_{2p} = \frac{P}{\rho_p}; \quad V_2 = V_{2,m} + V_{2,p} \tag{13.43}$$

Alternatively, one may calculate the monomer distribution between the two phases according to the Flory–Huggins theory. The activity coefficient of monomer in the gel phase is:

$$a = (1 - \varphi_2) e^{(\varphi_2 + \chi \varphi_2^2)} \tag{13.44}$$

where

$$\varphi_2 = \frac{V_{2,p}}{V_2} \tag{13.45}$$

$$\chi = \frac{1286.4}{T} - 3.02 \tag{13.46}$$

When the monomer phase is still present, the activity coefficient of the monomer dissolved in the gel phase is unity. Therefore, the volumetric fraction of polymer in the gel phase is calculated by solving, for the unknown φ_2, the following equation:

$$a(\varphi_2, T) - 1 = 0 \tag{13.47}$$

If this approach is taken, Eq. (13.38) for calculating the critical conversion is no longer necessary.

For the initiator, we denote by κ the partition coefficient between the two phases. This leads to the following relations to be applied before the critical conversion:

$$I_2 = \frac{\kappa V_2/V_1}{1 + \kappa V_2/V_1} I; \quad I_1 = I - I_2 \tag{13.48}$$

where the partition coefficient is assumed to be constant:

$$\kappa = \frac{I_2/V_2}{I_1/V_1} = 0.77 \tag{13.49}$$

After the critical conversion is reached, the initiator can be found only in the polymer-rich phase.

Concentrations of free radicals in the two phases are given by:

Before the critical conversion:

$$c_{R,1} = \sqrt{\frac{2 f_1 k_{d,1} c_{I,1}}{k_{t,1}}}; \quad c_{R,2} = \sqrt{\frac{2 f_2 k_{d,2} c_{I,2}}{k_{t,2}}} \tag{13.50}$$

After the critical conversion:

$$c_{R2} = \sqrt{\frac{2 f_{2,2} k_{2,2} c_{I,2}}{k_{2,2}}} \tag{13.51}$$

where the concentrations of initiator in the two phases are:

$$c_{I,1} = \frac{I_1}{V_1}; \quad c_{I,2} = \frac{I_2}{V_2} \tag{13.52}$$

13.6.2
Molecular-Weight Distribution

The following kinetic model is used to derive the molecular-weight distribution:

Initiation: $\quad I \xrightarrow{k_{d,j}} 2I^*$
$\quad I^* + M \xrightarrow{k_{p,j}} R_1$

Propagation: $\quad R_x + M \xrightarrow{k_{p,j}} R_{x+1}$

Chain transfer to monomer: $\quad R_x + M \xrightarrow{k_{tm,j}} P_x + R_1$

Termination by disproportionation: $\quad R_x + R_y \xrightarrow{k_{t,j}} P_x + P_y$

The rate expressions for the macroradical and dead polymer of each chain-length are [16]:

$$\frac{dR_{1,j}}{dt} = 2f_j k_{d,j} I_j + k_{tm,j} M_j \sum_y \frac{R_{y,j}}{V_j} - k_{p,j} \frac{M_j}{V_j} R_{1,j} - k_{tm,j} \frac{M_j}{V_j} R_{1,j} - k_{t,j} R_{1,j} \sum_y \frac{R_{y,j}}{V_j} \tag{13.53}$$

$$\frac{dR_{x,j}}{dt} = -k_{p,j} \frac{M_j}{V_j} R_{x,j} - k_{tm,j} \frac{M_j}{V} R_{x,j} - k_{t,j} R_{x,j} \sum_y \frac{R_{y,j}}{V_j} + k_{p,j} \frac{M_j}{V_j} R_{x,j} \tag{13.54}$$

$$\frac{dP_x}{dt} = \sum_j \left(k_{tm,j} \frac{M_j}{V_j} R_{x,j} \right) + \sum_j \left(k_{t,j} R_{x,j} \sum_y \frac{R_{y,j}}{V_j} \right) \tag{13.55}$$

The moments of live and dead polymer distribution are defined as:

$$\lambda_{i,j} = \sum_x x^i R_{x,j} \tag{13.56}$$

$$\mu_{i,j} = \sum_x x^i P_{x,j} \quad i = 0, 1, 2 \text{ in } j = 1, 2 \text{ phase} \tag{13.57}$$

The set of differential equations (13.53) and (13.54) describing the evolution of each radical chain length can be reduced to the following set of differential equations for the moments:

$$\frac{d\lambda_{0,j}}{dt} = 2f_j k_{d,j} I_j - k_{t,j} \frac{\lambda_{0,j}}{V_j} \tag{13.58}$$

$$\frac{d\lambda_{1,j}}{dt} = 2f_j k_{d,j} I_j - k_{t,j} \frac{\lambda_{0,j}\lambda_{1,j}}{V_j} + k_{tm,j} \frac{M_j}{V_j} (\lambda_{0,j} - \lambda_{1,j}) + k_{p,j} \frac{M_j}{V_j} \lambda_{0,j} \tag{13.59}$$

$$\frac{d\lambda_{2,j}}{dt} = 2f_j k_{d,j} I_j - k_{t,j} \frac{\lambda_{0,j} \lambda_{2,j}}{V_j} + k_{tm,j} \frac{M_j}{V_j} (\lambda_{0,j} - \lambda_{2,j}) + k_{p,j} \frac{M_j}{V_j} (\lambda_{0,j} + \lambda_{2,j}) \quad (13.60)$$

If we assume the quasisteady approximation for the radical concentration, from Eqs. (13.58) to (13.60) we can express the radical moments in the following form:

$$\lambda_{0,j} = \sqrt{\frac{V_j(2f_j k_{d,j} I_j)}{k_{t,j}}} \quad (13.61)$$

$$\lambda_{1,j} = \frac{2f_j k_{d,j} I_j V_j + (k_{tm,j} + k_{p,j}) M_j \lambda_{0,j}}{k_{t,j} \lambda_{0,j} + k_{tm,j} M_j} \quad (13.62)$$

$$\lambda_{2,j} = \frac{2f_j k_{d,j} I_j V_j + (k_{tm,j} + k_{p,j}) M_j \lambda_{0,j} + k_{p,j} M_j (\lambda_{0,j} + 2\lambda_{1,j})}{k_{t,j} \lambda_{0,j} + k_{tm,j} M_j} \quad (13.63)$$

The set of differential equations (13.55) describing the evolution of each polymer chain length can be reduced to the following set of differential equations for the moments:

$$\frac{d\mu_0}{dt} = \sum_j k_{tm,j} \frac{M_j}{V_j} \lambda_{0,j} + \sum_j k_{t,j} \frac{\lambda_{0,j}^2}{V_j} \quad (13.64)$$

$$\frac{d\mu_1}{dt} = \sum_j k_{tm,j} \frac{M_j}{V_j} \lambda_{1,j} + \sum_j k_{t,j} \frac{\lambda_{0,j} \lambda_{1,j}}{V_j} \quad (13.65)$$

$$\frac{d\mu_2}{dt} = \sum_j k_{tm,j} \frac{M_j}{V_j} \lambda_{2,j} + \sum_j k_{t,j} \frac{\lambda_{0,j} \lambda_{2,j}}{V_j} \quad (13.66)$$

Equations (13.64) to (13.66) are integrated together with Eqs. (13.58) and (13.60) or (13.61) to (13.63).

The number- and weight-average molecular weights can be expressed in terms of moments as:

$$M_n = MW_m \frac{\mu_1}{\mu_0} \quad (13.67)$$

$$M_w = MW_m \frac{\mu_2}{\mu_1} \quad (13.68)$$

where MW_m is the molecular weight of the monomer.

13.6.3
Heat Balance

The energy balance of the reactor contents can be written as:

$$m_s c_{p,s} \frac{dT_r}{dt} = \frac{(-\Delta H)}{62.5} \cdot \left(-\frac{dM}{dt}\right) - Q_{trans} \quad (13.69)$$

m_s is the total mass of the suspension, c_{ps} is defined in Table 13.6.

In Eq. (13.69), we neglect the energy generated by mixing. Q_{trans} includes the energy removed by the cooling jacket, by an external condenser, and needed to bring the water added in the system to the reaction temperature (see Section 13.7)

Intensive heat transfer may be obtained with a channel-like jacket, where the cooling water circulates at a high constant flow rate. This can be described by the following model that neglects the dynamics of the cooling jacket:

$$\frac{dT_j}{d\xi} = \frac{Uh_j L_j}{F_j c_{p,j}}(T_r - T_j), \quad T_j(0) = T_{j,0} \quad \text{and} \quad \xi \in (0, 1) \tag{13.70}$$

U is defined in Eq. (13.84).

$T_{j,0}$ is the jacket-inlet temperature, which is the manipulated variable in a PID temperature control loop, as will be discussed in Section 13.6.7. Integration of Eq. (13.70) gives the distribution of temperature along the jacket. In particular, the temperature at outlet is given by

$$T_j(1) = T_r - \exp\left(-\frac{Uh_j L}{F_c c_{p,j}}\right)(T_r - T_j(0)) \tag{13.71}$$

Then, the amount removed by jacket cooling is:

$$Q = F_c c_{p,j}(T_j(1) - T_j(0)) \tag{13.72}$$

13.6.4
Heat-Transfer Coefficients

For the mixing system a flat-blade-type turbine is adopted. The heat-transfer coefficient reactor contents–wall may be calculated by the following relation:

$$\frac{1}{U_{r\text{-wall}}} = \frac{1}{h_{r\text{-wall}}} + \frac{1}{h_{\text{scale},1}} \tag{13.73}$$

For the $h_{r\text{-wall}}$, a Sieder–Tate correlation is used:

$$\text{Nu} = C \times \text{Re}^{0.66} \, \text{Pr}^{0.33} \tag{13.74}$$

$$\text{Re} = \frac{\rho_s N_{\text{imp}} D_{\text{imp}}^2}{\eta_s} \tag{13.75}$$

$$\text{Pr} = \frac{\eta_s c_{p,s}}{\lambda_s} \tag{13.76}$$

$$h_{r\text{-wall}} = \text{Nu}\frac{\lambda_s}{d_{ws}} \quad \text{with} \quad d_{ws} = D \tag{13.77}$$

The value of the constant C depends on the type of the agitator. For a flat-blade turbine, Kiparissides et al. [12] recommends $C = 0.56$ valid for simple mixed vessels. Nagy and Agachi [14] use $C = 0.38$. More intensive heat transfer to the wall can be obtained by using internal baffles, for which Wieme et al. [13] consider $C = 0.75$, the value that was used in our study.

Heat transfer between the reactor wall and the cooling jacket considers turbulent flow and is calculated using the following relations:

$$\frac{1}{U_{\text{wall-j}}} = \frac{1}{h_{\text{wall-j}}} + \frac{1}{h_{\text{scale},2}} \tag{13.78}$$

$$h_{\text{wall-j}} = \text{Nu}\frac{\lambda_j}{d_{h,j}} \tag{13.79}$$

$$\text{Nu} = 0.023 \times \text{Re}^{0.8} \times \text{Pr}^{0.33} \tag{13.80}$$

$$\text{Re} = \frac{\rho_j v_j d_{h,j}}{\lambda_j} \tag{13.81}$$

$$v_j = \frac{F_j}{\rho_w} \cdot \frac{1}{A_{fc}} \tag{13.82}$$

$$\text{Pr} = \frac{\eta_j c_{p,j}}{\lambda_j} \tag{13.83}$$

Finally, the global heat-transfer coefficient includes the resistance of the jacket wall:

$$\frac{1}{U} = \frac{1}{U_{\text{wall-j}}} + \frac{1}{U_{\text{r-wall}}} + \frac{\lambda_{\text{wall}}}{\delta_{\text{wall}}} \tag{13.84}$$

13.6.5
Physical Properties

The temperature-dependent physical properties are calculated as shown in Tables 13.5 and 13.6.

13.6.6
Geometry of the Reactor

Calculation of various heat-transfer areas, jacket volume, and hydraulic diameter are detailed in Table 13.7.

13.6.7
The Control System

The inlet jacket temperature $T_{j,0}$ is adapted continuously to the heat-balance requirements by injecting a thermal agent (steam for heating period, chilled water for cooling period). This operation is realized by means of a controller whose

Table 13.5 Relationships used for calculating the physical properties [13].

Monomer

Density, kg/m³	$\rho_m = \dfrac{94.469}{0.2702^{\left(1+\left(1-\frac{T}{432}\right)^{0.2716}\right)}}$	
Specific heat, J/kg/K	$c_{p,m} = 5.165T - 165.1$	
Viscosity, Pa·s	$\eta_m = 10^{-3} \exp\left(9.373 - \dfrac{648.32}{T_r} - 4.294 \times 10^{-2}T + 4.316 \times 10^{-5}T^2\right)$	
Conductivity, W/m/K	$\lambda_m = 0.172\left(0.229 + 1.529\left(1 - \dfrac{T}{425}\right)^{2/3}\right)$	

Polymer

Density, kg/m³	$\rho_p = 1394 - 0.203T - 2.19 \cdot 10^{-3}T^2$ (with T in °C)
Specific heat, J/kg/K	$c_{p,m} = 1220(0.64 + 1.2 \cdot 10^{-3}T)$
Conductivity, W/m/k	$\lambda_p = 0.1164 + 1.373 \cdot 10^{-4}T$

Water

Density, kg/m³	$\rho_w = \dfrac{98.349}{0.3054^{\left(1+\left(1-\frac{T}{647.13}\right)^{0.081}\right)}}$
Specific heat, J/kg/K	$c_{p,w} = 1.534 \times 10^4 - 116T + 0.451T^2 - 7.835 \times 10^{-4}T^3 + 5.201 \times 10^{-7}T^4$
Viscosity, Pa·s	$\eta_w = 10^{-3} \exp\left(-24.71 + \dfrac{4209}{T} + 4.5271 \times 10^{-2}T - 3.3761 \times 10^{-5}T^2\right)$
Conductivity, W/m/k	$\lambda_w = -0.3838 + 5.254 \times 10^{-3}T - 6.369 \cdot 10^{-6}T^2$

Wall

Density, kg/m³	$\rho_{wall} = 7930$
Specific heat, J/kg/K	$c_{p,wall} = 460.5$
Conductivity, W/m/k	$\lambda_{wall} = 16.3$

action has to be introduced explicitly in the model. In this case, a cascade configuration (Figure 13.7) is appropriate because it gives a fast reaction to disturbances. The first controller (called "secondary") adjusts the cooling-water make-up in order to reject disturbances due to variability of the recycled or fresh coolant. The secondary controller receives its setpoint from the reactor temperature controller (called "primary"). The dynamics of the secondary control loop is fast, and will not be modeled in this study. The primary loop is slower, the PID control law being implemented as shown in Eqs. (13.85) to (13.88):

$$\varepsilon(t) = T_r^{SP} - T_r \tag{13.85}$$

$$\frac{dI_\varepsilon(t)}{dt} = \varepsilon(t) \tag{13.86}$$

$$D_\varepsilon(t) = \frac{dT_r}{dt} \tag{13.87}$$

$$T_{j,0} = T_{j,0}^* + K_p\left(\varepsilon(t) + \frac{1}{T_I}I_\varepsilon(t) + T_D D_\varepsilon(t)\right) \tag{13.88}$$

Table 13.6 Relationships used for averaging the physical properties [13].

Density	
Specific heat	$c_{p,s} = \omega_m c_{p,m} + \omega_p c_{p,p} + \omega_w c_{p,w}$
Viscosity	$\varphi_p = \dfrac{V_p}{V_1 + V_2}$
	$\varphi_{drop} = \dfrac{V_1 + V_2}{V_r}$
	$\eta_{drop} = \eta_m(1 + 2.5\varphi_p + 7.54\varphi_p^2)$
	$\eta_s = \eta_w\left(1 + \varphi_{drop}\dfrac{\eta_w + 2.5\eta_{drop}}{\eta_w + \eta_{drop}} + \varphi_{drop}^2\dfrac{\eta_w + 7.54\eta_{drop}}{\eta_w + \eta_{drop}}\right)$
Conductivity	$\lambda_{drop} = (1 - \varphi_p)\lambda_m + \varphi_p\lambda_p$
	$\lambda_s = \lambda_w\dfrac{2 + \dfrac{\lambda_{drop}}{\lambda_w} - 2\varphi_{drop}\left(1 - \dfrac{\lambda_{drop}}{\lambda_w}\right)}{2 + \dfrac{\lambda_{drop}}{\lambda_w} + \varphi_{drop}\left(1 - \dfrac{\lambda_{drop}}{\lambda_w}\right)}$

Table 13.7 Geometric data of the polymerization reactor.

Volume	$V = \dfrac{\pi D^2}{4} H$
The volume occupied by the reaction mixture (variable)	$V_r = \dfrac{V_w}{\rho_w} + \dfrac{V_m}{\rho_m} + \dfrac{V_p}{\rho_p}$
The height of the reaction mixture	$H_r = \dfrac{V_r}{\pi D^2/4}$
Heat transfer area	$A = \dfrac{4V_r}{D} = h_j L_j$
Cross-sectional area available for coolant flow	$A_f = \delta_j \cdot h$, with $\delta_j = h_j = 0.09\,m$
Equivalent (hydraulic) diameter of the jacket	$d_{h,j} = \dfrac{4A}{P} = \dfrac{4\delta_j h_j}{2(\delta_j + h_j)}$

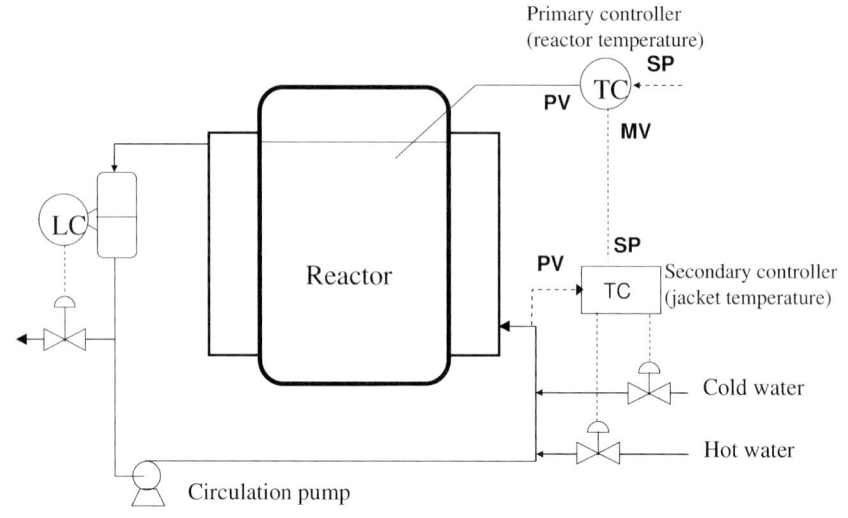

Figure 13.7 Configuration of the heating/cooling system for VCM polymerization.

13.7
Design of the Reactor

We start the design of the reactor by choosing a moderate value for the volume, namely $V = 75\,m^3$ and the polymerization temperature of 50 °C. For an initial ratio water / monomer = 1.2 (weight), the reactor is charged with 32 654 kg of monomer. The initial initiator concentration is 0.002 kmol/m³. For the H/D ratio of 1.75, approximate geometry values are $D = 3.75$ m and $H = 6.56$ m. The operating conditions are similar to those used for the laboratory reactor described by Xie et al. [7].

The model described in Section 13.6 is used to simulate the transient behavior of the reactor operated isothermally at 50 °C.

Figure 13.8 presents the polymerization rates in the two phases, r_1 and r_2, and the global rate r. The maximum polymerization rate of 0.63 %/min corresponds to 205.7 kg/min.

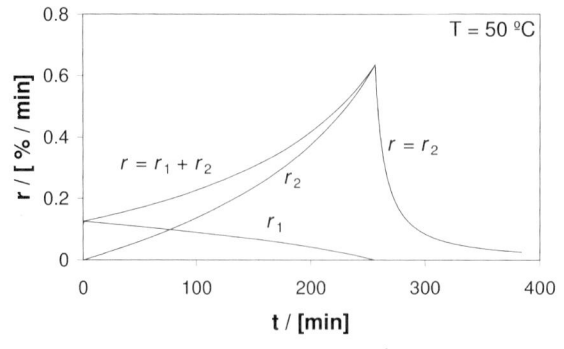

Figure 13.8 Polymerization rates in a 75-m³ reactor operated isothermally at 50 °C.

13.7 Design of the Reactor

The cooling system must ensure isothermal operation. Therefore, the target for the design is the capacity of removing the heat generated by the polymerization reaction at the critical conversion, when the generation rate reaches the maximum value:

$$Q_{max} = r_{max}(-\Delta^r H) = (0.0063 \times 32654/60) \times (1.025 \times 10^8 /62.5) \approx 5.62 \times 10^6 \text{ W}$$

A 75-m³ reactor has a heat-transfer area of about $A = 78 \text{ m}^2$. We also assume a heat-transfer coefficient $U = 800 \text{ W/m}^2\text{K}$. The driving force for the heat transfer is the temperature difference between the reactor contents and coolant. To achieve the required heat-transfer capacity, the temperature difference must be at least:

$$\Delta T_{min} = \frac{Q_{max}}{UA} = \frac{5.62 \times 10^6}{800 \times 78} = 90 \text{ K}$$

Obviously, when the reactor is operated at 50 °C, the temperature difference ΔT_{min} between the reactor contents and the coolant cannot be realized by using normal cooling water, for example with a temperature of 5 °C,

13.7.1
Additional Cooling Capacity by Means of an External Heat Exchanger

During polymerization, a certain amount of polymer evaporates. One design option giving additional cooling capacity is to condense, in an external heat exchanger, the polymer from the vapor phase and to return the liquid to the reactor bulk. We assume that, under the most demanding conditions, 30% of the heat generated by the polymerization reaction can be removed in the condenser. Under these conditions, the minimum temperature difference becomes:

$$\Delta T'_{min} = 0.7 \times \Delta T_{min} = 63 \text{ K}$$

For the condenser, we assume that water with $T = 20$ °C is used as coolant, and that the heat-transfer coefficient is $U_c = 2000 \text{ W/m}^2\text{K}$. In these conditions, the required area is:

$$A_c = \frac{Q_c}{U_c \Delta T_c} = \frac{0.3 \times 5.62 \times 10^6}{2000 \times 30} = 28.1 \text{ m}^2$$

This is a small condenser, which can be easily built, even for lower values of the heat-transfer coefficient or of the temperature difference.

Still, the temperature difference $\Delta T'_{min}$ is too large. In practice, the cooling capacity of the external condenser might be as high as 50% of the total duty. However, the polymerization reaction also takes place inside the condenser. Because the reaction conditions are different compared to the bulk of the reactor, the polymer obtained in this way might not satisfy the quality requirements. Therefore, we will look for other ways to increase the cooling capacity.

13.7.2
Additional Cooling Capacity by Means of Higher Heat-Transfer Coefficient

A solution may come from the ability to realize a large heat-transfer coefficient, for example $U = 1000\,\text{W/m}^2/\text{K}$. This means that special attention must be paid to scaling and the thermal resistance of the metal wall must be reduced, as discussed in Section 13.2.1. The minimum temperature difference becomes $\Delta T''_{\min} = 0.8 \Delta T'_{\min} = 50.4\,\text{K}$. This might offer a feasible solution, but the robustness of the reactor is questionable.

13.7.3
Design of the Jacket

Assuming that the coolant-inlet temperature is 5 °C, and the average coolant temperature is 10 °C, the coolant-outlet temperature is 15 °C. Then, the flow rate of water can be calculated:

$$F_c = \frac{Q}{c_p \times \Delta T} = \frac{3.6 \times 10^6}{4185 \times 10} = 86\,\text{kg/s}$$

When the flow-area of the coil-like jacket is about $0.08\,\text{m}^2$, the velocity of the water is:

$$u = \frac{86\,\text{kg/s}}{1000\,\text{kg/m}^3} \cdot \frac{1}{0.008\,\text{m}^2} \approx 10\,\text{m/s}$$

This is a high value; therefore the jacket will be split into four segments running in parallel.

13.7.4
Dynamic Simulation Results

Figure 13.9 presents dynamic simulation results for a 75-m³ reactor, cooled by means of the jacket and external condenser previously designed. The heat-transfer coefficients are rigorously calculated according to Section 13.6.4. Scaling due to the polymer deposition and heat-transfer resistance of the metal wall are neglected. Figure 13.9 shows that the reactor cannot be controlled. After 200 min, the heat-generation rate is high and cannot be removed using cooling water of 5 °C. Therefore, the reactor temperature increases to more than 85 °C.

Dynamic simulation provides the evolution in time of several process variables. Among them, we note that the volume of the suspension decreases, because of

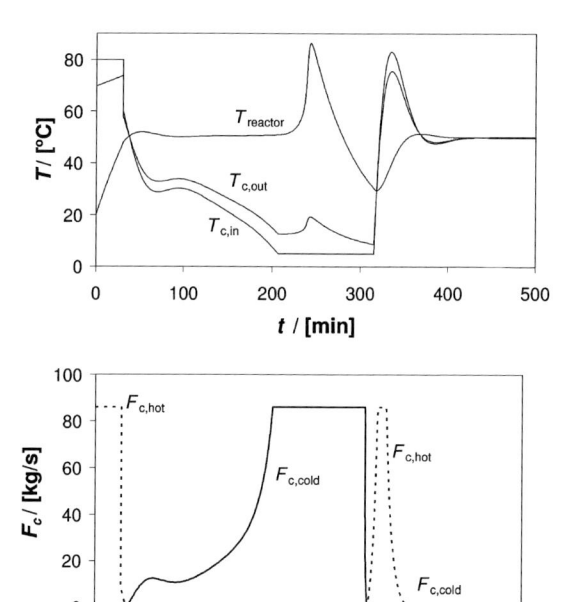

Figure 13.9 Evolution of the temperature and cooling water flow rate in a 75-m³ polymerization reactor. For about 100 min, the maximum cooling capacity is reached.

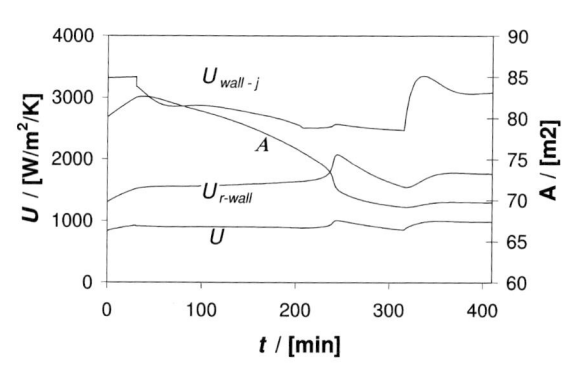

Figure 13.10 Evolution of the heat-transfer coefficients and heat-transfer area in a 75-m³ polymerization reactor. Due to volume contraction, the heat-transfer area decreases from 85 m² to 70 m².

the higher density of the polymer (compared to the monomer). Because the suspension fills a smaller part of the reactor, the heat-transfer area between the reactor contents and the wall decreases, meaning that our assumption of 80 m² might be optimistic (Figure 13.10).

13.7.5
Additional Cooling Capacity by Means of Water Addition

To account for the volume contraction, cold water may be added to the reactor. An additional benefit is the cooling provided by the water introduced in the system in this way. Around the critical conversion, the volume change is:

$$\Delta V = r_{max}\left(\frac{1}{\rho_m} - \frac{1}{\rho_p}\right) = \frac{0.0063 \times 32654}{60}\left(\frac{1}{862} - \frac{1}{1378}\right) = 0.0015\,\mathrm{m^3/s}$$

If this volume is replaced by cold water with a temperature of about 10 °C, bringing this amount to reaction temperature contributes with

$$Q_w = 0.0015\,\mathrm{m^3} \times 1000\,\mathrm{kg/m^3} \times 4185\,\mathrm{J/kg/K} \times 40\,\mathrm{K} = 2.5 \times 10^5\,\mathrm{W}$$

This is, approximately, 5% of the required duty. In terms of minimum temperature difference, this could be reduced to $\Delta T'''_{min} = 0.95\Delta T'''_{min} = 48\,\mathrm{K}$, corresponding to a duty of the jacket cooling $Q = 3.7 \times 10^6\,\mathrm{W}$.

Figure 13.11 presents simulation results for the case of cold water being added to the reaction mixture. Compared to the previous case, the maximum cooling

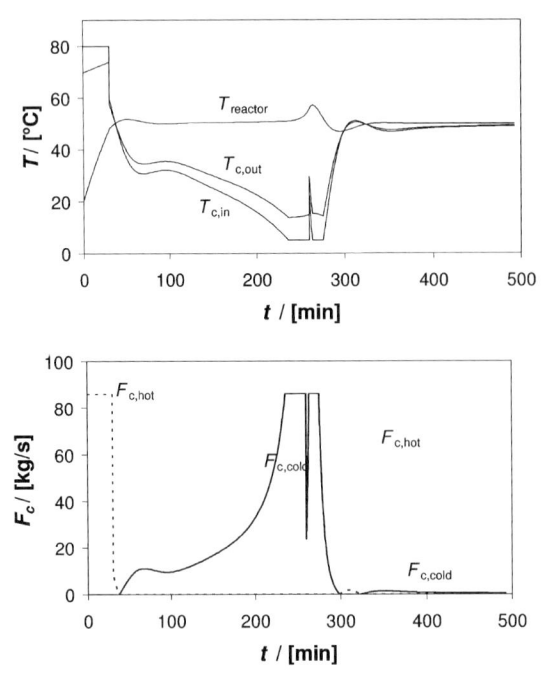

Figure 13.11 Evolution of the temperature and cooling water flow rate in a 75-m³ polymerization reactor where the suspension volume is kept constant by water addition.

13.7 Design of the Reactor | 393

capacity is reached for a shorter period. Despite this, the large thermal inertia of the 75-m^3 reactor prevents the temperature from rising too much.

13.7.6
Improving the Controllability of the Reactor by Recipe Change

The evolution of the temperature in the reactor designed so far is barely controllable. We expect the controllability to deteriorate due to factors that have been neglected, such as scaling due to polymer deposition and resistance to heat transfer of the reactor wall. Moreover, we must be concerned about the robustness of the reactor when other parameters used during design are uncertain.

One solution would be to reduce the reaction rate by using a smaller amount of initiator. The side effect of this change is a larger duration of the polymerization batch.

A closer look at Figure 13.8 shows the exponential increase of the polymerization rate near the critical conversion. Our target is to flatten the reaction profile, by increasing the initial rate. This can be achieved by using a cocktail of initiators. In this work we will use a mixture of perkadox and tert-butyl peroxyneodecanoate (TBPD). It turns out that using the slower initiator TBPD allows not only better temperature control, but also higher reaction temperature and a much larger reactor volume. In the following, we will present simulation results for a 150-m^3 reactor, operated at 57 °C. The amounts of initiator used are 0.075 kmol perkadox and 0.045 kmol TBPD. The duty of the external condenser is increased to 40% of the heat generated. The simulation includes the resistances to heat transfer of a polymer deposit of 50 μm thickness (4000 W/m^2/K) and of the 10 mm jacket wall (3000 W/m^2/K). The global, jacket and reactor-bulk heat-transfer coefficients are around 800 W/m^2/K, 3000 W/m^2/K and 1400 W/m^2/K, respectively.

Table 13.8 summarizes the design and operating procedures of the 75-m^3 and 150-m^3 reactors.

Figure 13.12 describes the evolution of conversion and pressure. The critical conversion is obtained in 300 min, while 500 min are necessary to obtain the final conversion of 85%. After the critical conversion is reached, the pressure quickly decreases. It can be also observed that, compared to Figure 13.8, the polymerization rate is more constant and the exponential increase around the critical point has disappeared. This can be explained by the combined action of the two initiators: Perkadox is a fast initiator that supplies the radicals during the first period of the polymerization process. The rate at which the second initiator, TBPD, produces radicals is almost constant. The result is the more constant reaction rate profile.

Evolutions of the reactor, jacket-inlet and jacket-outlet temperatures are depicted in Figure 13.13, together with the coolant flow rate. It can be observed that the control of the temperature is good. The maximum cooling capacity is never reached and therefore the temperature stays close to the desired value.

Figure 13.14 presents the evolution of the average molecular weight. Initially, the molecular weight is high due to fast initiation and slow termination. As the

Table 13.8 Summary of design and operating procedure for two polymerization reactors.

	Reactor 1	Reactor 2
Volume / [m^3]	75	150
Diameter / [m]	3.75	4.18
Height / [m]	6.55	10.45
H/D	1.75	2.5
Stirrer diameter / [m]	1.4	3.9
Number of jacket segments running in parallel	4	4
Length of one jacket segment / [m]	222	398
Heat transfer area of one jacket segment	20	35.8
Flow rate of cooling water in one segment / [kg/s]	21.5	21.5
Initial amount of monomer / [kg]	32 645	65 289
Initial amount of water / [kg]	39 175	78 347
Initial amount of perkadox / [kmol]	0.1435	0.075
Initial amount of TBPD / [kmol]	0	0.045
Setpoint of the temperature controller / [°C]	50	57
Temperature when cooling begins / [°C]	48	55
Temperature of the fresh cooling water / [°C]	5	5
Condenser duty / heat-generation rate	0.3	0.4

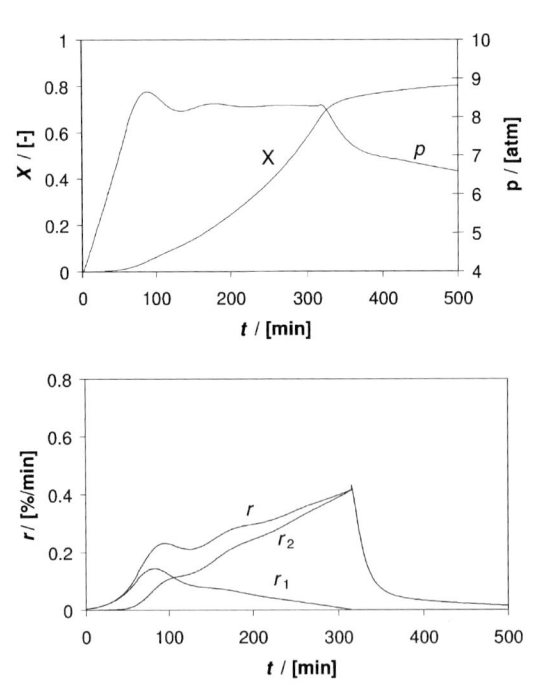

Figure 13.12 Evolution of conversion, pressure (left) and reaction rates (right) in a 150-m^3 polymerization reactor operated at 57 °C.

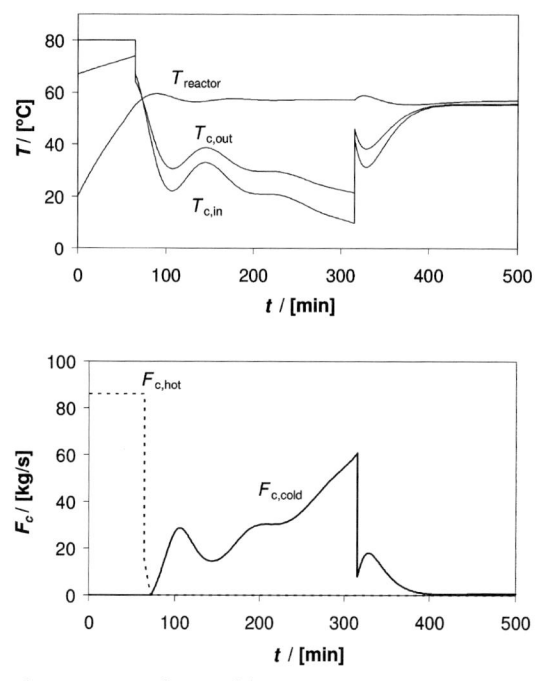

Figure 13.13 Evolution of the temperature and cooling water flow rate in a 150-m^3 polymerization reactor operated at 57 °C.

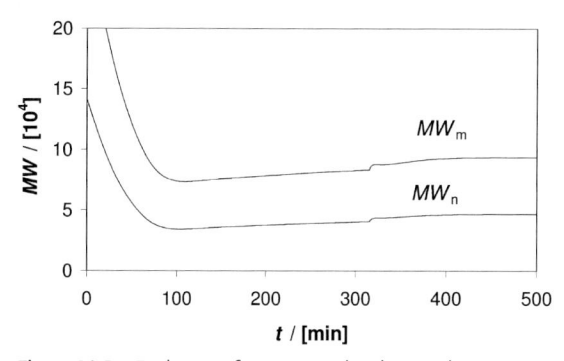

Figure 13.14 Evolution of average molecular weights in a 150-m^3 reactor operated at 57 °C.

reaction proceeds, the average molecular weight rapidly drops and then slowly increases to its final value.

The operating procedure presented in this section can be further optimized. The objective function should include not only the productivity of the reactor, but also costs related to investment, utilities and manpower. Examples of decision variables that can be considered are the temperature profile, the amounts of initiator that are added, and the timing of initiator addition. We leave this optimization as an exercise for the reader.

13.8
Conclusions

The selected case study illustrates a complex product and process design. High productivity and good product quality can be achieved simultaneously only if the reaction conditions are well controlled. However, in a batch process the degrees of freedom for reactor sizing are quite limited. The optimality of the design arises from a better operation procedure, but mainly by changing drastically the chemistry. This characteristic is shared by other batch processes.

The design starts from the recipe discovered at a laboratory-scale reactor. This recipe cannot be used in a large industrial reactor (75 m^3), because insufficient heat transfer capabilities even in the most clean and intensive conditions. Detailed modeling and simulation help identifying the limiting factors. Improvements are suggested, such as the addition of water to compensate volume contraction and increasing the heat transfer capacity by an external condenser. However, the most significant improvement arises from changing the recipe of polymerization, namely switching from mono-initiation to multi-initiation systems and by changing the reaction temperature to an optimal one. A much uniform reaction rate profile can be achieved. In this way the polymerization can be carried out in a much larger reactor of 150 m^3, the reactor heat transfer capacity offering this safe and robust operation.

References

1 Saeki, Y., Emura, T., Technical progresses for PVC production, *Prog. Polym. Sci.*, 2002, 2055–2131
2 Talamini, G., Heterogeneous bulk polymerization of vinylchloride, *J. Polym. Sci.*, A-2, 1966, 4, 535
3 Talamini, G., Visentini, A., Kerr. J., Bulk and suspension of vinyl chloride: the two-phase model, *Polymer*, 1998, 39, 1879–1891
4 Talamini, G., Visentini, A., Kerr. J., Bulk and suspension of vinyl chloride at high conversion, *Polymer*, 1998, 39, 4379–4384
5 Kiparissides, C., Daskalakis, G., Achilias, D. S., Sidiropoulou, E., Dynamic simulation of polyvinylchloride batch suspension polymerisation reactors, *Ind. Eng. Chem. Res.*, 1997, 36, 1253–1267
6 Xie, T.Y, Hamielec, A.E., Wood P.E., Woods, D.R., Experimental investigation of vinyl chloride polymerisation at high conversion: mechanism, kinetics and modeling, *Polymer*, 1991, 32(6), 537–557
7 Xie, T.Y, Hamielec, A.E., Wood P.E., Woods, D.R., Experimental investigation of vinyl chloride polymerisation at high conversion - Reactor dynamics, *J. Appl. Polym. Sci.*, 1991, 43, 1259–1269
8 Xie, T. Y., Hamielec, A. E., Wood, P. E., Woods, D. R., Suspension, bulk and emulsion of vinyl chloride-mechanism, kinetics and reactor modeling, *J. Vinyl Technol.*, 1991, 13(1), 2–25
9 Dimian, A. C., van Diepen, D., Van der Wal, G. A., Dynamic simulation of a PVC suspension reactor, *Comput. Chem. Eng.*, 1995, S427–432
10 Krallis, A., Kotoulas, C., Papadopoulos, S., Kiparissdes, C., Bousquet, J., Bonardi, C., A comprehensive kinetic model for the free-radical polymerisation of vinyl chloride in the presence of monofunctional and bifunctional initiators, *Ind. Eng. Chem. Res.*, 2004, 43, 6382–6399
11 Pinto, J. M., Giudici. R., Optimisation of cocktail of initiators for suspension

polymerisation of vinyl chloride in batch reactors, *Chem. Eng. Sci.*, 2001, *56*, 1021–1028

12 Kiparissides, C., Daskalakis, G., Achilias, D.D., Sidiropoulou, Dynamic simulation of industrial poly(vinyl chloride) batch suspension polymerization reactors, *Ind. Eng. Chem. Res.*, 1997, *36*, 1253–1267

13 Wieme, J., de Roo, T., Marin, G., Heynderickx, G., Simulation of pilot- and industrial-scale vinyl chloride batch suspension polymerization reactors, *Ind. Eng. Chem. Res.*, 2007, *46*, 1179–1196

14 Nagy, Z., Agachi, S., Model predictive control of a PVC batch reactor, *Comput. Chem. Eng.*, 1997, *21*, 571–591

15 Mejdell, T., Pettersen, T., Naustdal, C., Svendsen, H., Modeling of industrial S-PVC reactor, *Chem. Eng. Sci.*, 1999, *54*, 2459–2466

16 Chung, S.T., Jung, S.H., Kinetic modeling of commercial scale mass polymerization of vinyl chloride, *J. Vinyl Additive Technol.*, 1996, *2*, 295–303

14
Biodiesel Manufacturing

14.1
Introduction to Biofuels

Biofuels are valuable sources of sustainable energy. Interest in them will rise tremendously in the coming years due to the perspective of a dramatic shortage of oil and gas reserves, simultaneously with an accelerated worldwide demand in fuels, namely in the large emergent economies. Biofuels have at least five merits. They:

1. reduce the energetic dependence of nations with respect to fossil resources acting as stabilizing factors in a global market environment;
2. reduce the global pollution by less CO_2 emissions on a lifecycle basis;
3. reduce the local pollution in terms of CO, CO_2, sulfur and fines particles;
4. enable recycling various potentially energetic industrial and domestic wastes, as cooking oil and fats, as well as agricultural residues;
5. ensure a better balance between industry and agriculture, creating new jobs in rural areas and sustainable economic growth.

The biofuels also raise concerns with respect to a possible shortage of agricultural resources needed for food. In return, it could be said that the biofuels aim to valorize primarily nonfood raw materials and nonused land resources. In addition, the biofuels should be considered as only a component, an important one, of the diversity of renewable energetic resources.

14.1.1
Types of Alternative Fuels

Table 14.1 presents a classification of alternative transportation fuels in terms of origin and ecological efficiency with respect to classical gasoline and petroleum diesel. Some features are briefly commented.

Natural gas for vehicles (NGV) is a methane-based fuel that presents superior combustion efficiency, reduced CO and NO_x emissions, and no sulfur. The proven natural-gas reserves are superior to those of oil. NGV usage is widespread in countries possessing large resources, such as Argentina, Columbia and Venezuela, but is very limited in Europe and the USA.

Chemical Process Design: Computer-Aided Case Studies. Alexandre C. Dimian and Costin Sorin Bildea
Copyright © 2008 WILEY-VCH Verlag GmbH & Co. KGaA, Weinheim
ISBN: 978-3-527-31403-4

Table 14.1 Alternative fuels, origin and ecological efficiency.

Name	Acronym	Origin	Use	Ecological efficiency
Natural gas for vehicles	NGV	Natural methane	Compressed (CNG)	A
Liquefied petroleum gas	LPG	Oil and gas processing	Fuel	B
Synfuels				
– Gas to liquid	GTL	Rich methane gas	Gasoline or diesel substitutes	C
– Coal to liquid	CTL	Coal liquefaction		
– Biomass to liquid	BTL	Syngas		
Ethanol	BE	Sugar cane, corn, ...	Gasoline blend	B
	ETBE	Etherification C4 =	Gasoline additive	
Biodiesel	BD	Oils and fats	Diesel blend	B
Hydrogen		CH_4, hc, H_2O	Fuel cell	C
Methanol		syngas	Fuel cell	B

LPG is a blend of propane and butane obtained as secondary products of natural-gas exploration or during crude oil processing. For this reason LPG is included usually in alternative fuels. Although the combustion properties of LPG are excellent, its wider usage is prevented by supply limitations and safety problems.

Synfuels are synthetic hydrocarbons obtained from syngas, a mixture of CO and H_2. The composition of synfuels is comparable to either gasoline or diesel, depending on the manufacturing conditions. The technology, based on the well-known Fischer–Tropsh synthesis, is in practice very demanding. Syngas can be obtained from gas, coal and biomass, the liquid products being named accordingly GTL, CTL and BTL. The technology of GTL was developed by Shell in Malaysia in the decade 1980–90. Today, new large-scale plants are being erected in the Persian Gulf region. The liquefaction of coal (CTL) has been running at large scale for decades at Sasol in South Africa. New CTL plants are in construction in China and the USA. Despite progress in technology, the CTL has a low ecologic efficiency. The valorization of biomass as a renewable resource is much more advantageous and generates significant interest today, namely in Europe. The efficiency of synfuels production depends largely on the gas-reforming technology. From this viewpoint the use of a methane-rich feedstock has a definite advantage over coal.

Bioethanol is experiencing a fast growth on the global scale, particularly in Brazil and the USA. Ethanol is already one of the largest chemical commodities. A major advantage of bioethanol is its blending capacity with normal gasoline, between 5 and 85%. Up to 5% bioethanol may be tolerated without particular problems, but higher concentrations require an adaptive motor engine known as "flex technology". Another interesting valorization of bioethanol is as ETBE (ethanol-

Table 14.2 Emissions generated by biodiesel compared with normal diesel (in %).

Emission	Biodiesel 100	Biodiesel 20
Carbon monoxide	−48	−12
Unburned hydrocarbon	−67	−20
Fines	−47	−12
NO_x	+10	+2
Sulfates	−100	−20
Air toxics	−60 to −90	−12 to −20

tert-butyl-ether), a gasoline additive for better combustion. Unlike MTBE (methyl-tert-butyl-ether) banned today because of the pollution threat to ground water, the ETBE has a low solubility in water and is consequently environmental acceptable. Older MTBE plants are revamped today to ETBE processes.

Biodiesel consists of a mixture of fatty-acid esters. Fatty-acid methyl esters (FAME) are the most involved because methanol is the cheapest alcohol, but other alcohols, namely ethanol, may be employed as well. The manufacturing process is based on the transesterification of triglycerides by alcohols to fatty esters, with glycerol as a byproduct. In this way, highly viscous triglycerides are converted in long-chain monoesters presenting much lower viscosity and better combustion properties. Homogeneous or heterogeneous catalysis are used to enhance the reaction rate. Raw materials are vegetable oils, preferably nonedible, but also different wastes, such as used frying oils or animal fats (tallow).

The biodiesel can be used alone or blended, as B20 in 20 wt% with petrodiesel. As shown in Table 14.2, the biodiesel has remarkable combustion properties reflected in a drastic reduction of all emissions, excepting for a small increase in NO_x. Over a life cycle the CO_2 reduction is about 65% [13].

Intensive research is devoted to fuel-cell vehicles (FCVs). Hydrogen-powered engines can be viable if a cost-effective hydrogen process can be found. At the present time the main hydrogen source is methane reforming, including from biogas. Another possibility is coal gasification, but with a substantial CO_2 penalty. Water electrolysis might be interesting if the price of nonfossil electricity becomes competitive. Methanol fuel cells are in rapid progress.

14.1.2
Economic Aspects

Economic incentives and private interest in new opportunities are boosting the biofuels market at exceptionally high rates. The European Union gives a strong example. By 2010 the share of alternative fuels should rise to 5.75% by energy value and 6% by volume [14, 15]. Biodiesel is the preferred option in Europe. In contrast, in the USA and Brazil bioethanol is leading, but biodiesel has good prospects too [9, 15, 35]. As Figure 14.1 shows, the growth of biodiesel capacity in Europe is impressive: 6.1 Mt in 2006 or 65% increase compared with 2005, by a ratio of six with respect to 2001. Germany is the leading country (2.7 Mt in 2006), followed by France and Italy. In recent years large-scale production capacities emerged in the UK, Spain, Poland and Check Republic. In 2006 there were biodiesel plants in 21 of the 25 European countries involving around 3 million hectares of arable land. The use of biodiesel as a transport fuel does not require any changes in the distribution system, therefore avoiding expensive infrastructure changes. Biodiesel is also used as an efficient heating oil.

14.2
Fundamentals of Biodiesel Manufacturing

14.2.1
Chemistry

Fatty esters are currently manufactured by the transesterification of triglycerides with light alcohols. The triglycerides are found in vegetable oils and animal fats, more generally known as lipids. The transesterification reaction takes place in the presence of a suitable catalyst, acid or base. The fatty ester is released simultaneously with the reformation of the OH group in glycerol. The overall reaction occurs in three stages is controlled by chemical equilibrium, as expressed by the reactions below:

$$\text{Triglyceride} + CH_3\text{-}OH \underset{}{\overset{K_1}{\rightleftharpoons}} \text{Diglyceride} + R_1\text{-}COO\text{-}CH_3$$

$$\text{Diglyceride} + CH_3\text{-}OH \underset{}{\overset{K_2}{\rightleftharpoons}} \text{Monoglyceride} + R_1\text{-}COO\text{-}CH_3$$

$$\text{Monoglyceride} + CH_3\text{-}OH \underset{}{\overset{K_3}{\rightleftharpoons}} \text{Glycerol} + R_1\text{-}COO\text{-}CH_3 \text{ (Methyl esters)}$$

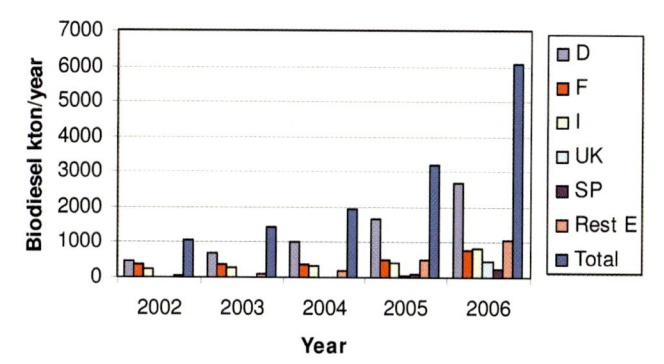

Figure 14.1 Evolution of biodiesel manufacturing capacity in Europe.

At full conversion of intermediates the overall reaction is:

$$\begin{array}{c} H_2C\text{---}OOCH_2\text{-}R_1 \\ | \\ CH\text{---}OOCH_2\text{-}R_1 \\ | \\ H_2C\text{---}OOCH_2\text{-}R_1 \end{array} + 3\,CH_3\text{-}OH \rightleftharpoons \begin{array}{c} H_2C\text{---}OH \\ | \\ HC\text{---}OH \\ | \\ H_2C\text{---}OH \end{array} + 3\,R_1\text{-}COO\text{-}CH_3$$

Triglyceride   *Glycerol*   *Methyl esters*

Thus, three molecules of fatty esters are produced for each molecule of triglyceride that needs three molecules of alcohol. The yield in biodiesel is about 90%. Full profit could be taken by converting glycerol to fuel by an appropriate chemistry.

The maximum achievable composition depends on the equilibrium constants of the individual steps. As an order of magnitude, they are of the order of about 10, 3 and 10 at 60 °C. In consequence, a substantial excess of alcohol is necessary to achieve higher ester yield close to 99%, as well as lower content in di- and monoglycerides.

Since the reaction rate is not fast enough at low temperature the transesterification makes use of catalysts. In fact, they make the difference between technologies. The catalysts can have acid or base character, and be homogeneous or heterogeneous. Base catalysts are preferred since they are faster. Homogeneous catalysts manifest higher activity, but need expensive postprocessing stages. Therefore, technologies based on continuous operation and solid catalyst would be preferable.

The lipid feedstock may contain variable proportions of free fatty acids (FFA), which should be converted in esters before transesterification. Otherwise, the formation of soaps occurs by reaction with the hydroxide catalyst, as follows:

$$R_1\text{-COO-CH} + NaOH \rightarrow R_1\text{-COONa} + H_2O$$

Another source of soaps is the saponification of ester dissolved in the glycerol phase:

$$R_1\text{-COO-CH}_3 + \text{NaOH} \rightarrow R_1\text{-COONa} + \text{CH}_3\text{OH}$$

The soaps can be reconverted to FFA by treatment with mineral acid:

$$R_1\text{-COO-CNa} + \text{acid} \rightarrow R_1\text{-COOH} + \text{salt}$$

The above simple reactions involve, in practice, costly operations, such as neutralization, washing, liquid phase and solid separations. These can be substantially reduced or even suppressed if superactive heterogeneous catalysts are employed.

For the esterification of FFA, superacid solid catalysts can be applied, such as ion-exchange resins (Amberlist, Nafion) or sulfated zirconia [22–24]. The preferred environment is a reactive-distillation column.

Regarding the transesterification, the research of heterogeneous catalysts is less advanced because of the difficulty of finding superbase catalysts. This issue will be presented later in this chapter. To date the most interesting innovation comes from the French Petroleum Institute where a heterogeneous catalyst based on zinc and aluminum oxides was developed and is currently being applied in commercial plants [2].

14.2.2
Raw Materials

A remarkable feature of lipids, either vegetal or animal, is that they share the same fatty acids in triglycerides in the range C12–C20 (Table 14.3). However, there are significant differences in composition. Thus, soybean, sunflower and rapeseed oils are all based on C18 acids, the first two being richer in unsaturated linoleic acid, which could introduce a problem of stability with respect to oxidation. The palm oil has an important amount of C16 acid. Coconut oil is given as an example of C12–C14 rich oil. As in palm oil the composition of tallow spreads over C16–C18 acids.

Table 14.3 Typical composition in fatty acids of some lipids.

Fatty acid	Formula	Soybean	Rapeseed	Sunflower	Palm	Coconut	Tallow
Lauric	12:0	0.1			0.1	46.5	0.1
Myristic	14:0	0.1			1.0	19.2	2.8
Palmitic	16:0	10.2	3.49	6.08	23.6	9.8	23.3
Stearic	18:0	3.7	0.85	3.26	14.2	3.0	19.4
Oleic	18:1	22.8	64.4	16.93	44.2	6.9	42.4
Linoleic	18:2	53.7	22.3	73.73	10.7	2.2	2.9
Linolenic	18:3	8.6	8.23		0.4	0	0.9

14.2.3
Biodiesel Specifications

As engine fuel, the long-chain unbranched fatty esters of biodiesel behave similarly to higher n-alkanes from petrodiesel. The biodiesel exhibits cleaner burning because of its oxygenated components, although the heat of combustion is only 90% from petrodiesel. Table 14.4 presents the quality specifications of biodiesel following the German norm 14214, which is typical of European standards. The most important combustion characteristic is the cetane number (CN), around 49. The CN number increases with the chain length, but decreases with the unsaturation of the fatty ester, and as a result depends on the raw-material composition. For example, CN is approximately 54 for methyl rapeseed ester, 50 for methyl palm ester, but only 46 for methyl soybean and sunflower esters.

The flash point indicates the temperature above which the fuel will ignite when exposed to a spark. This value, of about 110 °C, lower than for petrodiesel, is safer for transport purposes.

With respect to viscosity, which controls the fuel injection, the biodiesel shows somewhat higher values compared with petrodiesel, but this can be kept under 5 mm^2/s by controlling the feedstock composition or by blending.

The cold filter plugging point (CFPP) indicates the possibility of using the fuel in low-temperature conditions. Similar information is given by the "pour point", as well as by the "cloud point". Biodiesel shows higher CFPP values, namely at larger content in saturated esters. The rapeseed methyl ester (pour point at −9 °C) exhibits good behavior compared with the palm methyl ester (pour point at 10 °C).

Carbon residue is a measure of deposit formation in the long run. Biodiesel manifests a more pronounced coke formation than conventional diesel. For this reason the content of mono-, di-, and triglycerides, should each be kept below 0.4 wt%. The total bounded and free glycerol should be below 1.5%.

Table 14.4 Specification of biodiesel following the german norm DIN EN 14214.

Analysis	Units	Min.	Max.
Density	g/ml	0.875	0.900
Cetane number (CN)		49	
Flash point	°C	110	
Viscosity (15 °C)	mm^2/s	3.5	5.0
CFPP (cold filter plugging point)	°C	−10	−20
Water	ppm	–	300
Acidity number	mg KOH/g	–	0.5
Mono-/Di-/Triglycerides	wt%	–	0.4/0.4/0.4
Glycerol	wt%	–	0.25
Oxidation stability (110 °C)	h	5	

Higher alcohol content may cause the degradation of rubber gaskets and should be limited to under 1%. The acidity number reflects the presence of free fatty acids, which can contribute to accelerated engine aging. For the same reasons the water content should be kept quite low, at max. 300 ppm. Finally, in order to ensure proper storage the oxidation stability of the fuel should be at least 5 h at 110 °C.

The above-described specifications are supported by suitable analytical methods, based on chromatography and spectroscopy [30].

14.2.4
Physical Properties

Saturated fatty acids are solids characterized by melting points well above room temperature, as well as by high boiling points (Table 14.5). These properties show monotonic evolution with the carbon number. If a double bond appears an abrupt change in properties takes place. Higher free energy is reflected in lower melting and boiling points. For example, there is a significant difference in the melting points of stearic acid (70 °C), oleic acid (16 °C) and linoleic acid (−5 °C), which are saturated, with one and two double bounds, respectively. The same trend is exhibited by the methyl fatty esters, which in general have much lower melting and boiling points than the corresponding fatty acids.

Group contribution methods can be applied for property prediction of fatty-acid systems. For vapor-pressure prediction the following equation has been recently proposed and tested successfully against experimental data [6, 40]:

$$\ln P_{v,i} = \sum_k N_k \left(A_{1k} + \frac{B_{1k}}{T^{1.5}} - C_{1k} \ln T - D_{1k} T \right) + \left[M_i \sum_k N_k \left(A_{2k} + \frac{B_{2k}}{T^{1.5}} - C_{2k} \ln T - D_{2k} T \right) \right] + Q \quad (14.1)$$

Table 14.5 Properties of fatty acids, methyl esters and triglycerides [34].

Acid	Carbon	Melting point (°C)					Nbp (°C)	
		Acid	Methyl ester	MG	DG	TRIG	Acid[a]	Methyl ester
Lauric	12:0	44					298.7	261.85
Myristic	14:0	54	18.8	70.5	66.8	57.0	326.2	294.85
Palmitic	16:0	63	30.6	77.0	76.3	63.5	351	337.85
Stearic	18:0	70	39.1	81.5	79.4	73.1	375.2	351.85
Oleic	18:1	16	−19.8	35.2	21.5	5.5	359.85	348.85
Linoleic	18:2	−5	−35	12.3	−2.6	−13.1	354.85	365.85

a) From Aspen Plus database

with T in Kelvin and $P_{v,i}$ in Pascal. N_k is the number of group k in the molecule and M_i the molecular weight. A, B, C, and D are group parameters obtained by regression of experimental data, as shown in Table 14.6. The correction term Q is given by:

$$Q = \xi_1 q + \xi_2 \tag{14.2}$$

The factor q is calculated by:

$$q = \alpha + \beta/T^{1.5} - \gamma \ln(T) - \delta T \tag{14.3}$$

Table 14.6 Parameters for predicting the vapor pressure of long-chain fatty acid and derived oxygenated components.

Group	A_{1k}	B_{1k}	C_{1k}	D_{1k}
CH_3	−117.5	7232.3	−22.7939	0.0361
CH_2	8.4816	−10987.8	1.4067	−0.00167
COOH	8.0734	49152.6	0.0359	−0.00207
CH = cis	2.4317	1410.3	0.7868	−0.004
CH = trans	1.843	526.5	0.6584	−0.00368
COO	7.116	49152.6	2.337	−0.00848
OH	28.4723	−16694	3.257	0
$CH_2-CH_2-CH_2$	688.3	−349293	122.5	−0.1814

Group	A_{2k}	B_{2k}	C_{2k}	D_{2k}
CH_3	0.00338	−63.3963	−0.00106	0.000015
CH_2	−0.00091	6.7157	0.000041	−0.00000126
COOH	0.00399	−63.9929	−0.00132	0.00001
CH = cis	0	0	0	0
CH = trans	0	0	0	0
COO	0.00279	10.0396	−0.00034	0.00000295
OH	0.00485	0	0	0
$CH_2-CH_2-CH_2$	−0.00145	0	0	0

Compound	f_0	f_1	s_0	s_1
Esters	0.2773	−0.00444	−0.4476	0.0751
Acylglycerols	0	0	0	0
Fatty acids	0.001	0	0	0
Alcohols	0.7522	−0.0203	0	0

Calculation q	α	β	γ	δ
	3.4443	−499.3	0.6136	−0.00517

The parameters α, β, γ and δ have been obtained by the regression of the databank as a whole. Note that the term Q takes into account the effect of functional groups by means of two parameters ξ_1 and ξ_2, in turn given by the following expressions:

$$\xi_1 = f_0 + N_c f_1 \qquad (14.4)$$

$$\xi_2 = s_0 + N_{cs} s_1 \qquad (14.5)$$

N_c is the number of carbon in the molecule, while N_{cs} is the number of carbon in the alcoholic part. Table 14.6 shows the values of the parameters involved in the above correlations.

As an example, let us consider propyl laurate ester. The groups are CH_3 (2), CH_2 (12) and COO (1). In addition $N_c = 15$, $N_{cs} = 3$ and $M_i = 242$. The vapor pressure (Pa) is:

$$\ln P_v = \{[2(-117.5) + 12(8.4816) + 1(7.116) + 242[2(0.00338) + 12(0.00091) \\ + 1(0.00279)]\} + \{[2(7232.3) + 12(-10987.8) + 1(49152.6) \\ + 242[2(-63.3963) + 12(6.7157) + 1(10.0396)]\}/T^{1.5} - \{[2(-22.7939) \\ + 12(1.4067) + 1(2.337) + 242[2(-0.00106) + 12(0.000041) + 1(-0.00034)]\} \ln T \\ - \{[2(0.0361) + 12(-0.00167) + 1(-0.00848) + 242[2(0.000015) \\ + 12(-0.00000126) + 1(2.98E - 6)T + (0.2773 + 15(-0.00444))[3.4443 \\ - 499.3/T^{1.5} - (0.6136) \ln T + 0.00517T + [-0.4476 + 3(0.0751)]$$

By consequence, the following dependence vapor pressure/temperature is generated:

T	(K)	396.85	423.15	453.15	473.15	523.15	593.34
P_v	(Pa)	261.4	1034.4	3749	7680	18798	100000

The method can be applied for saturated fatty acids, unsaturated fatty acids, fatty esters, fatty alcohols and acyl-glycerols. The regression is based on 1200 data points. The absolute deviation in predicting vapor pressure is 6.82%. Another advantage of Eq. (14.1) is the capability of predicting the VLE of mixtures of fatty acids and esters by using the UNIFAC model for liquid activity. The comparison with experimental data shows good accuracy not achieved by other methods [40].

As mentioned above, the viscosity is an important physical parameter. Table 14.7 presents comparatively the values recommended by the standards for diesel fuels in the USA and Europe. It can be observed that they are in a small range, but the upper limit is somewhat higher for biodiesel with respect to petrodiesel. Actually, the fatty esters exhibit larger variation of viscosity due to their chemical nature compared with hydrocarbon molecules. Influencing factors are chain length, position, number, and nature of double bonds, as well as the nature of the

Table 14.7 Standards for diesel fuels in the USA and Europe.

Fuel	Standard	Kinematic viscosity (mm^2/s)
Petrodiesel	ASTM/D975–USA	1.3–2.4 (No. 1 diesel)
		1.9–4.1 (No. 2 low sulfur)
Petrodiesel	EN 590/ISO 3104–Europe	2.0–4.5
Biodiesel	ASTM/D6751–USA	1.9–6.0
Biodiesel	EN 14214/ISO 3104–Europe	3.5–5.0

Table 14.8 Kinematic viscosity (mm^2/s) at 40 °C of species implied in biodiesel [26].

Species	C18:0	C18:1	C18:2	C18:3	C18:1;12–OH
Triglyceride	nd	32.94	24.91	17.29	nd
Acid	nd	19.91	13.46	nd	nd
Alcohol	nd	17.53	11.94	nd	142.21
Methylester	5.85	4.51	3.65	3.14	15.44
Ethylester	5.92	4.78	4.25	3.42	nd
Propylester	6.78	5.44	4.39	nd	nd

oxygenated groups [26]. As illustrated by Table 14.8, the introduction of a double bond contributes significantly to lowering the kinematic viscosity of all concerned species. The effect of oxygenated moieties is approximately COOH ≈ C–OH > COOCH$_3$ ≈ C=O > C–O–C > no oxygen. It is interesting to note that the introduction of a second —OH group in the C18:1 chain, as in methyl ricinoleat issued from ricinoleic acid (C18:1–12—OH), has a dramatic impact on viscosity, which rises by an order of magnitude. For example, the unblended biodiesel from castor oil exceeds the viscosity specification.

14.3
Manufacturing Processes

14.3.1
Batch Processes

Older biodiesel processes are essentially batchwise. The oil is submitted to transesterification in a stirred-tank reactor in the presence of a large amount of methanol, and base catalyst, mostly NaOH or KOH. An excess of methanol is necessary chiefly to ensure full solubility of triglyceride and keep the viscosity of the reaction mixture low, but also for shifting the chemical equilibrium. A minimum molar ratio methanol:triglyceride of 6:1 is generally accepted [16, 17, 29]. The reaction

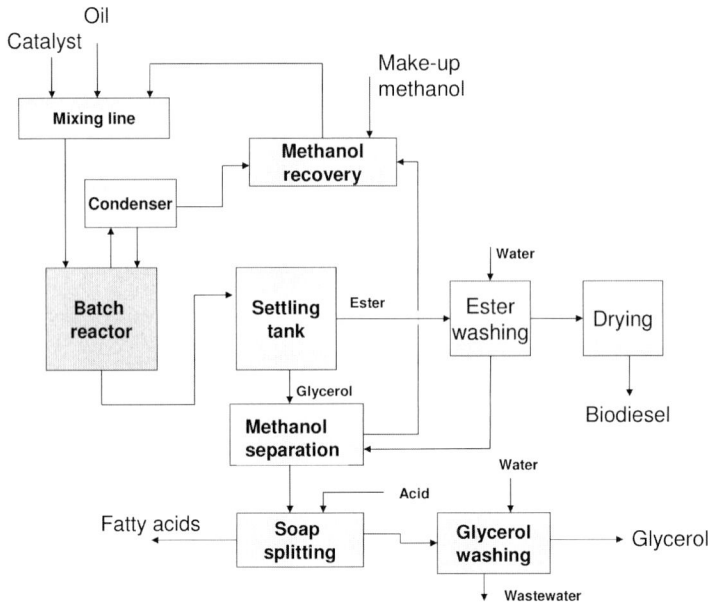

Figure 14.2 Batch-process flow diagram for biodiesel manufacturing.

takes place at temperatures from 60 to 80 °C, slightly below the mixture boiling point at the operating pressure. Previously, the oil should be neutralized by treatment with aqueous sodium hydroxide for the removal of free fatty acids. These can be found between 0.5 and 5% in the vegetable oils, somewhat more in animal fats, but can rise to up to 30% in used cooking oil. High FFA content needs special pretreatment by esterification. The transesterification reaction may be considered finished when the conversion reaches 98.5%. However, the mixture composition should respect the quality biodiesel specifications. The excess methanol is recovered for the next batch. The remaining mixture is submitted to the separation of esters from glycerol. This can take place either by decantation or by centrifugation. Water may be added to improve the phase split. The oil phase containing fatty esters is sent to finishing by neutralization with acid, followed by washing and drying. Phosphoric acid is frequently for neutralization used since Na_3PO_4 or K_3PO_4 can be recovered and sold as fertilizers. The water phase from washing is returned to glycerol separation.

After mixing of glycerol streams the result is about 50% glycerol–water solution with some methanol, residual base catalyst and soaps. Firstly, the methanol recovery takes place by flash distillation or film evaporation. Then, by adding acid the soaps are transformed in free fatty acids, which separate from glycerol as a top oily phase. Next, the FFA can be recovered and valorized by esterification with methanol. Finally the glycerol should have a purity of about 85% and be sold to specialized refiners. Purity of 99.5–99.7% can be achieved by applying vacuum distillation or ion-exchange process.

The batch process allows high flexibility with respect to the composition of the feedstock. In turn, the economic indices are on the lower side because of lower equipment productivity and higher operation costs, such as manpower and automation. The use of a large excess of methanol is reflected in higher energy consumption if no heat-integration measures are taken. Large amounts of wastewater formed by acid-base neutralization need costly treatment.

The productivity can be greatly improved by the implementation of continuous operations and the use of process-intensification techniques, such as reactive distillation. The replacement of a homogeneous catalyst by a heterogeneous one is highly desirable. These aspects will be discussed in greater detail in the next section.

14.3.2
Catalytic Continuous Processes

Figure 14.3 depicts the conceptual scheme of a continuous process working at low pressure that is capable of processing a feedstock with a larger amount of free fatty acids, such as unrefined nonedible vegetable oils, tallow fat and used cooking oil. For this reason in the first reactor R-1 the esterification of free fatty acids with methanol is carried out, preferably in a reactive-distillation device with solid catalyst. The amount of FFA should be reduced to below 1%, but preferably under 0.5%. Then the transesterification reaction follows in the unit R-2. A homogeneous catalyst is currently used, either as alkaline hydroxide or alkaline methoxide. The conversion should be high enough, in general over 98.5%. To ensure high yield in monoester and minimum amounts of mono-/di-/triglycerides minimum two reactors in series with glycerol intermediate separation should be employed. The

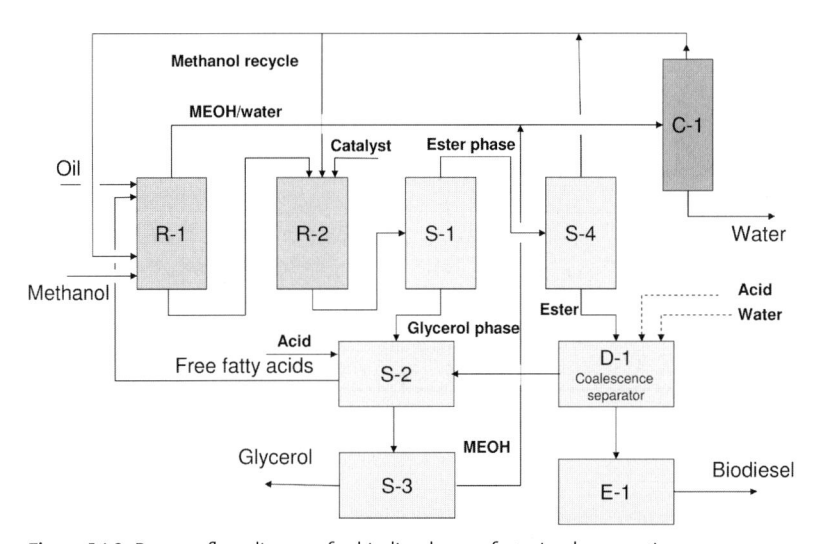

Figure 14.3 Process flow diagram for biodiesel manufacturing by a continuous process.

reaction mixture is then submitted to phase separation in crude ester and glycerol in the unit S-1. The separation can take place by decanting or by centrifugation. The glycerol phase is treated with acid for soap removal and recovery as FFA. Then, the methanol is recovered by evaporation and recycled. The crude ester follows the route of methanol separation in the unit S-4, the neutralization of the entrained catalyst, as well as the conditioning of biodiesel by washing and drying. The material balance loop is closed by the recovery of excess methanol from water solution by distillation.

The scheme in Figure 14.3 shows a modern method of ester/glycerol separation, namely by the technique known as "coalescence separator" [21]. This is applied in Figure 14.3 to the unit D-1. Such equipment avoids the use of water for phase separation. In this way, much more concentrated glycerol can be obtained with water only from the neutralization operation of low salt content. Another advantage is low energy consumption for methanol recovery. Such technology is applied by Henkel in Germany by the production of fatty-acid methyl esters [7].

Using a solid catalyst in the transesterification phase allows a substantial simplification of the process flowsheet. Figure 14.4 illustrates the process ESTERFIP-H™ developed by the French Petroleum Institute [2]. Two reactors are employed with intermediate glycerol separation. Excess methanol is recovered by multistage flash evaporation represented by the units S-1, S-2, S-3 and S-4. Phase separation of ester and glycerol are carried out in the units D-1, D-2 and D-3 by coalescence separation or centrifugation. It can be seen that the neutralization and washing steps are absent. Methanol can be recycled as vapor, with the result of low energy consumption.

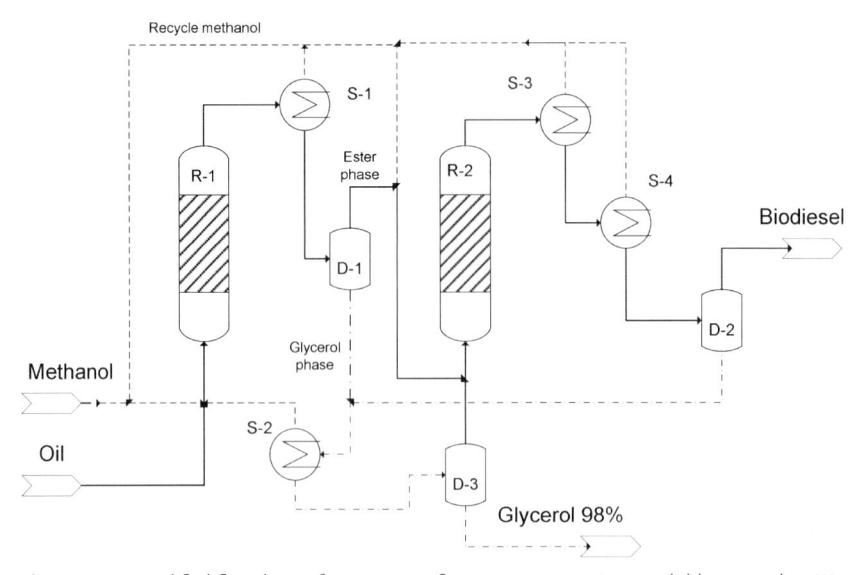

Figure 14.4 Simplified flowsheet of a transesterification process using a solid base catalyst [2].

14.3.3
Supercritical Processes

Performing the esterification in supercritical conditions has been studied initially as a method to solve the problem of miscibility of oil and methanol that hinders the kinetics in normal conditions. Since the critical coordinates of methanol are $T_c = 239\,°C$ and $P_c = 80\,bar$, raising the temperature and pressures at sufficiently high values is necessary. Studies conducted in Japan demonstrated the feasibility of producing biodiesel by the esterification of rapeseed with methanol without a catalyst working around 350 °C and 200 bar at molar ratio methanol:oil of 42:1 for reaction times below 4 min [10, 27, 28, 38]. The advantage of avoiding a catalyst is obvious. However, the conditions of pressure and temperature are severe and need special equipment. Recent research showed the real yield can be reduced by thermal degradation of biodiesel, namely of unsaturated fatty esters [20]. For this reason, lowering the reaction temperature and pressure is highly desirable.

The assessment of the critical region of a mixture triglyceride/methanol can be made by applying the approach explained next. The critical properties of individual triglycerides components can be estimated by a suitable group-contribution method, and then using mixing rules for averaging the parameters function of composition. As an example, let us consider coconut oil, based on lauric and myristic acids (see Table 14.3). The critical properties of the oil calculated by Lydersen's method [34] are $T_c = 606.8\,°C$ and $P_c = 6.2\,atm$. Table 14.9 presents the critical values obtained for different molar ratio methanol/oil starting with 6, corresponding to a low solubility limit, and ending with 42, the highest practical value tested in laboratory experiments [3]. It can be seen that the critical temperature drops rapidly with increased methanol/oil ratio to about 280 °C, while the critical pressure rises up to about 70 atm, close to methanol. In practice, operating parameters of 350 °C and 190 bar were employed at a residence time of 400 s.

The addition of cosolvent in combination with supercritical conditions seems to be an efficient means to reduce significantly the operating temperature [4]. For example, soybean oil could be converted with methanol into biodiesel with 98% yield by using propane, at least in 0.05 molar ratio to methanol, at 280 °C and 12.8 MPa. Similar results have been reported with CO_2 in a molar ratio of 0.1 with respect to methanol. In both cases the optimal ratio methanol/oil was 24 and residence time of 10 min [5].

Table 14.9 Critical properties of mixtures oil/methanol at different molar ratios R [3].

Properties	MeOH	Coconut oil	R = 6	R = 12	R = 24	R = 42
T_c, °C	239.5	606.8	395.9	345.9	305.8	282.2
P_c, atm	79.9	6.2	37.2	50.3	61.7	68.4
V_c, l/mol	118	2.366	0.33	0.22	0.22	0.22

Due to the absence of the catalyst the process flowsheet employing the supercritical technology should be much simpler, but in exchange the manufacture of hardware is much more demanding. Effective energy integration is also necessary. Despite these advantages the industrial implementation of supercritical esterification has not been reported.

14.3.4
Hydrolysis and Esterification

A simpler manufacturing procedure would consist in first performing the hydrolysis of triglycerides and isolating the fatty acids followed by esterification employing the robust technology of a solid heterogeneous catalyst. Significant advantages would be the possibility of extracting high value fatty acids from the lipid material, as well as obtaining high-purity glycerol. The hydrolysis reaction can be carried out without a catalyst working in milder conditions compared to full esterification. A temperature close to 270 °C and pressures from 70 to 200 bar have been found applicable [31]. Another advantage is that the overall yield can be increased by suppressing the back reaction of glycerol with the methyl ester. The reaction exhibits an autocatalytic effect due to the fatty acid produced, from which a small recycle can be provided.

Figure 14.5 presents a conceptual flowsheet. Oil and water are brought at high pressure, homogenized in a static mixer and heated. A volumetric ratio water/oil 1 : 1 is appropriate. The hydrolysis takes place in the reactor R-1 in slightly subcritical conditions at 270 °C and 100 bar. The yield in fatty acids is around 90% for a residence time about 40 to 60 min [28]. Therefore, a simple long coil can be used as the chemical reactor. After cooling and pressure reduction, the reaction mixture is separated into two phases in S-1. The oily phase containing a large majority of fatty acids is sent directly to esterification, or optionally to fatty-acid separation in the unit S-2 by vacuum distillation. The esterification reactor R-2 is preferably a reactive distillation using a solid acid catalyst on structured packing. The heavies

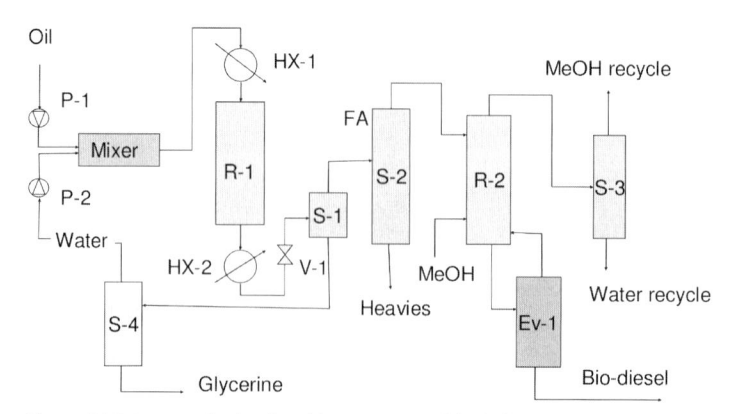

Figure 14.5 Process for biodiesel by supercritical hydrolysis and esterification.

from S-2 containing glycerides can be recycled to R-1, or disposed of as combustible waste. The unit R-2 delivers in bottom fatty-acid methyl esters diluted with methanol, from which biodiesel with fuel specifications is obtained from the evaporator Ev-1. The top stream from R-2 is sent to the distillation column S-3, from which water and methanol are recovered and recycled to R-1 and R-2, respectively. The glycerol phase from S-1 goes to the unit S-4 from which high-purity glycerol is obtained.

Summing up, the process based on the hydrolysis of triglycerides seems very attractive, despite the fact that supercritical operation raises a technical challenge. By making use of recycles the process can be designed to achieve material consumption close to stoichiometric requirements. Pumping liquids at high pressures requires moderate energy. By heat integration the utility consumption could be kept at low level.

14.3.5
Enzymatic Processes

The transesterification reaction can be catalyzed by enzymes, the most common being the lipase. The reaction takes place at normal pressure and temperatures 50 to 55 °C with low energy consumption. The yield of methanolysis depends on several factors as temperature, pH, type of micro-organism producing the enzyme, the use of cosolvents, *etc.* However, low yields in methyl esters and very long reaction times make the enzymatic processes not competitive enough at this time [9, 11, 17].

14.3.6
Hydropyrolysis of Triglycerides

A fundamentally different chemical way of converting biotriglycerides to fuels is hydrogenation followed by pyrolysis in the presence of a suitable catalyst. The process invented by the Finnish company Neste Oy [www.nesteoil.com] is known as NExBTL (biomass to liquid). Figure 14.6 shows a simplified reaction scheme. The fuel produced is essentially a mixture of long-chain hydrocarbons instead of long-chain esters. In addition, the whole feedstock is valorized, including glycerol converted to propane. It is claimed that the resulting fuel has superior combustion

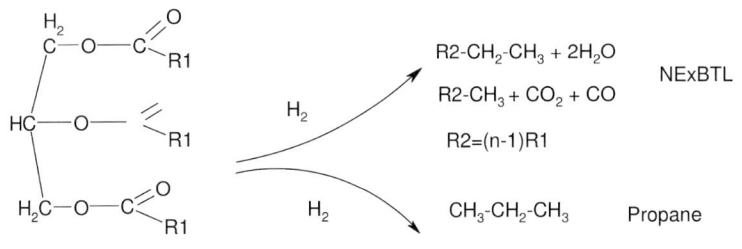

Figure 14.6 Reaction scheme describing the NExBTL process.

properties compared to ester biodiesel, such as for example higher cetane number (84 to 99), cloud point down to −30 °C and better storage stability. For this reasons NExBTL is occasionally called "second-generation" biodiesel. However, this process implies the availability of a low-cost hydrogen source, as well as more complex and expensive equipment.

14.3.7
Valorization of Glycerol

Glycerol is a high added value byproduct when it can be isolated of high purity. This is the case when solid catalyst or supercritical hydrolysis is employed. Because of biodiesel large amounts of low-quality glycerol become available and the price is pushed down. For example, 10 million tons per year biodiesel supplies glycerol of 1 million tons per year. Converting the glycerol to chemicals and/or fuels becomes imperative. One alternative is etherification with alcohols (e.g. methanol or ethanol) or alkenes (e.g. isobutene) to produce branched oxygenated components. Various catalysts can be employed such as zeolites, ion-exchange resins and acidic homogeneous catalysts. Tert-butyl ethers of glycerol can be used as ingredients in biodiesel and gasoline, offering an alternative to oxygenates additives such as ETBE.

14.4
Kinetics and Catalysis

14.4.1
Homogeneous Catalysis

Homogeneous catalysis remains largely employed today because ensuring simple and robust technology, as well as high reaction rates, despite some important economical and environmental disadvantages [7–9, 11, 16, 17, 37]. For the production of biodiesel by transesterification both acid and base catalysis can be applied, but the latter is much more efficient. The difference can be explained by the reaction mechanism, as explained by Figures 14.7 and 14.8. The glycerol part of the triglyceride is designated by R1 and the fatty acid by R2. In homogeneous acid catalysis (Figure 14.7) the first step consists of triglyceride activation by protonation (1) at the CO group where the oxygen is more active, followed by the formation of a carbocation complex (2). By nucleophilic attack with methanol (3) a tetrahedral carbon complex forms, which by losing the proton decomposes further (4) in a new methyl fatty ester and diglycerides. The methanolysis proceeds similarly with diglyceride and monoglyceride. It can be seen that if water is present it will produce fatty acid by hydrolysis in the step (3) and as a consequence it will decrease the yield of fatty ester. For this reason the water amount in the triglicerides should be reduced below 0.5%.

Base catalysis involves a completely different mechanism, as explained by Figure 14.8. The active species this time is an alkoxide, namely the methoxide $^-O-CH_3$.

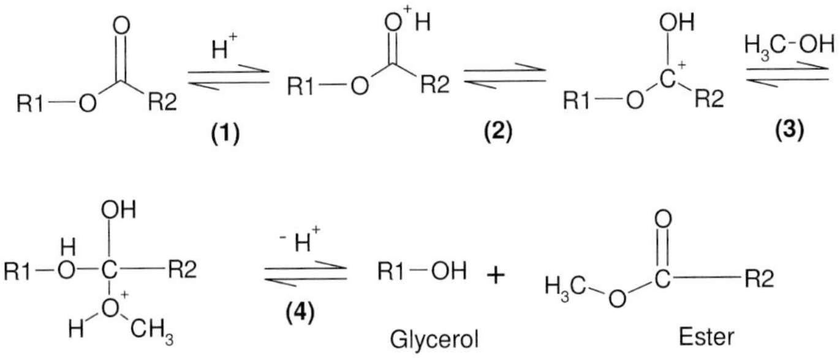

Figure 14.7 Mechanism of acid-catalyzed transesterification: R1–glyceride fragment, R2–fatty-acid carbon chain.

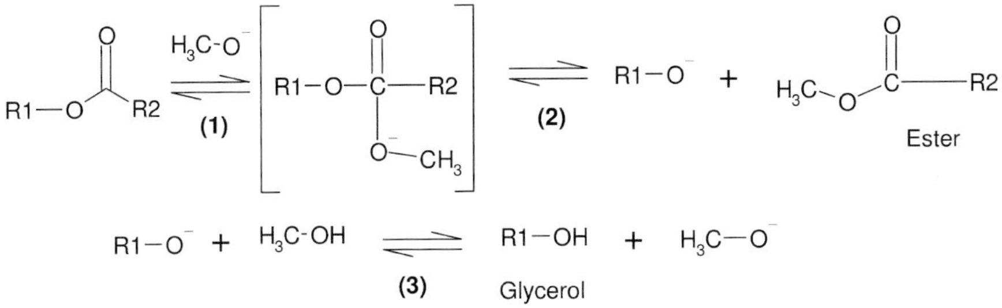

Figure 14.8 Mechanism of base-catalyzed transesterification of triglicerides [30].

This can be produced *in-situ* by the reaction of methanol with hydroxide liberating water:

$$NaOH + CH_3-OH \leftrightarrow Na^+ {}^-OCH_3 + H_2O$$

The methoxide can be introduced as preprepared from alcohol and an alkali metal. In this case, the absence of water favors the reaction rate, as well as easier post-processing. The first step consists of the nucleofilic attack of the methoxide to the carbonyl group, which leads to the formation of a tetrahedral carboanionic complex. Next, the transition complex decomposes into a fatty ester and a diglycerol anion, which reacts with an alcohol molecule, reforming the catalytic species. The other transesterification stages take place similarly. The whole reaction process is controlled by chemical equilibrium.

As mentioned above, the reaction rate by base catalysis is much faster than by acid catalysis by three orders of magnitude. The above presentation of the reaction mechanism can deliver an explanation: the formation of carboanionic tetrahedral intermediate results directly by the nucleophilic attack of the substitution species, while the carbocationic complex needs an intramolecular arrangement.

The transesterification of triglycerides with methanol to fatty methyl esters and glycerol implies three reaction stages passing through the formation of diglyceride and monoglyceride intermediates. For each stage the equilibrium constant can be expressed as the ratio of forward and backward reactions:

$$K_1 = k_1/k_{-1};\ K_2 = k_2/k_{-2};\ K_3 = k_3/k_{-3} \tag{14.6}$$

Kinetic experiments have been reported in literature for the treatment of different types of lipidic feedstock with methanol, such as for soybean [16, 33], rapeseed [25] and sunflower oil [1, 36]. The aim is to optimize the reaction conditions, namely the amount of catalyst, the excess of methanol and the reaction temperature. The results depend largely on the composition of the raw materials, but some trends can be distinguished:

1. Optimal amount of base catalyst is 0.5 to 1.5% w/w. Methoxides are more active than alkali hydroxides. More catalyst than needed favors soap formation and makes phase separation difficult.
2. The optimal molar ratio of methanol/triglyceride is between 6 and 9. Adding small amounts of cosolvents, such as propane and tetrahydrofuran, can reduce the excess of methanol.
3. The optimal reaction temperature is rather low, between 60 and 80 °C.

Since the reaction network consists of three-step series-parallel reactions, an important aspect in the kinetic modeling is the relation to the end-product specifications, more precisely the content in tri-, di- and monoglycerides. Table 14.10 presents the results of a recent study regarding the methanolysis of sunflower oil by using NaOH catalyst [1] and second-order kinetics based on the stoichiometric coefficients. The standard run measurements were done at a molar reactant ratio of 6, 0.5% catalyst, 60 °C and mixing speed of 400 rpm. The results draw attention to a complex reaction mechanism in the sense that the kinetic constants depend on the operation conditions. The first reaction leading to diglyceride is the slowest step. The equilibrium constants of the first and last stage are close to 10, but about 3 for the intermediate step. As a consequence, even if the triglyceride conversion can reach high values over 97.5% the amount of monoglyceride will remain substantial, around 2%, well above the maximum of 0.4% tolerated by quality specifications. In contrast, the amount of di- and tri-glycerides may drop below 0.5%. Using an excess of methanol does not help to reduce the concentration of monoglyceride. Another strategy is needed, as shown later in this chapter.

Table 14.10 Kinetic constants for the methanolysis of sunflower oil [1].

	k_1 [a]	k_{-1}	k_2	k_{-2}	k_3	k_{-3}
Standard run	0.0895	0.0094	0.3480	0.1285	0.4884	0.0380
Mixing at 600 rpm	0.1316	0.0195	0.3227	0.2470	0.8611	0.0606
Reaction at 40 °C	0.0218	0.0029	0.0651	0.0319	0.2280	0.0128
1.0% catalyst	0.2314	0.0166	0.4488	0.1068	0.8770	0.0632
10:1 mole ratio	0.0738	0.0067	0.0811	0.0798	0.3472	0.0537
Activation energy [b]	14 040	10 739	16 049	13 907	7173	10 997

a) l/mol/min
b) cal/mol

14.4.2
Heterogeneous Catalysis

The replacement of homogeneous catalysis by solid catalysts brings obvious economical and technological advantages. For this reason, a considerable research effort is being devoted in this area.

A first application regards the esterification reaction. Here, solid catalysts with acidic character can be used, such as zeolites, ion-exchange resins, sulfated metal oxides, sulfated carbon fibers, *etc.* However, only few are suitable for handling long-chain complex molecules. Figure 14.9 presents a comparison of acidic heterogeneous catalysts that can be used in biodiesel manufacturing [23]. Some can achieve super acidity comparable with sulfuric acid. Ion-exchange resins, such as Nafion and Amberlyst are capable of achieving high reaction rate at moderate temperatures below 130 °C, but their chemical stability at longer operation seems to raise concerns. On the contrary, sulfated zirconia and tin oxides can be used at higher temperatures, 140 to 180 °C, and ensure high reaction rates, but are sensitive to deactivation by sulfur-group leaching if free water is present. Since the water produced by esterification limits also achieving high conversion because of the chemical equilibrium, a good solution for solving both problems is employing reactive distillation.

The second area of heterogeneous catalysis in biodiesel manufacturing is the transesterification reaction. Here again, the base catalysts exhibit typically much higher activity than the acidic ones, but finding effective catalysts is still an open problem. Some solid metal oxides, such as those of tin, magnesium, and zinc could be used directly, but they actually act by a homogeneous mechanism

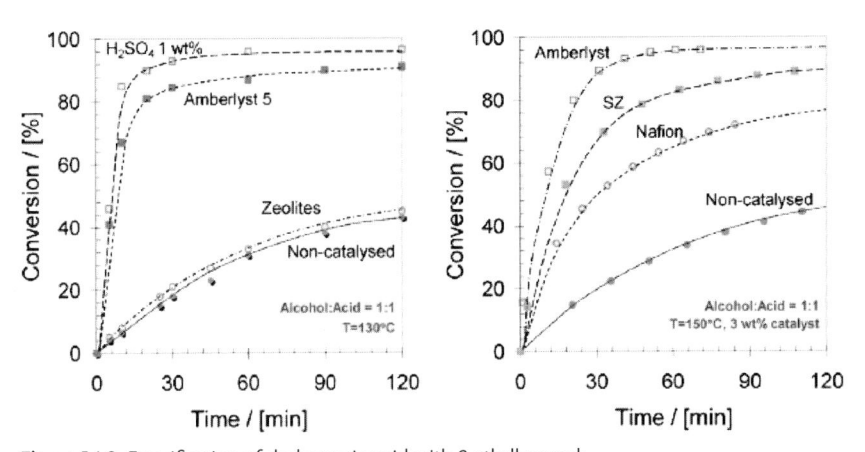

Figure 14.9 Esterification of dodecanoic acid with 2-ethylhexanol: comparison of homogeneous and heterogeneous acid catalysts at 130 °C (left); comparison of Amberlyst, Nafion and sulfated zirconia (SZ) at 150 °C (right). The amount of solid catalyst, 3 wt%, refers to the total mass of reactants [23, 24].

producing a significant amount of soaps or glycerates. A variety of synthetic catalysts based on alkaline-earth metal oxides (Ca, Mg, Sn, Zn), as well as alkali metals (Na, K) hydroxides or salts impregnated on alumina have been studied to date, but their activity and robustness remains insufficient [11–13, 39]. As mentioned, a successful industrial application of base heterogeneous catalysis has been reported [2], but not confirmed by other studies so far.

14.5
Reaction-Engineering Issues

In older processes based on a batch reactor the research from the laboratory can be scaled up without much difficulty. However, the overall productivity is seriously diminished by the time of additional operations, such as charge loading, heating, cooling and discharging. Switching on continuous reactors offers a number of alternatives, as will be presented in this section. As a modeling basis we consider the kinetics of sunflower oil catalyzed by NaOH, as described by the data in Table 14.11. The simulation is done in Aspen Plus™, the chemical species being assimilated to those corresponding to oleic acid. The target is to obtain the biodiesel quality as close as possible to the specifications presented in Table 14.5.

Firstly, the relation composition–time in a batch operation is computed (Figure 14.10). The profile corresponds to series-parallel equilibrium reactions. At 60 °C the reaction rate catalyzed by NaOH is fast: conversion over 98.5% can be reached in 20 min and chemical equilibrium at 99.8% in about 30 min. However, the content in mono- and diglycerides remains relatively high at 6.5 and 2.2% mol, or 2.6 and 1.6 wt%, much higher than the required specifications. Lowering the

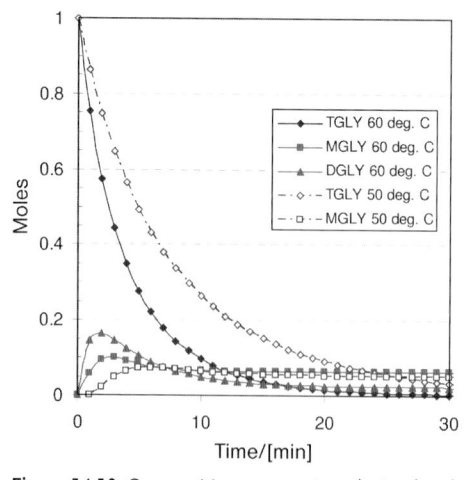

Figure 14.10 Composition versus time during batch methanolysis of oleic triglyceride using NaOH 0.5% catalyst and molar ratio methanol/oil 6:1 at 60 and 50 °C.

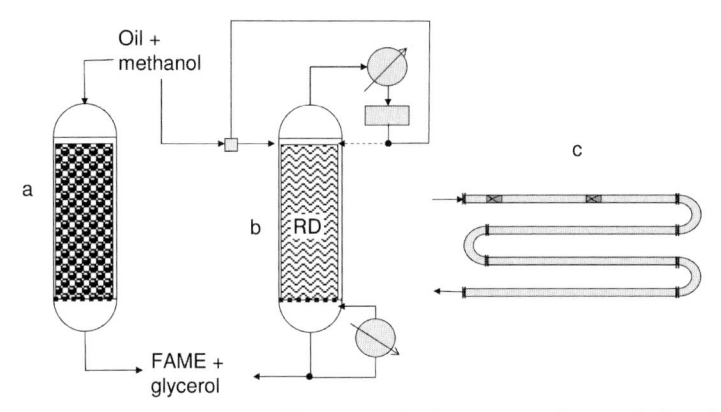

Figure 14.11 Continuous PFR-like reactors for the transesterification of glycerides.

temperature to 50 °C can slightly shift the equilibrium to lower DGLY and MGLY, still above specifications, but the reaction time becomes longer. Another possibility is raising the amount of methanol with the disadvantage of supplementary costs. Therefore, an effective method is performing the reaction in minimum two steps, with intermediate removal of glycerol so as to limit the backward reactions as much as possible.

Figure 14.11 presents alternative continuous plug-flow reactors (PFR) for the transesterification reaction. An isothermal plug-flow reactor has identical performance with a batch reactor in terms of conversion and product distribution. A PFR can be built either as a tubular reactor, or as series of a sufficient number of perfectly stirred zones hosted in the same shell. Considering homogeneous or heterogeneous catalysis generates subsequent alternatives. A packed-bed column (Figure 14.11a) is suitable when a superactive heterogeneous catalyst is available. PFR as long reaction tube (Figure 14.11b) can be used with homogeneous or heterogeneous catalyst, at least as a prereactor. Static mixing elements can be inserted to ensure homogeneous phase reaction.

An interesting possibility for performing the transesterification reaction is offered by reactive distillation (RD). This has been applied at the laboratory scale with a homogeneous catalyst demonstrating superior productivity [19]. Solid catalyst imbedded in structured packing can be used if its activity is high enough to cope with the constraints set by the hydraulics. A key advantage of reactive distillation is the possibility of ensuring much higher local ratio methanol/glycerides than in a PFR by means of the internal reflux, while the initial feed can be close to stoichiometric requirements. The energy consumption can be kept reasonable low by appropriate heat-integration measures, namely operating the column at higher pressure and temperatures.

Continuous stirred-tank reactors (CSTR) are currently used for biodiesel manufacturing. Obviously, a single CSTR would require a huge volume to achieve the performance of a batch or PFR-like reactor, but the use of several CSTRs in series can improve the productivity considerably (Figure 14.12a). Intermediate

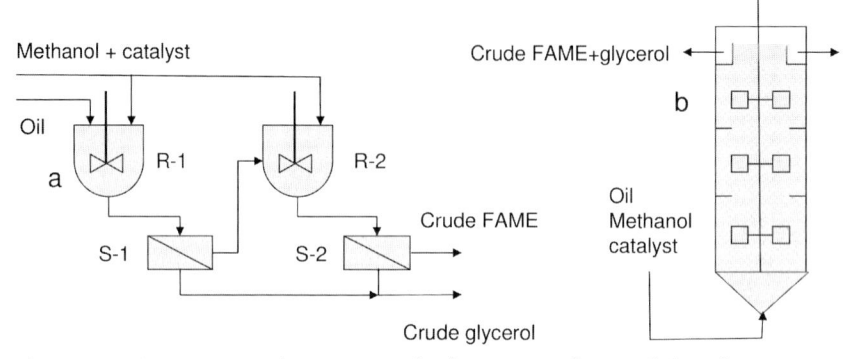

Figure 14.12 Continous stirred reactor setup for the transesterification of glycerides.

Table 14.11 Liquid–liquid equilibrium for the mixture methanol (MeOH -glycerol (G) methyl oleat (ME) at 60 °C and 135 °C [32].

Fatty phase			Glycerol phase		
x'_{MeOH}	x'_G	x'_{ME}	x''_{MeOH}	x''_G	x''_{ME}
0.043	0.002	0.955	0.105	0.895	0.000
0.098	0.008	0.894	0.331	0.669	0.000
0.163	0.007	0.830	0.446	0.554	0.000
0.267	0.008	0.725	0.598	0.402	0.000
0.303	0.008	0.689	0.669	0.330	0.001
0.016[a]	0.018[a]	0.966[a]	0.020[a]	0.98[a]	0.000[a]

a) Data at 135 °C

separation of glycerol will help achieving lower monoglycerides content. Another possibility for increasing the productivity is a multiagitated reaction column (Figure 14.12b), applicable for both homogeneous and heterogeneous catalysts. In addition, PFR and CSTR-like reactors can be combined, the first being more productive at higher conversions.

Summing up, from the reaction-engineering viewpoint there are a considerable number of alternatives. The most critical factor in design is the availability of an active catalyst and its performance with respect to the raw materials.

14.6
Phase-Separation Issues

A key operation in biodiesel manufacturing is the separation of glycerol and FAME from the reaction mixture by liquid–liquid decanting. Thermodynamic studies on this subject are scarce despite the industrial interest. Table 14.11 presents data for

Table 14.12 Stoichiometric material balance for trioleine methanolysis.

	Input			Output		
	Trioleine	Methanol	Total	Methyl ester	Glycerol	Total
Mass, kg	884	96	980	888	92	980
Volume, l	1040	122.2	1162.2	1082.9	83.6	1166.5

the system methanol/methyl oleate/glycerol that could be seen as representative [32]. It can be observed that the reciprocal solubility of glycerol and methyl ester at 60 °C is very low, the ester being practically insoluble in glycerol. Methanol is distributed between the phases but preferentially in glycerol. The same remains valid at higher temperature. The mono-olein has a low solubility in the glycerol phase. Note that the LLE prediction by UNIFAC or UNIFAC-Dortmund for the system methanol-glycerol-methyl ester is in satisfactory agreement with the experimental data, but deviates considerably for the mixture involving mono-olein. The data from Table 14.12 can serve for identifying the binary interaction parameters of a convenient thermodynamic model, such as NRTL.

14.7
Application

As a simulation example we treat the production of biodiesel from rapeseed in a plant capacity of 200 ktonne per year. The feedstock has a high content of oleic acid triglyceride, around 65%, such that the kinetic data from Section 14.6 can be used for sketching the design of the reaction section. For simplification, we consider that the oil was pretreated for removing impurities and gums, as well as FFA by esterification over solid catalyst. The free fatty acids and water content in oil feed should be less than 0.5%w. NaOH and KOH in 0.5 to 1.5% w/w are used as catalysts.

It is useful to examine a simple material balance in terms of the stoichiometric requirements, as presented in Table 14.12. It can be seen that the amount of lipid largely dominates the manufacturing rate. The glycerol obtained as a byproduct is approximately equal to the amount of methanol introduced in the reaction. The same large imbalance is also valid for the volumetric feeds.

Figure 14.13 presents a process flow diagram built up with the conceptual elements examined so far in this chapter. The flowsheet aims to illustrate different reaction and separation techniques. Note that the heat-integration elements are not presented in order to keep the flowsheet simple. The feed of oil and methanol, including recycle, ensures a molar ratio methanol:oil of 6:1. Good mixing of reactants is necessary to ensure a homogeneous reaction phase, otherwise an induction period will affect the conversion. The mixing of reactants can be done preferably in a static device.

424 | *14 Biodiesel Manufacturing*

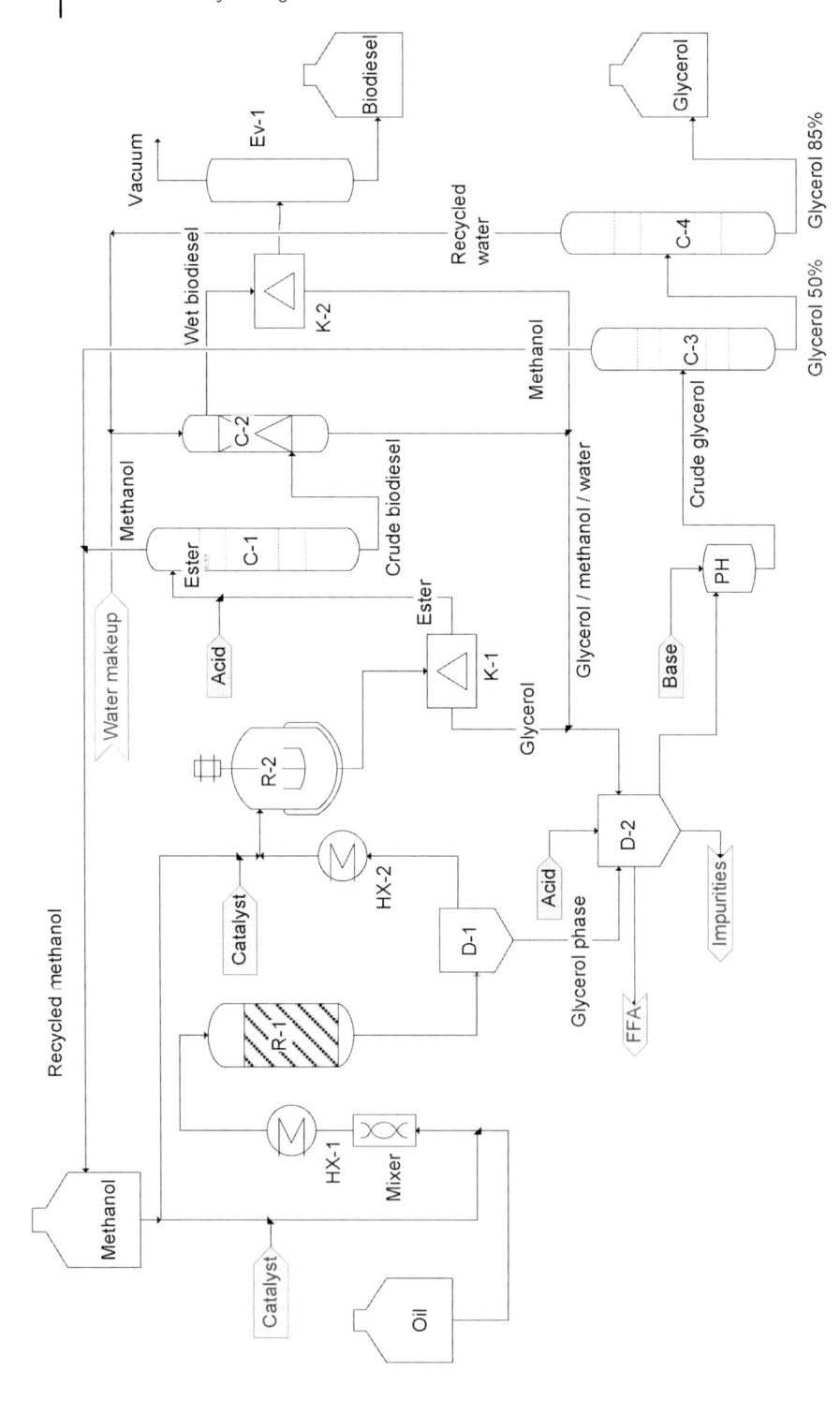

Figure 14.13 Process flow diagram for biodiesel manufacturing.

After heating at about 65 °C the esterification starts in the reactor R-1, which can be a CSTR, but preferably a PFR or a combination PFR/CSTR. The first reactor should ensure a conversion slightly above 90%. Intermediate removal of glycerol takes place to shift the equilibrium and get lower content of monoglyceride. A simple phase split by decanting can be applied at temperatures of 40 to 60 °C. The decanting time could be very variable, between a few minutes and 1 h. The presence of soaps and monoglycerides hinders the phase separation, while more neutral pH and lower methanol content helps. Modern coalescence separators can ensure a relatively smooth separation if the amount of soap is not excessive.

After makeup in methanol and catalyst the oil phase is submitted to a second transesterification step in the reactor R-2. The conversion should rise to a minimum of 98.5%. The phase separation is done this time by means of the centrifuge K-1. It is worth noting that centrifugal phase separation is becoming a state-of-the-art method in biodiesel technology. Centrifugation can be applied in all separation steps, including neutralization, washing and soap removal. Centrifuges of various capacities are available, such as from 60 to 1600 tonne/day supplied by the German company GEA.

After neutralization with acid, H_3PO_4, the resulting mixture is submitted to methanol stripping. In this way, about 90% from the excess methanol can be recycled directly to reaction section. The next stage is a wash with hot water at 50 °C for deep purification from methanol, glycerol, soaps and salt. This operation can be done in a countercurrent column, or directly in a centrifugal separator. Because of the presence of double bounds and ester groups the oil phase will contain a significant amount of water, well above the threshold of 300 ppm set by the specification norm (see Table 14.5). The removal of water and residual solids from the biodiesel is done by centrifugation in the unit K-2, and the final polishing before shipping by a vacuum flash evaporator.

The glycerol recovered in the previous stages is treated with acid for catalyst neutralization. The soaps are converted to fatty acids that precipitate at the top of the glycerol phase, and are sent back to the esterification stage.

After pH adjustment the crude glycerol is submitted to methanol recovery by vacuum distillation. The bottom product is usually a glycerol–water solution of about 50%. Its concentration can be increased up to 85% by vacuum evaporation. The water is recycled to the washing step. Normally, the glycerol is shipped to a specialized refiner, where the purity can be increased further to 99.5 and 99.7% by ion-exchange techniques and vacuum distillation.

The above flowsheet can be simulated by means of an appropriate simulation package. In the absence of a comprehensive kinetic model and of fundamental thermodynamic data the results will be only approximate, namely with respect to satisfying the quality specifications. However, the simulation allows the designer to obtain an overall view of streams, utilities and equipment, needed for an economic assessment.

Haas et al. [18] developed a computer model to estimate the capital and operating costs of a moderately sized industrial biodiesel production facility with a capacity of 33.5 ktonne (10 million gallons) using degummed soybean oil as feedstock. The major process operations were continuous transesterification, as well as ester and glycerol recovery, the process flow diagram being close to those presented in Figure 14.13. The investment costs inside battery limits were calculated to be 11.3 million US$. The largest contributors, accounting for nearly one third of the expenditures, were storage tanks of feedstock and products, sized for 25-day period. At a value of US$ 0.52/kg for feedstock soybean oil, a biodiesel production cost of US $0.53/l was predicted. The oil feedstock accounted for 88% of total estimated production costs. These are linearly dependent on the cost of oil. The sale of glycerol 80% could reduce the production costs by 6%.

14.8
Conclusions

Biodiesel is an alternative renewable fuel that has seen rapid development in recent years, namely in Europe. It can be manufactured from vegetable or animal fats by reaction with light alcohols, namely with methanol and ethanol. In most cases, the biodiesel is a mixture of fatty-acid methyl esters (FAME). From the chemistry viewpoint the raw materials for biodiesel are quite homogeneous, being based on triglycerides. These are esters of glycerol with fatty acids involving both saturated and unsaturated long carbon chains from C14 to C20. Another possibility is converting the triglycerides with hydrogen to hydrocarbons. The fuel obtained has superior combustion features and good CO_2 balance being sometimes called "second-generation" biodiesel.

The key merit of converting triglycerides into fatty esters is a drastic reduction in viscosity, about one order of magnitude, at a level compatible with the fuel-injection devices. The biodiesel is environmental friendly. Better combustion allows the level of greenhouses gases to be reduced, while sulfur is practically absent.

The quality of biodiesel is regulated by standards. The most important regards the content of free and bound glycerol. This cannot be modified by effective separation techniques. Therefore, the major element in design should be to obtain a composition of the mixture leaving the reaction system capable of matching the biodiesel specifications. This is difficult to achieve in view of the variety of raw materials.

Several technologies can be employed. The most widespread today makes use of homogeneous catalysts, in batch or in continuous-flow environments. Both reaction and separation steps can create bottlenecks. The availability of heterogeneous catalysis allows the suppression of neutralization and washing steps, leading to a simpler and more efficient process. However, the research of super active and robust catalysts is still an open problem. Supercritical hydrolysis and transesterification can be conducted without a catalyst, but in extreme conditions of pressure and temperature.

The design of the reaction section offers several alternatives. Plug-flow-like reactor type gives the best productivity with resonable reaction times from 10 to

20 min. Two-stage reactions with intermediate glycerol removal is necessary for pushing the equilibrium composition to low mono- and di-glycerides content. A reactive distillation environment is particularly efficient for both esterification and transesterrification when a superactive and robust solid catalyst is available.

Evaluating the profitability of biodiesel manufacturing reveals that this is dominated by the cost of the raw materials by more than 80%. The storage tanks account for more than one third of the equipment costs. Therefore, further progress can be achieved in simplifying the process by adopting heterogeneous catalysis or supercritical processing.

References

1 Bambase, M.E., Jr., Nakamura Naka, J., Matsumara, M., Kinetics of hydroxide-catalysed methanolysis of crude sunflower oil for the production of fuel-grade methyl esters, *J. Chem. Technol. Biotechnol.*, 82, 273–280, 2007

2 Bournay, L., Casanave, D., Delfort, B., Hillion, G., Chodorge, J.A., New heterogeneous process for biodiesel production, *Catal. Today*, 106, 190, 2005

3 Bunyakiat, K., Makmee, S., Sawangekeaw, R., Ngamprasertsith, S., Continuous production of biodiesel via transesterification from vegetable oils in supercritical methanol, *Energy Fuels*, 20, 812–817, 2006

4 Cao, W., Han, H., Zhang, J., Preparation of biodiesel from soybean using supercritical methanol and co-solvent, *Fuel*, 84, 347–351, 2005

5 Cao, W., Han, H., Zhang, J., Preparation of biodiesel from soybean using supercritical methanol and CO_2, *Process Biochem.*, 40, 3148–3151, 2005

6 Ceriani, R., Meirelles, A., Predicting vapor–liquid equilibria of fatty systems, *Fluid Phase Equil.*, 215(2), 227–36, 2004

7 Christ, C. (ed.), Production-Integrated Environmental Protection and Waste Management in the Chemical Industry, Reduction of waste production and energy consumption in the production of fatty-acid methyl esters (Henkel), Wiley-VCH, Weinheim, Germany, 1999

8 Darnoko, D., Cherayan, M., Kinetics of palm oil transesterification in a batch reactor, *J. Am. Oil Chem. Soc.*, 77(12), 1263–1267, 2000

9 Demirbas, M.F., Balat, M., Recent advances on the production and utilization trends of bio-fuels, *Energy Conver. Manag.*, 47, 2371–2381, 2006

10 Dembiras, A., Biodiesel production via non-catalytic SCF method and biodiesel fuel characteristics, *Energy Conver. Manag.*, 47, 271–2282, 2006

11 Dembiras, A., Comparison of transesterification methods for production of biodiesel, *Energy Conver. Manag.*, 2007

12 Dembiras, A., Biodiesel from sunflower oil in supercritical methanol with calcium oxide, *Energy Conver. Manag.*, 48, 937–941, 2007

13 Dossin, T.F., Reyniers, M.F., Berger, R.J., Marin, G.B., Simulation of heterogeneously MgO-catalyzed transesterification for fine-chemical and biodiesel industrial production, *Appl. Catal. B: Environmental*, 67, 136–148, 2006

14 European Union, Biofuels Progress Report 2006

15 European Union, An energy policy for Europe 2006

16 Freedman, B., Buterfield, R., Pryde, E., Transesterification kinetics of soybean oil, *J. Am. Oil. Chem. Soc.*, 77(12), 1375–380, 1986

17 Van Gerpen, J., Biodiesel processing and production, *Fuel Proc. Techn.*, 86, 1097–1107, 2005

18 Haas, M.J., Mc Aloon, A.J., Yee, W.C., Foglia, T.A., A process model to estimate biodiesel production costs, *Bioresource Technol.*, 97, 671–678, 2007

19 He, B.B., Singh, A.P., Thompson, J.C., A novel continuous-flow reactor using

20 He, H., Wang, T., Zhu, S., Continuous production of biodiesel fuel from vegetable oil using supercritical methanol process, *Fuel*, 86, 442–447, 2007
21 Henkel, A. G., German patent 3 9111 538, 1991
22 Lotero, E., Liy, Y., Lopez, D., Suwannakararn, K., Bruce, D.A., Goodwin. J.G., Jr, Synthesis of biodiesel via acid catalysis, *Ind. Eng. Chem. Res.*, 44, 5355–5363, 2005
23 Kiss, A.A., Dimian, A.C., Rothenberg, A., Solid acid catalyst for biodiesel production – towards sustainable energy, *Adv. Synth. Catal.*, 368, 75–81, 2006
24 Kiss, A.A., Dimian, A.C., Rothenberg, A., The heterogeneous advantage: biodiesel by catalytic reactive distillation, *Top. Catal.*, 40, 141–150, 2006
25 Komers, K., Skopal, F., Stloukal, R., Machek, J., Kinetics of the KOH-methanolysis of rapeseed oil for biodiesel production, *Eur. J. Lipid. Sci.*, 104, 728–737, 2002
26 Knothe, G., Steidley, K.R., Kinematic viscosity of biodiesel fuel components and related compounds, *Fuel*, 84, 1059–1065, 2005
27 Kusdiana, D., Saka, S., Kinetics of transesterification in rapeseed oil to biodiesel fuel as treated in supercritical methanol, *Fuel*, 80, 693–698, 2001
28 Kusdiana, D., Saka, S., Two-step preparation for catalyst-free biodiesel fuel production: Hydrolysis and methyl esterification. *Appl. Biochem. Biotechnol.*, 115, 781–92, 2004
29 Ma, F., Hanna, M.A., *Biodiesel Prod., Bioresource Technol.*, 70, 1–15, 1999
30 Meher, L.C., Vidya Sagar, D., Naik, S.N., Technical aspects of biodiesel production by transesterification, *Renew. Sustain. Energy Rev.*, 10, 248–268, 2006
31 Minami, E., Saka, S., Kinetics of hydrolysis and methyl esterification for biodiesel production in two-step supercritical methanol process. *Fuel*, 85, 2479–2483, 2006
32 Negi, D.S., Sobotka, F., Kimmel, T., Wozny, G., Schomaker, R., *Ind. Eng. Chem. Res.*, 45, 3693–3696, 2006
33 Noureddini, H., Zhu, D., Kinetics of transesterification of soybean oil, *J. Am. Oil Chem. Soc.*, 74(11), 1457–1463, 1997
34 Poling, B.E., Prausnitz, J.M., O'Connell, J.P., The Properties of Gases And Liquids, McGraw-Hill, 5th edn, 2001
35 Pousa, G., Santos, A., Suarez, P., History and Policy of Biodiesel in Brazil, *Energy Policy*, 2007
36 Vicente, G., Martinez, M., Aracil, J., A comparative study for vegetable oils for biodiesel production in Spain, *Energy Fuels*, 20, 394–398, 2006
37 Vicente, G., Martinez, M., Aracil, J., Integrated biodiesel production: a comparison of different homogeneous catalyst systems, *Bioresource Technol.*, 92, 297–305, 2004
38 Warabi, Y., Kusdiana, D., Saka, S., Reactivity of triglycerides and fatty acids of rapeseed oil in supercritical alcohols, *Bioresource Technology*, 91, 283–287, 2004
39 Xie, W., Li, H., Alumina-supported potassium iodide as heterogeneous catalyst for biodiesel production from soybean oil, *J. Molec. Catal. A: Chemical*, 205, 1–9, 2006
40 Yuan, W., Hansen, A.C., Zhang, Q., Vapour pressure and normal boiling point predictions for pure methyl esters and biodiesel fuels, *Fuel*, 84, 943–950, 2005

reactive distillation for biodiesel production, *Trans. ASABE*, 49(1), 107–112, 2006

15
Bioethanol Manufacturing

15.1
Introduction

Bioethanol is the most important biofuel today with a worldwide output of about 32 million tons in 2006 [1, 2]. Currently the bioethanol is produced by the fermentation of sugars derived from various crops including sugar cane, corn and sugar beet. In the near future, technologies based on the abundant and low-cost lignocellulosic biomass will become cost efficient such that the conflicts raised by the shortage of crops for food will be avoided. The competitiveness of bioethanol against fossil fuels will be greatly enhanced both by the technological progress and the continuous rise of the crude oil price.

The purpose of this case study is to examine the challenges raised by the bioethanol manufacturing as an economically profitable biofuel from a system perspective. First, economic and ecological aspects are discussed. The biorefinery concept is presented as a strategic evolution from low-cost, high-volume fuels to high-cost, low-volume chemicals. Then, the technological features of manufacturing processes are briefly reviewed. Particular attention is given to innovative technologies based on integrated processes that maximize the value of raw materials and the recovery of energy. The case study itself deals with the design of a process for manufacturing bioethanol starting from lignocellulosic biomass. Emphasis is placed on developing a comprehensive material balance on the basis of performances of fermentation reactors. Since the operation costs are important penalties in manufacturing the bioethanol, particular attention is given to the key measures with respect to saving energy and water.

15.2
Bioethanol as Fuel

When Henry Ford designed the famous Model T he assumed ethanol as the fuel for it, as it was readily available from the agricultural resources of the Midwest of the USA. However, cheaper gasoline emerged as the dominant automotive fuel, despite many disadvantages, such as lower octane rating, higher toxicity

Table 15.1 Some physical properties of fuels with impact on combustion [3].

Property	Ethanol	Gasoline	Biodiesel	Petrodiesel
Formula	C_2H_6O	Iso-C_8H_{18}	$C_{18}H_{36}O_2$	$C_{14}H_{30}$
Molecular weight	46.07	114.20	296.5	198.40
Sp. gravity	0.7939	0.7021	0.8724	0.7667
Normal boiling point (°C) (range of true boiling points)	78.65	117.6 (36–204)	343.85	253.6 (125–400)
ΔH_v (kJ/kg)	845.12	294.43	214.59	241.90
Combustion heat (kJ/kg)	−2.68E + 4	−4.44E + 04	−3.74E + 04	−4.40E + 04
Carbon content (wt.%)	52.2	85.5	76.9	87

(particularly when blended with tetra-ethyl lead), being more dangerous in storage and handling, showing less clean combustion and being more polluting. For a long time the domination of petroleum fuels was favored by the relative low cost and abundance of resources, while until recently the ecological concerns were largely ignored. This situation is rapidly changing. However, the huge investments in the oil and auto industries in capital and technology make the acceptance of a new cost-competitive industry difficult.

Table 15.1 presents some physical properties of interest when evaluating the ethanol versus different fuels. Reference components are iso-octane (methylheptane) for gasoline, methyl-oleate for biodiesel, and tetradecane for petrodiesel. The combustion heat of ethanol is clearly below that of gasoline and diesel, more ethanol being necessary to ensure the same engine power. Pure ethanol contains approximately 1/3 less energy per mass unit than gasoline. In contrast, the equivalent octane number is 113, superior to gasoline of only 87 to 92. As a result, ethanol-fuelled cars can use higher compression ratios and develop more power. Due to a higher heat of vaporization the temperature peak inside the combustion chamber is lower, resulting in less NO_x emissions. Ethanol burns cleaner than gasoline producing less carbon monoxide and hydrocarbon emissions. However, a negative effect seems to be an increased aldehyde content that enhances the occurrence of photochemical smog.

In most cases the ethanol is blended with gasoline. The vast majority of cars and trucks can use a 10% blend with gasoline (E10) in USA. A 22% blend of ethanol with gasoline (E22) is the most sold fuel in Brazil. Flex-fuel vehicles (FFV) are capable of running on variable blends up to 85% bioethanol (E85). However, when using E85 the fuel consumption increases by 20–30% due to differences in energy per unit volume.

Blends of ethanol with diesel are a valuable car fuel too. The formation of NO_x is drastically reduced even by small amounts of ethanol. In addition, the temperatures of the exhaust gas and lubricating oil are lower, while the engine can be started normally both hot and cold.

Table 15.2 Fuel ethanol specifications.

	ASTM D4806-98	EN 15376
Ethanol, vol.%	92.1 min	98.70 min
Methanol, vol.%	0.5 max	1.0
Saturated alcohols (C3–C5)	–	2.0
Water, vol.%	1.0 max	max 3.0
Solvent-washed gum, mg/100 ml	1.0	10
Acidity, as acetic acid, ppm	70	70 ppm
Chloride ion, mg/l	40 max	20 max
Copper content, mg/kg	0.1 max	0.1 max
Denaturant, vol.%	1.96 to 4.76 gasoline	none

Table 15.2 presents the specifications for ethanol fuel following American and European norms. The ethanol concentration should be a minimum 92.1 and 98.7 vol.%, respectively, while the amount of water is kept below 1%.

15.3
Economic Aspects

Worldwide ethanol production in 2005 was about 30 million tons (10 Ggal), of which 90% was from only two countries, Brazil and USA, each with 13.5 million tons. Brazil gets more than 30% of its transport fuels from sugar-cane ethanol. The production in the EU was much lower, about 0.5 million tonnes, mainly for manufacturing ETBE. The EU target in biofuels is 5.75% in 2010, composed of 2/3 ethanol and 1/3 biodiesel [1].

In the reference scenario of the International Energy Agency [2] the world output of biofuels is projected to climb to 92 Mtoe in 2030 at an average annual rate of 7% covering 4% of world road-fuel demand. In an alternative scenario, the production should rise even faster, at 9% per year, reaching 147 Mtoe or 7% of road-fuel use. The tone of oil equivalent (toe) is a unit of energy, defined as the amount of energy released by burning one tone of crude oil, approximately 42 G.

At the level of 2006 the vast majority of bioethanol originates from glucose sugars and starch. The manufacturing processes based on sugar-crop fermentation belong to the "first-generation bioethanol". Diverting an important part of the food crops to biofuels could destabilize the food supply chain. To avoid this, the agricultural surface allocated to biofuel crops is in many countries restricted by law. Improving the cropland productivity has only a limited effect on increasing the ethanol production. For these reasons a lot of R&D effort is devoted to developing the "second-generation bioethanol" based on the use of low-cost and nonedible lignocellulosic materials. However, the technologies for converting biomass into ethanol necessitate higher investment and operation costs.

Table 15.3 Comparison of production costs for biofules at 2006 level [1].

	EtOH	EtOH	EtOH	Biodiesel	Gasoline	Diesel
	Europe	Brazil	USA	Europe	$60/bbl	$60/bbl
Euros/l	0.4–0.6	0.2	0.3	0.35–0.65	0.32	0.36
Euros/GJ	19–29	10	14	10.5–20	9	10

Table 15.4 Total product cost for ethanol in Europe in 2004 [3].

Plant capacity	50 million litres		200 million litres	
Raw materials	Wheat	Sugar beet	Wheat	Sugar beet
Feedstock cost	0.28	0.35	0.28	0.35
Coproduct credit	0.07	0.07	0.07	0.07
Net feedstock cost	0.21	0.28	0.21	0.28
Labor cost	0.04	0.04	0.01	0.01
Operating and energy	0.20	0.18	0.20	0.17
Net investment	0.10	0.10	0.06	0.06
Total product cost	0.55 USD	0.59 USD	0.48 USD	0.52 USD
Total-gasoline-per-equivalent liter	0.81	0.88	0.71	0.77

Source: International Energy Agency 2004.

Clearly the cost is the key factor for the long-term future of biofuels. Table 15.3 presents production costs in Europe, Brazil and USA [1]. It can be seen that bioethanol is competitive with gasoline when produced from low-cost sugar cane in Brazil, or taking profit from high-productivity farming as in the USA. In Europe, the production price of ethanol is higher by a factor of two because of relatively expensive raw materials, wheat and sugar beet. At present, subventions are practised in Europe and the USA to stimulate the biofuels sector, but at term these should be suppressed. The steep rise of oil price makes biofuels more and more competitive.

As an illustration of the cost structure, Table 15.4 presents the situation in Europe at the level of 2003 for two plant capacities, from wheat and sugar beet, respectively [4]. The raw materials account for about half of the final cost. The contribution of operating and energy costs is significant too, although the effect of scale seems minor. Larger plants need lower specific investment. The conclusion is that progress is necessary in two directions: (1) using low-cost raw materials, such as lignocellulosic biomass and biowaste, and (2) reducing drastically the energy costs by improving the fermentation technology and by process integration.

The manufacturing of bioethanol needs large resources of land, but there is no strong evidence that this would disrupt the food market. In the USA in 2006 about 7% of the corn crop was used for ethanol, 58% for domestic livestock feed, 11% for food and 22% exported. In Europe in 2004 the bioethanol consumed 0.4% of cereals and 0.8% of sugar beets. The cultivable surface was 23 Mha for wheat (23 member states) and 2.2 Mha for sugar beets, with a mean yield of 5.4 and 57.4 t/ha, respectively. It is estimated that the maximum land potential for biofuel crops in Europe is about 8.2 Mha, or 10% from the cultivable land, which could supply up to 10 Mtoe of biodiesel/bioethanol, provided adequate crop rotation. The 5.75% target in 2010, which implies 16.6 Mtoe biofuels, seems difficult to achieve. It is clear that beyond 2010 the implementation of second-generation ethanol is necessary. The potential of biomass in Europe represents about 100 Mtoe dry materials, from which 16 Mtoe biofuels could be produced [1]. A substantial import of biofuels from countries with more favorable conditions will be necessary to fulfil the greenhouse-gases reduction targets.

Switching to large-scale cellulosic ethanol production is necessary to solve the apprehension raised by using the edible raw materials. A recent analysis [5] estimates that cellulosic ethanol production costs at the level of 2006 could be a minimum of 10% lower than that of gasoline. For a plant of 172 000 tonnes/y capacity (the units for mass are "tonnes") the total investment would be 250 million USD shared as follows: feedstock handling (wood or switchgrass) 12.0, pretreatment 41.9, xylose fermentation 10.9, cellulase production 5.0, simultaneous saccharification and fermentation 37.0, ethanol recovery 7.1, offsite tankage 7.2, environmental systems 7.0, utilities (steam, electricity, water) 90.0, miscellaneous 8.5, fixed capital investment 227.3, start-up costs 11.4, working capital 11.3 (all numbers are million USD), total capital investment 250.0. The total production costs would be 57 c/L of which feedstock accounts for 18.2 c/L, enzymes 5.3 c/L, other raw materials (sulfuric acid, lime, glucose, nutrients) 2.6 c/L, gypsum disposal 0.26 c/L, electricity 2.2 c/L, water 0.11 c/L, labor 1.06 c/L, maintenance 3.49 c/L (13.2 c/L), direct overhead 0.63 c/L, general overhead 3.17 c/L, insurance and property taxes 1.74 c/L. Thus the total cash costs of 34.37 c/L plus annualized capital charge 22.59 c/L gives a total production cost of 56.96 c/L ("c" means USD cent).

15.4
Ecological Aspects

The ecological behavior of biofuels is an issue of maximum importance but the subject of much controversy. An up-to-date paper [6] tries to examine this problem by means of a life-cycle assessment (LCA) on the basis of 47 published studies. The environmental sustainability is evaluated from three perspectives: (1) reducing dependence on fossil fuels through energy-balance assessments; (2) reducing greenhouse gases (GHGs); and (3) reducing health and environmental impacts. Figure 15.1 illustrates a general framework of a LCA for biofuels evaluation. This account for the following stages: production of inputs, agriculture and harvesting, transport, conversion processes and combustion, and consumer output. Note that

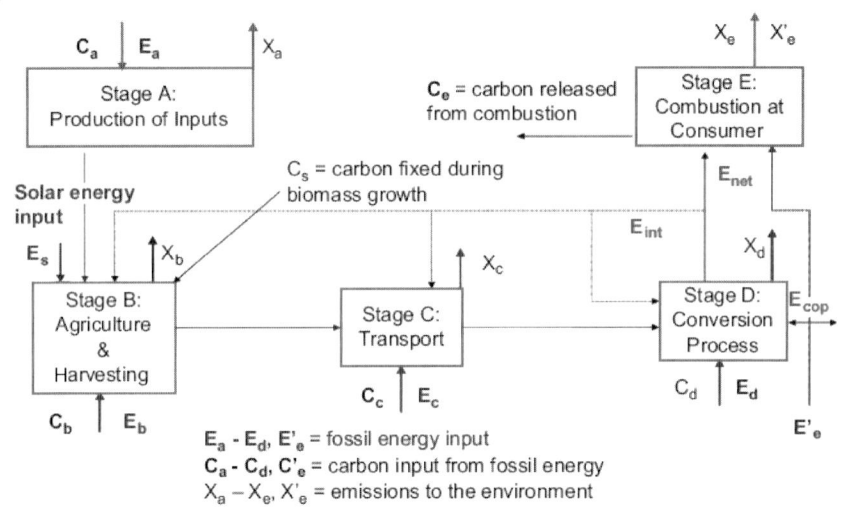

Figure 15.1 Material flow and environmental interventions across the life-cycle stages in a biofuel system [6].

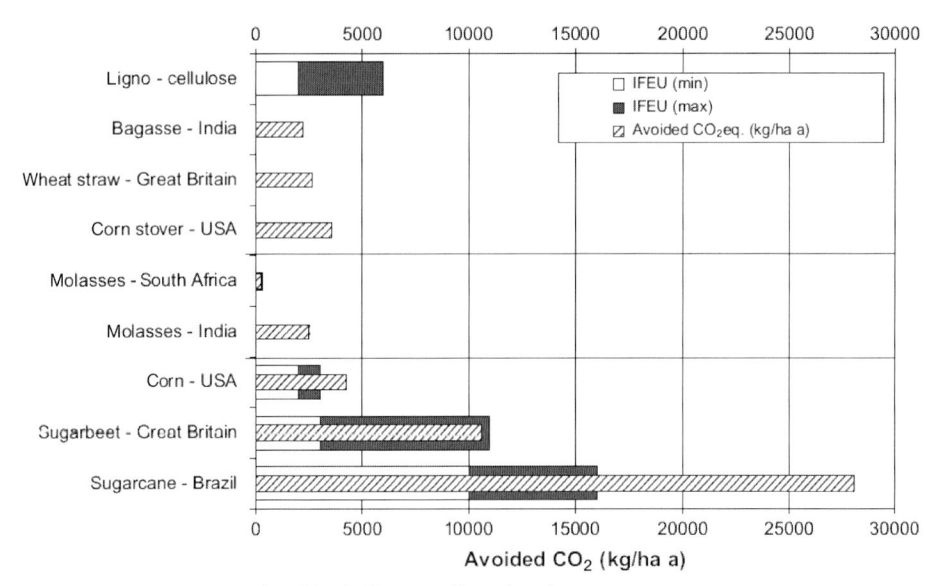

$E_a - E_d$, E'_e = fossil energy input
$C_a - C_d$, C'_e = carbon input from fossil energy
$X_a - X_e$, X'_e = emissions to the environment

Figure 15.2 Agricultural land efficiency of bioethanol in replacing fossil energy for transportation [6, 7].

greenhouse gases include carbon dioxide (CO_2), methane (CH_4), nitrous oxides (N_xO), hydrofluorocarbons (HFCs), perfluorocarbons (PFCs), etc. The warming potential is expressed in units of CO_2 equivalents.

Results of specific situations are presented in Figure 15.2, from the viewpoint of the authors and of the Institute for Energy and Environmental Research (IFEU)

Table 15.5 Gains of biofuels versus reference oil-based automotive fuel [8].

Fuel	GHG emissions	Nonrenewable energy
Ethanol from wheat	30%	22%
Ethanol from sugar beat	32%	24%
Ethanol from lignocellulosic	78%	76%
Ethanol from sugar cane	88%	91%
Biodiesel from rapeseed	53%	64%

in Germany [7]. It should be stressed that these should be not extrapolated being representative to a certain geographical region. Each case has specific characteristics regarding the supply chain of raw materials, as well as the ethanol manufacturing technology. The best performance is for tropical sugar cane in Brazil. However, good results are seen for the ethanol from sugar beet in Great Britain, being superior to corn and lignocellulosic materials. Most of the reports lead to the conclusion that bioethanol is beneficial with respect to reduction in fossil resource use and global warming. However, the impacts on land degradation and ecological toxicity, occurring mainly during the growing and processing of biomass, were more often unfavorable than favorable. Further work is needed.

The environmental performance of biofuels is analyzed in some reports issued from European industry [1, 8]. Table 15.5 gives some results from a well-to-wheels analysis of future automotive fuels and power trains in the European context issued from the Joint Research Study group of the EU at the level of May 2006 [8]. It can be seen that, depending of the pathways considered, the GHG gains can range from 30 up to 88% and the nonrenewable saved energy from 22 to 91%. At the low bound one finds the ethanol from wheat and sugar beet, while at the upper bound sugar cane and lignocellulosic materials dominate. With respect to tailpipe emissions the reduction is 61% for NO_x and 50% for CO, as well as benzene and 1,3-butadiene, the absolute values being lower than the future Euro 5 standard for automotive fuels. Even if the real effect on the environment is much lower, since the biofuels are used in blending, the effect is significant at the large scale of the transportation sector.

15.5
Raw Materials

Bioethanol can be produced from a large variety of natural renewable materials, such as agricultural crops, land and forest products, as well as from industrial and domestic waste, such as paper, textile and beverages. The choice is determined by the need and availability, which in turn depends largely on climate, geographical location and economic profile of each country. From the ethanol manufacturing viewpoint the raw materials can be classified in the following categories [9]:

1. readily fermentable carbohydrates to ethanol,
2. starch-based raw materials,
3. lignocellulosic materials.

The potential of each resource is given in Table 15.6, which is a compilation from different sources.

Carbohydrates are readily available from various crops, such as sugar cane, sugar beet and fruits, which contain primarily sucrose. Sugar cane is very prolific and can give a yield of about 63 t/ha. With an average sucrose content of 13.2% a total sucrose yield of 8.32 t/ha can be obtained. From this, about 7 t/h crystal sugar can be produced, while the residual molasses can give 0.675 t/ha of ethanol at 100% yield. If all sugar-cane juice is transformed into ethanol 4800 l/ha or 4.46 t/ha can be obtained [9]. The vegetal residue, known as begasse, has a high content of lignin and is difficult to alcoholize. However, it can be used to produce the energy needed, the eventual energy surplus being exported.

Sugar beet is a versatile crop with good potential, the land productivity ranging from 50 to 150 t/ha. In Europe the mean productivity is of 64 t/ha. The sugar content can vary from 17 to 19%. On average, about 13.5 t/ha sugar can be obtained, from which about 7.23 t/ha ethanol could be theoretically produced. A recent feasibility study in Australia demonstrates that an ethanol plant producing annually 40 million litres ethanol would need 6000 hectares of sugar-beet crop. Additional value can be placed on the tops and fermented mash for livestock feed. This combination appears to offer an attractive return for farmers [10].

Starch can be found typically in corn and potatoes. Corn is a preferred raw material to make ethanol in the Midwest region of USA, and in some parts of Europe. The crop productivity in modern farming is about 8–10 t/ha. The yield of converting corn to ethanol is 2.5 (wet milling) to 2.8 (dry milling) gallon per bushel corn (equivalent with 24.4 kg), which results in a yield of 0.417 l/kg, or 3400 l/ha. [11].

Table 15.6 Potential of the most common raw materials for bioethanol production.

Raw material	Substrate content	Productivity ton/ha	Ethanol l /ton (dry)	Yield l/ha
Sugar cane	13.2% sugar	63	75	4800
Sugar beet	18% sugar	57.4	86	4936 (7980 France)
Corn		8.160	417	3400
Wheat		5.4	359	1917 (2996 Ireland)
Potatoes	12–21% starch			
Cassave	20–35% starch	10–30	165–180	1650–4800
Sweet sorghum	18–20% sucrose	100–120	400	3500–6000
Jerusalem artichokes	16	40	90	3600
Biomass	cellulose 20–50% hemicellulose 10–40% lignin 0–35%	500 m^3/ha		

The lignocellulosic materials consist mainly of cellulose, hemicellolose and lignin. *Cellulose* is a polysaccharide with the formula $(C_6H_{10}O_5)_n$ derived exclusively from β-glucose condensed through β(1 → 4)-glycosidic bonds. *Hemicellulose* contains glucose combined with other monomers, such as xylose, mannose, galactose, rhamnose, and arabinose. Hemicelluloses contain most of the D-pentose sugars. *Lignin* is a large, crosslinked macromolecule, hydrophobic and aromatic in nature. Only cellulose and hemicelulose can be hydrolyzed chemically or by bacteria to sugars. Biomass pretreatment for removing lignin is necessary. Therefore, low-content lignin feedstock is preferable, such as in wheat straw, leaves, switch grass, waste paper. On the other hand, lignin can be used for producing energy.

15.6
Biorefinery Concept

15.6.1
Technology Platforms

A biorefinery is an integrated processing facility that converts biomass to fuels, power, and value-added chemicals. Besides high-volume, low-value (HVLV) biofuels, low-volume, high-value (LVHV) biochemicals are produced, with much higher economic profitability [12–15]. In addition, the carbon-rich residues can generate heat and power to be used onsite or exported. Figure 15.3 presents the concept. The plant input consists of various renewable and waste materials, such as lignocellulose, cereals and maize grains, sugar beet, sugar cane and starch, lipids: vegetable oils and animal fats, and municipal solid waste (MSW).

In the first step, the biomass is submitted to *preprocessing* with the goal of extracting added-value products. For example, by wood pyrolysis various chemicals can be isolated, such as alcohols, esters, phenols, but also complex organic molecules, such as guaiacol or syringol, which otherwise would require complicated organic synthesis.

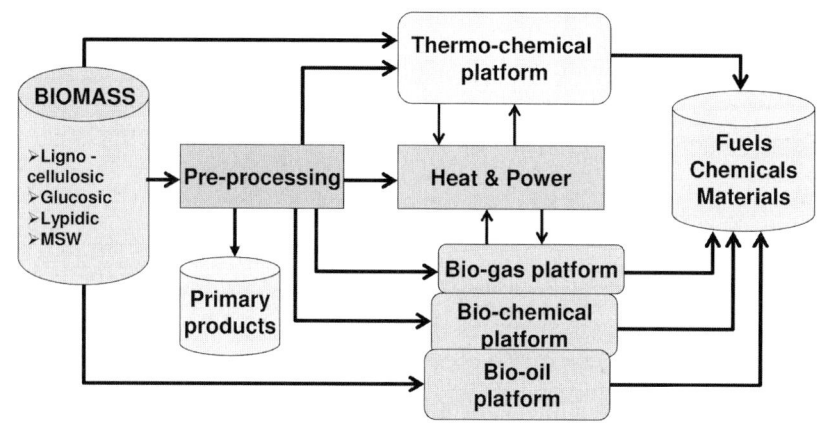

Figure 15.3 Technology platforms in biorefinery.

15 Bioethanol Manufacturing

The biorefineries can be classified into several *technology platforms* taking into account the core conversion-to-fuel technology applied:

1. Thermochemical refinery or biomass to liquid (BTL): conversion of syngas to fuels and chemicals by Fisher–Tropsh synthesis.
2. Biochemical refinery: conversion of lignocellulosic biomass by fermentation to bioethanol and chemicals.
3. Biogas platform: production of methane-rich gas from MSW.
4. Long-chain carbon refinery: conversion of oils and fats to biodiesel and chemicals.

The most promising biorefinery is based on a lignocellulosic feedstock (LCF). By pretreatment the feed is separated into three main streams: cellulose, hemmicellolose and lignin. Cellulose can be converted to glucose and further to bioethanol, solvents and organic chemicals based on lactic and levulinic acids. Hemicellulose can be transformed into hexose and pentose, and further furfural from which resins and nylon-type polymers can be manufactured. From lignin, adhesive and bituminous products can be obtained. The carbonaceous wastes are burned to produce energy.

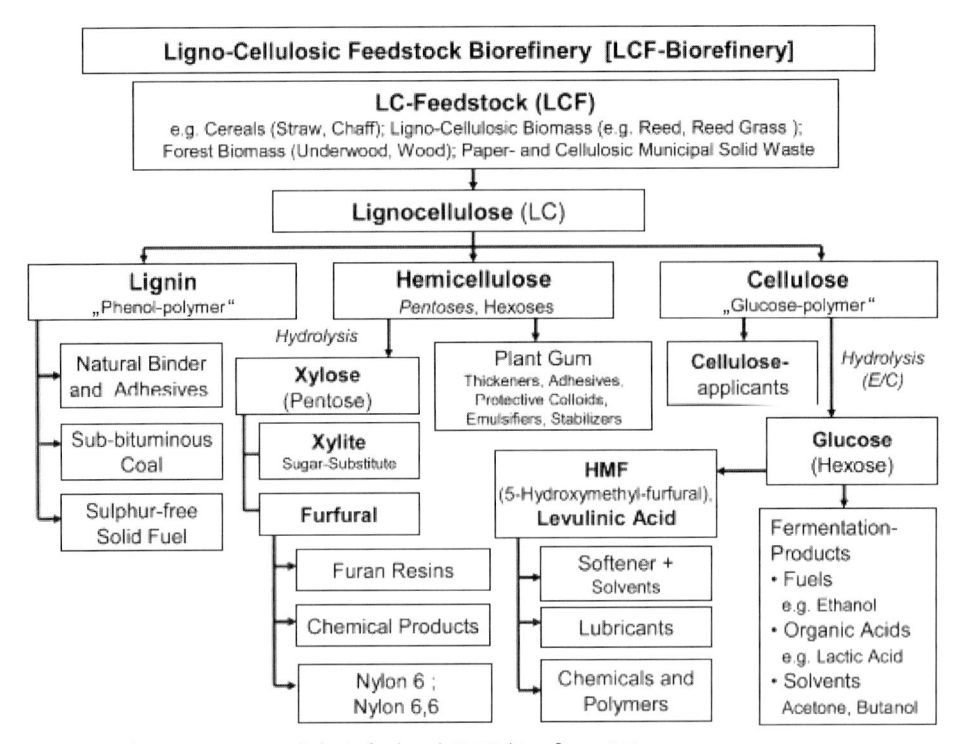

Figure 15.4 Ligno-cellulosic feedstock (LCF) biorefinery [14].

15.6.2
Building Blocks

Building blocks are simple molecules on which further diversification is possible by applying organic synthesis methods. The whole range of petrochemical products is built around C1, C2, C3, C4, C5 and BTX components. As opposed to petrochemistry, where large hydrocarbons are chopped into small molecules by spending large amounts of energy, biotechnology can offer the same building blocks with much better exergetic yield. Figure 15.5 shows key biochemical building blocks. The C1 chemistry can be developed on methane obtained from biogas, C2 chemistry on ethanol and acetic acid, C3 chemistry on glycerol and lactic acid, C4 chemistry on succinic and fumaric acids, C5 chemistry on furfural and levulinic acids, C6+ on lysine and sorbitol. Large amounts of BTX are not available, but some natural complex aromatic molecules can be extracted with high benefit.

Some polymers can be manufactured with good economic efficiency from renewable raw materials. Thus, polylactic acid (PLA) obtained from corn is

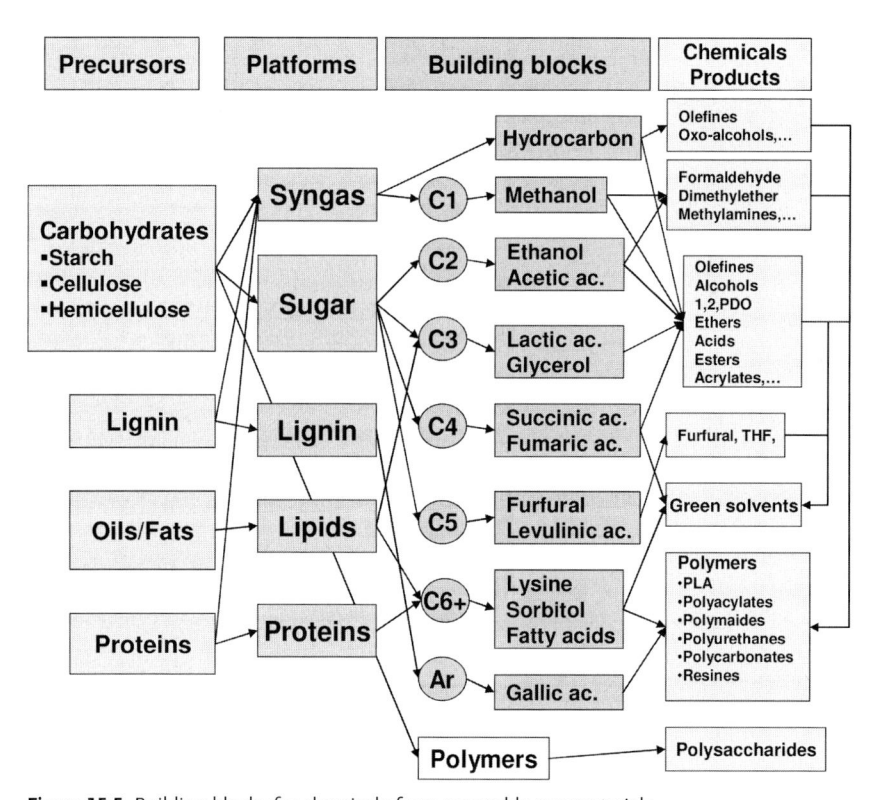

Figure 15.5 Building blocks for chemicals from renewable raw materials.

biodegradable and can replace PET in food packaging and medical-care applications. The PLA process results in about 60% GHG reduction compared with the polyester PET. 1,3-propandiol got from corn can replace ethylene glycol in polyesters, such as SORONA®. The polyamide Rilsan 11® that can be produced from castor oil has outstanding resistance against ethanol, being used in manufacturing parts for fuel-injection systems.

These examples demonstrate that the valorization of renewable raw materials should go beyond the production of biofuels, to added-value chemicals and products, as well as to energy generation, by combined heat and power cycles.

15.7
Fermentation

15.7.1
Fermentation by Yeasts

Yeasts are unicellular uninucleate fungi that have the ability to ferment sugars for the production of ethanol. The most widely used are *Saccharomyces cerevisiae*, *S. uvarum* (formerly *S. carlsbergensis*), *Candida utilis* and *Kluyveromyces*. Each micro-organism is suited for a given type of sugar and operation conditions. The main chemical reaction in fermentation consists of converting the basic monomer hexose into ethanol and two carbon dioxide molecules:

$$C_6H_{12}O_6 \rightarrow 2C_2H_5OH + 2CO_2$$

The process goes through the formation of the intermediate pyruvate $CH_3-CO-COO-$, which further decarboxylates into acetaldehyde, and is reduced finally to ethanol. The complex reaction mechanism is described elsewhere [9]. The above reaction takes place only in anaerobic conditions (without oxygen), otherwise the complete oxidation to carbon dioxide and water takes place.

The production of ethanol by yeasts is characterized by high selectivity and low formation of byproducts. The main variables describing the fermentation process are substrate (sugar) concentration, tolerance to ethanol, temperature, pH and oxygen. These effects are briefly reviewed.

Increasing the concentration of substrate has a positive effect on the reaction rate up to a maximum limit, when inhibition by the product occurs. The tolerance limit depends on the yeast nature, in most cases 110 g ethanol/l, but the best are resistant up to 20% [9].

The temperature should be kept in a range defined by optimum and maximum values. In most cases the maximum yeast growth is obtained at 39–40 °C, with yields of alcohol above 90%. Since the fermentation process is exothermic (−586 J per g glucose) cooling facilities should be provided.

The pH affects the ethanol production rate, yeast growth, byproduct formation and bacterial contamination. Sugar fermentation by yeast is relatively insensitive to pH values between 3.5 and 6. To suppress the bacterial growth the pH should be kept below 5.

Although theoretically the oxygen should be absent in anaerobic fermentation, the presence of small and controlled amounts is necessary to ensure the survival and the growth of the yeast. Gaseous oxygen may be replaced by oxygenated organic compounds, such as fatty acids or ergosterol. In this way, an ethanol concentration up to 15 wt.% can be achieved with an ethanol yield above 95%.

15.7.2
Fermentation by Bacteria

Another possibility of producing ethanol on an industrial scale is by using bacteria. One of the most widely employed is *Zymomonas mobilis*, which can ferment glucose to ethanol with high yield, similar to *S. carlsbergensis* [9]. In addition to ethanol other fermentation products can occur, such as lactic acid, acetic acid, formic acid, acetone and hydrogen. Unlike yeast, *Z. mobilis* does not require oxygen for growth and greatly simplifies the technology. However, the fermentation process is sensitive to contamination with other bacteria that could produce inhibiting species. Most bacteria can resist up to 10–20 g/l ethanol, but *Z. mobilis* can tolerate alcohol concentrations of 120 g/l.

The interest in using bacteria in fermentation is the possibility of modifying their metabolic behavior by a combination of strains or by genetic engineering, so as to accept several substrates. This is the case in the fermentation of pentose, abundant in lignocellulosic materials such as xylose and arabinose, which cannot be treated by yeasts. The fermentation of pentose can be ensured by a bacteria called *Escherichia coli*, which can be combined with the efficient ethanol-forming activity of *Z. mobilis*. Moreover, pentose active strains of *Z. mobilis* have been developed by genetic engineering.

15.7.3
Simultaneous Saccharification and Fermentation

Prior to fermentation the raw materials containing polysaccharides (cellulose or starch) have to be hydrolyzed to glucose. Simultaneous saccharification and fermentation (SSF) combines enzymatic hydrolysis of polysaccharides with glucose fermentation to ethanol. In this way, the formation as an intermediate of glucose does not inhibit the cellulose hydrolysis, resulting in higher overall ethanol yield. As mentioned, cocultures of microbes can be used, each one capable of hydrolysis and fermentation, but compatible with respect to optimal activity. The most promising ethanologenic bacteria are *E. coli*, *Klebsiella oxytoca* and *Z. mobilis* [16]. Particular attention has been paid to genetically modified *Z. mobilis* in the field of lignocellulosic biomass. This approach is adopted in the case-study developed next.

15.7.4
Kinetics of Saccharification Processes

The kinetics of biochemical processes may be simple or complex, depending on the number of variables having a significant influence. One of the simplest approach but with wide applications is the Michaelis–Menten model. The enzymatic reaction takes place by a two-step mechanism:

$$S + E \xleftrightarrow{K_m} ES \quad \text{(fast)}$$

$$ES \xleftrightarrow{k_r} P + E \quad \text{(slow)}$$

The process rate becomes:

$$r_p = \frac{k_r c_{E0} S}{K_m + S} \tag{15.1}$$

K_m is the Michaelis constant representing the equilibrium enzyme-substrate intermediate, k_r the rate-determining constant, c_{E0} the initial enzyme concentration, and S the concentration of substrate. The physical significance is that the reaction rate is proportional to the substrate concentration at low values, but tends to maximum at higher values. If initial rate experiments are conducted then the limit of the reaction rate will correspond to a maximum initial rate $V_{max} = k_r c_{E0}$. By replacing it into Eq. (15.1) the following kinetic expression is obtained:

$$r_p = \frac{V_{max} S}{K_m + S} \tag{15.2}$$

In the case of enzymatic saccharification of lignocellulosic biomass the reaction mechanism and the kinetic modeling is much more complex. Multireaction modeling is necessary including adsorption and inhibition steps. As an illustration, one can give the modeling of a saccharification of corn stover [17], as illustrated in Figure 15.6. The assumptions are: (1) enzyme adsorption follows a Langmuir-type isotherm with first-order reactions occurring on the cellulose surface, (2) the cellulose matrix is uniform in terms of susceptibility to enzymatic attack, (3) the enzyme activity is constant, (4) the conversion of cellobiose to glucose follows the Michaelis–Menton kinetics. Let S, G, G_2 and X denote the concentrations of cellulose, glucose, cellobiose and xylose, respectively. Then, the kinetic equations are:

Enzyme adsorption

$$E_{iB} = \frac{E_{imax} K_{iad} E_{iF} S}{1 + K_{iad} E_{iF}} \tag{15.3}$$

Cellulose to cellobiose with competitive inhibition by glucose, cellobiose and xylose

$$r_1 = \frac{k_{1r} E_{1B} R_S S}{1 + G_2/K_{1IG2} + G/K_{1IG} + X/K_{1IX}} \tag{15.4}$$

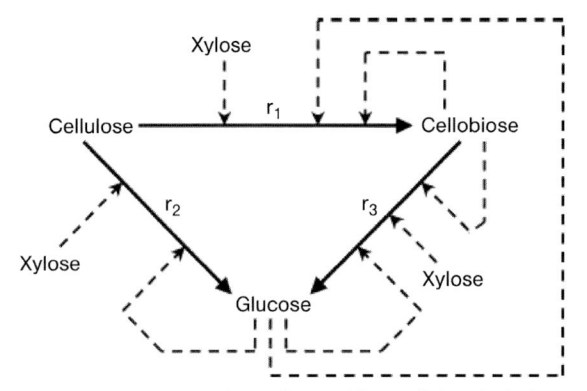

Figure 15.6 Reaction scheme for modeling cellulose hydrolysis [17].

Cellulose to glucose with competitive inhibition by glucose and xylose:

$$r_2 = \frac{k_{2r}(E_{1B} + E_{2B})R_S S}{1 + G_2/K_{2IG2} + G/K_{2IG} + X/K_{2IX}} \tag{15.5}$$

Cellobiose to glucose with competitive inhibition by glucose and xylose:

$$r_3 = \frac{k_{3r} E_{2F} G_2}{G_2 + K_{3M}(1 + G/K_{3IG} + X/K_{3IX})} \tag{15.6}$$

with the mass balance:

$$\text{cellulose:} \quad \frac{dS}{dt} = -r_1 - r_2 \tag{15.7}$$

$$\text{cellobiose:} \quad \frac{dG_2}{dt} = 1.056 r_1 - r_3 \tag{15.8}$$

$$\text{glucose:} \quad \frac{dG}{dt} = 1.111 r_2 - 1.053 r_3 \tag{15.9}$$

$$\text{enzyme:} \quad E_{Ti} = E_{Fi} + E_{Bi} \tag{15.10}$$

The temperature dependence is described by the equation

$$k_{ir(T2)} = k_{ir(T1)} \exp(-E_{ai}/R[1/T_1 - 1/T_2]) \quad \text{with} \quad T_1 = 303.15\,\text{K} \quad \text{and} \quad T_2 = 328.15\,\text{K}.$$

Model parameters are given in Table 15.7.

Table 15.7 Model parameters for the saccharification of lignocellulose.

Dissociation constant (g protein/g substrate)	Inhibition constant (g/kg)	Reaction constant (g/mg h)
$K_{1ad} = 0.4$	$K_{11G} = 0.1$ $K_{11G2} = 0.015$ $K_{11X} = 0.1$	$k_{1r} = 22.3$
$K_{2ad} = 0.1$	$K_{21G} = 0.04$ $K_{21G2} = 132.0$ $K_{21X} = 0.2$	$k_{2r} = 7.18$
$E_{1max} = 0.06$	$K_{31G} = 3.9$ $K_{3M} = 24.3$ $K_{31X} = 201$	$k_{3r} = 285.5$
$E_{2max} = 0.01$		$E_a = 5540$ cal/mol
$R_s = S/S_0$		

Enzymes involved in r_1: endo-β-1,4-glucanase, exo-β-1,4-cellobiohydrolase. Enzymes involved in r_2: exo-β-1,4-cellobiohydrolase, exo-β-1,4-glucan glucohydrolase. Enzymes involved in r_3: β-bludosidse or cellobiases.

15.7.5
Fermentation Reactors [9]

Batch Reactor   Bach reactors are easy to scale-up. The conversion of sugars with yeast in a simple batch reactor can achieve 75 to 95% and final ethanol concentrations of 10–16 vol.%. The productivity is usually 1.8 to 2.5 g of ethanol/l h. Recycling the yeast can reduce the initial yeast growth period and increase the productivity. Due to low productivity and high operation costs, the batch reactors were replaced by continuous reactors.

CSTR   Substrate and nutrients including oxygen are continuously fed so as to achieve the desired steady state. The beer containing ethanol, biomass and unconsumed nutrients is removed continuously. The productivity can reach 6 g/l h, about three times of a batch reactor. The yield of ethanol is limited by the inhibition effect. Total productivity is also limited by low biomass concentration (10–12 g/l). A sugar concentration of 10% in the feed gives the highest productivity. Two or more CSTR in series can be used.

Separating the yeast from the ethanol by centrifugation and recycling it can improve significantly the reactor productivity. For example, concentrations of yeast up to 83 g/l can be achieved with productivities of 30 to 51 g/l h, or tenfold over fermentation without cell recycle.

Tower Reactor   The tower reactor is convenient when working with flocculating yeast cells. The reactor consists of a cylinder provided with bottom and upper zones for feeding substrate and cells and solid/liquid separation. The overall aspect ratio is of 10:1, with 6:1 for the reaction zone. A tower reactor does not use mechanical mixing, and is simpler to build. Cell concentrations up to 100 g/l can be achieved with productivities 30–80 times higher than in batch reactors. The residence time is below 0.4 h and the yield up to 95% of the theoretical one. A design procedure is available [18].

Combined Reaction and Separation Devices The continuous removal of ethanol from the reaction zone favors building up a higher yeast concentration and accordingly a higher reaction rate. The separation of ethanol can be achieved by a membrane, vacuum and solvent extraction.

The static membrane reactor has the disadvantage of fouling. In a modern setup a rotating microporous cylinder is used, through which the ethanol is continuously removed. Larger yeast cells are sent back into the annular zone by the centrifugal force. In this way, the membrane is self-cleaning but allows diffusion through it. The ethanol production rate can reach good values, about 30 g/l h, but the reactor can be confronted with serious mechanical and operating problems.

When vacuum is used the bioreactor operates at a total pressure compatible with the yeast, of 6.7 kPa, at the boiling point, such that the heat of reaction can be taken off by evaporation. A high ethanol production rate of 82 g/l h is achieved with a yeast density of 120 g/l and an ethanol concentration at 35 g/l. Alternatively, the reactor can be operated at atmospheric pressure and CO_2 released directly. Beer is sent to a flash vessel where the ethanol is stripped out under vacuum.

The application of such techniques can reduce considerably the energetic costs in the next step, the distillation of ethanol from beer.

Immobilized Cell Reactors An immobilization technique consists of attaching the cells to a gel matrix. A high cell concentration can be obtained. Agitation, cell separation and recycling are not needed. Immobilized cells can be used in fixed-bed and fluidized-bed reactors. The substrate solution flows continuously through the reactor, while the immobilized cells convert the sugar to ethanol. A maximum ethanol productivity of 53.8 g/l was achieved at a dilution rate of $4.6\,h^{-1}$ and an initial glucose concentration of 127 g/l.

15.8
Manufacturing Technologies

15.8.1
Bioethanol from Sugar Cane and Sugar Beets [9]

The scheme of manufacturing ethanol from sugar cane is presented in Figure 15.7. In the first stage, milling, the washed, chopped, and shredded sugar cane is repeatedly treated with water and crushed between rollers. The collected cane juice contains 10–15% sucrose. The remaining fibrous solid, called "bagasse", is a valuable material that can be used as a solid fuel for producing onsite steam and electricity, as animal food, for paper manufacture, or to produce supplementary ethanol by a lignocellulosic process. Typically, the cane juice is split into two streams, for sugar production and for ethanol. The residue obtained from the crystal sugar processing, called molasses, is mixed with fresh cane juice to form a viscous liquid called "mash". This is further sterilized by heating and adjusted to the desired sugar content at 14 to 22 vol%. The next step, fermentation, can be

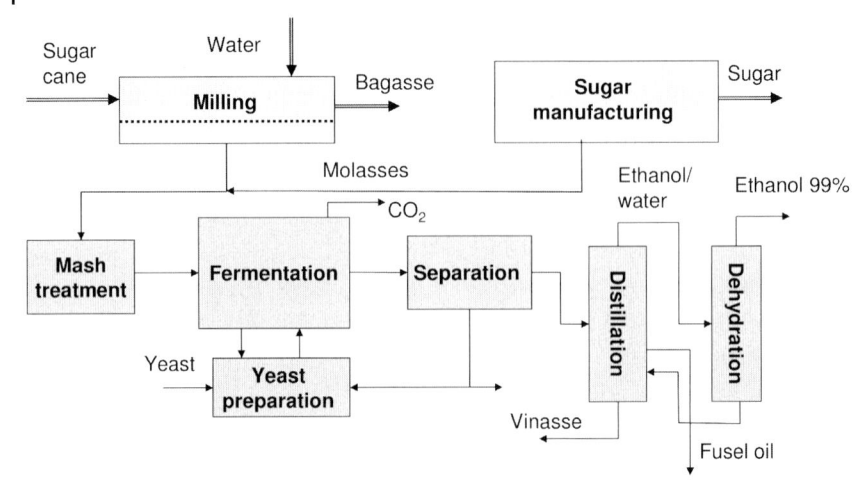

Figure 15.7 Manufacturing scheme of ethanol from sugar cane.

carried out in batches or continuously. The yeast is initially produced and activated in a separate reactor. The outflow from fermentation is submitted to liquid/solid separation by filtration or by centrifugation. The solid cake containing yeast is recycled to fermentation. The mother liquor goes to distillation, from which a rich alcohol solution is obtained in top, byproducts are fuel oil as a side stream, and a liquid residue (vinasse) in bottom. The latter can be used for obtaining biogas or as a fertilizer. The last step in the process is the alcohol dehydration, which can take place by azeotropic distillation with an entrainer (benzene, cyclohexane), by extractive distillation (di-ethyleneglycol), or by means of molecular sieves. The above scheme shows that the whole raw material can be transformed efficiently into ethanol, useful byproducts and energy. There are potentially no ecologically unfriendly residues.

15.8.2
Bioethanol from Starch

The manufacturing of ethanol from corn can follow two paths, dry grind and wet mill as shown in Figure 15.8. Dry grind, which is employed in two thirds of US plants, takes advantage from a simple flowsheet that maximizes the return of capital. Wet milling is more capital intensive but allows the separation of valuable grain components before fermentation, such as corn oil and gluten, approaching in this way the concept of a biorefinery.

Getting ethanol from corn consists of five steps: grain grinding, mash cooking, starch liquefaction, saccharification and fermentation, followed by ethanol separation and dehydration. Before fermentation, the starch has to be converted to soluble dextrin by liquefaction in the presence of specific enzymes, such as α-amylase. This process takes place at temperatures above 100 °C at suitable pH and residence time. Then, another enzyme glucoamylase is added that transforms the

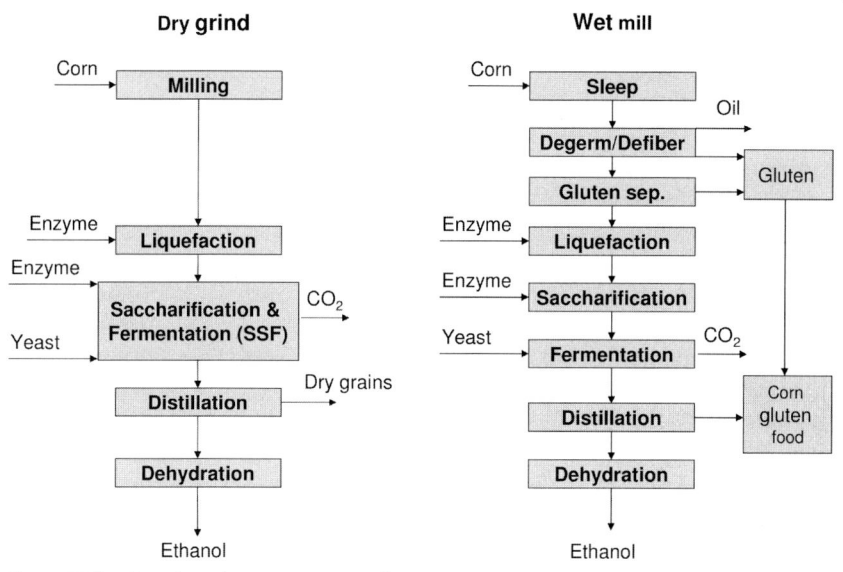

Figure 15.8 Ethanol production processes from corn [11].

dextrine into glucose, from which ethanol can be obtained by fermentation with yeast. The last two processes can take place simultaneously.

15.8.3
Bioethanol from Lignocellulosic Biomass

Cost-efficient manufacturing of bioethanol from abundant lignocellulosic biomass is the key challenge for the future of biofuels. Intensive R&D is taking place at the present time in the USA, Canada, Europe and Japan [11, 12, 15, 16, 19–25]. The construction of six pilot plants each with a capacity of 700 tonnes biomass per day was started in the USA in 2007 with governmental support. In this way, various raw materials and adapted technologies can be studied. The key problem in lignocellulosic ethanol is finding cost-efficient and ecological technologies for feedstock pretreatment, namely for removing lignin and increasing the fraction of hydrolyzed hemicelluloses. Figure 15.9 presents a generic block diagram. By pretreatment the biomass containing cellulose (C), hemicellulose (H) and lignin (L) is split in a solid fraction (C + L) and a liquid fraction, containing pentose (P) and inhibitors (I). The last stream (P + I) has to be submitted to detoxification in order to remove the inhibitors. The core section is the consolidated bioprocessing (CBP). By a classical approach, cellulose hydrolysis and hexose fermentation can be carried out sequentially (SHF). Alternatively, both steps can be carried out by simultaneous saccharification and fermentation (SSF) by using suitable micro-organisms. On the contrary, pentose fermentation should be carried out separately, since the micro-organisms are more sensitive to inhibitors and manifests slower reaction rates when hexoses are present [27]. However, by using genetically

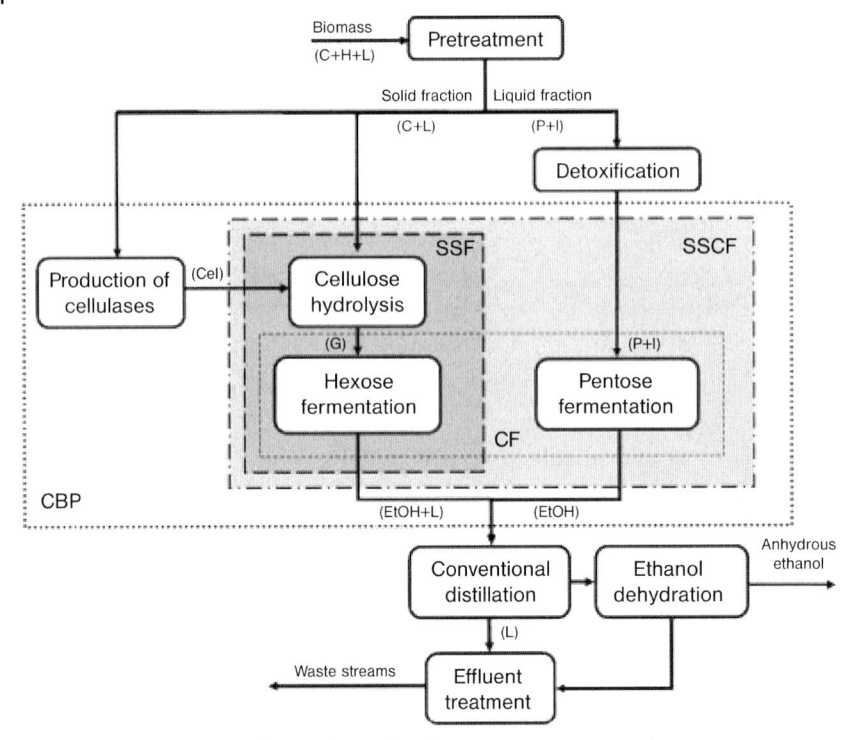

Figure 15.9 Block diagram for bioethanol from lignocellulosic biomass [27]. CBP = consolidated biprocessing, CF = cofermetation, SSF = simultaneous saccharification and fermentation, SSCF = simultaneous saccharification and cofermentation. Components: C = cellolose, H = hemicellulose, L = lignin, Cel = cellulases, G = glucose, P = pentose, I = inhibitors, EtOH = ethanol.

modified bacteria, it is possible to perform the steps together by simultaneous saccharification and fermentation, SSCF. Finally, the ethanol is separated by distillation and purified by dehydration.

Process-integration issues should receive particular attention [15, 24, 26]. The pretreatment step may imply the use of aggressive chemicals, such as acids and bases. Liquid water and steam are used for extracting the useful substrates from solids. Accordingly, waste streams are produced, from which the water should be recovered and recycled in an efficient manner. Distillation of water solution is energy intensive. Fortunately, a substantial part of the feedstock is formed by lignin, which can be used as a fuel in a combined heat and power cycle. In this way, the energetic needs of the plant could be entirely covered, eventually exporting the excess. The above scheme can be exemplified by a number of process flowsheets [27].

15.9
Process Design: Ethanol from Lignocellulosic Biomass

15.9.1
Problem Definition

The goal of this design project is to realize the manufacturing of ethanol from lignocellulosic biomass by using advanced pretreatment and fermentation procedures. The feedstock is based on yellow poplar hardwood. Poplar is a perennial tree well adapted for moderate climate regions, cheap to grow, with low input costs, low tillage and fertilization requirements, and easy to store. The design will use many elements from an extensive study performed at the National Renewable Energy Laboratory (NREL) [24].

Table 15.8 presents the feedstock composition expressed in key components: cellulose, hemicellulose (xylan, arabinan, mannan, galactan) and lignin.

The manufacturing process can be captured in blocks, as displayed in Figure 15.10. Ethanol fermentation technologies start with removal of large or unsuitable

Table 15.8 Feedstock composition.

Component	Flow rate (kg/h)	% dry basis
Cellulose	46140	43.10
Hemicellulose		
• Xylan	20600	19.24
• Arabinan	850	0.79
• Mannan	4250	3.97
• Galactan	260	0.24
Lignin	29930	27.96
Acetate	5020	4.69
Moisture	51800	48.39
Total	158850	

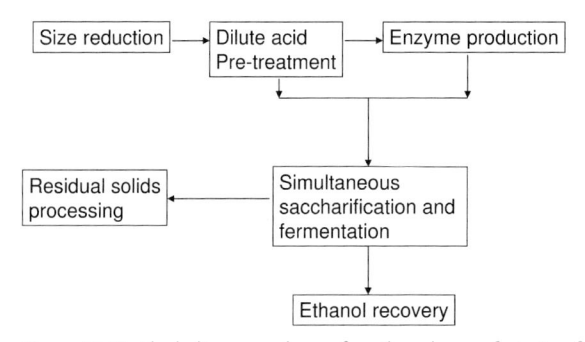

Figure 15.10 Block-diagram scheme for ethanol manufacturing from lignocellulosic biomass.

materials, followed by mechanical preprocessing and shredding of the material. Then, the material is processed to promote hydrolysis. Afterwards, the slurried material is fermented to produce alcohol, which is finally purified through filtration and distillation.

When cellulose is used as a raw material, the activity of cellulase (the enzyme catalyzing cellulose hydrolysis) is inhibited by glucose and short cellulose chains. One way to overcome this inhibition is to combine enzymatic hydrolysis with glucose fermentation to ethanol, as the accumulation of ethanol in fermenter does not inhibit cellulase.

15.9.2
Definition of the Chemical Components

For the purpose of conceptual design of the bioethanol plant, Aspen Plus will be used as the flowsheet simulator. However, most of the key components involved in the process are not defined in the standard Aspen Plus property databases, and therefore their physical property data are not available. The National Renewable Energy Laboratory (NREL) has developed a database that includes a complete set of properties for the currently identifiable compounds in the ethanol process [28].

The components included in the database can be classified as components involved in vapor–liquid equilibrium, and identifiable solids. For compounds involved in vapor–liquid equilibrium, the simulator requires a set of properties to allow flash calculations, even though the compound may be a very high boiler and will be found in the liquid phase exclusively. Materials that are commonly solids but that in the process are found exclusively in aqueous solution are treated as liquids. This category includes glucose, xylose, arabinose, mannose and galactose. In addition, we included in this class the oligomers that occur during the saccharification process. The properties specified in the database are molecular weight, heat and free energy of formation, critical parameters, Antoine parameters, heat of vaporization, liquid molar volume, ideal gas capacity, and liquid heat capacity.

The second class comprises conventional solids, defined by a chemical formula, but whose property requirements are very minimal. In this class we included lignin, cellulose, mannan, galactan, xylan, arabinan and the biomass. The properties specified in the database include molecular weight, heat of formation, solid molar volume, and solid heat capacity.

15.9.3
Biomass Pretreatment

The purposes of pretreatment of lignocellulosic materials are the removal of lignin and hemicellulose, reduction of cellulose crystallinity, and increase in the porosity of the materials. Among the physical methods, *mechanical* treatments, such as chipping, grinding and milling are used to reduce cellulose crystallinity. The size of the materials is 10–30 mm after chipping, and 0.2–2 mm after milling or grinding.

The most commonly used physicochemical method is *steam explosion*. Here, chipped biomass is treated with high-pressure saturated steam and then the pressure is swiftly reduced. This makes the material undergo an explosive decompression. Steam explosion is initiated at 160–260 °C (0.7–4.83 MPa), for several minutes before the material is exposed to atmospheric pressure. The process causes hemicellulose degradation and lignin transformation due to the high temperature. Addition of H_2SO_4 in steam explosion can improve enzymatic hydrolysis, decrease the production of inhibitory compounds, and lead to more complete removal of hemicellulose. The advantages of steam-explosion pretreatment include low energy requirement and no recycling or environmental costs. Limitations of steam explosion include a destruction of a portion of the xylan fraction, incomplete disruption of the lignin–carbohydrate matrix and generation of compounds that may be inhibitory to micro-organisms. Because of the formation of degradation products that are inhibitory to microbial growth, enzymatic hydrolysis and fermentation, pretreated biomass needs to be washed by water, which decreases the overall saccharification yields due to removal of soluble sugars.

Ammonia fiber explosion (AFEX) is another physicochemical pretreatment in which lignocellulosic materials are exposed to liquid ammonia at high temperature and pressure for a period of time, and then the pressure is swiftly reduced. A typical dosage is 1–2 kg ammonia/kg dry biomass at 90 °C and residence time of 30 min. To reduce the cost and protect the environment, ammonia must be recycled after the pretreatment.

Acid hydrolysis has been successfully employed for pretreatment of lignocellulosic materials. Dilute sulfuric acid, used at either low or high temperature, achieves high xylan to xylose conversion. This is favorable to the overall economics, as xylan accounts for a large part of the total carbohydrates in the lignocellulosic materials.

In *alkaline hydrolysis*, bases such as NaOH and NH_3 are used for pretreatment. The effects include increased porosity, larger internal surface area, a decrease of the degree of polymerization, separation of structural linkages between lignin and carbohydrates, and disruption of the lignin structure.

In this project, we will use acid hydrolysis as the pretreatment method. The milled wood chips are heated to about 100 °C using low-pressure steam. Then, steam and sulfuric acid are added to the mixture in a pretreatment reactor. The reactor temperature is 190 °C, while the pressure is 12.2 atm. The concentration of acid is 0.5% (weight), while the concentration of solids in the outlet stream is 22%.

Table 15.9 presents the main reactions that occur in the pretreatment reactor, modeled as a stoichiometric reactor, and the corresponding conversions.

The reactor outlet is flashed to 1 atm and 100 °C. This removes in the vapor stream (53 650 kg/h) about 25% of the water, and 15% of the acetic acid.

The next step is separation of the solid and liquid phases, which is modeled as a single-stage solid washer (Swash). The separation will leave the solid portion

Table 15.9 Reactions in the pretreatment reactor.

Reaction	Conversion	Reference reactant
$(Cellulose)_n + n\ H_2O \rightarrow n$ Glucose	0.065	Cellulose
$(Cellulose)_n + m\ H_2O \rightarrow n$ Glucose oligomer	0.007	Cellulose
$(Cellulose)_n + 0.5n\ H_2O \rightarrow 0.5n$ Cellobiose	0.007	Cellulose
$(Mannan)_n + n\ H_2O \rightarrow n$ Mannose	0.75	Mannan
$(Mannan)_n + m\ H_2O \rightarrow n$ Mannose oligomer	0.05	Mannan
$(Mannan)_n \rightarrow n$ HMF $+ 2n\ H_2O$	0.15	Mannan
$(Galactan)_n + n\ H_2O \rightarrow n$ Glucose	0.75	Galactan
$(Galactan)_n + m\ H_2O \rightarrow n$ Glucose oligomer	0.05	Galactan
$(Galactan)_n \rightarrow n$ HMF $+ 2n\ H_2O$	0.15	Galactan
$(Xylan)_n + n\ H_2O \rightarrow n$ Xylose	0.75	Xylan
$(Xylan)_n + m\ H_2O \rightarrow n$ Xylose oligomer	0.05	Xylan
$(Xylan)_n \rightarrow n$ Furfural $+ 2n\ H_2O$	0.1	Xylan
$(Arabinan)_n + n\ H_2O \rightarrow n$ Arabinose	0.75	Arabinose
$(Arabinan)_n + m\ H_2O \rightarrow n$ Arabinose oligomer	0.05	Arabinose
$(Arabinan)_n \rightarrow n$ Furfural $+ 2n\ H_2O$	0.1	Arabinose
Acetate $\rightarrow$ Acetic acid	1	Acetate

with about 40% insoluble solids. This means that not the entire amount of the toxic materials (acetic acid) will be found in the liquid stream. One way to remove more of the toxic materials is to wash with water. In this project, we choose to wash with an amount of water (10^5 kg/h) which is about 50% of the initial amount of liquid (2.11×10^5 kg/h). In this way, about 65% of the acetic acid will be found in the liquid stream. Another option is remixing the liquids and solids after conditioning, followed by another separation and recycling.

From the liquid stream, about 88% of the acetic acid is removed in a continuous ion-exchange unit, simultaneously with the entire amount of sulfuric acid.

After ion exchange, the material is overlimed. First, the liquid pH is increased to 2 by addition of sulfuric acid (90 kg/h). Then, lime is added (68.5 kg/h) to raise the pH to 10, and the mixture is heated by steam injection to 50 °C, at a residence time of 1 h. The liquid is then adjusted to pH 4.5 and held for 4 h. In this way, large gypsum crystals are formed, which can be separated by means of hydrocyclone and rotary drum filtration. The amount of gypsum produced is about 200 kg/h, in which the solid content is 80%.

The flowsheet of the pretreatment and hydrolyzate conditioning is presented in Figure 15.11.

15.9.4
Fermentation

The fermentation section is modeled by means of two stoichiometric reactors placed in parallel, followed by a flash for separation of the vapor products (see Figure 5.12). The first reactor (SSCF) describes the saccharification and

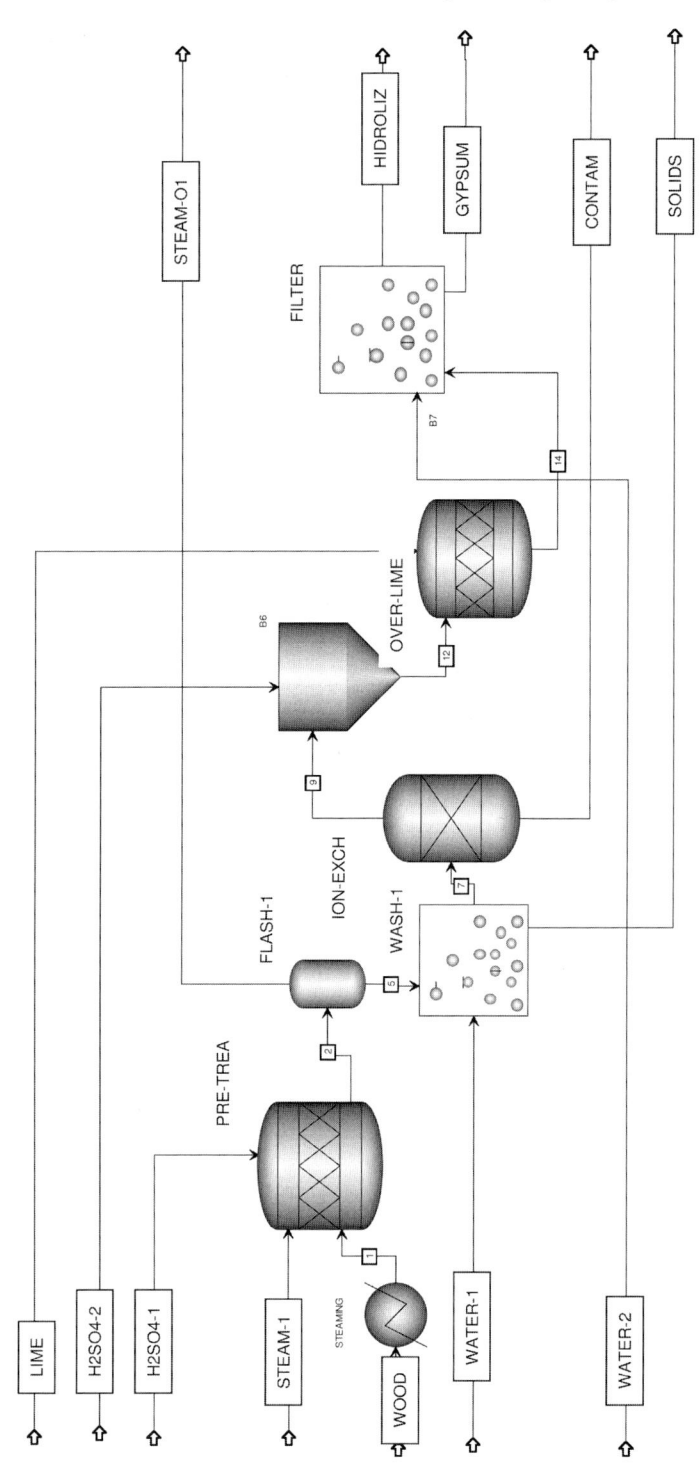

Figure 15.11 Pretreatment and hydrolyzate conditioning.

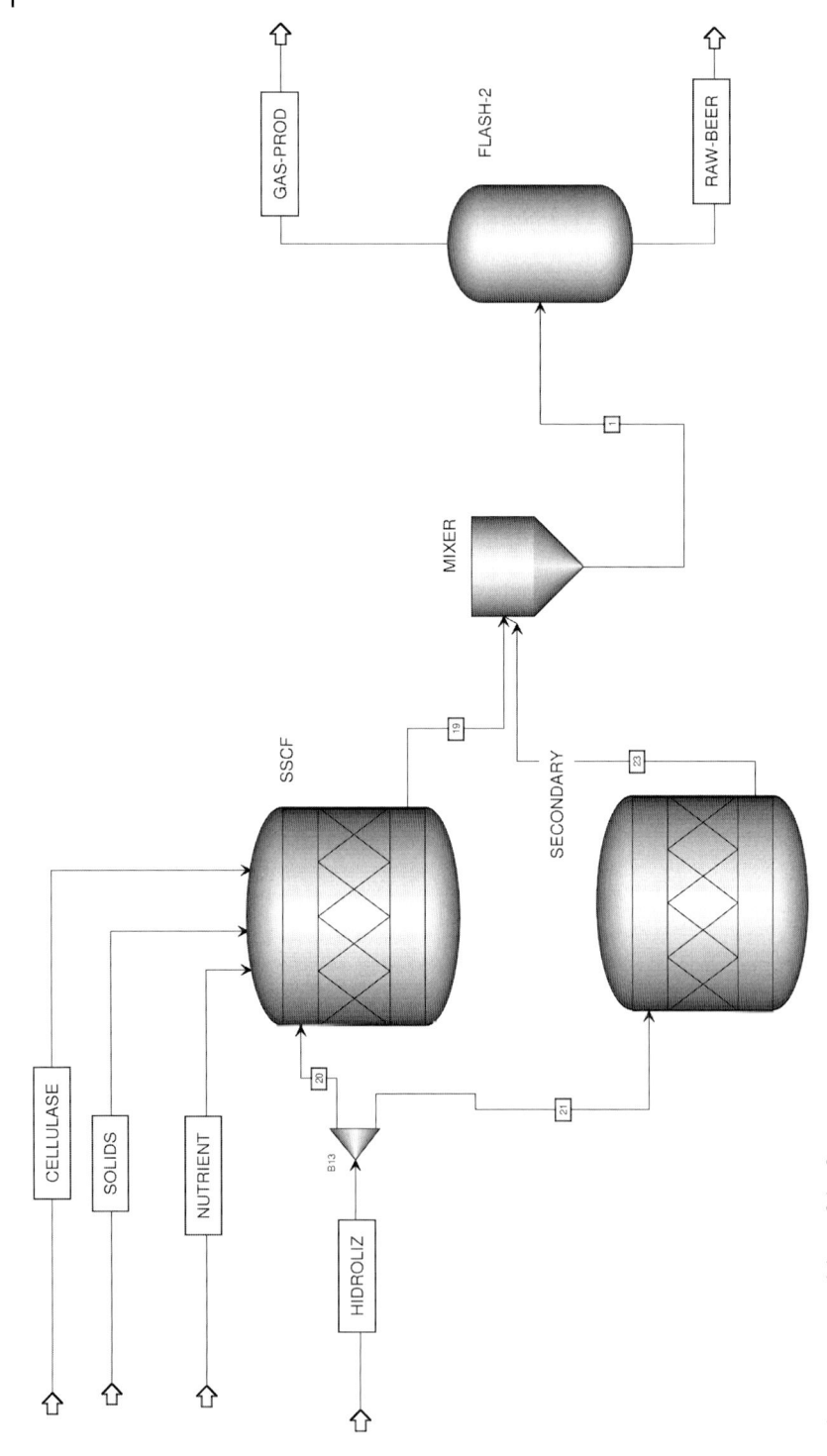

Figure 15.12 Modeling of the fermentation section.

15.9 Process Design: Ethanol from Lignocellulosic Biomass

fermentation processes. Saccharification requires cellulase enzymes, which are produced in another area of the plant. For fermentation, the recombinant *Z. mobilis* is used as an ethanologen. It is assumed that *Z. mobilis* will ferment glucose and xylose to ethanol, but not other sugars. The ethanologen must be grown in a seed-fermentation train. Detoxified hydrolizate and nutrients are combined with an initial seed inoculum, grown in the laboratory. The result of one batch is used as a seed for the next batch, until the amount of biomass obtained is large enough to support the production fermentation.

The chemical reactions and their conversions are shown in Tables 15.10 and 15.11. A typical value for the residence time is 7 days, giving a total fermentation volume of about $60\,000\,m^3$. This can be arranged in 3 trains, with 5 reactors of $4000\,m^3$ in each train.

In addition to sacharification and fermentation, loss to other product occurs (Table 15.12). This is modeled by a side stream bypassing the SSCF reactor that reacts to lactic acid. A total of 7% of the sugars available for fermentation are considered lost to contamination.

The fermentors are cooled by pump-around loops and external heat exchangers. The total duty required is $9.5 \times 10^6 \, kcal/h$.

Table 15.10 SSCF Saccharification reactions and conversions.

Reaction	Conversion	Reference reactant
$(Cellulose)_n + m\,H_2O \rightarrow m$ Glucose oligomer	0.068	Cellulose
$(Cellulose)_n + 0.5n\,H_2O \rightarrow 0.5n$ Cellobiose	0.012	Cellulose
$(Cellulose)_n + n\,H_2O \rightarrow n$ Glucose	0.8	Cellulose
Cellobiose + $H_2O \rightarrow 2$ Glucose	1.0	Cellobiose

Table 15.11 SSCF Fermentation reactions and conversions.

Reaction	Conversion	Reference reactant
Glucose $\rightarrow 2$ Ethanol + 2 CO_2	0.92	Glucose
Glucose + 1.2 $NH_3 \rightarrow 6$ Z. mobilis + 2.4 H_2O + 0.3 O_2	0.027	Glucose
Glucose + 2 $H_2O \rightarrow 2$ Glycerol + O_2	0.002	Glucose
Glucose + 2 $CO_2 \rightarrow 2$ Succinic acid + O_2	0.008	Glucose
Glucose $\rightarrow 3$ Acetic acid	0.022	Glucose
Glucose $\rightarrow 2$ Lactic acid	0.013	Glucose
3 Xylose $\rightarrow 5$ Ethanol + 5 CO_2	0.85	Xylose
Xylose + $NH_3 \rightarrow 5$ Z. mobilis + 2H_2O + 0.25 O_2	0.029	Xylose
3 Xylose + 5 $H_2O \rightarrow 5$ Glycerol + 2.5 O_2	0.029	Xylose
3 Xylose + 5 $CO_2 \rightarrow 5$ Succinic acid + 2.5 O_2	0.009	Xylose
2 Xylose $\rightarrow 5$ Acetic acid	0.024	Xylose
3 Xylose $\rightarrow 5$ Lactic acid	0.014	Xylose

Table 15.12 SSCF Contamination loss reactions.

Reaction	Conversion	Reference reactant
Glucose → 2 Lactic acid	1.0	Glucose
3 Xylose → 5 Lactic acid	1.0	Xylose
3 Arabinose → 5 Lactic acid	1.0	Arabinose
Galactose → 2 Lactic acid	1.0	Galactose
Mannose → 2 Lactic acid	1.0	Mannose

A large quantity of vapor products result from the fermentation: about 25 000 kg/h, mostly CO_2, but also 300 kg/h of ethanol that is recovered by scrubbing with water.

15.9.5
Ethanol Purification and Water Recovery

Distillation and molecular-sieve absorption are used to recover ethanol from the raw fermentation beer. The flowsheet of this section is presented in Figure 15.13.

First, the solids are separated. The liquid is mixed with the aqueous solution of ethanol recovered from the gas product of the fermentation section. The resulting mixture consists of about 3.23×10^5 kg/h, of which about 8% (weight) is ethanol.

The first distillation column has 32 trays, with the feed on tray 12, being operated at a reflux rate of 50 000 kg/h. The column has a partial condenser and delivers 1000 kg/h vapor distillate (sent to the absorption column) and 32 000 kg/h liquid distillate. The liquid distillate contains mainly ethanol (77% weight) and water (20% weight), with furfural the main impurity. The recovery of ethanol is almost 100%. The bottoms stream (2.9×10^5 kg/h) consists mainly of water, the concentration of dissolved organic compounds being around 7% (weight).

The second distillation column has 30 trays, with the feed on tray 15. It delivers 26 730 kg/h distillate where the concentration of ethanol is 92.5% (weight). The reflux ratio is 2. The distillate is free of furfural, the ethanol recovery being 99.5%. The bottoms are mixed with the bottoms of the first column and sent to water purification.

The residual water contains about 9% organics. The wastewater is treated by anaerobic digestion, where 90% of organics are converted to biogas (methane and CO_2). The biogas can be sent to a burner. Subsequently, the water is treated by aerobic digestion, which removes 90% of the remaining organics.

An additional source of energy is the solids stream separated after the fermentation. This contains mainly lignin (30 000 kg/h), cellulose (7800 kg/h), xylan (2000 kg/h) and biomass (126 kg/h). The solids can be burned, and the energy recovered as high-pressure steam.

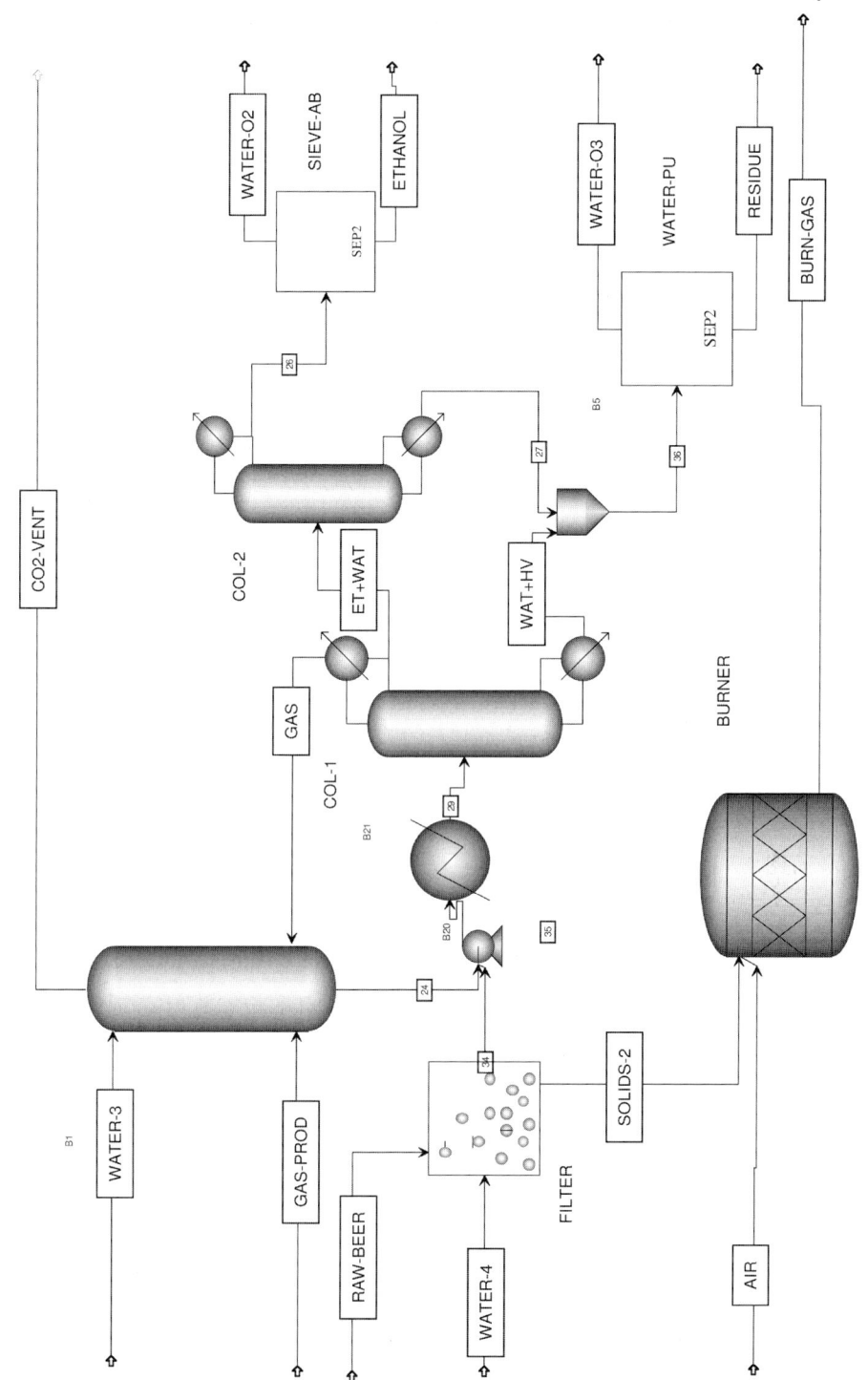

15.9 Process Design: Ethanol from Lignocellulosic Biomass | 457

Figure 15.13 Ethanol separation and dehydration.

Table 15.13 Summary of the mass and energy balance for the bioethanol plant.

Raw material	158 850 kg/h wood
	102 030 kg/h dry matter
Production of ethanol	25 000 kg/h
Energy produced	293×10^6 kcal/h
Energy required	8×10^6 kcal/h for steaming
	80×10^6 kcal/h for pretreatment
	28×10^6 kcal/h first distillation column
	19×10^6 kcal/h second distillation column
CO_2 produced	23 760 kg/h from fermentation
	94 675 kg/h from lignin burning
Gypsum produced	158 kg/h

Table 15.13 presents a summary of the mass and energy balance for the bioethanol plant. When the entire amount of cellulose and hemicellulose is converted to ethanol, the maximum achievable yield is about 35% kg/kg dry matter. The process designed in this chapter achieves an ethanol yield of about 25%. The energy produced by burning the non-fermentable part is enough to satisfy the requirements of the pre-treatment and purification sections, and the excess could be used to generate electricity. Large quantities of CO_2 are produced, the amount resulting from fermentation being almost pure.

15.10
Conclusions

The production of biofuels is growing sharply worldwide. Bioethanol is a valuable ecological fuel that can be used in blending with gasoline and diesel. In Brazil it represents already about 30% of the total transport fuel due to very favorable economics for manufacturing from sugar cane, as well as in the USA where it takes profit from the large land surface and highly productive farming methods. The majority of studies based on life-cycle analysis from well-to-wheel conclude that the impact of using biofuels, particularly of bioethanol, is positive to reduce GHG emissions and nonrenewable energy. However, there are serious concerns regarding the negative effects on the food supply chain, as well as about soil damaging by intensive exploitation.

A promising solution to the constraints raised by the use of edible raw materials is the advent of the second-generation bioethanol based on abundant and cheap lignocellulosic biomass. However, the capital and operation costs are much higher than with plants handling sugars and starch. Here, the key problem is finding an environmentally friendly and cost-effective method for feed pretreatment by which fermentable cellulose and hemicellulose are separated from lignin. Drastically reducing the cost of enzymes is also imperative.

The case study handles the design of a bioethanol plant based on lignocellulosic biomass following the NREL technology and making use of simulation in Aspen Plus. Emphasis is set on getting a realistic and consistent material and energy balance over the whole plant so as to point out the impact of the key elements on the investment and operation costs. To achieve this goal the complicated biochemistry is expressed in term of stoichiometric reactions and user-defined components. Acid hydrolysis was chosen as the pretreatment method. The conversion process is based on simultaneous saccharification and fermentation. The ethanol product is separated by distillation and molecular sieves. The nonfermentable materials are burned in order to generate steam or electricity.

The development of pretreatment technologies that are tuned to the characteristics of the biomass is still needed. Ideally, lignocellulose should be fractionated into multiple streams that contain valuable compounds in concentrations that make purification, utilization, and/or recovery economically feasible. Predictive pretreatment models should enable the design of this step to match both the biomass feedstock and the fermentation technology.

A very important step in the process is the simultaneous saccharification and fermentation. This requires enzymes that can effectively break the cellulosic and hemicellulosic material into sugar components. Additionally, micro-organisms that can use a wide range of sugars are desired. These are challenges to be solved by future research in the field of biotechnology.

From the viewpoint of energy consumption, the process is self-sustained (Table 15.13). However, because large quantities of energy are involved, any saving achieved by process integration is important. The process also implies very large flow rates that can be handled only by large equipment, for which a good mechanical design becomes very important. Moreover, special attention should be paid to control and operation of the bioreactors where undesired micro-organisms mutations or contamination should be avoided. Additional challenges arise from the need for manipulating large amounts of solids, or liquid–solid mixtures.

The process uses large quantities of water, therefore purification of wastewater and recycle is imperative. In this context, the accumulation of trace elements can become a problem.

We conclude that, although significant progresses have been achieved in the technology of bioethanol production, development of large-scale competitive processes requires an integrated approach, where the most recent results from various field of research are brought together.

References

1 French Petroleum Institute, Panorama reports 2007, 1. Biofuels and their environmental performance; 2. Biofuels worldwide; 3. Biofuels in Europe
2 International Energy Agency, World Energy Outlook 2006: fact sheet biofuels
3 Aspen Plus™ release 12.1, Aspen Technologies, Massachusetts USA, 2006
4 USDA, The economics of bioethanol production in the EU 2006
5 Solomon, B. D., Barnes, J. R., Halvorsen, K. E., Grain and cellulosic ethanol:

History, economics, and energy policy, *Biomass Bioenergy 31*, 415–425, 2007
6 Blottnitz V. H., Curran, M. A., A review of assessments conducted on bioethanol, *J. Cleaner Prod.*, 15, 607–619, 2007
7 Quirin, M., Gaertner, O., Pehnt, M., Reinhardt G., CO_2 mitigation through biofuels in the transport sector, Institute for Energy and Environmental Research (IFEU), Heidelberg, Germany, 2004
8 European Commission JRC, Well-to-wheals analysis of future fuels and power trains in the European context, version 2c, March 2007
9 Ullmann's Encyclopaedia of Industrial Chemistry, Ethanol, Wiley-VCH, Weinheim, Germany, 2003
10 Rural Industries Research and Development Corporation, Australia, Sugar beet report, 2005
11 Bothast, B. J., Schlicher, M. A., Biotechnological process for conversion of corn into ethanol, *Appl. Microbiol Biotechnol.*, 67, 19–26, 2005
12 Dale, B., Greening the chemical industry: research and development priorities for biobased industrial products, *J. Chem. Technol. Biotechnol.*, 78, 1093–1103, 2003
13 Kamm, B., P.R. Gruber, M. Kamm, Biorefineries (eds.) – Industrial Processes and Products, Wiley-VCH, Weinheim, Germany, 2006
14 Kamm B., M. Kamm, Principles of biorefineries, *Appl. Microbiol. Biotechnol.* 64, 137–145, 2004
15 Wyman, C., Potential synergies and challenges in refining cellulosic biomass to fuels, chemical, and power, *Biotechnol. Prog.*, 19, 254–262, 2003
16 Liu, Y., Tanaka, S., Ethanol fermentation from biomass resources, *Appl. Microbiol. Biotechnol.*, 69, 627–642, 2006
17 Kadam, K. L., Rydolm, E. C., McMillan, J. D., Development and validation of a kinetic model for enzymatic saccarification of lignocellulosic biomass, *Biotechnol. Prog.*, 20, 699–705, 2004
18 Olivera. S. C., Castro, H. F., Visconti, Q., Giudici, R., Continuous ethanol fermentation in a tower reactor, *Bioprocess Eng.*, 20, 525–530, 1999
19 Lynd. L. R., Weimer, P. J., van Zyl, W., Pretorius, Y., Microbial Cellulose Utilization: Fund. *Biotechnol.*, 66 (3), 506–577, 2002
20 Mielenz, J. R., Ethanol production from biomass, *Curr. Opin. Microbiol.*, 4, 324–329, 2001
21 Mosier, N., Wyman, C., Dale, B., Elander, R., Lee, Y., Holzapple, M., Ladisch, M., Features of promising technologies for pretreatment of lignocellulosic biomass, *Bioresource Technol.*, 96, 673–686, 2005
22 Rabinovitch. M. L., Ethanol production from material containing cellulose: the potential of Russian R&D, *Appl. Biochem. Microbiol.*, 42, 1, 1–26, 2006
23 Schell, D. J., Riley, C., Dowe, N., Ibsen, K., Ruth, M., Toon, S., Lumpkin, R., A bioethanol process development unit with a corn fiber feedstock, *Bioresources Technol.*, 91, 179–188, 2004
24 Wooley, R., Ruth, M., Glassner, D., Sheehan, J., Ibsen, K., Majedski, H., Galvez, A., Lignocellulosic biomass to ethanol process design and economics. Report NREL/TP-580-26157, 1999
25 Klinke, H., B., Thomsen, A. B., Ahring, B. K., Inhibition of ethanol-producing yeast and bacteria by degradation products during pretreatment of biomass, *Appl. Microbiol. Technol.*, 66, 10–26, 2004
26 Pfeffer, M., Wukovits, W., Beckmann, G., Friedl, A., Analysis and decrease of the energy demand of bio-ethanol production by process integration, *Appl. Thermal Eng.*, 27, 2657–2664, 2007
27 Cardona, C. A., Sanchez, O. J., Fuel ethanol production: process design trends and integration opportunities, *Biores. Technol.*, 98, 2415–2457, 2007
28 Wooley, R.J. and Putsche, V., Development of an ASPEN PLUS Physical Property Database for Biofuels Components, NREL/MP-425-20685, 1996

Appendix A
Residue Curve Maps for Reactive Mixtures

A.1
Coupling Reaction and Distillation

Useful insights into the thermodynamics of a multiphase reacting system can be obtained by analyzing the situation when chemical and phase equilibrium (C&PE) are achieved simultaneously. The Gibbs rule can be used to find the degrees of freedom F:

$$F = c + 2 - P - R \tag{A.1}$$

where P, c, and R are the number of phases, components and independent reactions, respectively. If only vapor–liquid equilibrium exists then $P = 2$, $R = 0$ and $F = c$, usually selected among $(c - 1)$ molar fractions, plus one state variable, P or T, but when one independent chemical reaction takes place then $R = 1$ and F has one dimension less. Some examples below will illustrate this fundamental rule.

Insightful solution of a C&PE problem can be found by graphical representation, particularly using ternary diagrams. Let us consider the general stoichiometric reaction:

$$v_A A + v_B B + \ldots \leftrightarrow v_P P + v_R R + \ldots \quad \text{or} \quad \sum_{i=1}^{c} v_i A_i = 0 \quad \text{with} \quad v_t = \sum_{i=1}^{c} v_i \tag{A.2}$$

The composition can be expressed with respect to a reference species k as follows [1]:

$$x_i = \frac{x_{io}(v_k - v_t x_k) + v_i(x_k - x_{ko})}{v_k - v_t x_{ko}} \tag{A.3}$$

The relation (A.3) describes the so-called *stoichiometric lines* converging into a *pole* π, whose location can be determined by the following relation:

$$x_i = \frac{v_i}{\sum_i v_i} = \frac{v_i}{v_t} \tag{A.4}$$

Chemical Process Design: Computer-Aided Case Studies. Alexandre C. Dimian and Costin Sorin Bildea
Copyright © 2008 WILEY-VCH Verlag GmbH & Co. KGaA, Weinheim
ISBN: 978-3-527-31403-4

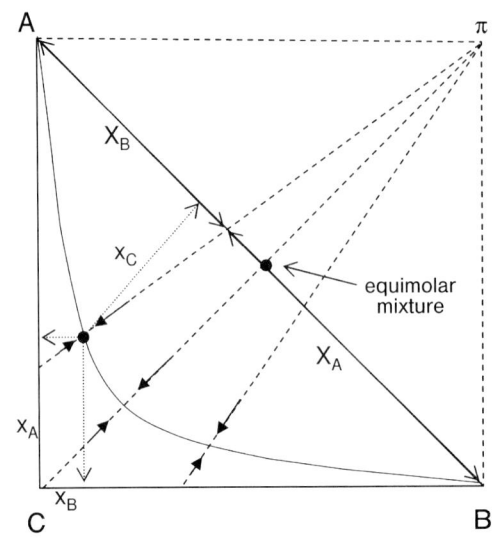

Figure A.1 Conventional and transformed composition variables in a ternary diagram.

Figure A.1 illustrates the graphical interpretation in a right-angle triangle. Pure components are marked in the vertices A, B, C. The molar fractions x_A and x_B are represented on the edges CA and CB, while x_C is visualized by the height of the point representing the ternary mixture with respect to AB. In Figure A.1 the chemical equilibrium curve is displayed too drawn by means of the relation:

$$K_{eq} = \frac{a_P^{v_P} a_R^{v_R} \ldots}{a_A^{v_A} a_B^{v_B} \ldots} = \frac{x_P^{v_P} x_R^{v_R} \ldots}{x_A^{v_A} x_B^{v_B} \ldots} \cdot \frac{\gamma_P^{v_P} \gamma_R^{v_R} \ldots}{\gamma_A^{v_A} \gamma_B^{v_B} \ldots} = \prod_i (x_i \gamma_i)^{v_i} = K_x K_\gamma \quad (A.5)$$

The mass-balance problem can be solved graphically. The median connecting the vertex C with the AB edge corresponds to the transformation of an equimolar AB mixture into C. Extending this line with an equal segment gives the position of the pole π of coordinates (0,1) and (1,0). From this point, stoichiometric lines can be drawn for any initial composition of the reaction mixtures. When the reaction preserves the number of moles ($v_t = 0$) the stoichiometric lines are parallel.

A.1.1
Transformed Variables

For more species a change of variables appears useful in order to reduce the dimensionality of a graphical representation. For simplicity, let us consider the reaction $A + B = C$ taking place in one phase containing initially only A and B. Before the reaction the number of moles is $N_0 = N_A + N_B$. After the reaction the number of moles becomes N_A^*, N_B^* and N_C^*, the total number being

$N_t = N_A^* + N_B^* + N_C^* = N_0 - N_C^*$. The actual molar fractions may be written as: $x_A = N_A^*/N_t$, $x_B = N_B^*/N_t$, $x_C = N_C^*/N_t$ with $N_t = N_0/(1 + x_C)$. The composition can be expressed only as function of A and B using C used as reference, since the three species are linked by a stoichiometric equation. Since $N_A = N_A^* + N_C^*$ and $N_B = N_B^* + N_C^*$ one gets the expressions:

$$\overline{X}_A = \frac{x_A + x_C}{1 + x_C} \quad \text{and} \quad \overline{X}_B = \frac{x_B + x_C}{1 + x_C} \tag{A.6}$$

X_A and X_B are new composition variables X_i that verify the relation $\Sigma X_i = 1$.

More generally, the composition of a reacting system characterized by c molar fractions can be reduced to $(c - 1)$ new composition variables by the following transformation [1, 2]:

$$X_i = \frac{\nu_k x_i - \nu_i x_k}{\nu_k - \nu_t x_k} \tag{A.7}$$

The reference k component should preferably be a product.

For example, the AB mixture expressed in Figure A.1 by X_A and X_B mole fractions on the AB edge leads at equilibrium to a mixture (x_A, x_B, x_C) obtained by intersecting the equilibrium curve with the stoichiometric line passing through the initial mixture. Conversely, a ternary mixture where a chemical reaction at equilibrium takes place may be described only by two transformed composition variables.

A.2
Reactive Residue Curve Map Analysis

Residue curve (RCM) and distillation curve (DCM) maps are today standard tools for designing distillation systems dealing with nonideal mixtures involving azeotropes. A *residue curve* characterizes the evolution of the liquid composition in a vessel during a batchwise distillation experiment. The whole compositional space may be spanned by residue curves considering different initial mixture compositions. For nonreactive mixtures the RCM is obtained by solving the component dynamic material balance expressed by the following differential equation:

$$\frac{dx_i}{d\xi} = x_i - y_i \tag{A.8}$$

in which $\xi = H/V$ is a "warped-time" variable defined as the ratio of molar liquid holdup H by the molar vapor rate V, while x_i and y_i are vapor and liquid compositions, respectively. Getting reliable RCMs, their classification and use in the

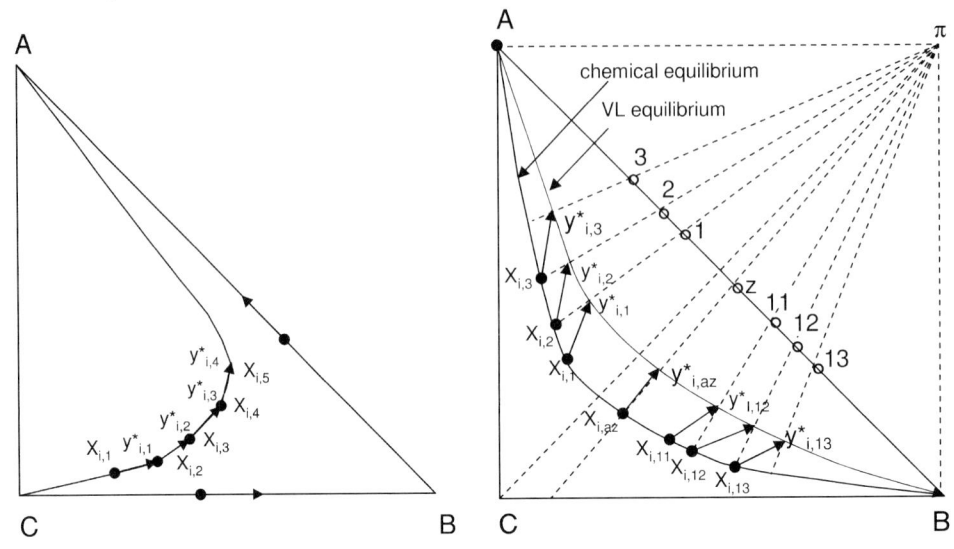

Figure A.2 Construction of the distillation lines for nonreactive (left) and reactive mixtures (right).

design of distillation systems is discussed thoroughly in the book of Doherty and Malone [2].

A similar representation is based on *distillation lines* [1], which describe the composition on successive trays of a distillation column with an infinite number of stages at infinite reflux (∞/∞ analysis). In contrast with relation (A.8) the distillation lines may be obtained much easier by algebraic computations involving a series of bubble and dew points, as follows:

$$x_{i,1} \to y_{i,1}^* = x_{i,2} \to y_{i,2}^* = x_{i,3} \to y_{i,3} \ldots \tag{A.9}$$

Figure A.2 (left) shows the construction of a distillation for an ideal ternary system in which A and C are the light (stable node) and the heavy (unstable node) boilers, while B is an intermediate boiler (saddle). The initial point $x_{i,1}$ produces the vapor $y_{i,1}^*$ that by condensation gives a liquid with the same composition such that the next point is $x_{i,2} = y_{i,1}^*$, etc. Accordingly, the distillation line describes the evolution of composition on the stages of a distillation column at equilibrium and total reflux from the bottom to the top. The slope of a distillation line is a measure of the relative volatility of components. The analysis in RCM or DCM leads to the same results.

Similarly, when a chemical reaction takes place the residue curves can be found by the equation [2]:

$$\frac{dx_i}{d\xi} = x_i - y_i + \mathrm{Da}(v_i - v_T x_i)R \tag{A.10}$$

In the above Da denotes the Damköhler number as the ratio of the characteristic process time H/V to the characteristic reaction time $1/r_o$. The reaction rate r_o is a reference value at the system pressure and an arbitrary reference temperature, as the lowest or the highest boiling point. For catalytic reactions r_o includes a reference value of the catalyst amount. R is the dimensionless reaction rate $R = r/r_o$. The kinetics of a homogeneous liquid-phase reaction is described in general as function of activities:

$$r = k\left[\left(\prod a_i^{v_i}\right)_{pr} - \left(\prod a_j^{v_j}\right)_{eq}/K_{eq}\right] \tag{A.11}$$

Thus, the parameter Da is a measure of the reaction rate, but its absolute value cannot be taken as a basis for comparing different systems.

For sufficient high values of the Da number the reaction can reach an equilibrium state. This situation, in which phase and chemical equilibrium are reached simultaneously is very useful in analyzing real systems. In this case the RCM can be computed by using the following equation [2]:

$$\frac{d\overline{X}_i}{d\xi} = \overline{X}_i - \overline{Y}_i \tag{A.12}$$

Equations (A.8) and (A.12) are similar, but the last makes use of transformed variables.

Analogous with the procedure presented before, *reactive distillation lines* can be obtained by computing a series of dew and bubble points incorporating a chemical equilibrium term, as follows:

$$x_{i,1} \rightarrow y^*_{i,1} \leftrightarrow x^{eq}_{i,2} \rightarrow y^*_{i,2} \leftrightarrow x^{eq}_{i,3} \rightarrow y^*_{i,3} \ldots \tag{A.13}$$

Graphical construction of reactive distillation lines at equilibrium is shown in Figure A.2 (right) for the reversible reaction $A + B \leftrightarrow C$. The initial point $x_{i,1}$, at chemical equilibrium produces a vapor $y^*_{i,1}$, which by condensation and equilibrium reaction gives the liquid with the composition $x_{i,2}$. This is found by crossing the stoichiometric line passing through $y^*_{i,1}$ with the chemical equilibrium curve. Then, the liquid $x_{i,2}$ produces the vapor $y^*_{i,2}$, etc. Similarly, the points 11, 12, 13, etc. shows the situation in which the mixture becomes richer in B and poorer in A and C.

Figure A.2 (right) emphasizes a particular position where phase equilibrium and stoichiometric lines are collinear. In other words the liquid composition remains unchanged because the resulting vapor, after condensation, is converted into the original composition. This point is a *potential reactive* azeotrope, but when the composition satisfies chemical equilibrium too it becomes a true *reactive azeotrope*. Some examples of residue curve maps are presented below. Ideal mixtures are used to illustrate the basic features, which may be applied to some important industrial applications.

A.2.1
Three Reactive Components Mixtures

Let us consider the reversible reaction $A + B \leftrightarrow C$ for which the relative volatilities are in the order 3/2/1 and the equilibrium constant $K_x = 6.75$. The physical and reactive distillation lines may be obtained simply by computation in Excel™ by applying the relations (A.13), as illustrated in Figure A.3. Note that in this case the starting point, liquid with composition (0.1, 0.1, 0.8), is not at chemical equilibrium. The co-ordinates of the triangle are in normal mole fractions. It may be observed that after a short straight path the reactive distillation line superposes the chemical equilibrium curve. Starting from other points shows the same trend. Inverting the volatilities to 3/1/2 does not change the situation. This result is consistent with the Gibbs phase rule. Since $c = 3$, $P = 2$ and $R = 1$, we have $F = 3 - 2 - 1 + 2 = 2$. If one variable, pressure or temperature, is fixed in order to determine the phase state, it remains that only one composition variable is needed. This may be X_A or X_B as transformed variables. Such situations designate one degree-of-freedom systems.

Figure A.3 also illustrates graphically the formation of a reactive azeotrope as the point where a particular stoichiometric line becomes tangential to the nonreactive residue curve and intersects simultaneously the chemical equilibrium curve.

Let us now consider the reaction $A + B \leftrightarrow C$ conducted in the presence of an inert I, which is the lightest component. The number of degrees of freedom becomes $F = 4 - 2 - 1 + 2 = 3$. One specification as P or T leaves two composition variables. It is convenient to start with mixtures of C and I that enables us to span the whole composition space. The representation of four components by means of a tetrahedron is possible, although not comfortable. As can be seen from

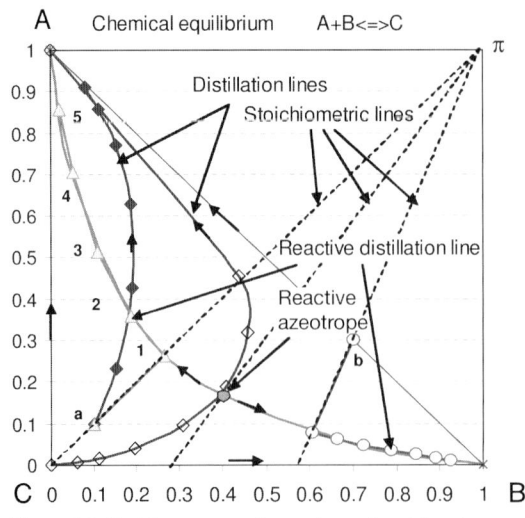

Figure A.3 Residues curves for a mixture involving three components, reaction $A + B \leftrightarrow C$.

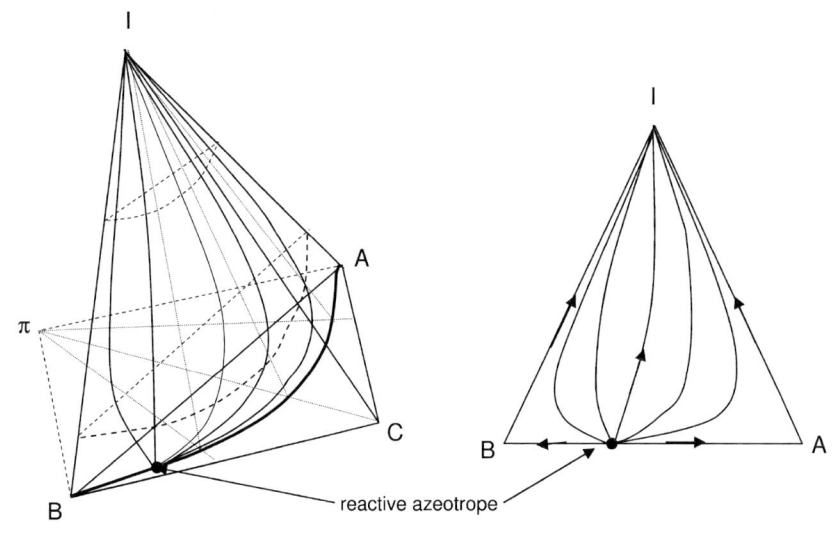

Figure A.4 Reactive residue curve maps for a ternary system containing inert, reaction $A + B + I \leftrightarrow C + I$.

Figure A.4 (left), the equilibrium curve from the previous case becomes a surface of conical shape limited by the inert-free equilibrium curve and the edges AI and BI. It is to be noted that all the reactive residue curves start at the reactive azeotrope but with different paths towards the vertex I.

Another possibility is the representation in a two-dimensional diagram, as in Figure A.4 (right). The component C being chosen as the reference, the relation (A.3) gives the transformed co-ordinates: $X_A = (x_A + x_C)/(1 + x_C)$ and $X_B = (x_B + x_C)/(1 + x_C)$. The residue curves run from the reactive azeotrope to the vertex of component I. This situation is denoted by "two degrees of freedom systems".

A.2.2
Four Reactive Components Mixtures

The advantage of bidimensional representation is evident if four reactive components are involved, as in the class of reversible reactions $A + B \leftrightarrow C + D$. This situation covers an important number of industrial applications, as the esterification of acids with alcohols. Selecting C as the reference, the transformed variables are $X_A = x_A + x_C$, $X_B = x_B + x_C$ and $X_D = x_D - x_C$ since $v_t = 0$. The transformed variables sums to one, but only two are used as co-ordinates. Accordingly, the pure components may be placed in the corner of a square diagram, reactants or products on the same diagonal. Figure A.5 displays the reactive distillation map traced as before for the relative volatilities 4/2/6/1 and the equilibrium constant $K_{eq} = 5$.

The above behavior has been verified experimentally in a number of cases. Figure A.6 illustrates the RCM for the reactive system methanol / acetic acid / methyl acetate / water at physical and chemical equilibrium [3] in which experi-

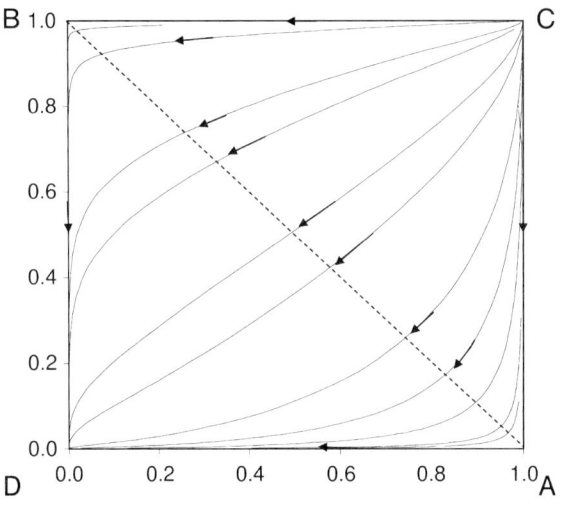

Figure A.5 Residue curve map for an ideal reactive mixture with relative volatilities 4/6/2/1 and reversible reaction $A + B \leftrightarrow C + D$ with $K_x = 5$.

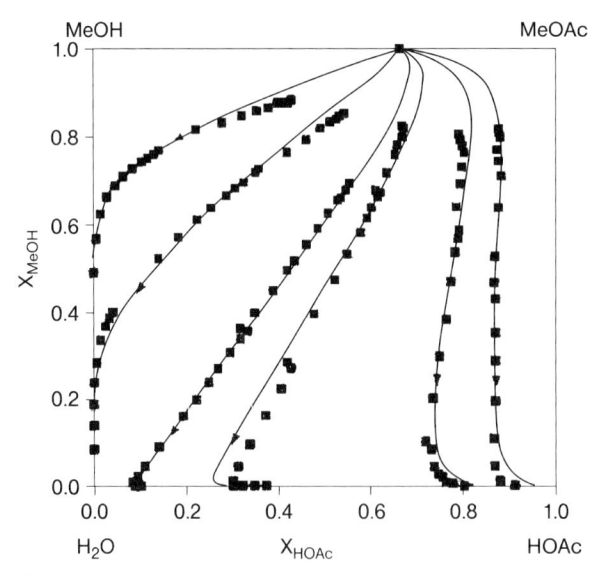

Figure A.6 Comparison of computed (solid line) and experimentally measured equilibrium reactive RCM for the methyl acetate system at atmospheric pressure [3].

mental and calculated data are compared. The liquid activities were computed by using UNIQUAC. Note the existence of a minimum-boiling azeotrope of methanol/methyl acetate, which is the unstable node. The trajectories run from the azeotrope to the acetic acid vertex, which is the stable node. The other components, methyl acetate, methanol and water are saddles. The agreement is good, taking into account the sensitivity of modelling to experimental errors: the equilibrium constant K_x was in the range of 4.6–8.2, the correction for liquid activities $K = 3.4$–4.5, leading to a true equilibrium constant K_{eq} in the range 20–30.

A.3
Influence of Reaction Kinetics on the Residue Curve Map

The kinetics of a reaction rate has a substantial influence on residue curve maps. Distillation boundaries and physical azeotropes can vanish, while other singular points due to kinetic effects might appear. The influence of the kinetics on RCM can be studied by integrating Eq. (A.10) for finite Da numbers. In addition, the singular points satisfy the relation:

$$Da R = \frac{x_i - y_i}{x_i - \nu_T x_i} \tag{A.14}$$

As an illustration, Figure A.7 presents the evolution of RCM for the mixture isobutene/methanol/MTBE at a pressure of 8 bar from pure physical equilibrium to simultaneous chemical and phase equilibrium [4]. Note that the absolute values of Da from this work cannot be compared against other papers because of different scaling of variables. Plot (a) shows the physical phase equilibrium (Da = 0) in which there are two physical azeotropes, MeOH/MTBE ($x = 0.55$) and MeOH/IB ($x = 0.08$), the first being saddle and the second unstable node, both linked by a distillation boundary. Consequently, two distillation fields appear above and below the separatrix in which MeOH and MTBE are stable nodes. Plot (b) shows the situation at Da = 10^{-4} when the reaction is slow. The upper distillation region is not affected by reaction, unlike near to the MTBE corner. The stable point moves from MTBE to a new position containing about 76.5% MTBE, 20% MeOH and 3.5% IB. Raising Da further to 2×10^{-4} (plot c) leads to a situation in which the saddle point disappears and all trajectories point out to the methanol vertex; the reaction rate becomes dominant. For Da = 1, when the reaction rate is high enough (plot d), the trajectories are collected by the chemical equilibrium curve, the RCM being similar to the Figure A.3.

Summing up, the influence of the kinetics of a chemical reaction on the vapor–liquid equilibrium is very complex. Physical distillation boundaries may disappear, while new kinetic stable and unstable nodes may appear. As result, the residue curve map with chemical reaction could look very different from the physical plots. As a consequence, evaluating the kinetic effects on residue curve maps is of great importance for conceptual design of reactive distillation systems. However, if the reaction rate is high enough such that the chemical equilibrium is reached quickly, the RCM simplifies considerably. But even in this case the analysis may be complicated by the occurrence of reactive azeotropes.

470 | *Appendix A Residue Curve Maps for Reactive Mixtures*

- ● Stable node
- ○ Unstable node
- □ Saddle point
- – – – – Pure chemical equilibrium
- ········ Separatrix

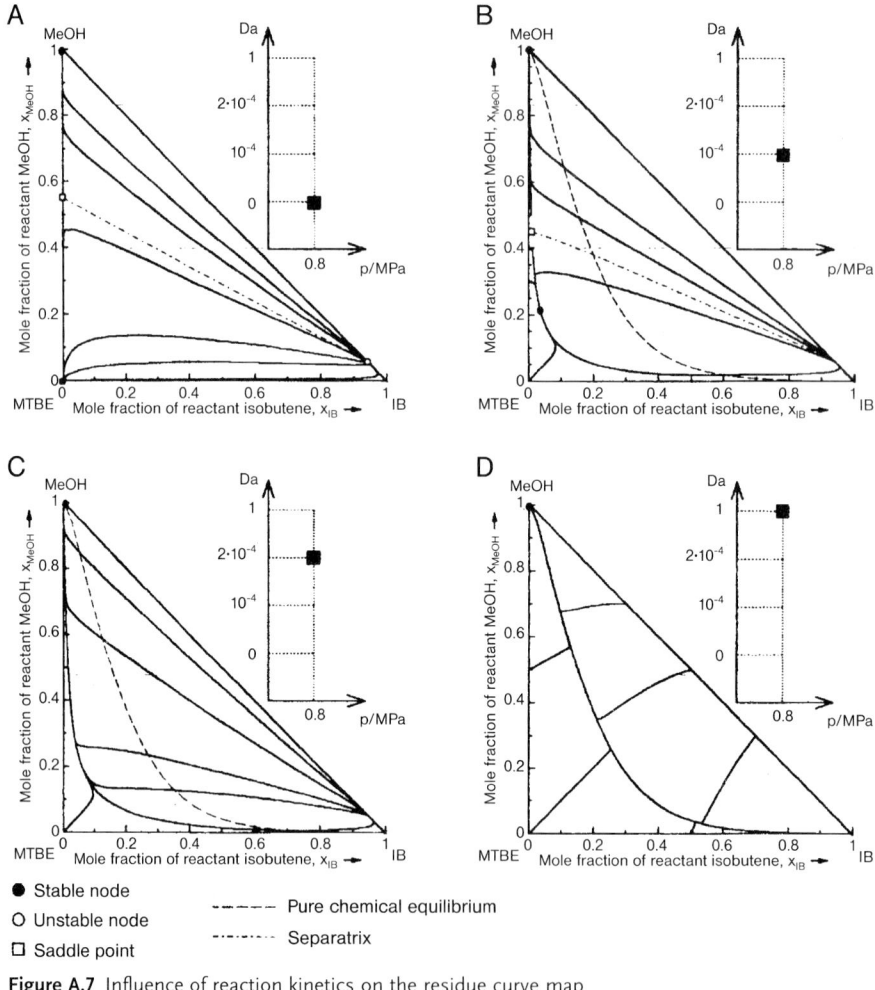

Figure A.7 Influence of reaction kinetics on the residue curve map (a) no reaction Da = 0, (b) slow reaction Da ~ 10^{-4}, (c) intermediate reaction Da = 2 × 10^{-4}, (d) fast reaction Da = 1 [4].

A.4
Influence of the Accuracy of Thermodynamic Data

The accuracy of the thermodynamic data has a significant effect on RCM computation. In the case of slow reactions both kinetics and phase equilibrium should be modelled accurately. If the reaction is fast enough the chemical reaction prevails. In many cases chemical equilibrium may be taken as the reference. Consequently, accurate knowledge of the chemical equilibrium constant is needed. When reactive azeotropes and/or phase splitting might occur accurate modelling of phase equilibrium is also needed.

It should be kept in mind that the calculation of the equilibrium constant is subject to high sensitivity to small errors in thermochemical data due to its exponential dependency to the standard Gibbs free energy variation, as expressed by the relation:

$$K_{eq} = \exp(-\Delta G^0(T, P)/RT) = \exp[-(\sum_i v_i \Delta G^0_{f,i}(T, P)/RT]$$

ΔG^0 is the standard Gibbs free energy change of reaction obtained from the standard energy of formation of pure components $\Delta G^0_{f,i}$ at 25 °C, $P = 1$ atm and a suitable aggregation state. Once having determined K_{eq} in standard conditions, the van't Hoff equation may be used to calculate K_{eq} at other temperatures, as follows:

$$\left(\frac{d \ln K_{eq}}{dT}\right)_P = \frac{\Delta H^0_R(T)}{RT^2}$$

where ΔH^0_R is the standard enthalpy change of reaction. Alternatively, ΔG^0 may be calculated from enthalpy and entropy changes as expressed by the relation $\Delta G^0 = \Delta H^0 - T\Delta S^0$. Whichever method is used the accuracy of data should be very good in order to avoid false predictions.

For example, for a liquid-phase reaction at 400 K the product RT has the value 0.792 cal/mol. For $\Delta G^0 = -2$ kcal/mol one gets $K_{eq} = 15$. For an absolute error of ±1 kcal/mol the values are 44 and 3.5, respectively, obviously leading to very different results. An error of ±1 kJ/mol would be acceptable.

Therefore, experimental validation of the chemical equilibrium constant is recommended for accurate design. In addition, for nonideal mixtures the equilibrium constant has to be expressed in term of activities. Accurate modelling of phase equilibrium is required particularly for binaries with a strong influence on K_γ.

A.5
Reactive Azeotropes

From a mathematical viewpoint the necessary and sufficient condition for azeotropy is:

$$\frac{y_1 - x_1}{v_1 - v_t x_1} = \frac{y_i - x_i}{v_i - v_t x_i} \quad i = 2, \ldots, c-1 \tag{A.15}$$

The mathematical solution of Eq. (A.15) is tedious. An elegant graphical solution has been proposed by Stichlmair and Fair [1]. The occurrence of a reactive azeotrope is expressed geometrically by the necessary condition that the tangent to the residue (distillation) curve be collinear with the stoichiometric line. Such points form the locus of *potential reactive azeotropes*. In order to become a *true reactive azeotrope* the intersection point must also belong to the chemical equilibrium

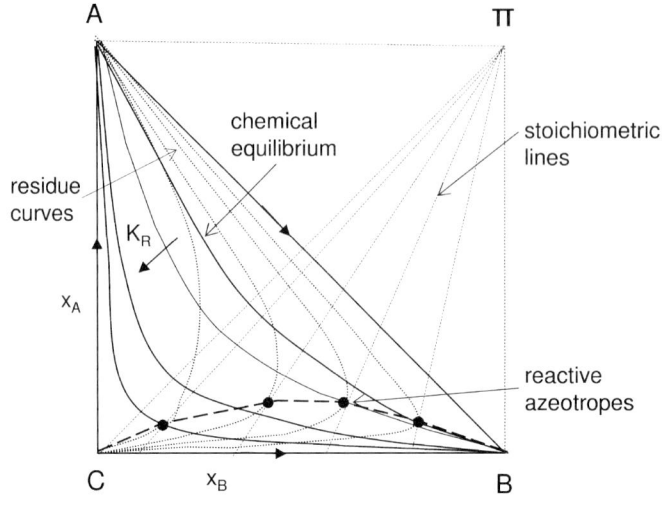

Figure A.8 Graphical identification of reactive azeotropes for the reaction $A + B \leftrightarrow C$.

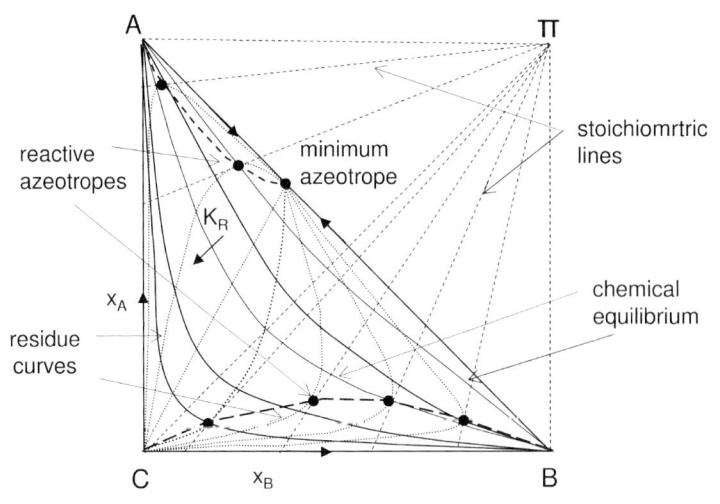

Figure A.9 Reactive azeotropes for the reaction $A + B \leftrightarrow C$ when A and B form a minimum azeotrope.

curve. Thus, reactive azeotropes can be identified by a graphical construction, as illustrated in Figure A.8 for the reaction $A + B \leftrightarrow C$. Residues curves emerge from the unstable node C, turn to the saddle B and terminate into the stable node C. Stoichiometric lines tangent can be traced from the pole π, as previously explained. The graphical construction is able to find points that satisfy the conditions of reactive azeotropy, which form an arched curve around the edge BC. It may be observed that when the residue curves are concave they cannot cross the stoichiometric lines, and as a result there are no reactive azeotropes. This would happen if B is the highest boiler and C and intermediate.

This simple graphical illustration allows the formulation of a practical rule: reactive azeotropes may occur for ideal mixtures having "segregated volatilities" (reactants either lighter or heavier with respect to products), but should not form in the case of "mixed volatilities".

A similar construction can be imagined when A and B form a minimum boiling azeotrope, as illustrated in Figure A.9. This time a second reactive azeotrope curve appears in the upper distillation region, between the AB-minimum azeotrope and the reactant A. On the other hand, when the volatility order is changed, as for example B becomes the heaviest and C the intermediate, no reactive azeotropes can be found since the residue curves and the equilibrium curves are aligned in the same direction.

References

1 Stichlmair, J. G., Fair, J. R., Distillation: Principles and Practice, Wiley-VCH, Weinheim, Germany, 1998
2 Doherty M. F., Malone M. F., Conceptual design of distillation systems, McGraw-Hill, New York, USA, 2001
3 Song, W., Venimadhavan, G., Manning, J., Malone, M. F., Doherty, M. F, *Eng. Chem. Res.*, 37, 1917–1928, 1998
4 Thiel, C., Sundmacher, K., Hoffmann, U., Residue curve maps for heterogeneously reactive distillation of fuel ethers MTBE and TAME, *Chem. Eng. Sci.*, 52, no. 6, 993–1005, 1997

Appendix B
Heat-Exchanger Design

The information given below is useful in preliminary conceptual design and flowsheeting. More theory and data can be found by consulting specialized references [1–5].

B.1
Heat-transfer Fluids

Water Critical pressure and temperature values of water are 220 bar and 373.14 °C. Steam is a valuable heating agent below 200 °C, where the saturation pressure is about 24 bar. Superheated steam can be used to enlarge the temperature range. Liquid water is excellent for cooling, but also for heating at mild temperatures below 100 °C. For higher temperatures thermal fluids are more suitable.

Salt Brines Salt brines are water solutions of inorganic salts. Aqueous $CaCl_2$ solutions of maximum 25% are recommended down to −20 °C. Salt brines are low cost but expensive in operation. Antifreezes described below are preferable.

Glycol Solutions Ethylene glycol can be used in principle down to −35 °C, but in practice is limited to −10 °C because of high viscosity. Propylene glycol has the advantage of being nontoxic. Other antifreeze fluids, such as methanol and ethanol solutions raise safety and toxicity problems.

Refrigerants Refrigerants remove heat from a body or process fluid by vaporization. Ammonia (R717) seems to be popular again after years of decline in favor of chlorinated hydrocarbons (CFCs). Because of damage to the ozone layer, the CFCs are being replaced by refrigerants based on hydrochlorofluorocarbons (HCFC), although these are not completely inoffensive. Thermodynamic properties of new HCFC can be found in the Perry's Handbook (1997). One of the most recommended is R134a for replacing R12.

Thermal Fluids By using thermal fluids the heat-transfer operations can be carried out over a larger temperature interval but at reasonable operating

Chemical Process Design: Computer-Aided Case Studies. Alexandre C. Dimian and Costin Sorin Bildea
Copyright © 2008 WILEY-VCH Verlag GmbH & Co. KGaA, Weinheim
ISBN: 978-3-527-31403-4

Table B.1 Properties of some thermal fluids.

Fluid	Composition	Temperature range	P_{sat} max. temp.	$C_p^\#$	$\rho^\#$	$\eta^\#$
		°C	bar	kJ/kgK	kg/m³	cP
Dowtherm A	$(C6H5)_2O / (C6H5)_2$	L 15–400	10.6	1.556	1062	5
		V 257–400		2.702	680	0.13
Dowtherm J	Alkylated aromatics	−80–315	11.9	1.571	933.6	9.98
				3.012	568.2	0.16
Dowtherm Q	Alkylated aromatics	L −35–330	3.4	1.478	1011	46.6
				2.586	734	0.2
Syltherm 800	Siloxane	−40–400	13.7	1.506	990	51
				2.257	547	0.25
Syltherm XLT	Siloxane	−100–260	5.2	1.343	947	78
				2.264	563	0.18

\# = values at minimum and maximum temperatures.

pressures. Table B.1 shows the properties of some thermal fluids produced by Dow Chemicals. The best known Dowtherm A, is a mixture of diphenyl oxide/diphenyl capable of working as liquid or vapor up to 400 °C at a maximum pressure of 10 bar. Other fluids are based on mixtures of heavy hydrocarbons. Silicones are excellent liquid heating/cooling media over a wide temperature range, as for example between −100 °C and 400 °C. More information can be found on the Internet sites of producers.

Inorganic Salts Several formulations are known but the most widely used salt is a molten mixture of the eutectic $NaNO_2$ (40%)/$NaNO_3$ (7%) /KNO_3 (53%), for operation between 146 °C (melting point) and 454 °C.

B.2
Heat-transfer Coefficients

The overall heat-transfer coefficient between two fluids separated by a wall is the reciprocal of the sum of the individual resistances. Partial heat-transfer coefficients depend on the hydrodynamic regime and physical properties of fluids, particularly viscosity and thermal conductivity. Table B.2 shows typical values for partial heat-transfer coefficients that can be used in preliminary design. The assumed values have to be checked by rigorous calculation in final design. Particularly attention should be given to two-phase mixtures, hydrogen-rich gases and condensation of vapors with noncondensable gases.

Table B.3 gives values of thermal resistance due to fouling. Thicker walls of stainless steel should also be included in the overall heat-transfer coefficient. The

Table B.2 Partial heat-transfer coefficients.

Fluid	h (W/m² K)	Fluid	h (W/m² K)
Gases		*Boiling liquids*	
Gases, low pressure	20–80	Boiling water	1500–2000
Gases, high pressure	100–300	Boiling organics	800–1300
Hydrogen-rich gases	80–150		
Liquids		*Condensing vapor*	
Water, turbulent regime	1500–3000	Condensing steam	4000–5000
Dilute aqueous	1000–2000	Thermal fluids	2000–3000
Solutions			
Light organic liquids	1000–1500	Organics	800–2000
Viscous organic liquids	500–800	Organics with NC	500–1500
Heavy-ends	200–500	Refrigerants	1500
Brines	800–1000		
Molten salts	500–700		

NC: noncondensables.

Table B.3 Fouling as the equivalent heat-transfer coefficient.

Fluid	Fouling (W/m² K)
Cooling water (towers)	3000–6000
Organic liquids & light hydrocarbon	5000
Refrigerated brine	3000–5000
Steam condensate	3000–5000
Steam vapor	4000–10000
Condensing organic vapors	5000
Condensing thermal fluids	5000
Aqueous salt solutions	1000–3000
Flue gases	2000–5000

combination of different situations leads to the overall heat-transfer coefficients listed in Tables B.4 to B.7. Note that fouling is included.

B.3
Shell-and-tubes Heat Exchangers

Figure B.1 shows the American TEMA (Tubular Exchanger Manufacture Association) standards that are largely accepted. The codification makes use of three letters that indicate the type of stationary head, shell and rear head, respectively. One of the most common types is AES or AEL that designate removable channel and cover (A), one-pass shell (E) and fixed tube sheet (L) or floating-rear (S). A similar

Table B.4 Overall heat-transfer coefficients for shell-and-tubes heat exchangers.

Hot fluid	Cold fluid	U (W/m²K)
Heat exchangers		
Water	Water	800–1500
Organic solvents	Organic solvents	200–500
Light oils	Light oils	100–400
Heavy oils	Heavy oils	50–300
Gases	Gases	10–50
Coolers		
Organic solvents	Water	250–750
Light oils	Water	350–900
Heavy oils	Water	60–300
Gases	Water	20–300
Water	Brine	600–1200
Organic solvents	Brine	150–500
Gases	Brine	15–250
Water	Natural gas mixture with hydrogen	500–800
Water or brine	Gases, moderate pressures	100–200
Heaters		
Steam	Organic solvents	500–1000
Steam	Light oils	300–900
Steam	Heavy oils	60–450
Steam	Gases	30–300
Dowtherm	Heavy oils	50–300
Dowtherm	Gases	20–200
Flue gases	Steam or hydrocarbon vapors	30–100
Condensers		
Steam	Water	1000–1500
Organic vapors	Water	700–1000
Organics vapors, high NC, A	Water	100–500
Organics vapors, low NC, V	Water	250–600
Thermal fluid vapors	Tall oil	300–400
Tall oil, vegetable oil vapors	Water	100–250
Vaporizers		
Steam	Aqueous diluted solutions	1000–2000
Steam	Light organics	1000–1500
Steam	Heavy organics	600–900
Evaporators		
Steam	Sea water (long tube falling film)	1500–3000
Steam	Sea water (long tube rising film)	700–2500
Steam	Sugar solution (agitated film)	1000–2000

Table B.5 Overall heat-transfer coefficients for air-cooled heat (bare tube basis).

Process fluid	U (W/m^2 K)
Water cooling	500
Light-organics cooling	400–500
Fuel-oil cooling	150
High viscous liquid cooling	40–100
Hydrocarbon gases, 3–10 bar	60–200
Hydrocarbon gases, 10–30 bar	300–400
Condensing hydrocarbons	400–600

Table B.6 Overall heat-transfer coefficients for jacketed vessels.

Jacket	Vessel	U (W/m^2 K)
Steam	Aqueous solutions	500–1000
	(glass-lined CS)	(300–500)
Steam	Light organics	250–800
	(glass-lined CS)	(200–400)
Steam	Viscous solutions	50–300
	(glass-lined CS)	(50–200)
Water, brine, thermal fluid	Aqueous solutions	250–1500
	(glass-lined CS)	(150–450)
Water, brine, thermal fluid	Light organics	200–600
	(glass-lined CS)	(150–400)
Water, brine, thermal fluid	Viscous solutions	100–200
	(glass-lined CS)	(50–150)

Table B.7 Overall heat-transfer coefficients for immersed coils in agitated vessels.

Coil	Pool	U (W/m^2 K)
Steam	Diluted aqueous solutions	500–1000
Steam	Light oils	250–500
Steam	Heavy oils	150–400
Water or brine	Aqueous solutions	400–700
Light oils	Aqueous solutions	200–300

NC: noncondensables.

type is BEM but with bonnet-type cover. The type A is preferred when fouling in tubes is likely. The types NEN designates channel integral with tube sheets and removable cover.

In preliminary design the problem is the selection of the right type of exchanger and its sizing that complies with design specifications. Conversely, the design

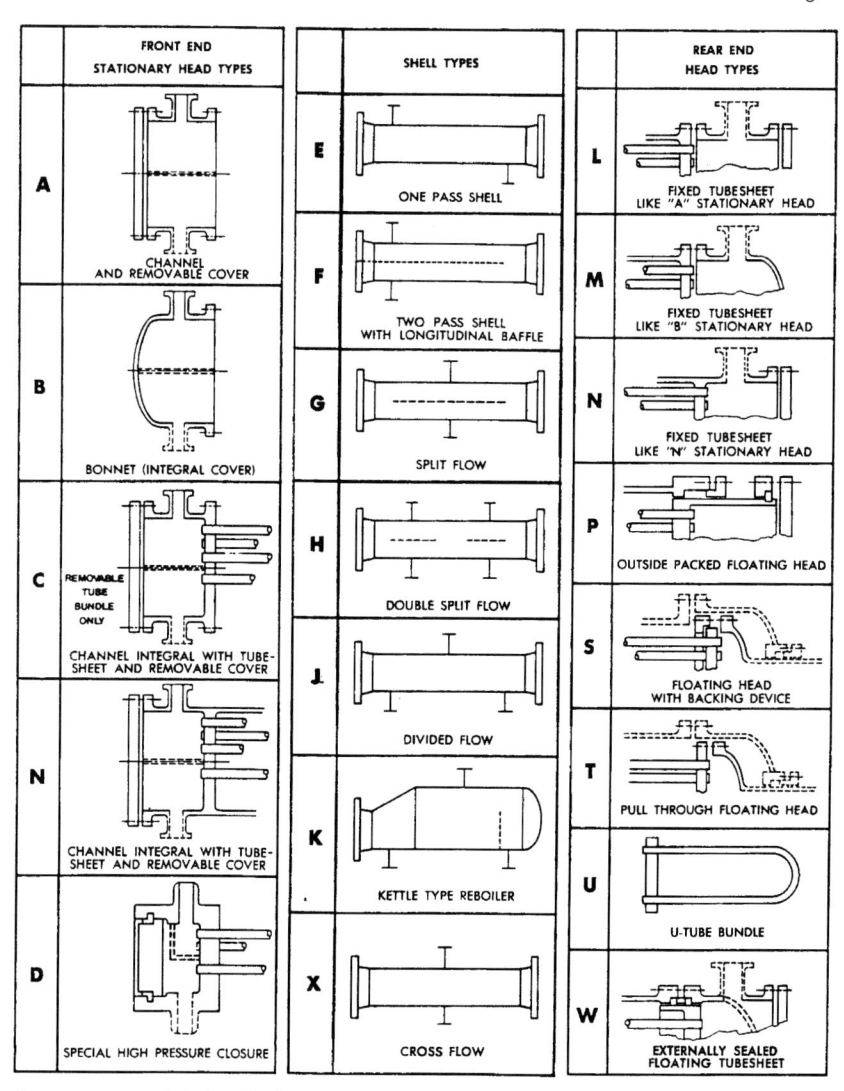

Figure B.1 Type of shell-and-tubes heat exchangers following TEMA standards [1].

should be developed so as to use standard heat exchangers as much as possible. The designer should decide which side, shell or tube, is appropriate for each fluid, and find a compromise between heat-transfer intensity and maximum pressure drop. For example, cooling water usually passes through tubes in low-pressure condensers. When the flow velocity cannot ensure high transfer then 2, 4, or 6 passes are recommended. Note that at higher pressures the tubes are more appropriate for condensing, while the cooling water is better fed in the shell side, where the fluid velocity can be manipulated by means of baffles.

Rules for fluid side selection are:

1. Corrosion: most corrosive fluid to the tube side.
2. Fouling: fouling fluids in tubes.
3. Fluid temperatures: high-temperature fluid in tubes.
4. Pressure: high-pressure fluids in tubes.
5. Pressure drop: lower pressure drop can be obtained in one or two passes.
6. Condensing steam and vapor at low pressures: shell side.
7. Condensing gas–liquid mixtures: tube side with vertical position.
8. Stainless and special steels: corresponding fluid in tubes.

Allowable pressure drop is the key design parameter. This is in general 0.5 to 0.7 bar for liquids, occasionally larger for tube-side flow, and of 0.1 bar for gases.

The shell diameter depends on the number of tubes housed, as well as the limitations set by pressure and temperature. The diameter may vary between 0.3 and 3 m. High values are valid for fixed-tube sheet. If a removable bundle is necessary then the shell diameter is limited to 1.5 m.

Tube size is designated by outside diameter (O.D.) × thickness × length. Diameters are normalized in inch or mm. Examples are 1/4, 3/8, 1/2, 5/8, 3/4, 1, $1^1/_4$, and $1^1/_2$ inches. Tubes of 3/4 in or schedule 40 (19/15 mm), as well as 1 in (25/21 mm) are the most widely used. Tube lengths may be at any value up to 12 m, the more common values being of 6, 9, and 10 m.

Triangular layout of tubes is the most encountered. The tube pitch is 1.25 times the outside diameter. Exact tube counting can be obtained by means of specialized design programs. For preliminary calculations the number of tubes can be found by means of relations based on the factor $C = (D/d) - 36$, where D and d are the bundle and outside tube diameters, respectively. The total number of tubes N_t can be calculated by means of the following relations [1]:

1 tube pass: $\quad N_t = 1298 + 74.86C + 1.283C^2 - 0.0078C^3 - 0.0006C^4$
2 tube pass: $\quad N_t = 1266 + 73.58C + 1.234C^2 - 0.0071C^3 - 0.0005C^4$
4 tube pass: $\quad N_t = 1196 + 70.79C + 1.180C^2 - 0.0059C^3 - 0.0004C^4$
6 tubes pass: $\quad N_t = 1166 + 70.72C + 1.269C^2 - 0.0074C^3 - 0.0006C^4$

As an illustration, Table B.8 shows some layouts for $3/4$ and 1 inch O.D. tubes.

Double-pipe heat exchangers are widely used for smaller flow rates. When the heat-transfer coefficient outside is too low, a solution consists of using longitudinal finned tubes as an extended surface.

B.4
Air-cooled Heat Exchangers

Air-cooled heat exchangers are employed on a large scale as condensers of distillation columns or process coolers. The approach temperature – the difference between process outlet temperature and dry-bulb air temperature – is typically of 8 to 14 °C above the temperature of the four consecutive warmest months. By air-

Table B.8 Shell and tubes layout.

Shell I.D., in	One pass		Two pass		Four pass	
	3/4 in O.D.	1 in O.D.	3/4 in O.D.	1 in O.D.	3/4 in O.D.	1 in O.D.
8	37	21	30	16	24	16
12	92	55	82	52	76	48
15 1/4	151	91	138	86	122	80
21 1/4	316	199	302	188	278	170
25	470	294	452	282	422	256
31	745	472	728	454	678	430
37	1074	674	1044	664	1012	632

humidification this difference can be reduced to 5 °C. Air-cooled heat exchangers are manufactured from finned tubes. A typical ratio of extended to bare tube area is 15:1 to 20:1. Finned tubes are efficient when the heat-transfer coefficient outside the tubes is much lower than inside the tubes. The only way to increase the heat transferred on the air side is to extend the exchange area available. In this way the extended surface offered by fins significantly increases the heat duty. For example, the outside heat-transfer coefficient increases from 10–15 W/m² K for smooth tubes to 100–150 or more when finned tubes are used. Typical overall heat-transfer coefficients are given in Table B.5. The correction factor F_T for LMTD is about 0.8.

B.5
Compact Heat Exchangers

Compact heat exchangers (CHEs) are characterized by high efficiency in reduced volume, but much higher cost. If area density of the shell-and-tubes heat exchangers can achieve 100 m²/m³, the compact heat exchangers have significantly higher values, between 200 and 1500 m²/m³. However, because of higher cost, the CHEs are employed only in special applications. Some common types are briefly presented.

B.6
Plate Heat Exchangers

Plate heat exchangers are intensively used in the food and pharmaceutical industries, but less so in chemical industries. Because of the small cross section, intensive heat transfer can be realized, as for example from 400 W/m² K with viscous fluids up to 6000 W/m² K for water. Gasket plate devices are the most common. The effective area per plate can be larger than 1 m². Up to 400 plates can be assembled in a frame. However, the operation is limited to 30 bars and 250 °C. Welded plate heat exchangers are similar. The operation can rise to 80 bars and 500 °C, but cleaning is problematic.

Plate–fin heat exchangers are manufactured by assembling plates separated by corrugated sheets, which form the fins. The plates are made from aluminum sealed by brazing. The operation of these devices requires clean fluids. The main applications can be found in cryogenic and natural-gas liquefaction for pressures and temperatures up to 60 bars and 150 °C.

B.7
Spiral Plate Exchangers

Spiral plate exchangers are usually of two types: spiral-plate and spiral-tubes. Very intensive heat transfer can be achieved, with transfer area per unit up to 250 m^2. The operation is limited to 20 bars and 400 °C. However, the cost of such devices is high.

References

1 Perry's, Chemical Engineers Handbook, 7th edn, McGraw-Hill, New York, USA, 1997
2 Sinnot, R.K, Coulson & Richardson's Chemical Engineering, vol. 6, Butterworth-Heinemann, 2nd revision, 1996
3 Linnhoff, B., D. W. Townsend, D. Boland, G. F. Hewitt, B. Thomas, A. R. Guy, R. H. Marsland, User Guide on Process Integration for the Efficient Use of Energy, The Institution of Chemical Engineers, 2nd edn, 1994
4 McCabe, W. L., Smith, J. C., Harriott, P., Unit Operations of Chemical Engineering, 7th edn, McGraw-Hill, New York, USA, 2005
5 Ludwig, E. E., Applied Process Design, vol. 3, 3rd edn, Butterworth-Heinemann, 1999

Appendix C
Materials of Construction

Selection of materials is an important issue in equipment design and costing evaluation. More information can be found in dedicated engineering handbooks [1–3]. The following criteria should be kept in view:

1. Resistance to process conditions, such as corrosion, erosion, stress and temperature.
2. Mechanical processing properties of manufacturing equipment parts, such as sheets, piping, profiles, including welding properties.
3. Contamination regarding the interaction between material and fluid, and its consequence on process, such as for example catalyst deactivation.
4. The cost of materials and of processing.

The materials of construction can be classified as follows:

1. Ferrous metals and alloy steels.
2. Nonferrous metals and alloys.
3. Inorganic nonmetallic, as glass and glassed steel, porcelain and stone.
4. Plastic and thermoplastic materials.

Here we refer only to the first two categories that dominate process equipment manufacture, more particularly to ferrous metals and alloys.

Resistance to fluid corrosion is by far the most important aspect. This topic is discussed in detail in the mentioned references, where extensive corrosion tables are given. From this point of view process fluids can be classified as follows:

1. Hydrocarbon mixtures, usually nonaggressive for carbon and low-alloy steel, but aggressive for plastics.
2. Nonoxidizing and reducing media
   a) Acid solutions excluding hydrochloric, phosphoric, sulfuric.
   b) Neutral solutions, such as nonoxidizing salts, chlorides, sulfates.
   c) Ammonium hydroxide and amines.
3. Oxidizing media
   a) Acid solutions, e.g. nitric acid,
   b) Neutral or alkaline solutions, e.g. persulfates, peroxides, chromates,
   c) Pitting media, such as acidic ferric chloride solutions.

4. Neutral waters
   a) Fresh water supply, slow moving or turbulent.
   b) Seawater, low moving or turbulent.
5. Gases
   a) Steam, dry or wet.
   b) High-temperature (furnace) gases with oxidizing effect.
   c) Hydrogen-rich gases with pitting effect.
   d) Halogens and halide acids.

Besides fluid corrosion particular attention has to be given to extreme temperature conditions in which the equipment should work or ensure safety operation. Temperatures higher than 600 °C require special high-temperature steels. The same is valid for cryogenic conditions, where material still should have good toughness at low temperatures. The mechanical strength of steels degrades significantly with increasing temperature, particularly near to the creep limit.

When wall thickness is calculated, care should be given to correct estimation of the tensile strength (stress) at the design temperature. This aspect is important when selecting high-alloy and special steels that are much more expensive than carbon steel and mild alloy (see Table C.2). Note that the thermal conductivity of stainless steels is considerably lower than of carbon steel, implying much higher heat-transfer resistance. For example, the thermal conductivity of carbon steel is about 50 W/m K, but only 10–15 W/m K for stainless steels.

Common metallic materials of constructions are discussed briefly below. Specific information about properties and applications can be found in quality standards practiced in each country, such as for example SAE, AISI, ASTM (USA), BS (UK), DIN (Germany), NF (France), UNI (Italy), NEN (The Netherlands), etc.

C.1
Iron and Carbon Steel

Low carbon steel (CS) is the most used and cheapest material. It is suitable for hydrocarbon and organic solvents, except chlorinated solvents. Concentrated sulfuric acid and caustic alkalis can be stored occasionally in CS. High-silicon irons can be used for storing acids. Surface treatment (quenching, tempering, nitriding, age hardening) can be used to improve the resistance to corrosion or higher temperatures. Coating with protected sheets of more resistant materials can also be applied, leading to important cost reduction.

C.2
Stainless Steels

A wide range is available in a great variety of compositions that can be tailored for specific applications. According to their microstructure the stainless steels can be classified into three main categories:

1. Ferric steels, with 13–20% Cr, <0.1% C, but no nickel.
2. Austenitic steels, with 18–29% Cr, >7% Ni.
3. Martensitic, with 10–12%Cr, 0.2–0.4%C, and up to 2% Ni.

The class of austenitic stainless steels is by far the most widely used. Adding nickel considerably improves the resistance to corrosion, which can be enhanced further by adding molybdenum. Both ferric and austenitic alloys cannot be hardened by heat treatment. Among different types of stainless steels the most used in process industries are of austentic types. The main are briefly described.

1. Type 18Cr/8Ni SS type (304 in USA, 801B in UK, and by X5-CrNi 18/9 in Germany) is the most used in process industries. It is suitable for most applications except furnace gases, see water and pitting media. It has a fair to good resistance to acidic solutions. The low-carbon variant 304L is recommended for thicker welded parts. The version 321 stabilized with titanium is more suitable for higher temperatures.
2. Type 18Cr/8Ni/2Mo (316 in USA, 845B in UK, X5-CrNiMo 18/12 in Germany) where molybdenum has been added to improve the resistance to reducing media, and dilute acidic solutions, is almost a universal steel. However, mechanical processing is more difficult. For welding thick parts the low-carbon version 316L is more suitable.

Table C.1 shows some typical stress values for mechanical design as a function of temperature. A higher safety factor is taken into account at higher temperatures compared with the nominal tensile strength. Although the corrosion resistance of SS 304 and 316 types is good, their thermal resistance above 600 °C is not satisfactory. Heat-resistant steels for furnaces and cracking units are discussed in Ullmann [1].

C.3
Nickel Alloys

Three classic nickel alloys are the best known in process industries. Monel, a nickel-copper alloy in the ratio 2:1, is, after stainless steel, the most popular. It

Table C.1 Typical design stress [3].

Material	Tensile strength (N/mm^2)	Design stress at temperature °C in N/mm^2				
		100	200	300	400	500
Carbon steel	360	125	105	85	70	
Low alloys steel	550	240	240	235	220	170
Stainless steel series 340 (18Cr/8Ni)	510	145	115	105	100	90
Stainless steel series 316 (18Cr/8Ni/2Mo)	520	150	120	110	105	95

Table C.2 Relative cost of material of construction.

Carbon steel	1
Low alloy steels (Cr–Mo)	2
Austentic steel type 304 (18/8 CrNi)	2–6
Austentic type 316	2.5–8
Copper	5.7
Monel	25
Titanium	26

stands better than stainless steel against ferric chloride, acidic solutions, alkalis, reducing media and sea water. Inconel (76%Ni, 7% Fe, 15% Cr) can be used for high-temperature applications and furnace gases (sulfur free). Hastelloy alloys are high resistant to strong mineral acids, such as HCl, and strong oxidizing media. However, all these materials are more expensive than the stainless steel, so their use should be restricted as strictly necessary.

Table C.2 gives a price scale with reference to carbon steel. The price of steel is very variable depending on country, process, shape, etc. A reasonable composite steel price for process industries purposes at 2005 year level was about 600€/tonne.

References

[1] Ullmann's Encyclopaedia of Industrial Chemistry, Construction Materials in Chemical Industry, by H. Gräfen, Wiley-VCH, Weinheim, Germany, 2002

[2] Perry's, Chemical Engineers Handbook, 7th edn, McGraw-Hill, New York, USA, 1997

[3] Sinnot, R.K, Coulson & Richardson's Chemical Engineering, vol. 6, Butterworth-Heinemann, 2nd revision, 1996

Appendix D
Saturated Steam Properties

		Specific volume			Enthalpy		Entropy		
T °C	P kPa	V_l m³/kg	V_v m³/kg	H_l kJ/kg	ΔH_v kJ/kg	H_v kJ/kg	S_l kJ/kg K	ΔS_v kJ/kg K	S_v kJ/kg K
0	0.6106	0.000999	206.42	−0.017	2501.3	2501.3	−0.0031	9.1572	9.1541
5	0.8724	0.001000	147.09	20.432	2490.2	2510.6	0.07178	8.9528	9.0246
10	1.2287	0.001000	106.3	41.105	2478.8	2520.0	0.14592	8.7545	8.9005
15	1.7071	0.001001	77.846	61.946	2467.2	2529.2	0.21916	8.5624	8.7815
20	2.3414	0.001002	57.726	82.909	2455.5	2538.4	0.29143	8.3761	8.6676
25	3.1728	0.001003	43.313	103.96	2443.5	2547.5	0.36265	8.1957	8.5583
30	4.2505	0.001004	32.862	125.06	2431.5	2556.6	0.43278	8.0208	8.4536
35	5.6330	0.001006	25.196	146.18	2419.4	2565.6	0.50180	7.8513	8.3532
40	7.3890	0.001008	19.511	167.32	2407.2	2574.5	0.56971	7.6871	8.2568
50	12.355	0.001012	12.028	209.56	2382.7	2592.2	0.70220	7.3733	8.0755
60	19.946	0.001017	7.6696	251.71	2358.0	2609.7	0.83041	7.0779	7.9083
70	31.196	0.001023	5.0418	293.76	2333.1	2626.9	0.95460	6.7991	7.7537
80	47.404	0.001029	3.4067	335.73	2308.0	2643.7	1.07510	6.5354	7.6105
90	70.169	0.001036	2.3600	377.66	2282.5	2660.2	1.19220	6.2853	7.4774
100	101.33	0.001044	1.6722	419.61	2256.5	2676.1	1.3062	6.0473	7.3535
110	143.38	0.001052	1.2095	461.65	2230.0	2691.6	1.4175	5.8201	7.2376
120	198.70	0.001061	0.8912	503.85	2202.7	2706.5	1.5263	5.6026	7.1289
130	270.34	0.001070	0.6680	546.26	2174.5	2720.7	1.6328	5.3937	7.026.S
140	361.64	0.001080	0.5084	588.93	2145.2	2734.2	1.7373	5.1924	6.9297
150	476.30	0.001091	0.3924	631.92	2114.8	2746.8	1.8399	4.9979	6.8378
160	618.38	0.001102	0.3068	675.25	2083.2	2758.4	1.9408	4.8093	6.7501
170	792.32	0.001114	0.2426	718.96	2050.0	2769.0	2.0401	4.626	6.6661
180	1002.9	0.001127	0.1939	763.07	2015.4	2778.4	2.1379	4.4474	6.5853
190	1255.2	0.001141	0.1564	807.60	1979.0	2786.6	2.2343	4.2729	6.5072
200	1554.7	0.001156	0.1273	852.59	1940.8	2793.3	2.3295	4.1018	6.4312
210	1907.3	0.001172	0.1044	898.05	1900.5	2798.6	2.4235	3.9337	6.3572
220	2319.1	0.001189	0.0861	944.05	1858.2	2802.2	2.5166	3.768	6.2845
230	2796.6	0.001208	0.0715	990.63	1813.4	2804.0	2.6088	3.6041	6.2129
240	3346.6	0.001228	0.0597	1037.9	1766.0	2803.9	2.7005	3.4415	6.1420
250	3976.2	0.001251	0.0501	1086.0	1715.7	2801.6	2.7918	3.2795	6.0714
260	4692.8	0.001275	0.0421	1135.0	1662.1	2797.1	2.8831	3.1175	6.0006

Chemical Process Design: Computer-Aided Case Studies. Alexandre C. Dimian and Costin Sorin Bildea
Copyright © 2008 WILEY-VCH Verlag GmbH & Co. KGaA, Weinheim
ISBN: 978-3-527-31403-4

Appendix D Saturated Steam Properties

T °C	P kPa	Specific volume			Enthalpy		Entropy		
		V_l m³/kg	V_v m³/kg	H_l kJ/kg	ΔH_v kJ/kg	H_v kJ/kg	S_l kJ/kg K	ΔS_v kJ/kg K	S_v kJ/kg K
270	5504.3	0.001302	0.0356	1185.1	1604.8	2789.9	2.9745	2.9546	5.9291
280	6418.7	0.001332	0.0301	1236.6	1543.3	2779.8	3.0665	2.7899	5.8565
290	7444.6	0.001366	0.0255	1289.6	1476.8	2766.4	3.1595	2.6224	5.7819
300	8591.0	0.001404	0.0216	1344.7	1404.6	2749.3	3.2538	2.4507	5.7045
310	9867.7	0.001447	0.0183	1402.0	1325.6	2727.6	3.3499	2.2732	5.6232
320	11286	0.001499	0.0155	1462.0	1238.4	2700.4	3.4484	2.0879	5.5362
330	12858	0.001560	0.0130	1525.4	1140.7	2666.1	3.55	1.8913	5.4413
340	14599	0.001637	0.0108	1593.7	1028.6	2622.2	3.6572	1.6775	5.3347
350	16527	0.001740	0.0088	1669.8	894.27	2564.0	3.7744	1.4351	5.2094
360	18668	0.001895	0.0069	1760.9	720.17	2481.1	3.9133	1.1374	5.0507
370	21052	0.002219	0.0049	1895.5	438.99	2334.5	4.1202	0.6826	4.8028
374	22090	0.003155	0.0031	2099.3	0	2099.3	4.4298	0	4.4298

Appendix E
Vapor Pressure of Some Hydrocarbons

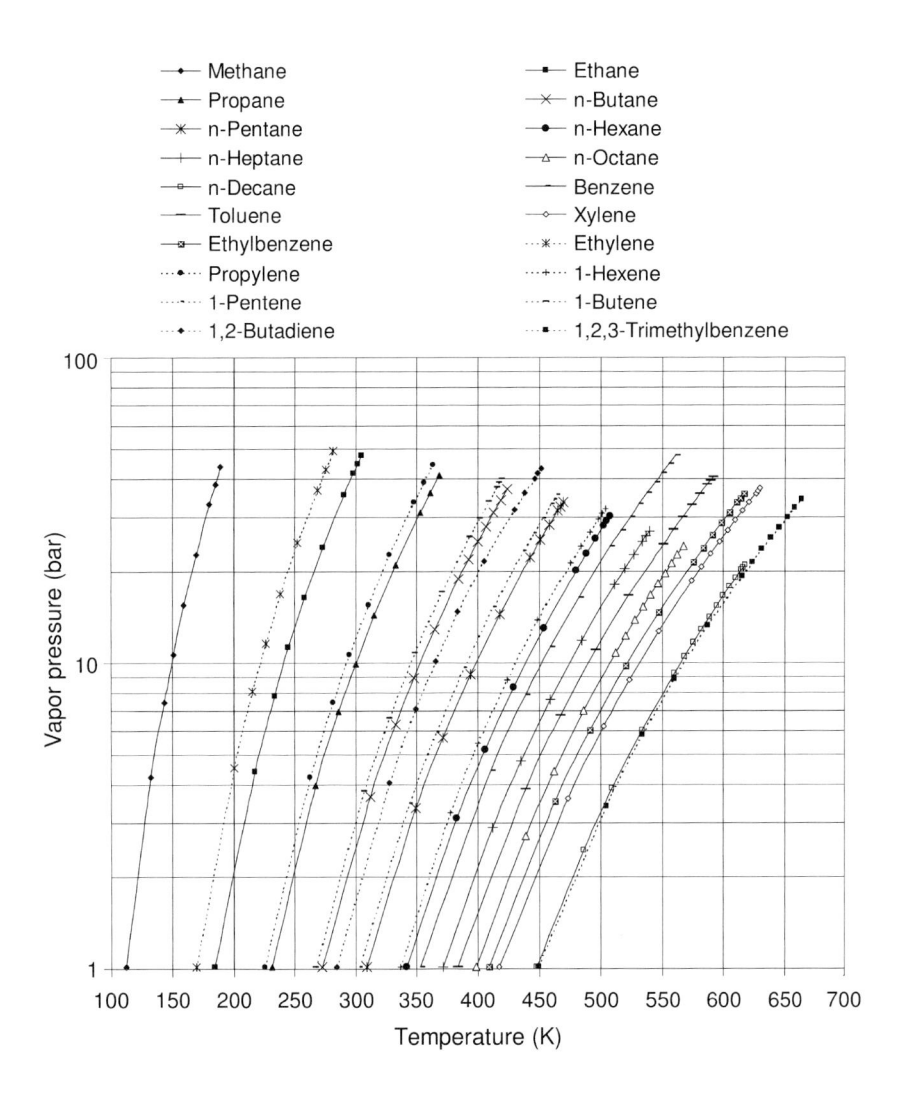

Appendix F
Vapor Pressure of Some Organic Components

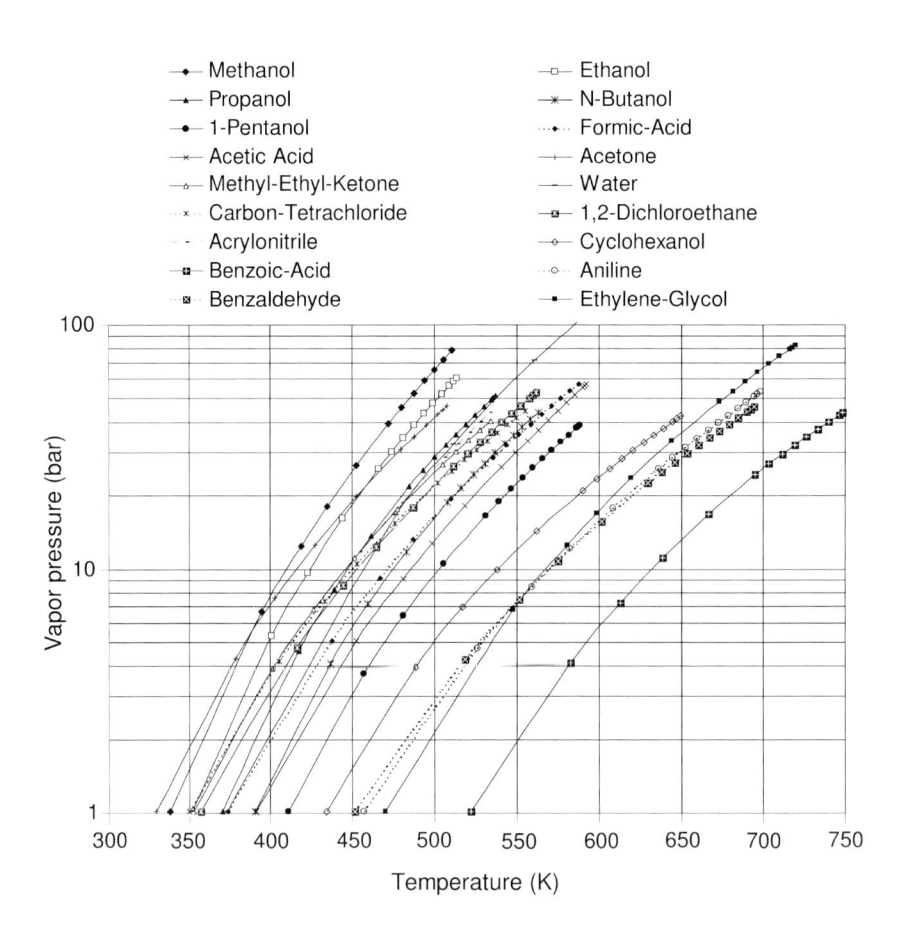

- Methanol
- Propanol
- 1-Pentanol
- Acetic Acid
- Methyl-Ethyl-Ketone
- Carbon-Tetrachloride
- Acrylonitrile
- Benzoic-Acid
- Benzaldehyde
- Ethanol
- N-Butanol
- Formic-Acid
- Acetone
- Water
- 1,2-Dichloroethane
- Cyclohexanol
- Aniline
- Ethylene-Glycol

Chemical Process Design: Computer-Aided Case Studies. Alexandre C. Dimian and Costin Sorin Bildea
Copyright © 2008 WILEY-VCH Verlag GmbH & Co. KGaA, Weinheim
ISBN: 978-3-527-31403-4

Appendix G
Conversion Factors to SI Units

To convert from unit	To SI unit	Multiply by
Length		
in	m	2.5400×10^{-2}
ft	m	0.3048
Mass		
lb	kg	0.45359
Force		
lbf	N	4.4482
kgf	N	9.81
Temperature		
°C	K	°C + 273.15
°F	K	(°F + 459.67)/1.8
°R	K	°R/1.8
Pressure		
in. of water (60 °F)	Pa	2.4884×10^{2}
atm	Pa	1.0133×10^{5}
psi	Pa	6.8948×10^{3}
torr (mmHg, 0 °C)	Pa	1.3332×10^{3}
Volume		
ft^3	m^3	2.8317×10^{-2}
in^3	m^3	1.6387×10^{-3}
gal	m^3	3.7854×10^{-3}
bbl (42 gal)	m^3	0.15899
Density		
lb/in^3	kg/m^3	2.7680×10^{4}
lb/ft^3	kg/m^3	16.018

Chemical Process Design: Computer-Aided Case Studies. Alexandre C. Dimian and Costin Sorin Bildea
Copyright © 2008 WILEY-VCH Verlag GmbH & Co. KGaA, Weinheim
ISBN: 978-3-527-31403-4

Appendix G Conversion Facturs to SI Units

To convert from unit	To SI unit	Multiply by
Energy		
Btu	J	1.0544×10^3
Btu/lb	J/kg	2.3244×10^3
Btu/lb/°F	J/kg/K	4.1840×10^3
kcal	J	4180
Power		
HP (550 ft.lbf/s)	W	7.457×10^2
Btu/hr	W	0.2931
kcal/h	W	1.1622
ton of refrigeration	kW	3.517
Heat-transfer coefficient		
Btu/h/ft^2/°F	W/m^2/s	5.6783
kcal/h/m^2/°C	W/m^2/K	1.162
Pressure drop		
psi/ft	kPa/m	2.2621×10^1

Index

a

absorber–bioreactor–recycle model   354 f
absorption
– ethanol from lignocellulosic biomass   456
– NO_x removal   342 f, 351 f, 356
– separation system   50
acetaldehyde   202
acetaldehyde/acetic anhydride route   287
acetate   449
acetic acid, sustainability metrics   12
acetic acid/acetylene route   287
acetone/chloroform/toluene separation   92
acetone manufacturing   173
acetylene   202
acid-catalyzed transesterification   417
acid concentration   281
acid hydrolysis   451
acid-soluble oil (ASO)   263
acid-to-alcohol ratio   242
acidic quench   321
acidification potential (AP)   167
acidity number   405
acroleine   316
acrylonitrile (ACN)
– propene ammoxidation   313–338
– propylene ammoxidation   37
– sustainability metrics   12
activation energy   140, 266
active carbon catalysts   137
active centers   315
additional cooling capacity   389, 392
adiabatic PFR
– cumene synthesis   184
– ethylene chlorination   212
– phenol hydrogenation   145
adiabatic temperature change
– cumene synthesis   180
– reactor-design   47

adiabatic tubular reactor   123
adipic acid production   129
adsorption
– liquid-separation system   73
– phenol hydrogenation   140
– separation system   64 ff, 69
aerobic digestion   456
agglomerate   369
agricultural land efficiency   434
agricultural residues   399
air emissions   334
alcohol–water–entrainer   256
alcohols
– biodiesel manufacturing   406
– fatty-ester synthesis   232
– lauric acid esterification   249, 256
alkaline hydrolysis   451
alkanes   256
alkylation
– benzene   173–200
– cumene synthesis   176
– isobutane   262
alternative fuels types   399
alumina catalysts   137
Amberlyst   231, 404, 419
ammonia   314 ff
ammonia fiber explosion (AFEX)   451
ammoxidation   313–338
animal lipids   404
antipollution measures   7
Antoine equation   236
applications
– biodiesel manufacturing   423
– PVC   364, *see also* examples
aquatic toxicity potential (ATP)   167
arabinan   449
aromatics
– alkylation   178
– lauric acid esterification   256

Chemical Process Design: Computer-Aided Case Studies. Alexandre C. Dimian and Costin Sorin Bildea
Copyright © 2008 WILEY-VCH Verlag GmbH & Co. KGaA, Weinheim
ISBN: 978-3-527-31403-4

– separation system   67, 80 f
aspect ratio, isobutane alkylation   281
Aspen Plus
– acrylonitrile production   318, 324
– biofuel manufacturing   420, 449
– cumene synthesis   195
– EDC pyrolysis   212
– fatty-ester synthesis   235
– hydrodealkylation   123
– isobutane alkylation   280
– lauric acid esterification   251
– phenol hydrogenation   134
– separation system   92
asymmetry index   377
auxiliary materials   6, 33
azeotropes
– acrylonitrile production   318, 324
– fatty-ester synthesis   234
– lauric acid esterification   255
– phenol hydrogenation   131, 140
– ternary systems   86
– vinyl acetate monomer process   297
azeotropic distillation
– liquid-separation system   73
– separation system   82 ff, 88, 95
– vinyl acetate monomer process   293

b

Bacillus azotoformans   354
backbiting   372
bacteria fermentation   441
bagasse   445
balance equations, NO_x removal   348
balance yields   8
balanced vinyl chloride monomer process   202
base-catalyzed transesterification   417 f
basic flowsheet structures (BFSs)   3, 103 ff, *see also* flowsheet
basic quench   321
basis of design
– benzene alkylation   173
– isobutane alkylation   263
– NO_x removal   341
– phenol hydrogenation   129
– process synthesis   27
– vinyl chloride monomer process   201, 287
batchwise biodiesel manufacturing   409, 444
batchwise suspension polymerization (S-PVC)   364
Bayer-type catalyst   290
Benson method   135

benzene
– alkylation to cumene   173–200
– physical properties   133
– separation system   67
– toluene conversion   52, 82
best available technique (BAT)   334
beta factor   139
binary interaction parameters, phenol hydrogenation   141
biochemical ethanol refinery   438
biochemical NO_x removal   339–362
biochemical oxygen demand (BOD5)   335
BioDeNO$_x$ process   343, 356
biodiesel manufacturing   399–430
bioethanol manufacturing   400, 429–460
biofuels   399
biogas   67
biomass   436, 450
bioreactor   354 f
biorefinery   437
biotriglycerides   415
bleed streams   34
blending   262, 430
block diagram, *see* flowsheet
boilers   89
boiling points
– acrylonitrile production   317
– biodiesel manufacturing   406
– cumene synthesis   176
– fatty-ester synthesis   235
– hierarchical approach   34
– hydrocarbon   81
– isobutane alkylation   265
– lauric acid esterification   255
– phenol hydrogenation   140
– split sequencing   64
– vinyl acetate monomer process   293
– vinyl chloride monomer process   205
boundary conditions   344
boundary value method   98
BTX C6–C8 aromatics   80, 439
building blocks, bioethanol manufacturing   439
bulk aromatics separation   80
bulk gas/liquid   351 f
bulk liquid separation system   73
bulk PVC polymerization   364
butadiene   202
butene   265 ff, 275
tert-butyl peroxineodecanoate (TBPD)   379
byproducts   6, 32, 36
– isobutane alkylation   264
– physical properties   205

c

C–C bonds   209 ff
C1–C5 components   439
C4–C16 isoparaffins   280
C6–C8 aromatics   80, 439
C-hexanol/hexanone
– one-reactor material balance   170
– phenol hydrogenation   163
– physical properties   132, 140
C-hexene   170
Candida utilis   440
ε-caprolactam production   129
carbohydrates   436
carbon deposit formation   211
carbon dioxide
– acrylonitrile production   323
– bioethanol manufacturing   434
– emissions   6, 399
– PVC manufacturing   364
– separation system   67
– vinyl acetate monomer process   288
carbon fibers catalysts   137
carbon monoxide   364, 399
carbon residue   405
carbonyl   324
Carnot cycle   331
cascade configuration   386
catalysts
– acrylonitrile production   313
– aromatics alkylation   178
– biodiesel manufacturing   404, 414 ff, 418
– cumene synthesis   181
– fatty-ester synthesis   231 ff, 243
– hierarchical approach   30
– isobutane alkylation   263
– NO_x removal   341
– operating point   275
– phenol hydrogenation   130, 137 ff
– propene ammoxidation   335
– vinyl acetate monomer process   287, 297
– vinyl chloride monomer process   206
catalytic continuous processes   411
catalytic cracking   50
catalytic distillation
– cumene synthesis   196
– fatty-ester synthesis   231–260
catalytic hydrogenation   65
catalytic oxidation   65
catalytic reforming   80
cellobiose conversion   442
cellulose   436
cellulosic ethanol production   433
centrifugation   424
cetane number (CN)   405

chain transfer   288, 372, 382
chelating agents   339
chemical absorption   50
chemical components   450
chemical equilibrium   31
– benzene alkylation   181
– biodiesel manufacturing   402
– fatty-ester synthesis   236, 241
– phenol hydrogenation   133
chemical oxygen demand (COD)   335
chemical reactions   29
– benzene alkylation   176
– fatty-ester synthesis   231 ff
– phenol hydrogenation   132
– propene ammoxidation   319 f
– RSR structures   41
– vinyl chloride monomer process   205
chemical routes, phenol hydrogenation   130
chemical treatment   65
chemical/physical equilibrium (CPE)   47, 233, 236
chemical-based cleaning processes   339
chemically enhanced distillation   79
chemistry
– biodiesel manufacturing   402
– isobutane alkylation   264
– process synthesis   25, 29
– propene ammoxidation   314
chlorinated hydrocarbons   211
chlorine
– EDC pyrolysis   213
– physical properties   205
– PVC manufacturing   364
– vinyl chloride monomer process   202
chloroethane   67
chloroform   211
chloroform/toluene/acetone separation   92
chloromethyl-propane   256
climate conditions   27
closed-loop instability   107
cloud point   405
coal liquefaction (CTL)   400
coalescence separator   412
coconut   404
cocurrent flows   99
coil-like jacket   390
coke formation   405
cold filter plugging point (CFPP)   405
cold stream table, cumene synthesis   187
columns
– acrylonitrile production   324
– complex   77
– EDC pyrolysis   216
– ethanol from lignocellulosic biomass   456

– fatty-ester synthesis 246
– isobutane alkylation 280
– NO$_x$ removal 354
– reactive distillation 99
– vinyl chloride monomer process 207, see also distillation columns
combustion
– acrylonitrile production 313, 334
– bioethanol manufacturing 430
– oxychlorination 209
– PVC manufacturing 364
– vinyl acetate monomer process 288
component classification 35
component split matrix 326
composite curves
– biodiesel manufacturing 420
– cumene synthesis 188, 191
– isobutane alkylation reactor inlet 277
computer simulation, see simulation
concentration profiles
– lauric acid esterification 241, 253, 256
– NO$_x$ removal 348, 355 ff
– phenol hydrogenation 147
conceptual design 13, 18, 22
– reactive distillation columns 100
condenser
– PVC manufacturing 388
– separation system 64, 69 ff
– vinyl chloride monomer process 219
conductivity 386
consolidated bioprocessing (CBP) 447
contamination loss reactions 456
continuous PFR-like reactors 421
continuous stirred-tank reactors (CSTR) 44
– biofuel manufacturing 421, 444
– isobutane alkylation 270
– separator/recycle system 108
control 1, 17, 23
– one-reactor process 162
– phenol hydrogenation 161
– process synthesis 58
– PVC manufacturing 385, 393
– reactor-inlet stream fixing 112
– RSR systems 105 ff
– selectivity 45, see also plantwide control
convection-mass-transfer reaction 351
convergence requirements 217
conversion
– acrylonitrile production 313, 320
– biofuel manufacturing 403, 442, 455
– hydrodealkylation 124
– isobutane alkylation 279
– phenol hydrogenation 133
– PVC manufacturing 365, 375–394
– RSR systems 114, see also per-pass conversion
cooling
– heat exchanger 278, 389
– phenol hydrogenation 146
– PVC manufacturing 383, 388, 391, 395
copper catalyst 209
corn 436, 447
costs 10, 95
countercurrent flow 47, 99
CPE diagram, lauric acid esterification 252
cracking section control, vinyl chloride monomer plant 223
critical conversion, see conversion
critical manifolds 274
crosscurrent flows 47
cryogenic acrylonitrile production 331
cryogenic distillation 64, 69
cumene, benzene alkylation 173–200
cyanide 335
cyano–acroleine 324
cyclohexanone hydrogenation 129–172
cyclone system 335

d

Damköhler number 110 ff, 355
dead polymers 373, 373, 382
defect chain structure 372
degree-of-freedom analysis 105, 262
dehydration 457
dehydrogenation 135, 151
densities
– isobutane alkylation 266
– PVC manufacturing 386
design 3, 15, 35, 58
– acrylonitrile separation 328
– fatty-ester synthesis 251
– hierarchical approach 21
– phenol conversion 147 ff
– PVC polymerization reactors 394
– RSR systems 105
– separation system 98
destination code 35
detailed engineering 18
detergents manufacture 231
di(sec-butyl) peroxidicarbonate 379
diesel 261, 399
diffusion
– catalyst effectiveness 243
– cumene synthesis 182
– NO$_x$ removal 343 ff, 347
– phenol hydrogenation 139
– PVC manufacturing 374

di-isoalkyl sulfates   280
dilute separations   73
dimerization   177, 262
dimethyl sulfoxide (DMSO)   82
DIPB formation   174–198
direct chlorination   206
direct sequence separations   215
discretized model, NO_x removal   353
disproportionation   372, 382
dissociation
– lignocellulose saccharification   444
– PVC manufacturing   375
DIST–CAN, acrylonitrile production   330
distillation
– double-effect   189
– ethanol from lignocellulosic biomass   456
– isobutane alkylation   280
– liquid-separation system   71 ff
– phenol hydrogenation   130
– vinyl chloride monomer   215
distillation columns
– acrylonitrile production   324
– cumene synthesis   185
– EDC pyrolysis   216
– phenol hydrogenation   153
– separation system   50, see also columns
distillation line map (DLM)   86
divided wall column (DWC)   57, 82
dodecane   265
dodecanoic acid esterification
– biodiesel manufacturing   419
– lauric acid   252
– zirconia catalyst   242
Doherty–Malone method   98 f
Dow-Kellog technology   175
dynamic simulation
– isobutane alkylation   281
– PVC manufacturing   390
– vinyl acetate monomer process   310
– vinyl chloride monomer process   222, see also simulation

e

economics   4, 12 f, 27, 36, 49, 401
– bioethanol manufacturing   431 f
– hierarchical approach   21, 55
– PVC manufacturing   363 f
– vinyl chloride monomer process   202
Eerbeek   343, 354
efficient heat transfer   366
effluents   265
eigenvalues   349
electron donor   340

Eley–Rideal (ER) mechanism   182
emissions
– biodiesel manufacturing   399
– propene ammoxidation   334 ff
Emmen denitrifying   343
emulsion polymerization   364
end-of-pipe antipollution   7
endothermic reactions   47
energy balance
– isobutane alkylation   278
– PVC manufacturing   383
energy consumption
– acrylonitrile production   329
– cumene synthesis   189
– phenol hydrogenation   129
energy integration   10
– benzene alkylation   187
– biodiesel manufacturing   413
– cumene synthesis   194
– hierarchical approach   23 ff
– phenol hydrogenation   156
– process synthesis   56
– vinyl acetate monomer process   302
– vinyl chloride monomer process   219
enhanced distillation   79
enrichment   64
enthalpy   132
entrainers
– bioethanol manufacturing   446
– fatty-ester synthesis   255 f
– separation system   89
environmental protection   4 ff, 26 ff
– bioethanol manufacturing   434
– phenol hydrogenation   166
enzymatic processes   415, 442
equilibrium adsorption   65
equilibrium constant
– dodecanoic acid esterification   242
– lauric acid esterification   237
– NO_x removal   343
equilibrium design   48, 234, 238
equipment design   4
Escherichia coli   441
ESTERFIP–H process   412
esterification
– biodiesel manufacturing   414
– lauric acid   235, 251 ff, 256
ethanol   92, 400, 429–460
ethanol-*tert*-butyl-ether (ETBE)   400 f, 431
ethers   256
ethylene   205, 212, 288 f
ethylene dichloride (EDC)
– physical properties   205
– pyrolysis   212

– thermal cracking   210
– vinyl chloride monomer process   202 ff
ethylhexanol   235
eutectic salts   158
examples
– acetone/chloroform/toluene separation   92
– acrylonitrile/propylene ammoxidation   37
– aromatics separation   80 f
– landfill gas (LFG) separation   67
– phenone acetylation   11
– propyl laurate ester   408
– rapeseed   423
– toluene hydrodealkylation (HDA)   52, 122
exothermal reactions   45 ff
– acrylonitrile production   313
– propylene ammoxidation   328
– RSR systems   119
– vinyl chloride monomer process   203
explosion limit   28, 205
extractive distillation   79, 82
– acrylonitrile production   330

f

fast initiation systems   370
fatty-acid methyl esters (FAME)   401 f, 422
fatty-esters synthesis, catalytic distillation   231–260, 402
feasibility condition   1, 111
feed purification   33 f
feed streams control   163
feedback control   107
feed-effluent heat exchanger (FEHE)
– benzene alkylation   192
– cyclohexanol dehydrogenation   151, 156
– hydrodealkylation   124
– isobutane alkylation   279
– reactor-design   48
– vinyl acetate monomer process   297
– vinyl chloride monomer process   221
feedstock, bioethanol manufacturing   432, 438, 449
Fenske–Underwood–Gilliland (FUG) method   98
fermentation, bioethanol   429, 440–453
film model   343
fire risks   28
first separation step   50, 61
– propene ammoxidation   321
– vinyl acetate monomer process   299
– vinyl chloride monomer process   213
first-order reaction   118
Fisher–Tropsch synthesis   438
fixed recycle flow rate   44

fixed-flow equation   271
flashes sequence
– liquid-separation system   71 f
– separation system   51
– vinyl acetate monomer process   299
flat-blade-type turbine   384
Flory–Huggins theory   380
flow rates
– isobutane alkylation   270
– RSR systems   110 f
flowsheet
– acrylonitrile production   322, 333, 336
– benzene alkylation   190 f
– BioDeNO$_x$ process   357
– biodiesel manufacturing   410 f, 414, 423
– bioethanol from lignocellulosic biomass   448 ff
– catalytic cumene distillation   197
– cyclohexanol dehydrogenation   151
– Dow-Kellogg process   175
– ethanol from sugar cane   446
– 2-ethylhexyl dodecanoate synthesis   238
– fermentation section   454
– heat integration   220
– impurities/VCM process   226 f
– integrated process design   15
– one reactor phenol hydrogenation   159
– propene ammoxidation   335
– PVC manufacturing   366
– recycles/separation/heat integration   157
– vinyl acetate monomer   303, 306
– vinyl chloride monomer   204, 218
flue gases furnace   219
fluid properties   247
fluid-bed reactor   210, 320
N-formylmorpholine (NFM)   82
fossil resources   399
fouling   368
four-component distillation   75
free fatty acids (FFA)   403
free radicals   371 ff, 381
freezing point, acrylonitrile   314, 317
frequency factor   140
Frössling equation   244
fuel-cell vehicles (FCVs)   401
fuel manufacturing, bioethanol   429–460

g

galactan   449
Gani method   135, 239
gas cooling   321
gas dehydration setup   305
gas emissions   6

gas flows 247
gas–liquid column 207, 212
gas–liquid/solid systems 50 f, 343
gas-phase balance equations 351
gas separation system 64, 300
gas split manager (GSM) 61
gasoline
– biodiesel manufacturing 399, 430
– isobutane alkylation 261
– pyrolysis 80
gel effect 372 f
geometric characteristics, PVC reactor 368, 385 ff
Gibbs free energy 31
– fatty-ester synthesis 236
– phenol hydrogenation 132
Glitch ballast trays 159
global-warming potential (GWP) 167
glucose 442
glycerides 402 ff
glycerol 402–416
glycols 66, 82
grand composite curves 188, 191
greenhouse gases (GHGs) 6, 12, 433
group contribution methods 32

h

H_2SO_4 265
HAL 100 catalyst 263
halo-hydrocarbons 67, 256
Hatta number 342, 348
Hazop analysis 23, 26
HCl
– EDC pyrolysis 214
– physical properties 205
– PVC manufacturing 364
– vinyl chloride monomer process 204, 222
HCN 323 f, 328
health 28, 202, 289
heat effects
– PVC manufacturing 383
– reactor-design 47 f
– RSR systems 118
heat exchanger 389
heat exchanger network (HEN) 56
heat integration 26, 48
– alkylation reactor 190, 193
– propene ammoxidation 328
– vinyl chloride monomer process 219
heat transfer
– isobutane alkylation 279
– phenol hydrogenation 146
– process synthesis 56

– PVC manufacturing 363, 366, 384, 389 ff
– vinyl acetate monomer process 297
heavies 38 ff
– acrylonitrile production 323
– cumene synthesis 175, 184
– EDC pyrolysis 214
– hydrodealkylation 123 ff
– phenol hydrogenation 132–170
– separation systems 72 f, 76
– vinyl acetate monomer process 298
heavy-ends removing 72
heavy ions breakdown 264
hemicellulose 436, 449
Henry coefficients 347
Henry components 321
Henry law 299
heterogeneous azeotropic distillation 95, 255
heterogeneous catalysis 99, 419 ff
heterogeneous systems 46
heuristics
– azeotropic separation 77
– hierarchical approach 25
– integrated process design 15
– split generation 63
hexose fermentation 440, 447
hierarchical approach 16, 21–60, 262
high-temperature chlorination (HTC) 207
high-volume/low-value (HVLV) biofuels 437
homogeneous azeotropic distillation 88
homogeneous catalysis 416 ff
homogeneous systems 46
hot spot occurrence 146
hot stream table, cumene synthesis 187
human toxicity potential (HTPI/ HTPE) 167
hybrid separations 79
hydraulic correlations 245
hydraulic design 46, 234, 246
hydraulic diameter 385
hydrocarbons
– acrylonitrile production 335
– lauric acid esterification 256
– thermal cracking 211
– vinyl chloride monomer process 204
hydrodealkylation (HDA) 52, 83
hydrofluorocarbons (HFCs) 434
hydrogen 57
– biodiesel manufacturing 400
– one-reactor material balance 170
– phenol hydrogenation 129
– physical properties 132

– RSR systems 121
– separation system 52
hydrogen sulfide 67
hydrolysis 414
hydrolyzate conditioning 453
hydropyrolysis 415

i

immobilized cell reactors 444
impurities 6, 32
– acrylonitrile production 323
– oxychlorination 208
– phenol hydrogenation 130
– plantwide control 224
– separation system 63
– thermodynamic behavior 217
– vinyl acetate monomer process 288, 295
– vinyl chloride monomer process 201 ff
incineration process design 7
independent chemical equations 28
inhibitors
– lignocellulose saccharification 444
– PVC manufacturing 372
– vinyl acetate monomer process 288
initiators 211, 370–382
inlet temperatures 184
input/output analysis 25, 32–42, 294
input/output mass balance 109, 119
input/output structure
– isobutane alkylation 267
– phenol hydrogenation 141
– process synthesis 25, 33–42
inside battery limit (IBL) 33, 39
instabilities 125
integrated process design 1–20
intermediates 403
internal reactants recycle 99
internals selection 245
intramolecular transfer 372
inventory control 43, 107
ion-exchange resins 231, 404
iron-EDTA complex 339–362
isobutane alkylation 261–286
isomerization 177
isooctane 265, 430
isopropanol 85
isopropylbenzene (IPB), cumene synthesis 174–195
isothermal reactor 378

j

jackets 368, 383, 390
Jerusalem artichokes 436

k

KA-oil 129, 148, 158
Katapak-S 245 f, 253
ketones 130, 148, 256
key components 32
key mixtures 294, 318
kinetics
– benzene alkylation 181
– biodiesel manufacturing 409, 416
– dodecanoic acid esterification 242
– EDC pyrolysis 212
– fatty-ester synthesis 233, 241, 244
– hierarchical approach 31
– isobutane alkylation 266
– NO_x removal 347
– phenol hydrogenation 137
– PVC manufacturing 375, 378
– saccharification processes 442
– sunflower oil methanolysis 418
– suspension polymerization 371
– vinyl acetate monomer process 292
Klebsiella oxytoca 441
Kluyveromyces 440
Koch flexitrays 159

l

landfill gas (LFG) 67
large-scale reactor technology 365
large-volume organic chemicals (LVOC) 334
lauric acid 235, 254, 404 ff
lauroil peroxide (LPO) 374
Lewis-acid type catalysts 206
LHHW kinetic model 137
life-cycle assessment (LCA) 13, 17, 433
life-cycle inventory (LCI) 12
light-ends removing 71
lights 184, 298
lignin 436, 449
lignocellulosic biomass 429, 442–450
lignocellulosic feedstock (LCF) 438
limited reciprocal solubility 317
linoleic fatty acid 404 ff
lipase 415
lipid feedstock 403
liquefied petroleum gas (LPG) 261, 400
liquid activity 408
liquid holdup 246
liquid–liquid equilibrium 422
liquid–liquid separation 73 f, 79
– acrylonitrile production 317
liquid–liquid split
– acrylonitrile production 323, 332
– isobutane alkylation 280

– lauric acid esterification 255
liquid phase phenol hydrogenation 130
liquid separation 71, 82
– propene ammoxidation 324
– vinyl chloride monomer process 215, 300
liquid split manager (LSM) 61
liquid velocity 246
load point 247
long-chain carbon refinery 438
losses 161, 195
low-temperature chlorination (LTC) 207
low-volume/high-value (LVHV) biochemicals 437

m

macroradicals 382
make-up strategies 43
manifold projection 274 ff
mannan 449
manufacturing routes
– benzene alkylation 173
– biodiesel 409
– bioethanol 445
– vinyl acetate monomers 287
mass balance
– benzene alkylation 185
– cellulose hydrolysis 443
– ethanol from lignocellulosic biomass 458
– isobutane alkylation 268
– lauric acid esterification 250
– PVC manufacturing 379
– RSR systems 106
mass exchange networks 332
mass separation agent 51
mass transfer
– bulk 351
– NO_x removal 344, 347
– phenol hydrogenation 139
material balance 8, 35, 41
– acrylonitrile production 320
– cumene synthesis 175
– one-reactor process 170
– phenol hydrogenation 153, 163
– RSR structures 41
– trioleine methanolysis 423
– vinyl acetate monomer process 294, 302
– vinyl chloride monomer process 216
material flow analysis (MFA) 8, 434
material intensity 10
mathematical models 343
maximum uncertainty 276
Mayo equation 377
Mears factor 139, 243
mechanical treatments 450

mechanistic cycle 315
melt crystallization 73, 79
melting points 294, 406
membrane permeation 64, 69, 79
MESH equations 252
methane 67, 399, 434
methanol
– bioethanol manufacturing 431
– critical coordinates 413
– fuel cells 401
– sustainability metrics 12
methodology
– fatty-ester synthesis 232
– separation system 61
methyl acetate manufacturing 233
methyl-*tert*-butyl ether (MTBE) 261, 401
methyl esters 406
methyl oleate 430
N-metyl-pyrolidone (NMP) 82
metrics, sustainability 10
Michaelis–Menten model 442
MIMO methods 57
mineral acid 404
minimum energy requirements (MER) 56
minimum losses control 161
minimum reactor volume 45
minimum temperature difference 56, 389
mixed-integer linear programming (MILP) 17, 21, 78
mixing
– isobutane alkylation 284
– PVC manufacturing 369, 384
– RSR systems 109, 119
– VLE 294, 318
molar flow rates
– isobutane alkylation 270
– RSR systems 109
molar ratios
– biodiesel manufacturing 413, 418
– selectivity 181
molecular sieving 65, 456
molecular volumes, isobutane alkylation 266
molecular weight
– acrylonitrile 314, 317
– biodiesel manufacturing 407
– cumene synthesis 176
– isobutane alkylation 266
– PVC manufacturing 364, 376, 382, 395
– VCM process byproducts 205
– vinyl acetate monomer process 289, 294
moments, continuous/discrete distribution 376, 382
monomers 369, 386

Index

motor octane number (MON) 261
multieffect distillation 189
Multipak packing 245 f
multiple steady states 45
municipal solid waste (MSW) 67
myristic fatty acid 404 f

n

Nafion 231, 404, 419
natural gas for vehicles (NGV) 399
natural gas reforming 12
Nedalco NO_x removal 343
neutralization 411, 424
NExBTL process 415
nitrile impurities 323, 334
nitrogen 67
nitromethane 211
nitrous oxides (N_xO) 434
nominal operating points 275
nominal plant capacity 129
nominal production rate 106
normal-space approach, isobutane
  alkylation 274
NO_x
– biochemcial removal 339–362
– emissions 6, 399, 430
number of stages per meter (NSTM) 247

o

octane number 261
octane rating 429
olefin additions 264
oleic fatty acid 404 ff
oligomers 324
one distillation field 88
one reactor process 108, 117 f,158
operation conditions
– isobutane alkylation 271 ff
– phenol hydrogenation 148, 161
– PVC manufacturing 363, 394
– vinyl acetate monomer process 293
optimization 15, 25
– fatty-ester synthesis 233, 250
– isobutane alkylation 277
– material balance 54
– phenol hydrogenation 155
organic ion-exchange resin 231
outlet reactor mixtures 176
outlet streams number 34
overall approach 56
overall heat coefficient 367
overall mass balance 110 f
overall material balance 35, 144
oxidation stability 405

oxychlorination 203, 208
oxygen
– lauric acid esterification 256
– NO_x removal 346, 352
– separation system 67
– vinyl acetate monomer process 288
ozone-depletion potential (ODP) 167

p

packing, catalytic 245
palladium-type catalysts 130, 150, 158
palmitic fatty acid 404 ff
paraffins 264
parametric sensitivity 146
partial phenol conversion 147
particles morphology 364
partition ratio 371, 377
pasteurization 71
patents 28
payback time 54
Pd/Au catalysts 288
Peng–Robinson EOS 53
perfluorocarbons (PFCs) 434
Perkadox 16-W40 379
per-pass conversion
– acrylonitrile production 313
– isobutane alkylation 271
– phenol hydrogenation 150
– RSR systems 114
perselectivity 65
petrodiesel 399, 409, 430
phase equilibrium 32
– EDC pyrolysis 214
– fatty-ester synthesis 234 ff
phases
– biodiesel manufacturing 422
– NO_x removal 342
phenol
– cumene manufacturing 173
– one-reactor material balance 170
– physical properties 132, 140
phenol hydrogenation
– cyclohexanone 129–172
– RSR structure 144
– to cyclohexanone 137
phenone acetylation 11
phosphoric acid 12, 410
photochemical oxidation potential
  (PCOP) 167
physical absorption 50, 64
physical properties
– acrylonitrile 317
– benzene alkylation/cumene 175
– biodiesel manufacturing 406

- bioethanol manufacturing   430
- isobutane alkylation   265
- phenol hydrogenation   131
- PVC manufacturing   385
- vinyl acetate monomer process   293
- vinyl chloride monomer process   205
- zeolites   180
PID control law   386
pinch-point analysis (PPA)   16, 23, 26, 55
plant data   23, 27
- phenol hydrogenation   129, 148
plantwide control   17, 23
- hydrodealkylation   124
- impurities   224
- inlet flow rates   272
- isobutane alkylation   281
- phenol hydrogenation   162
- RSR systems   106
- vinyl acetate monomer process   305
- vinyl chloride monomer process   222
plasticizer adsorption   369
platinum catalysts   137
plug-flow reactors (PFR)   44
- glycerides transesterification   421
- RSR systems   114
- vinyl acetate monomer process   297
pollution   6, 11
- acrylonitrile production   334
- biodiesel manufacturing   399 f
polyalkylation   177
polychlorinated biphenyls (PCBs)   27
polydispersity index   377
polylactic acid (PLA)   439
polyvinyl acetate manufacture   287
polyvinyl chloride (PVC)
- suspension polymerization   363–398
- vinyl chloride monomer process   201
potatoes   436
potential environmental impact (PEI)   166, 170
power consumption   26
- acrylonitrile production   331
- PVC manufacturing   370
practical minimum-energy requirements (PME)   12
Prandtl number   139
prefractionator   77
preliminary chemical/phase equilibrium (CPE)   236
preliminary hydraulic design   246
preliminary material balance   36, 294
preprocessing   437, 452
pressure effect
- acrylonitrile production   320

- binary cyclohexanone/cyclohexanol VLE   143
pressure-swing distillation   79
primary raw materials (PRM)   8
process design
- BioDeNO$_x$ 357
- ethanol from lignocellulosic biomass   449
- phenol hydrogenation   161
process flow diagram (PFD)   18
- biodiesel manufacturing   423
process flowsheet
- alkylation/transalkylation/energy integration   194
- benzene alkylation   192, *see also* flowsheet
process intensification   15
process selection, NO$_x$ removal   341
process steps, vinyl chloride monomer process   202
process synthesis/integration   3, 14, 21–60
process utilities   27
process-control system   26
processing costs   37
product price   27
production costs, bioethanol manufacturing   432
production rate   4
- isobutane alkylation   266, 272
- lauric acid esterification   249
- phenol hydrogenation   161, 165
- RSR systems   109, 116
- vinyl chloride monomer plant   225
production-integrated environmental protection   6
project definition
- benzene alkylation   173
- bioethanol manufacturing   449
- fatty-ester synthesis   233 ff
- phenol hydrogenation   129
- propene ammoxidation   313
- PVC manufacturing   363
- vinyl acetate monomer process   288
- vinyl chloride monomer process   201
project-based learning   2
propagation   375, 382
propane
- acrylonitrile production   313
- butane blend   400
- cumene synthesis   176
- isobutane alkylation   265
propanol   255
n-propanol   85
propene ammonoxidation   12, 313–338
propene conversions   179

propyl laurate ester 408
propylene 176
propylene ammoxidation 37
proton transfer 265
pseudoalkylates 264
pseudohomogeneous process 266
pumping capacity 370
purge streams 34, 64
purification
– acrylonitrile 325
– EDC pyrolysis 215
– separation system 66
– vinyl chloride monomer process 206
purity
– biodiesel manufacturing 410
– cumene synthesis 173
– hierarchical approach 27
– lauric acid esterification 249
– phenol hydrogenation 161
– vinyl acetate monomer process 295
– vinyl chloride 202
pyrolysis
– EDC 212
– separation system 80
– vinyl chloride monomer process 203

q

quality specifications, phenol hydrogenation 130
quench, acrylonitrile production 321

r

Radfrac module
– cumene synthesis 187
– isobutane alkylation 281
– separation system 92
radicals 371 ff, 381
rank condition 111
rapeseed 404
raw materials 6 ff, 21, 27
– biofuel manufacturing 404, 435
– vinyl acetate monomer process 294
reactants 36
– excess 43
– injection strategy 47
– inventory 43
– makeup 41 ff
– recycling 33
– RSR systems 109
– vinyl acetate monomer process 293
reaction engineering 31
– benzene alkylation 176
– biodiesel manufacturing 420
reaction heat, RSR systems 118

reaction kinetics 6
– benzene alkylation 182
– ethylene chlorination 207
– isobutane alkylation 265
– lignocellulose saccharification 444
– oxychlorination 209
– phenol hydrogenation 132
– RSR systems 111
– vinyl acetate monomer process 289
reaction rates
– acrylonitrile production 313
– dodecanoic acid esterification 242
– phenol hydrogenation 138
– PVC manufacturing 370, 382, 394
– RSR systems 109
reaction scheme 30
– cellulose hydrolysis 443
– NExBTL process 415
– NO_x removal 340
– propene ammoxidation 314
– vinyl acetate monomer process 289
– vinyl chloride monomer process 202
reaction temperature
– biodiesel manufacturing 418
– EDC pyrolysis 212
– isobutane alkylation 279
reactive distillation (RD) 98 ff
– benzene alkylation 195
– biodiesel manufacturing 421
– cumene synthesis 195
– fatty-ester synthesis 231, 246
reactor design 31, 45 ff
– cyclohexanol dehydrogenation 151
– phenol hydrogenation 149
– PVC manufacturing 363, 367, 378, 388
– vinyl chloride monomer process 212
reactor inlet, phenol hydrogenation 163
reactor outlet
– EDC pyrolysis 214
– isobutane alkylation 265
– RSR systems 110
– vinyl chloride monomer process 219
reactor selection 45
reactor/separation/recycle (RSR) systems 103–128
– benzene alkylation 183
– hierarchical approach 23
– isobutane alkylation 268
– phenol hydrogenation 144
– process synthesis 25, 41, 103
– vinyl acetate monomer process 296
real balance yield 8
reboiled stripping 50, 71

reboiler
- cumene synthesis   189
- lauric acid esterification   249
- vinyl chloride monomer process   219
recipe changed PVC manufacturing   393
recovery
- acrylonitrile production   330
- first separation split   68
- isobutane alkylation   268
- separation system   51
rectifiers   77
recycle systems   6, 23 ff, 41 ff, 103
- isobutane alkylation   268, 272
- phenol hydrogenation   163
- RSR systems   110
- vinyl acetate monomer process   296
Redlich–Kwong model
- acrylonitrile production   318
- EDC pyrolysis   214
- phenol hydrogenation   140
redox potentials   341
redox properties, catalysts   313
reflux ratio
- lauric acid esterification   249
- phenol hydrogenation   155
refractive index, acrylonitrile   314
refrigeration systems   26, 56
regulation-by-feedback   43
research octane number (RON)   261 f
residence time
- cumene synthesis   183
- fatty-ester synthesis   244
residue curve maps (RCM)   16, 32
- fatty-ester synthesis   233
- lauric acid esterification   237, 255
- phenol/cyclohexanol/cyclohexanone mixture   143
- separation system   84
- vinyl acetate monomer process   295
resins   231, 404
resources optimization   57
revenues   10
reversible endo/exothermic reactions   47
Reynolds number   139, 146
Rilsan 11   440
robustness, isobutane alkylation   262, 271, 281

s

S. carlsbergensis   440
S. uvarum   440
saccharification/fermentation   441
Saccharomyces cerevisiae   440
safety   28

- isobutane alkylation   263
- vinyl acetate monomer process   289
- vinyl chloride monomer process   202
saponification   404
scaling   390
secondary raw materials (SRM)   8
segmentation   56
selection   245, 271
selective catalytic reduction (SCR/SNCR)   339
selectivity
- cumene synthesis   179
- EDC pyrolysis   213
- fatty-ester synthesis   242
- isobutane alkylation   266, 273 ff
- phenol hydrogenation   134 ff, 141
- vinyl acetate monomer process   288
selectors   61 ff, 70
self-regulation   43, 106
sensitivity   44
separation system   23 ff, 49, 61–102
- acrylonitrile production   320, 325
- biodiesel manufacturing   411, 424, 445
- ethanol from lignocellulosic biomass   457
- isobutane alkylation   280
- phenol hydrogenation   152
- propene ammoxidation   324 ff
- vinyl acetate plant   304
- vinyl chloride monomer process   206, 213, 298
separation/recycle structure   23 ff, 41, 103
- isobutane alkylation   268
- vinyl acetate monomer process   296
sequence optimization   78
sequential cellulose hydrolysis/hexose fermentation (SHF)   447
shell heat exchanger   148
shell-and-tube chemical reactor   159
Sherwood particle   139
shifts   43
shut-down control   161
side-chain aromatics   52
side-stream complex columns   77
Sieder–Tate correlation   384
silica gel catalysts   137
simulated moving bed (SMB) technology   84
simulation
- acrylonitrile purification   327
- benzene alkylation   185
- isobutane alkylation   281
- lauric acid esterification   248
- one-reactor phenol hydrogenation   160
- phenol hydrogenation   153

- propene ammoxidation   324
- vinyl acetate monomer process   310
- vinyl chloride monomer process   222
simultaneous saccharification/fermentation (SSF)   441
single-stage solid washer (Swash)   451
singularities   274
SISO methods   57
site data   27
site isolation   315
sizing   155, 355
slope condition   121
sloppy split   189
slurry gas/liquid reactor   49
snowball effect   44
- isobutane alkylation   262, 270
- RSR systems   105 f, 111
SO_2 emissions   6
social sustainability   5
SOHIO catalyst   315
solid catalysts   231 ff, 419
solid split manager (SSM)   61
solubility   64, 317
solvents   26, 57, 82
sorghum   436
SORONA   440
soybean   404, 418
space-time yield (STP)   292
specific balance yield   9
specific heat   366, 386
specifications   263
- acrylonitrile production   314
- biodiesel manufacturing   405
- vinyl acetate monomer process   288
spiral models   17
split sequences   61, 64 ff, 73, 215
SPYRO software   211
SR-Polar EOS   214, 299
stabilizers   71
starch   436, 445
start-up control   161
start-up heaters   48
state multiplicity   44, 107
static membrane reactor   445
steady-state model   118, 281
steam consumption   331
steam explosion   451
steam generator   48
stearic fatty acid   404 ff
stoichiometric parameters   8
- phenol hydrogenation   130
- RSR systems   111
- trioleine methanolysis   423

stoichiometric reactor
- EDC pyrolysis   212
- ethylene oxyacetylation   289
- phenol hydrogenation   153
storage facilities   27
stream table
- acrylonitrile production   323
- process synthesis   56
- vinyl acetate monomer process   297
strippers   73, 77
submixture analysis   63
sugars fermentation   429, 436, 445
sulfated zirconia (SZ)   404, 419
sulfolane   82
sulfuric acid alkylation   280 f
sunflower oil   404, 418
supercritical processes   413
superstructures   17, 51
supertargeting   56
supported liquid-phase catalysis (SLPC)   290
surface parameters   240
surfactants manufacture   231
suspension PVC polymerization   363–398
sustainability metrics   5, 10
sweet sorghum   436
synfuels   400
synthesis, separation systems   61–102
systematic approaches   1, 15, 103 ff

t
Talamini analysis   371 ff, 377
tallow   404
targeting   56, 192
task-oriented system   61
temperature
- bioethanol manufacturing   440
- lauric acid esterification   241
- liquid separation system   73
- PVC manufacturing   385 ff, 389 f, 395
- reactive distillation   99
- vinyl acetate monomer process   294
terephtalic acid   12
termination   372 ff, 382
ternary azeotropes   256
ternary systems
- distillation separation   75
- ethanol/water/entrainer   95
- residue curve maps (RCM)   86
terrestrial toxicity potential (TTP)   167
tetradecane   430
tetra-ethyl lead blending   430
theoretical balance yield   8
thermal coupling   57
thermal cracking   203, 210

thermal degradation 321
thermal design 47, 278
thermal effects 180
thermochemical bioethanol refinery 438
thermodynamic equilibrium
– NO$_x$ removal 344
– PVC manufacturing 371
thermodynamics
– acrylonitrile production 314, 318
– fatty-ester synthesis 233, 239
– integrated process design 15
– lauric acid esterification 254
– phenol hydrogenation 140
– process synthesis 25, 29, 32
– vinyl acetate monomer process 289
– vinyl chloride monomer process 202
Thiele modulus 139
toluene/acetone/chloroform separation 92
toluene hydrodealkylation (HDA) 52, 122
tortuosity 139
total cost assessment (TCA) 13
tower reactor 444
toxicity
– bioethanol manufacturing 429
– emissions 11
– fatty-ester synthesis 235
– hierarchical approach 28
– materials 6
– phenol hydrogenation 129
transalkylation 83, 177, 194
transesterification 402, 409
transient behavior 388
transitions time control 161
transportation fuels 399
trichlorethane (TCE) 206 ff
triglycerides
– biodiesel manufacturing 406
– hydropyrolysis 415, *see also* glycerides
trimethyl pentane (TMP) 262
tube heat exchanger 148
turndown ratio 343
turnover-frequency (TOF) 291
two distillation fields 89
two reactants/recycle processes 115
two-phase model, suspension PVC manufacturing 369

u
Underwood–Gilliland sequence optimization 78
UNIFAC model
– biodiesel manufacturing 409
– separation system 87

UNIQUAC model
– acrylonitrile production 318, 326
– fatty-ester synthesis 236 ff, 242 ff

v
V-cycle model 17
valorization, glycerol 416
value-added monetary unit 10
vapor–liquid equilibrium 99
vapor phase 130
vapor pressure 240, 406
vapor recovery 23, 64
Veendam 343, 354
vegetal lipids 404
vinyl acetate monomer (VAM) process 287–312
vinyl chloride monomer (VCM) process 201–230, 364 f
viscosity 386
VLE
– fatty-ester synthesis 239
– hierarchical approach 32
– key mixtures 294, 318
– separation system 91
void fraction 180, 245
volatile organic components (VOCs) 7, 27, 323
volatility
– cyclohexanone 152
– distillation separation 75
– split sequencing 64
volume parameters 240

w
walls 386
waste
– disposal costs 27
– process design 6
– propene ammoxidation 334
– reduction algorithm (WAR) 166
– treatment 10
– water 7, 334
water 10, 23, 26, 57
– acrylonitrile production 314, 318 f, 324, 332 ff
– azeotropes 236
– bioethanol manufacturing 431, 456
– burning 324
– cooling capacity 392
– ethanol from lignocellulosic biomass 456
– lauric acid esterification 255
– one-reactor material balance 170
– physical properties 133
– propene ammoxidation 332 ff

– PVC manufacturing   386
– residue curve maps   85
– separation systems   97
waterfall approach   17
Weber number   370
Weisz–Prater criterion   243
wheat   432, 436
Wilson/Redlich–Kwong model   141
Winn–Underwood–Gilliland method   280

x
xylan   449
xylenes   83
xylose   442

y
Y-type zeolites   179
yeasts fermentation   440
yields   8, 134

z
Z. mobilis   441, 455
zeolites
– biodiesel manufacturing   419
– cumene synthesis   175, 178 ff
– phenol hydrogenation   137
zeotropic mixtures   73 f
zirconia catalyst   242